Operator Theory: Advances and Applications
Vol. 143

Editor:
I. Gohberg

Reproducing Kernel Spaces and Applications

Daniel Alpay
Editor

Springer Basel AG

Editor:
Daniel Alpay
Department of Mathematics
Ben Gurion University of the Negev
P.O. Box 653
Beer Sheva 84105
Israel
e-mail: dany@math.bgu.ac.il

2000 Mathematics Subject Classification 46E22, 47B32, 47N20, 47N50

A CIP catalogue record for this book is available from the
Library of Congress, Washington D.C., USA

Bibliographic information published by Die Deutsche Bibliothek
Die Deutsche Bibliothek lists this publication in the Deutsche Nationalbibliografie; detailed bibliographic data is available in the Internet at <http://dnb.ddb.de>.

ISBN 978-3-0348-9430-2 ISBN 978-3-0348-8077-0 (eBook)
DOI 10.1007/978-3-0348-8077-0

Originally published by Birkhäuser Verlag, Basel, Switzerland in 2003

Member of the BertelsmannSpringer Publishing Group
Printed on acid-free paper produced from chlorine-free pulp. TCF ∞
Cover design: Heinz Hiltbrunner, Basel

9 8 7 6 5 4 3 2 1 www.birkhauser-science.com

Contents

Operator Theory:
Advances and Applications, Vol. 143, vii–xv

Editorial Introduction

Daniel Alpay

This volume contains a selection of papers that relate to the general topic of reproducing kernel Hilbert spaces, that is, Hilbert spaces of functions for which point evaluations are bounded. These spaces appear in a wide range of situations and they possess additional structure (thanks to the *reproducing kernel*) that other Hilbert spaces do not have. This gives, as is illustrated in the papers presented here, a supplementary point of view that allows one to state and solve new problems.

Let $\mathcal{H}$ be a Hilbert space of functions defined on a set Ω for which the point evaluations $f \mapsto f(w)$ are bounded for all $w \in \Omega$. By the Riesz representation theorem there exists a function $k(z, w)$ defined on $\Omega \times \Omega$ with the following two properties:

1. For every $w \in \Omega$ the function $k_w : z \mapsto k(z, w)$ belongs to $\mathcal{H}$.
2. For every $w \in \Omega$ and $f \in \mathcal{H}$

$$\langle f, k_w \rangle_{\mathcal{H}} = f(w).$$

The function $k(z, w)$ is uniquely defined and is called the reproducing kernel of the space.

An important feature in a reproducing kernel Hilbert space $\mathcal{H}$ is that one can compute the inner product explicitly on the dense set spanned by the functions $k_w : z \mapsto k(z, w)$ since

$$\langle k_w, k_v \rangle_{\mathcal{H}} = k(v, w).$$

The reproducing kernel is *positive* in the sense that all the matrices

$$\begin{pmatrix} k(w_1, w_1) & \cdots & k(w_1, w_N) \\ k(w_2, w_1) & \cdots & k(w_2, w_N) \\ \cdots & \cdots & \cdots \\ k(w_N, w_1) & \cdots & k(w_N, w_N) \end{pmatrix} \geq 0$$

for every choice of the integer N and of the points $w_1, \ldots, w_N \in \Omega$. Conversely each positive function defines a unique reproducing kernel Hilbert space of functions.

Two typical examples of positive kernels are $k(t, s) = \min\,(t, s)$ with $t, s \in \mathbb{R}^+$ (the covariance function of the Brownian motion) and $k(z, w) = \dfrac{1}{1 - zw^*}$ for z, w in the open unit disk (the reproducing kernel of the Hardy space of the open unit disk).

The notion of positivity can be traced back to the work of Mercer [47] at the beginning of the XXth century, the motivation arising from the work of Hilbert on integral equations; see Stewart's paper [61, p. 422–423] which contains a quite complete account of the history of positivity. We also send the reader to the notes and remarks on pages 84–85 of [20]. The notion of reproducing kernel seems to have appeared first in a paper of Zaremba and is more than 90 years old; see [66]. The papers of N. Aronszajn [14] and of L. Schwartz [59] are classics that can be studied repeatedly with much gain.

The main result of the theory, namely the one-to-one correspondence between positive functions and reproducing kernel Hilbert spaces, relates operator theory and the theory of functions. The notions of reproducing kernel and of positivity proved to be quite useful in numerous fields, of which we mention:

1. Partial differential equations. See [22].
2. Conformal mapping. See [21], [49].
3. Gaussian processes. Every positive function on a set T is the covariance function of a centered gaussian process indexed by T. See, e.g., [42, Theorem 8.2 p. 117] and [50, p. 79 and p. 93]. When T is a homogeneous tree one is lead to very interesting problems of harmonic analysis; see [43].
4. Harmonic analysis on semigroups. See [20].
5. The notion of hypergroup and harmonic analysis on hypergroups. See [25].
6. Operator theory. We refer in particular to the work of L. de Branges [27] and of L. de Branges and J. Rovnyak [28], [29] for work in the sixties and to [18].
7. Inverse problems. We refer in particular to the series of papers of D. Arov and H. Dym [15], [16], [17].
8. Orthogonal polynomials. See [30] for a recent account.
9. Zero counting problems. See [39], [36].
10. From the 1950' a new partner came into play: the theory of dissipative linear systems; see [3] for a review.
11. The Schur algorithm. See [7], [4].
12. Interpolation. See [35].
13. Inequalities. See [12].
14. Integral transforms. See [56], [57].
15. The theory of bounded symmetric domains. See [41], [13].
16. Generalizations of a formula of Kac (that in turn is a version of the Szegö formula for the growth of the determinants of blocks of a Toeplitz matrix) for the growth of the determinant (or traces) of finite sections of a convolution operator. See [34].
17. The basis problem. See [51, Chapter 4].
18. Approximation in reproducing kernel Hilbert spaces, see [33], and the theory of spline functions. See [24], [19].
19. Multiresolution analysis and the reproducing kernel particle method. We refer to the work of W.K. Liu; see [46], [45].

20. Pattern recognition and statistical learning theory (the theory of support vector machines). See [40], [58]. In this last volume we refer in particular to the papers [63] and [64]. Since this topic is maybe less known to the operator theory community we mention that *the support vector method is a general approach to function estimation problems.* See [63, p. 26].

We note that the above list and the given references are by no way exhaustive. We refer to the first section of the paper of S. Saitoh in the present volume for another (and mainly different) list of topics where reproducing kernel spaces appear.

Quite often a given question is best understood in a reproducing kernel Hilbert space (for instance when using Cauchy's formula in the Hardy space $\mathbf{H}_2$) and one finds oneself as Mr Jourdain of Molière' Bourgeois Gentilhomme speaking Prose without knowing it [48, p. 51]:

Par ma foi! il y a plus de quarante ans que je dis de la prose sans que j'en susse rien.[1]

The reproducing kernel setting allows one to attack more general situations and understand better the problem under study. As an example, let us mention the Beurling spaces $\mathcal{N} = s\mathbf{H}_2$ where s is an inner function. For such a space $\mathbf{H}_2 \ominus \mathcal{N} = \mathbf{H}_2 \ominus s\mathbf{H}_2$ is the reproducing kernel Hilbert space with reproducing kernel $k_s(z,w) = \frac{1-s(z)s(w)^*}{1-zw^*}$. Note that $\mathbf{H}_2 \ominus s\mathbf{H}_2$ is invariant under the backward shift operator $R_0f(z) = \frac{f(z)-f(0)}{z}$. Now the kernel $k_s(z,w)$ is positive for a wider class of functions than inner functions, namely for all functions analytic and contractive in the open unit disk. The characterization of reproducing kernel Hilbert spaces with reproducing kernel of the form $k_s(z,w)$ leads to a generalization of the Beurling–Lax theorem due to de Branges and Rovnyak when isometric inclusions are replaced by contractive inclusions. See [28], [6]. This approach led to recent progress in the study of the counterparts of the spaces with reproducing kernel $k_s(z,w)$ in the ball; see [5].

Another example of importance is taken from the theory of stochastic processes. If $x(n)$ is a second order real stationary process with correlation function $E(x(n)x(m)) = r(n-m)$ then

$$k_x(z,w) = \sum_{n,m=0}^{\infty} z^n w^{*m} E(x(n)x(m)) = \frac{\varphi(z)+\varphi(w)^*}{1-zw^*},$$

where E denotes the mathematical expectation, z, w are in the open unit disk, and

$$\varphi(z) = r(0) + 2\sum_{1}^{\infty} r(n)z^n.$$

[1] On my conscience, I have spoken prose above these forty years, without knowing anything of the matter. See [53, p. 234] for the translation.

If no hypothesis is made on the stochastic process (besides being second order) the function $k_x(z,w)$ is still positive but has no special additional structure. The associated reproducing kernel Hilbert space seems then of little use. Kailath and Lev-Ari introduced the family of α-stationary processes. For these the function $k_x(z,w)$ is of the form

$$k_x(z,w) = \alpha(z)\left\{\frac{\varphi(z)+\varphi(w)^*}{1-zw^*}\right\}\alpha(w)^*,$$

where α an φ are possibly matrix-valued functions. The case $\alpha(z)\equiv 1$ corresponds to classical stationary processes. See [44]. In this context, it is also worth mentioning that the theory of Hilbert spaces of entire functions of de Branges [27] serves to solve the prediction problem for continuous stationary processes. See [38].

Another and last example is the case of finite-dimensional spaces. Any finite-dimensional space of functions is trivially a reproducing kernel Hilbert space and one would think that this is the end of the story. Just the opposite is true, it is the beginning of a fascinating story, which has implications in the theory of fast algorithms (and connects then with the above example of α-stationary processes) and matrix theory. See [37].

Two recent new trends which connect with reproducing kernel spaces are *several complex variables* and *non commutativity*; examples of reproducing kernel Hilbert space of several complex variables exist for a long time (for instance the Hardy space of the ball, see [55], and kernels appearing in the more general theory of bounded symmetric domains, see [41], [13]) but we have in mind the work of Agler, see for example [1], [2], and Popescu, see [52]. We would like to mention also the theory of reproducing kernel spaces of hyperholomorphic functions. This is also a non commutative theory, which has been developed for a long time (see [32], [26]), but for which much remains to be done; see [10], [11]. We refer to [62] for an introduction to hypercomplex analysis.

Of course not every Hilbert space is a reproducing kernel Hilbert space, and even in a reproducing kernel Hilbert space, reproducing kernel methods are one way among others, and alternative routes exist. To quote Albert Camus on a different subject:

> *Il n'est pas possible, d'ailleurs, de fonder une attitude sur une émotion privilégiée. Le sentiment de l'absurde est un sentiment parmi d'autres. Qu'il ait donné sa couleur à tant de pensées et d'actions prouve seulement sa puissance et sa légitimité. Mais l'intensité d'un sentiment n'entraîne pas qu'il soit universel.*[2]

See [31, p. 20]. On the other hand, it is worth mentioning that, if one is willing to work with a weakened version of the axiom of choice (in particular in Solovay's

[2]Besides, it is not possible to base an attitude on a privileged emotion. The feeling of absurd is one feeling among others. That it gave its print to so many thoughts and actions proves only its power and its legitimacy. But the intensity of a feeling does not entail that it is universal.

model, in which all sets of real numbers are Lebesgue measurable and the axiom of choice for countable collections holds; see [60]), then all everywhere defined operators in Hilbert spaces are bounded (see [65]) and all Hilbert spaces of functions are then reproducing kernel Hilbert spaces. See [9] for an example of a Hilbert space of functions which is not a reproducing kernel Hilbert space.

The papers presented in this volume cover some of the various aspects of reproducing kernel spaces described above. More precisely:

1. **One complex variable theory:** The papers of Chang–Gilbert–Tie, Dym, Kriete and Saitoh deal with problems in operator theory in one complex variable. So does also the paper by Azizov, Dijksma, Langer, Wanjala and the editor of this volume. Topics touched upon include composition operators, Bergman spaces, geometrical aspects of reproducing kernel spaces, bitangential interpolation, and the Schur algorithm.
2. **Differential operators:** Reproducing kernel spaces and differential operators are closely related, in particular via the Weyl–Titchmarsh function. The paper of F. Gesztesy and L. Sakhnovich explores this direction. The paper of Mboup deals with the theory of self–similar systems and uses the theory of Hilbert spaces of entire functions of de Branges and the underlying string and differential operator.
3. **Several complex variables:** We first present a translation of a part of the thesis of M.F. Bessmertnyi [23]. A translation of another part of this thesis appears in the volume [8]. The paper of Agler–Yeh–Young explores one of the recent new directions in the passage from the one complex variable operator theory to the several complex variables operator theory, namely realization theory in the symmetrized bidisc. The paper of Vasilescu studies unitary dilations for tuples of commuting contractions via moment problems.
4. **The noncommutative case:** This is considered in the paper of Ball and Vinnikov (who consider the notion of reproducing kernel in the noncommutative setting) and in the paper of F. Szafraniec who considers operators of multiplication and their subnormality as a background for noncommutative complex analysis.

Acknowledgments: It is always a pleasure to acknowledge a debt and to thank a person. A few sentences will not be sufficient to express my gratitude to Harry Dym and Israel Gohberg for their help and collaboration over the years, but more than a few words would be more like a confession à *la Rousseau* [54]. So I will only thank Israel for inviting me to edit the present volume and Harry for introducing me to the theory of reproducing kernel spaces. If I recall correctly, the first assignment he gave me as a PhD student was to read and understand de Branges's book [27]. I am not sure that I have completed the assignment as of now.

Finally it is a pleasure to thank Professor Uri Abraham from Ben–Gurion University for discussions on Solovay's model.

References

[1] J. Agler. *On the representation of certain holomorphic functions defined on a polydisk*, volume 48 of *Operator Theory: Advances and Applications*, pages 47–66. Birkhäuser Verlag, Basel, 1990.

[2] J. Agler and J. McCarthy. Complete Nevanlinna-Pick kernels. *J. Funct. Anal.*, 175:111–124, 2000.

[3] D. Alpay. *The Schur algorithm, reproducing kernel spaces and system theory.* American Mathematical Society, Providence, RI, 2001. Translated from the 1998 French original by Stephen S. Wilson, Panoramas et Synthèses. [Panoramas and Syntheses].

[4] D. Alpay, T. Azizov, A. Dijksma, and H. Langer. The Schur algorithm for generalized Schur functions: Reproducing kernel spaces. To appear in Linear Algebra and Applications.

[5] D. Alpay, A. Dijksma, and J. Rovnyak. Un théorème de type Beurling-Lax dans la boule unité. *C. R. Math. Acad. Sci. Paris*, 334(5):349–354, 2002.

[6] D. Alpay, A. Dijksma, J. Rovnyak, and H. de Snoo. *Schur functions, operator colligations, and reproducing kernel Pontryagin spaces*, volume 96 of *Operator theory: Advances and Applications.* Birkhäuser Verlag, Basel, 1997.

[7] D. Alpay and H. Dym. *On applications of reproducing kernel spaces to the Schur algorithm and rational J-unitary factorization*, volume 18 of *Operator Theory: Advances and Applications*, pages 89–159. Birkhäuser Verlag, Basel, 1986.

[8] D. Alpay, I. Gohberg, and V. Vinnikov, editors. *The Harry Dym Anniversary Volume*, volume 134 of *Operator Theory: Advances and Applications.* Birkhäuser Verlag, Basel, 2002.

[9] D. Alpay and T. Mills. A family of Hilbert spaces which are not reproducing kernel Hilbert spaces. *Journal of Analysis and Applications.* To appear (2003).

[10] D. Alpay and M. Shapiro. Problème de Gleason et interpolation pour les fonctions hyper–analytiques. *Comptes Rendus Mathématiques, Académie des Sciences. Paris*, vol. 335 (2002) 889–894.

[11] D. Alpay and M. Shapiro. Reproducing kernel quaternionic Pontryagin spaces. To appear in *Integral Equations and Operator Theory.*

[12] T. Andô. Reproducing kernel spaces and quadratic inequalities. Lecture notes, Hokkaido University, Sapporo, 1987.

[13] J. Arazy. *A survey of invariant Hilbert spaces of analytic functions on bounded symmetric domains*, volume 185 of *Contemporary Mathematics*, pages 7–65. Amer. Math. Soc., Providence, RI, 1995.

[14] N. Aronszajn. Theory of reproducing kernels. *Trans. Amer. Math. Soc.*, 68:337–404, 1950.

[15] D. Arov and H. Dym. Some remarks on the inverse monodromy problem for 2×2 canonical differential systems. In *Operator theory and analysis (Amsterdam, 1997)*, volume 122 of *Oper. Theory Adv. Appl.*, pages 53–87. Birkhäuser, Basel, 2001.

[16] D. Arov and H. Dym. J-inner matrix functions, interpolation and inverse problems for canonical systems. IV. Direct and inverse bitangential input scattering problems. *Integral Equations Operator Theory*, 43(1):1–67, 2002.

[17] D. Arov and H. Dym. J-inner matrix functions, interpolation and inverse problems for canonical systems. V. The inverse input scattering problem for Wiener class and rational $p \times q$ input scattering matrices. *Integral Equations Operator Theory*, 43(1):68–129, 2002.

[18] D.Z. Arov. Stable dissipative linear stationary dynamical scattering systems. *J. Operator Theory*, 2(1):95–126, 1979.

[19] Marc Atteia. *Hilbertian kernels and spline functions*, volume 4 of *Studies in Computational Mathematics*. North-Holland Publishing Co., Amsterdam, 1992.

[20] Christian Berg, Jens Peter Reus Christensen, and Paul Ressel. *Harmonic analysis on semigroups*, volume 100 of *Graduate Texts in Mathematics*. Springer-Verlag, New York, 1984. Theory of positive definite and related functions.

[21] S. Bergman. *The kernel function and conformal mapping*. American Mathematical Society, 1950.

[22] S. Bergman and M. Schiffer. *Kernel functions and elliptic differential equations in mathematical physics*. Academic Press, 1953.

[23] M.F. Bessmertnyi. *Functions of several complex variables in the theory of linear systems*. PhD thesis, Kharkov, 1982.

[24] Anatoly Yu. Bezhaev and Vladimir A. Vasilenko. *Variational theory of splines*. Kluwer Academic/Plenum Publishers, New York, 2001.

[25] W. Bloom and H. Heyer. *Harmonic analysis of probability measures on hypergroups*, volume 20 of *de Gruyter Studies in Mathematics*. Walter de Gruyter & Co., Berlin, 1995.

[26] F. Brackx, R. Delanghe, and F. Sommen. *Clifford analysis*, volume 76. Pitman research notes, 1982.

[27] L. de Branges. *Espaces Hilbertiens de fonctions entières*. Masson, Paris, 1972.

[28] L. de Branges and J. Rovnyak. Canonical models in quantum scattering theory. In C. Wilcox, editor, *Perturbation theory and its applications in quantum mechanics*, pages 295–392. Wiley, New York, 1966.

[29] L. de Branges and J. Rovnyak. *Square summable power series*. Holt, Rinehart and Winston, New York, 1966.

[30] Adhemar Bultheel, Pablo González-Vera, Erik Hendriksen, and Olav Njåstad. *Orthogonal rational functions*. Cambridge University Press, Cambridge, 1999.

[31] Albert Camus. *L'homme révolté*. Gallimard, 1951.

[32] R. Delanghe and F. Brackx. Hypercomplex function theory and Hilbert modules with reproducing kernels. *Proc. London Math. Soc.*, 37:545–576, 1978.

[33] M. Duc-Jacquet. *Approximation des fonctionnelles linéaires sur les espaces hilbertiens autoreproduisants*. PhD thesis, Université scientifique et médicale de Grenoble, 1973.

[34] H. Dym. Trace formulas for a class of Toeplitz-like operators. *Israel J. Math.*, 27(1):21–48, 1977.

[35] H. Dym. *J–contractive matrix functions, reproducing kernel Hilbert spaces and interpolation*. Published for the Conference Board of the Mathematical Sciences, Washington, DC, 1989.

[36] H. Dym. On the zeros of some continuous analogues of matrix orthogonal polynomials and a related extension problem with negative squares. *Communications on pure and applied mathematics*, 47:207–256, 1994.

[37] H. Dym. *Shift, realizations and interpolation, Redux*, volume 73 of *Operator Theory: Advances and Applications*, pages 182–243. Birkhäuser Verlag, Basel, 1994.

[38] H. Dym and H.P. McKean. *Gaussian processes, function theory and the inverse spectral problem.* Academic Press, 1976.

[39] H. Dym and N. Young. A Schur–Cohn for matrix polynomials. *Proceedings of the Edinburgh Math. Soc.*, 33:337–366, 1990.

[40] Theodoros Evgeniou and Massimiliano Pontil. On the V_γ dimension for regression in reproducing kernel Hilbert spaces. In *Algorithmic learning theory (Tokyo, 1999)*, volume 1720 of *Lecture Notes in Comput. Sci.*, pages 106–117. Springer, Berlin, 1999.

[41] L.K. Hua. *Harmonic analysis of functions of several complex variables in the classical domains.* Translated from the Russian by Leo Ebner and Adam Korányi. American Mathematical Society, Providence, R.I., 1963.

[42] S. Janson. *Gaussian Hilbert spaces*, volume 129 of *Cambridge Tracts in Mathematics.* Cambridge University Press, Cambridge, 1997.

[43] G. Letac. Problèmes classiques de probabilité sur un couple de Gel′fand. In *Analytical methods in probability theory (Oberwolfach, 1980)*, volume 861 of *Lecture Notes in Math.*, pages 93–120. Springer, Berlin, 1981.

[44] H. Lev-Ari and T. Kailath. *Triangular factorisation of structured hermitian matrices*, volume 18 of *Operator theory: Advances and Applications*, pages 301–324. Birkhäuser Verlag, Basel, 1986.

[45] S. Li and W.K. Liu. Synchronized reproducing kernel interpolant via multiple wavelet expansion. *Comput. Mech.*, 21(1):28–47, 1998.

[46] Wing-Kam Liu, Shaofan Li, and Ted Belytschko. Moving least-square reproducing kernel methods. I. Methodology and convergence. *Comput. Methods Appl. Mech. Engrg.*, 143(1-2):113–154, 1997.

[47] J. Mercer. Functions of positive and negative type and their connection with the theory of integral equations. *Philos. Trans. Roy. Soc. London Ser. A*, 209:415–446, 1909.

[48] Molière. *Le bourgeois gentilhomme.* Nouveaux Classiques Larousse. Librairie Larousse, 1965. First published 1670.

[49] Zeev Nehari. *Conformal mapping.* McGraw-Hill Book Co., Inc., New York, Toronto, London, 1952.

[50] J. Neveu. *Bases mathématiques du calcul des probabilités.* Masson et Cie, Éditeurs, Paris, 1964.

[51] N.K. Nikolski. *Operators, functions, and systems: an easy reading. Vol. 2*, volume 93 of *Mathematical Surveys and Monographs.* American Mathematical Society, Providence, RI, 2002. Model operators and systems, Translated from the French by Andreas Hartmann and revised by the author.

[52] G. Popescu. Meromorphic interpolation in several variables. *Linear Algebra and Applications*, 357:173–196 (2002).

[53] Ernest Rhys, editor. *Selected Comedies by J.B. Poquelin Molière. Volume Two*, volume 831 of *Everyman's library*. J.M. Dent and Sons, London, 1929.

[54] J.J. Rousseau. *Les Confessions*. Moultou, 1782, 1789.

[55] W. Rudin. *Function theory in the unit ball of* $\mathbb{C}^n$. Springer–Verlag, 1980.

[56] S. Saitoh. *Theory of reproducing kernels and its applications*, volume 189. Longman scientific and technical, 1988.

[57] S. Saitoh. *Integral transforms, reproducing kernels and their applications*, volume 369. Longman scientific and technical, 1997.

[58] B. Schölkopf, C. Burges, and A. Smola, editors. *Advances in kernel methods. Support vector learning*. MIT Press, 1998.

[59] L. Schwartz. Sous espaces hilbertiens d'espaces vectoriels topologiques et noyaux associés (noyaux reproduisants). *J. Analyse Math.*, 13:115–256, 1964.

[60] R. Solovay. A model of set-theory in which every set of reals is Lebesgue measurable. *Ann. of Math. (2)*, 92:1–56, 1970.

[61] J. Stewart. Positive definite functions and generalization, an historical survey. *Rocky Mountain J. Math.*, 6:409–434, 1976.

[62] A. Sudbery. Quaternionic analysis. *Math. Proc. Cambridge Philos. Soc.*, 85(2):199–224, 1979.

[63] V. Vapnik. Three remarks on the support vector method of function estimation. In B. Schölkopf, C. Burges, and A. Smola, editors, *Advances in kernel methods. Support vector learning*, chapter 3, pages 25–41. MIT Press, 1998.

[64] G. Wahba. Support vector machines, reproducing kernel Hilbert spaces and randomized GACV. In B. Schölkopf, C. Burges, and A. Smola, editors, *Advances in kernel methods. Support vector learning*, chapter 6, pages 69–88. MIT Press, 1998.

[65] J.D. Maitland Wright. All operators on a Hilbert space are bounded. *Bull. Amer. Math. Soc.*, 79:1247–1250, 1973.

[66] S. Zaremba. L'équation biharmonique et une classe remarquable de fonctions fondamentales harmoniques. *Bulletin international de l'Académie des Sciences de Cracovie*, pages 147–196, 1907.

Daniel Alpay
Department of mathematics
Ben–Gurion University of the Negev
Beer-Sheva, Israel
e-mail: dany@math.bgu.ac.il

Operator Theory:
Advances and Applications, Vol. 143, 1–37

Realization of Functions into the Symmetrised Bidisc

J. Agler, F.B. Yeh and N.J. Young

Abstract. We give two realization formulae for analytic functions from the open unit disc to the open symmetrised bidisc G. We prove that the $ABCD$ matrices of the two realizations are equal, albeit with two different partitions. We illustrate the calculation of $ABCD$-matrices by finding explicitly the realizations of a certain class of extremal analytic functions from the disc to the domain G, to wit the *complex geodesics* of G.

G is defined to be the domain

$$G = \{(z_1 + z_2, z_1 z_2) : |z_1| < 1, |z_2| < 1\}.$$

The formulae permit the construction of analytic 2×2-matrix-valued functions F such that $F(\lambda)$ has spectral radius no greater than 1 for every λ in the disc.

Introduction

The idea of a *realization* of a function is a gift from engineering to pure mathematics. It provides a valuable addition to the repertoire of representations of functions, having some particular virtues in comparison with more traditional representations such as power series and infinite product expansions. It is well geared to numerical computation, and it often facilitates the generalization of results about scalar functions to matricial ones [9, 7]. A number of mathematicians have explored the scope of realizations in function theory over the past 15 years; the school of I.C. Gohberg has made extensive contributions, e.g., [9].

In this paper we establish a realization theory for a new class of analytic functions from the open unit disc $\mathbb{D}$ to $\mathbb{C}^2$; this class of functions arises in the context of the still-unsolved problem of μ-synthesis, which is relevant to the problem of robust stabilization in H^∞ control [13, 15].

An archetypal realization theorem asserts that a function f belongs to a class C if and only if there exist matrices (or operators) A, B, C, D with a particular

Received by the editors April 2002.
2000 *Mathematics Subject Classification.* 93B50,30E05,93B36,47A56.
Key words and phrases. Kobayashi extremal problem, spectral Nevanlinna-Pick problem, interpolation.
J. Agler's work is supported by an NSF grant in Modern Analysis. This work was also supported by EPSRC Visiting Fellowship GR/N 19250.

property $\mathcal{P}$ such that

$$f(z) = D + Cz(1 - Az)^{-1}B \tag{1}$$

for all relevant values of z. For example, if $\mathcal{C}$ is the Schur class of (operator- or scalar-valued) analytic functions on $\mathbb{D}$ then $\mathcal{P}$ is the property

$$\begin{bmatrix} A & B \\ C & D \end{bmatrix} \text{ is unitary.} \tag{2}$$

This is an elegant and useful connection between two of the most interesting objects in analysis – bounded analytic functions on $\mathbb{D}$ and bounded linear operators on Hilbert space. The utility of a realization result for the construction and analysis of functions in a class $\mathcal{C}$ depends on whether there is a manageable property $\mathcal{P}$ which corresponds to $\mathcal{C}$, and whether realizations of the given type can be calculated in practice. We shall prove a realization result for the class $\mathcal{O}(\mathbb{D}, G)$ of analytic functions from $\mathbb{D}$ to a certain domain $G \subset \mathbb{C}^2$ and show that it passes the two tests. Indeed, the relevant condition $\mathcal{P}$ is just the unitarity condition (2) above, D being of type 2×2 ; however, the formula for the function in terms of A, B, C, D is slightly more complicated than equation (1). In fact there are two natural approaches to realizing a function in $\mathcal{O}(\mathbb{D}, G)$. The first was established in [4] and is re-stated here as Theorem 1.2. The second is developed here and leads to the realization described in Theorem 1.9. Of the two approaches, the first gives a more explicit realization formula, while the second lends itself better to the *calculation* of realizations. It is a remarkable fact that the $ABCD$ matrices of the two realizations turn out to be equal (though with two different partitions). This (Theorem 1.10) is one of the main results of the paper, and it enables us to justify our claim that the realizations of the type in Theorem 1.2 can be computed in practice. To substantiate the claim we need to calculate realizations of a class of non-trivial members of $\mathcal{O}(\mathbb{D}, G)$. The functions we have chosen to realize are the solutions of extremal two-point interpolation problems for $\mathcal{O}(\mathbb{D}, G)$. We may regard such functions as higher-dimensional analogues of Möbius functions, on the grounds that the solutions of extremal two-point interpolation problems for $\mathcal{O}(\mathbb{D}, \mathbb{D})$ are precisely the Möbius functions. These extremal solutions in $\mathcal{O}(\mathbb{D}, G)$ are called *complex geodesics* of G (the terminology is explained further in Section 2), and they are found explicitly in [5]. They are rational functions of degree 1 or 2; the description is re-stated here in Theorems 2.1, 2.2 and 2.3. Section 3 is devoted to the computation of realizations of geodesics of degree 2; the result is stated as Theorem 3.5. After some lengthy algebra we arrive at a realization in which A, B, C and D are all explicitly-given 2×2 matrices. The case of degree 1 (in fact, affine) geodesics is a good deal simpler and is given in Theorem 1.6.

The fact that we are able to obtain such full and explicit results is a consequence of special features of the domain G that we study. G can be regarded as the space of monic polynomials of degree 2 that are stable, in the sense of having both zeros in $\mathbb{D}$. If we write such a polynomial as $z^2 - sz + p$ and its zeros as z_1, z_2

then we have

$$s = z_1 + z_2, \; p = z_1 z_2$$

and so, identifying the polynomial with the point $(s, \; p) \in \mathbb{C}^2$, we have

$$G = \{(z_1 + z_2, \; z_1 z_2) \; : \; z, \; z_2 \in \mathbb{D}\}.$$

We denote the closure of G by Γ, so that

$$\Gamma = \{(z_1 + z_2, \; z_1 z_2) \; : \; |z_1| \leq 1, \; |z_2| \leq 1 \}.$$

We call G, Γ the open and closed symmetrised bidiscs respectively. One can think of the sets as the images of the open and closed bidiscs under the "symmetrisation map"

$$(z_1, z_2) \mapsto (z_1 + z_2, z_1 z_2).$$

They are the symmetric products (of order 2) of the open and closed discs, naturally embedded in $\mathbb{C}^2$. The study of functions on G is the same as the study of symmetric functions on $\mathbb{D}$; however, we are concerned with the study of functions *into* G, which is quite a different matter. An easy way of generating analytic functions from $\mathbb{D}$ to G is to compose an analytic function $g : \mathbb{D} \to \mathbb{D}^2$ with the symmetrisation map, but many analytic functions $\mathbb{D} \to G$ do not factor continuously through $\mathbb{D}^2$ (e.g., $z \mapsto (0, z)$).

The significance of Γ for the spectral Nevanlinna-Pick problem is that a 2×2 matrix A has both eigenvalues in the closed unit disc if and only if $(\operatorname{tr} A, \det A) \in \Gamma$. There is thus a close connection between interpolation from $\mathbb{D}$ to Γ and interpolation from $\mathbb{D}$ to the space of 2×2 matrices with spectral radius at most 1 (see [2] for details).

Note that Γ and G are not convex. Two points of Γ are

$$w_1 = (1 + 1, 1 \cdot 1) = (2, 1) \text{ and } w_2 = (i + i, i \cdot i) = (2i, -1)$$

The mid-point of w_1 and w_2 is $(1 + i, 0)$, which does not belong to Γ.

The problem of analytic interpolation from $\mathbb{D}$ to G or Γ is roughly equivalent to the *spectral Nevanlinna-Pick problem* for 2×2 matrices, which has been an object of study for some years [10, 11, 13]. On the face of it G is not far removed from the bidisc $\mathbb{D}^2$, a much-studied domain, but the geometric properties of G and $\mathbb{D}^2$ are markedly different. G is inhomogeneous and non-convex, so that the property of mapping $\mathbb{D}$ into G is not immediately expressible by a norm inequality. Much classical function theory extends to mappings from $\mathbb{D}$ to a Cartan domain, but such domains are by definition homogeneous, and there appears to be no general theory which would apply to analytic functions from $\mathbb{D}$ to the domain G. It is heartening that there is nevertheless a successful realization theory for the class $\mathcal{O}(\mathbb{D}, \; G)$.

The *hyperbolic distance* d on $\mathbb{D}$ is given by

$$d(\lambda, \mu) = \tanh^{-1} \left| \frac{\lambda - \mu}{1 - \bar{\mu}\lambda} \right|$$

for all $\lambda, \mu \in \mathbb{D}$. The hyperbolic distance is invariant under the biholomorphic automorphisms of $\mathbb{D}$, which are the *Möbius functions*, i.e., the functions v of the form

$$v(z) = \omega \frac{z - \alpha}{1 - \bar{\alpha} z}$$

for some ω of unit modulus and $\alpha \in \mathbb{D}$. The *Schur class* of functions on a given domain Ω is the class of functions analytic in Ω and bounded by 1 in modulus. We denote the closed unit disc by Δ. This paper answers a question posed in [8].

1. The symmetrised bidisc and realizations

In this section we shall recapitulate some elementary properties of G and a realization theorem from [4], work out a motivating example which leads to a second realization formula and establish the relation between the two forms of realization of analytic functions from $\mathbb{D}$ to G.

We denote a typical point of G or Γ by (s,p) (think of "sum" and "product"). Given a point $(s,p) \in \mathbb{C}^2$ one may determine whether $(s,p) \in \Gamma$ by solving a quadratic equation to find $z_1, z_2 \in \mathbb{C}$ such that $z_1 + z_2 = s$ and $z_1 z_2 = p$. However, there are alternative criteria which are often easier to use [2, Theorem 1].

Theorem 1.1. *Let $s, p \in \mathbb{C}$. The following statements are equivalent:*
(1) $(s,p) \in \Gamma$;
(2) $|s| \leq 2$ *and* $|s - \bar{s}p| \leq 1 - |p|^2$;
(3) $|s| \leq 2$ *and, for all* $z \in \mathbb{D}$,

$$\left| \frac{2zp - s}{2 - zs} \right| \leq 1.$$

To present the realization formula for analytic Γ-valued functions we shall utilise some standard engineering notation. If H, U and Y are Hilbert spaces and

$$\begin{array}{llll} A : H & \to & H, & B : U & \to & H, \\ C : H & \to & Y, & D : U & \to & Y \end{array}$$

are bounded linear operators, then, for any $z \in \mathbb{C}$ for which $1 - Az$ is invertible we denote the operator

$$D + Cz(1 - Az)^{-1}B : U \longrightarrow Y$$

by

$$\left[\begin{array}{c|c} A & B \\ \hline C & D \end{array} \right] (z).$$

The following realization result is [4, Corollary 1.4].

Theorem 1.2. *An analytic function*

$$\varphi = (s,p) : \mathbb{D} \longrightarrow \mathbb{C}^2$$

maps $\mathbb{D}$ *into* Γ *if and only if there exist a Hilbert space* H *and a unitary operator*

$$\begin{bmatrix} A & B \\ C & D \end{bmatrix} : H \oplus \mathbb{C}^2 \longrightarrow H \oplus \mathbb{C}^2 \tag{3}$$

such that

$$s = \left[\begin{array}{c|c} A & B_1 \\ \hline C_1 & D_{11} \end{array}\right] - \left[\begin{array}{c|c} A & B_2 \\ \hline C_2 & D_{22} \end{array}\right]$$

and

$$p = \left(\frac{1}{4}\mathrm{tr}^2 - \det\right) \left[\begin{array}{c|c} A & B \\ \hline C & D \end{array}\right],$$

where

$$B = [B_1\ B_2] : \mathbb{C}^2 \to H, \quad C = \begin{bmatrix} C_1 \\ C_2 \end{bmatrix} : H \to \mathbb{C}^2$$

and

$$D = \begin{bmatrix} D_{11} & D_{12} \\ D_{21} & D_{22} \end{bmatrix} : \mathbb{C}^2 \to \mathbb{C}^2.$$

Furthermore, for any analytic function $\varphi : \mathbb{D} \to \Gamma$, *one may choose* H, A, B, C *and* D *in the realization* (3) *so that*

$$\mathrm{tr} \left[\begin{array}{c|c} A & B \\ \hline C & D \end{array}\right] \equiv 0,$$

and hence

$$s = 2 \left[\begin{array}{c|c} A & B_1 \\ \hline C_1 & D_{11} \end{array}\right] = -2 \left[\begin{array}{c|c} A & B_2 \\ \hline C_2 & D_{22} \end{array}\right]$$

and

$$p = -\det \left[\begin{array}{c|c} A & B \\ \hline C & D \end{array}\right].$$

Here is a reformulation of Theorem 1.2 which can be regarded as a quantitative refinement of a realization for analytic multifunctions due to Aupetit [6].

Theorem 1.3. *Let* $F : \mathbb{D} \times \mathbb{C} \to \mathbb{C}$ *be analytic. The following two statements are equivalent.*

(1) *For every* $\lambda \in \mathbb{D}$, $F(\lambda, \cdot)$ *is a monic polynomial of degree* 2 *whose zeros lie in the closed unit disc;*

(2) *there exists a* 2×2 *matricial inner function* ψ *such that, for all* $\lambda \in \mathbb{D}$ *and* $z \in \mathbb{C}$,

$$F(\lambda, z) = \det(zI_2 - \psi(\lambda)),$$

and the two entries on the main diagonal of ψ *are equal.*

Proof. It is trivial that (2) implies (1). Suppose that (1) holds, so that, for all $\lambda \in \mathbb{D}$ and $z \in \mathbb{C}$, we may write

$$F(\lambda, z) = z^2 - s(\lambda)z + p(\lambda)$$

for some analytic functions s, p on $\mathbb{D}$. Since the zeros of $F(\lambda, \cdot)$ lie in the closed unit disc, (s, p) maps $\mathbb{D}$ into Γ. Choose A, B, C, D as in Theorem 1.2, and let

$$\psi(\lambda) = \left[\begin{array}{c|c} A & B \\ \hline C & D \end{array}\right] \operatorname{diag}\{1, -1\}.$$

Since the $ABCD$ operator is unitary, ψ is a 2×2 inner function, and since

$$\operatorname{tr} \left[\begin{array}{c|c} A & B \\ \hline C & D \end{array}\right] = 0,$$

the two diagonal entries of ψ are equal, and we have

$$s(\lambda) = \operatorname{tr} \psi(\lambda), \quad p(\lambda) = \det \psi(\lambda).$$

Thus the characteristic polynomial of $\psi(\lambda)$ is $F(\lambda, \cdot)$, as required. □

Suppose we are given an analytic function $\varphi : \mathbb{D} \longrightarrow \Gamma$. How can we find a realization of φ of the type described in Theorem 2? The proof in [4] suggests a way. Write $\varphi = (s, p)$ and choose $f_1, f_2 \in H^\infty$ such that

$$f_1 f_2 = 4p - s^2$$

and

$$|f_1(z)| = |f_2(z)| \text{ for almost all } z \in \mathbb{T}$$

(the existence of such functions f_1, f_2 follows from the inner-outer factorization of $4p - s^2$). The function

$$\psi = \frac{1}{2} \begin{bmatrix} s & f_1 \\ f_2 & -s \end{bmatrix}$$

is a 2×2 matricial Schur function, that is, it is analytic in $\mathbb{D}$ and satisfies

$$\|\psi(z)\| \leq 1 \text{ for all } z \in \mathbb{D}.$$

The standard realization theorem for functions in the Schur class (e.g., [18]) then provides H, A, B, C and D as in the theorem.

Let us try out this procedure on a simple example. Let $\beta \in \mathbb{D}$ and let

$$\varphi_\beta(\lambda) = (\beta\lambda + \bar{\beta}, \lambda), \ \lambda \in \mathbb{D}.$$

We shall see in the next section that φ_β not only maps $\mathbb{D}$ into Γ, but also has a certain extremal property which makes it a natural object of study. According to the recipe, we need functions $f_1, f_2 \in H^\infty$ such that

$$\begin{aligned} f_1 f_2(\lambda) &= 4\lambda - (\beta\lambda + \bar{\beta})^2 \\ &= -\beta^2\lambda^2 + 2(2 - |\beta|^2)\lambda - \bar{\beta}^2 \end{aligned} \tag{4}$$

on $\mathbb{D}$, and

$$|f_1(\lambda)| = |f_2(\lambda)| \text{ for almost all } \lambda \in \mathbb{T}.$$

If $\beta = 0$ we may take $f_1(\lambda) = 2\lambda$, $f_2(\lambda) = 2$. Otherwise we may write equation (4) in the form

$$f_1 f_2(\lambda) = -\beta^2(\lambda - \gamma_+)(\lambda - \gamma_-)$$

where

$$\gamma_\pm = \frac{(1 \pm \sqrt{1 - |\beta|^2})^2}{\beta^2} \tag{5}$$

Observe that

$$|\gamma_-| < 1, \quad |\gamma_+| > 1,$$

so that we may define analytic branches of the functions

$$\lambda \mapsto (\lambda - \gamma_+)^{\frac{1}{2}}, \quad \lambda \mapsto (1 - \bar{\gamma}_- \lambda)^{\frac{1}{2}}$$

on Δ. Now let

$$f_1(\lambda) = \quad \beta B_{\gamma_-}(\lambda)(\lambda - \gamma_+)^{\frac{1}{2}}(1 - \bar{\gamma}_- \lambda)^{\frac{1}{2}},$$
$$f_2(\lambda) = -\beta(\lambda - \gamma_+)^{\frac{1}{2}}(1 - \bar{\gamma}_- \lambda)^{\frac{1}{2}}.$$

We have $|f_1| = |f_2|$ on $\mathbb{T}$, and $f_1 f_2 = 4p - s^2$. The function

$$\psi(\lambda) = \frac{1}{2}\begin{bmatrix} \beta\lambda + \bar{\beta} & f_1(\lambda) \\ f_2(\lambda) & -\beta\lambda - \bar{\beta} \end{bmatrix}$$

is in the Schur class, and so may be realized by known methods.

The calculation exhibits a disappointing feature: even though we started with an *affine linear* function φ_β, the realization we compute will have an infinite-dimensional state space, on account of the presence of square root functions in ψ. This can hardly be an essential property of realizations of φ_β: a simple dimension count suggests there should be a realization with a one-dimensional state space. It must accordingly be an artefact of our method of calculation, and we are led to introduce an alternative approach which will yield finite-dimensional realizations of rational functions from $\mathbb{D}$ to Γ.

Theorem 1.1 enables us to pass from a given analytic function $\varphi : \mathbb{D} \to \Gamma$ to a Schur function on the bidisc. Indeed, if we let

$$\Phi(z, s, p) \;=\; \frac{2zp - s}{2 - zs}$$

where $zs \neq 2$, then Theorem 1.1 tells us that Φ maps $\mathbb{D} \times \Gamma$ into Δ, and hence the function h defined by

$$h(\lambda, z) \;=\; \Phi(z, \varphi(\lambda)), \; z, \lambda \in \mathbb{D},$$

is analytic on $\mathbb{D}^2$ and maps into Δ; that is, it belongs to the Schur class on $\mathbb{D}^2$. It therefore has an Agler realization [1, Theorem 1]: there exist a Hilbert space H, complementary orthogonal projections P_1, P_2 on H, and a unitary operator

$$\begin{bmatrix} a & b \\ c & d \end{bmatrix} : H \oplus \mathbb{C} \longrightarrow H \oplus \mathbb{C}$$

such that, for all $\lambda_1, \lambda_2 \in \mathbb{D}$,

$$h(\lambda_1, \lambda_2) \;=\; d + c\Lambda(1 - a\Lambda)^{-1} b$$

where $\Lambda = \lambda_1 P_1 + \lambda_2 P_2$. Let us again illustrate the construction with the affine linear function φ_β. We define

$$\begin{aligned} h_\beta(\lambda_1, \lambda_2) &= \Phi(\lambda_2, \varphi_\beta(\lambda_1)) \\ &= \frac{2\lambda_2\lambda_1 - \beta\lambda_1 - \bar\beta}{2 - \bar\beta\lambda_2 - \beta\lambda_1\lambda_2}, \end{aligned}$$

which is an inner function on $\mathbb{D}^2$. Write $\mu = (\mu_1, \mu_2)$, $\lambda = (\lambda_1, \lambda_2)$ in $\mathbb{D}^2$. To compute an Agler realization of h_β we need to find positive semi-definite functions a_1, a_2 on $\mathbb{D}^2 \times \mathbb{D}^2$ such that

$$1 - \bar h_\beta(\mu) h_\beta(\lambda) = (1 - \bar\mu_1\lambda_1) a_1(\lambda, \mu) + (1 - \bar\mu_2\lambda_2) a_2(\lambda, \mu).$$

Agler's theorem guarantees that such functions a_1, a_2 exist, and after some trial and error we found the following identity.

Lemma 1.4. *Let $\beta \in \mathbb{D} \setminus \{0\}$ and let*

$$\begin{aligned} h_\beta(\lambda) &= \frac{2\lambda_1\lambda_2 - \beta\lambda_1 - \bar\beta}{2 - \bar\beta\lambda_2 - \beta\lambda_1\lambda_2}, \quad \lambda = (\lambda_1, \lambda_2) \in \mathbb{D}^2, \\ u_\beta(\lambda) &= 2 - \bar\beta\lambda_2 - \beta\lambda_1\lambda_2, \quad \lambda = (\lambda_1, \lambda_2) \in \mathbb{D}^2, \\ \eta &= \sqrt{1 - \sqrt{1 - |\beta|^2}}, \\ \xi &= 1 + \sqrt{1 - |\beta|^2}. \end{aligned}$$

For all $\lambda, \mu \in \mathbb{D}^2$,

$$1 - \bar h_\beta(\mu) h_\beta(\lambda) = \frac{1}{\bar u_\beta(\mu) u_\beta(\lambda)}$$
$$\times \left[(1 - \bar\mu_1\lambda_1) 2(\eta\bar\mu_2 - \eta^{-1}\bar\beta)(\eta\lambda_2 - \eta^{-1}\beta) + (1 - \bar\mu_2\lambda_2)(\xi\bar\mu_1 - \xi^{-1}\beta^2)(\xi\lambda_1 - \xi^{-1}\bar\beta^2)\right].$$

From this identity we easily obtain a realization. Define $f_1, f_2 : \mathbb{D}^2 \to \mathbb{C}$ by

$$\begin{aligned} f_1(\lambda) &= \sqrt{2}\, \frac{\eta\lambda_2 - \eta^{-1}\beta}{u_\beta(\lambda)}, \\ f_2(\lambda) &= \frac{\xi\lambda_1 - \xi^{-1}\bar\beta^2}{u_\beta(\lambda)}. \end{aligned}$$

The identity in Lemma 1.4 can be written

$$1 - \bar h_\beta(\mu) h_\beta(\lambda) = (1 - \bar\mu_1\lambda_1) \bar f_1(\mu) f_1(\lambda) + (1 - \bar\mu_2\lambda_2) \bar f_2(\mu) f_2(\lambda).$$

Re-write this relation without using minus signs:

$$\begin{aligned} &1 + \bar\mu_1\lambda_1 \bar f_1(\mu) f_1(\lambda) + \bar\mu_2\lambda_2 \bar f_2(\mu) f_2(\lambda) \\ &= \bar h_\beta(\mu) h_\beta(\lambda) + \bar f_1(\mu) f_1(\lambda) + \bar f_2(\mu) f_2(\lambda). \end{aligned}$$

This equation can be interpreted as the equality of the Gram matrices of two collections of vectors in $\mathbb{C}^3$

$$\left\langle \begin{bmatrix} f_1(\lambda) \\ f_2(\lambda) \\ h_\beta(\lambda) \end{bmatrix}, \begin{bmatrix} f_1(\mu) \\ f_2(\mu) \\ h_\beta(\mu) \end{bmatrix} \right\rangle_{\mathbb{C}^3} = \left\langle \begin{bmatrix} \lambda_1 f_1(\lambda) \\ \lambda_2 f_2(\lambda) \\ 1 \end{bmatrix}, \begin{bmatrix} \mu_1 f_1(\mu) \\ \mu_2 f_2(\mu) \\ 1 \end{bmatrix} \right\rangle_{\mathbb{C}^3}$$

for all $\lambda, \mu \in \mathbb{D}^2$. It easy to see that the vectors $\{(f_1(\lambda), f_2(\lambda), h_\beta(\lambda)) : \lambda \in \mathbb{D}^2\}$ span $\mathbb{C}^3$. Hence there exists a unitary matrix

$$\begin{bmatrix} a & b \\ c & d \end{bmatrix} : \mathbb{C}^2 \oplus \mathbb{C} \to \mathbb{C}^2 \oplus \mathbb{C}$$

such that, for all $\lambda \in \mathbb{D}^2$,

$$\begin{bmatrix} f_1(\lambda) \\ f_2(\lambda) \\ h_\beta(\lambda) \end{bmatrix} = \begin{bmatrix} a & b \\ c & d \end{bmatrix} \begin{bmatrix} \lambda_1 f_1(\lambda) \\ \lambda_2 f_2(\lambda) \\ 1 \end{bmatrix}. \tag{6}$$

Let $f(\lambda) = [f_1(\lambda)\ f_2(\lambda)]^T$ and $\Lambda = \operatorname{diag}\{\lambda_1, \lambda_2\}$. The last equation becomes the pair

$$\begin{aligned} f(\lambda) &= a\Lambda f(\lambda) + b, \\ h_\beta(\lambda) &= c\Lambda f(\lambda) + d. \end{aligned}$$

Eliminate $f(\lambda)$ from these two equations to obtain:

$$h_\beta(\lambda) = d + c\Lambda(1 - a\Lambda)^{-1} b.$$

It remains to find a, b, c and d. Multiply through by $u_\beta(\lambda)$ in equation (6) to obtain

$$\begin{bmatrix} \sqrt{2}(\eta\lambda_2 - \eta^{-1}\beta) \\ \xi\lambda_1 - \xi^{-1}\bar{\beta}^2 \\ 2\lambda_1\lambda_2 - \beta\lambda_1 - \bar{\beta} \end{bmatrix} = \begin{bmatrix} a & b \\ c & d \end{bmatrix} \begin{bmatrix} \sqrt{2}\lambda_1(\eta\lambda_2 - \eta^{-1}\beta) \\ \lambda_2(\xi\lambda_1 - \xi^{-1}\bar{\beta}^2) \\ 2 - \bar{\beta}\lambda_2 - \beta\lambda_1\lambda_2 \end{bmatrix}$$

for all $\lambda_1, \lambda_2 \in \mathbb{D}$. From this equation a, b, c and d are readily determined, with the following conclusion.

Theorem 1.5. *Let $\beta \in \mathbb{D} \setminus \{0\}$, let h_β be as in Lemma 1.4 and let*

$$\begin{aligned} \xi &= 1 + \sqrt{1 - |\beta|^2}, \\ \eta &= \sqrt{1 - \sqrt{1 - |\beta|^2}}, \end{aligned}$$

$$a = \begin{bmatrix} 0 & \frac{-\beta\eta}{\sqrt{2}\bar{\beta}} \\ \frac{-\xi\eta}{\sqrt{2}\beta} & \frac{\bar{\beta}}{2} \end{bmatrix}, \quad b = \begin{bmatrix} \frac{-\beta}{\sqrt{2}\eta} \\ \frac{-\bar{\beta}^2}{2\xi} \end{bmatrix},$$

$$c = \begin{bmatrix} \frac{\eta}{\sqrt{2}} & \frac{\xi}{2} \end{bmatrix}, \quad d = -\frac{\bar{\beta}}{2}.$$

Then $\begin{bmatrix} a & b \\ c & d \end{bmatrix}$ *is a unitary matrix and for all* $\lambda_1, \lambda_2 \in \mathbb{D}$,

$$h(\lambda_1, \lambda_2) \;=\; d + c\Lambda(1 - a\Lambda)^{-1}b,$$

where $\Lambda = \operatorname{diag}\{\lambda_1, \lambda_2\}$.

A lengthy calculation has led to a "realization" of φ_β which is more complicated than the original expression for φ_β; however, this example illustrates the nature of the computation required and the relation between an Agler realization of a Schur function of the form $\Phi(\lambda_2, \varphi(\lambda_1))$ on the bidisc and a realization of a function $\varphi : \mathbb{D} \to \Gamma$ in the sense of Theorem 2. What might this relation be? The most optimistic guess is that

$$\begin{bmatrix} A & B \\ C & D \end{bmatrix} = \begin{bmatrix} a & b \\ c & d \end{bmatrix} = \begin{bmatrix} 0 & \frac{-\beta\eta}{\sqrt{2\bar\beta}} & \frac{-\beta}{\sqrt{2}\eta} \\ \frac{-\xi\eta}{\sqrt{2}\beta} & \frac{\bar\beta}{2} & \frac{-\bar\beta^2}{2\xi} \\ \frac{\eta}{\sqrt{2}} & \frac{\xi}{2} & \frac{-\beta}{2} \end{bmatrix}$$

The guess is in fact correct; the following result is a matter of verification.

Theorem 1.6. *Let* $\beta \in \mathbb{D} \setminus \{0\}$ *and let*

$$\begin{aligned} \xi &= 1 + \sqrt{1 - |\beta|^2}, \; \eta \;=\; \sqrt{1 - \sqrt{1 - |\beta|^2}}, \\ A &= \begin{bmatrix} 0 \end{bmatrix}, \; B = \begin{bmatrix} B_1 & B_2 \end{bmatrix} = \begin{bmatrix} \frac{-\beta\eta}{\sqrt{2\bar\beta}} & \frac{-\beta}{\sqrt{2}\eta}, \end{bmatrix}, \\ C &= \begin{bmatrix} C_1 \\ C_2 \end{bmatrix} = \begin{bmatrix} \frac{-\xi\eta}{\sqrt{2}\beta} \\ \frac{\eta}{\sqrt{2}} \end{bmatrix}, \\ D &= [D_{ij}]_{i,j=1}^2 = \begin{bmatrix} \frac{\bar\beta}{2} & \frac{-\bar\beta^2}{2\xi} \\ \frac{\xi}{2} & \frac{-\beta}{2} \end{bmatrix}. \end{aligned}$$

Then $\begin{bmatrix} A & B \\ C & D \end{bmatrix}$ *is a unitary matrix, and is a realization of* $\varphi_\beta(\lambda) = (\beta\lambda + \bar\beta, \lambda)$ *in the sense of Theorem 1.2; that is, for all* $\lambda \in \mathbb{D}$,

$$\beta\lambda + \bar\beta \;=\; 2\left[\begin{array}{c|c} A & B_1 \\ \hline C_1 & D_{11} \end{array}\right](\lambda) \;=\; -2\left[\begin{array}{c|c} A & B_2 \\ \hline C_2 & D_{22} \end{array}\right](\lambda)$$

and

$$\lambda \;=\; -\det\left[\begin{array}{c|c} A & B \\ \hline C & D \end{array}\right](\lambda).$$

It may be checked that

$$\begin{aligned}\left[\begin{array}{c|c} A & B \\ \hline C & D \end{array}\right](\lambda) &= D + C\lambda B \\ &= \frac{1}{2}\begin{bmatrix} \beta\lambda + \bar\beta & \xi\lambda - \frac{\bar\beta^2}{\xi} \\ \frac{-\beta\eta^2}{\bar\beta}\lambda + \xi & -\beta\lambda - \bar\beta \end{bmatrix} \\ &= \frac{1}{2}\begin{bmatrix} \beta\lambda + \bar\beta & \xi(\lambda - \gamma_-) \\ \frac{-\beta\eta^2}{\bar\beta}(\lambda - \gamma_+) & -\beta\lambda - \bar\beta \end{bmatrix}\end{aligned}$$

where $\gamma_\pm$ are defined by equation (5).

For completeness we give a realization in the case $\beta = 0$, in which case $\varphi_0(\lambda) = (0, \lambda)$. Let

$$\psi(\lambda) = \begin{bmatrix} 0 & \lambda \\ 1 & 0 \end{bmatrix} = \left[\begin{array}{c|cc} 0 & 0 & 1 \\ \hline 1 & 0 & 0 \\ 0 & 1 & 0 \end{array}\right](\lambda);$$

then tr $\psi = 0$ and

$$\varphi_0 = (2\psi_{11}, -\det\psi).$$

We shall now establish the relation between the two notions of realization of a general analytic function from $\mathbb{D}$ to Γ. For any operator

$$P = \begin{bmatrix} P_{11} & P_{12} \\ P_{21} & P_{22} \end{bmatrix} : H \oplus U \longrightarrow H \oplus Y$$

we shall denote by $\mathcal{F}_P$ the linear fractional transformation

$$\mathcal{F}_P(X) = P_{22} + P_{21}X(1 - P_{11}X)^{-1}P_{12}$$

defined for any operator X on H such that $1 - P_{11}X$ is invertible. $\mathcal{F}_P(X)$, when defined, is an operator from U to Y. The following identity of standard type may be verified by straightforward expansion.

Lemma 1.7. *Let H, U and Y be Hilbert spaces and let $P = [P_{ij}]_{i,j=1}^2, Q = [Q_{ij}]_{i,j=1}^2$ be operators from $H \oplus U$ to $H \oplus Y$. For any pair of operators X, Y on H such that $1 - P_{11}X$ and $1 - Q_{11}Y$ are invertible,*

$$\begin{aligned}1 - \mathcal{F}_Q(Y)^*\mathcal{F}_P(X) = {} & Q_{12}^*(1 - Y^*Q_{11}^*)^{-1}(1 - Y^*X)(1 - P_{11}X)^{-1}P_{12} \\ & + [\, Q_{12}^*(1Y^*Q_{11}^*)^{-1}Y^* \;\; 1 \,](1 - Q^*P)\begin{bmatrix} X(1 - P_{11}X)^{-1}P_{12} \\ 1 \end{bmatrix}.\end{aligned}$$

Consider an analytic function $\varphi : \mathbb{D} \to \Gamma$. By [4, Theorem 3] there exists a 2×2 matricial function ψ in the Schur class such that tr $\psi = 0$ and

$$\varphi = (2\psi_{11}, -\det\psi).$$

Since $\|\psi\|_\infty \le 1$, the matricial kernel

$$(x, y) \mapsto \frac{1 - \psi(y)^*\psi(x)}{1 - \bar y x} : \mathbb{D}^2 \longrightarrow M_2(\mathbb{C})$$

is positive, and hence there exists a Hilbert space K and a mapping

$$F : \mathbb{D} \longrightarrow \mathcal{L}(\mathbb{C}^2, K)$$

such that, for all $x, y \in \mathbb{D}$,

$$\frac{1 - \psi(y)^*\psi(x)}{1 - \bar{y}x} = F(y)^*F(x) \tag{7}$$

We shall invoke the foregoing identity with $P = \psi(\lambda_1), Q = \psi(\mu_1)$.

Lemma 1.8. *Let ψ be a 2×2 matricial Schur function, let K be a Hilbert space and let $F : \mathbb{D} \to \mathcal{L}(\mathbb{C}^2, K)$ be a mapping such that (7) is satisfied. Define $h : \mathbb{D}^2 \to \Delta$ by*

$$\begin{aligned} h(\lambda_1, \lambda_2) &= \mathcal{F}_{\psi(\lambda_1)}(\lambda_2) \\ &= \psi_{22}(\lambda_1) + \psi_{21}(\lambda_1)\lambda_2(1 - \psi_{11}(\lambda_1)\lambda_2)^{-1}\psi_{12}(\lambda). \end{aligned}$$

For any pair of points $\lambda = (\lambda_1, \lambda_2), \mu = (\mu_1, \mu_2)$ in $\mathbb{D}^2$,

$$1 - \bar{h}(\mu)h(\lambda) = (1 - \bar{\mu}_1\lambda_1)\langle f(\lambda), f(\mu)\rangle_K + (1 - \bar{\mu}_2\lambda_2)\bar{g}(\mu)g(\lambda)$$

where

$$g(\lambda) = (1 - \psi_{11}(\lambda_1)\lambda_2)^{-1}\psi_{12}(\lambda_1)$$

and

$$f(\lambda) = F(\lambda_1)\begin{bmatrix} \lambda_2 g(\lambda) \\ 1 \end{bmatrix}.$$

Proof. In Lemma 1.7 take $H = U = Y = \mathbb{C}$, $P = \psi(\lambda_1)$, $Q = \psi(\mu_1)$, $X = \lambda_2$, and $Y = \mu_2$. For any pair $\lambda = (\lambda_1, \lambda_2), \mu = (\mu_1, \mu_2)$ of points of $\mathbb{D}^2$ we have

$$\begin{aligned} &1 - h(\bar{\mu})h(\lambda) \\ &= \bar{\psi}_{12}(\mu_1)(1 - \bar{\mu}_2\bar{\psi}_{11}(\mu_1))^{-1}(1 - \bar{\mu}_2\lambda_2)(1 - \psi_{11}(\lambda_1)\lambda_2)^{-1}\psi_{12}(\lambda_1) \\ &\quad + \left[\bar{\psi}_{12}(\mu_1)(1 - \bar{\mu}_2\bar{\psi}_{11}(\mu_1))^{-1}\bar{\mu}_2 \quad 1\right] \\ &\qquad \times (1 - \psi(\mu)^*\psi(\lambda))\begin{bmatrix} \lambda_2(1 - \psi_{11}(\lambda_1)\lambda_2)^{-1}\psi_{12}(\lambda_1) \\ 1 \end{bmatrix} \\ &= (1 - \bar{\mu}_2\lambda_2)\bar{g}(\mu)g(\lambda) + \begin{bmatrix} \bar{g}(\mu)\bar{\mu}_2 & 1 \end{bmatrix}(1 - \psi(\mu_1)^*\psi(\lambda_1))\begin{bmatrix} \lambda_2 g(\lambda) \\ 1 \end{bmatrix}. \end{aligned}$$

By equation (7),

$$1 - \psi(\mu_1)^*\psi(\lambda_1) = (1 - \bar{\mu}_1\lambda_1)F(\mu)^*F(\lambda).$$

On combining the last two equations we obtain the stated relation. □

The significance of the last lemma is that there is an Agler realization of $\Phi(\lambda_2, \varphi(\lambda_1))$ in which the portion of the state space corresponding to λ_2 is one-dimensional.

Theorem 1.9. *Let $\varphi = (s,p) : \mathbb{D} \to \Gamma$ be analytic and let $h : \mathbb{D}^2 \to \Delta$ be given by*

$$h(\lambda_1, \lambda_2) = \frac{2\lambda_2 p(\lambda_1) - s(\lambda_1)}{2 - \lambda_2 s(\lambda_1)}.$$

There exist a Hilbert space H and operators

$$a = \begin{bmatrix} a_{11} & a_{12} \\ a_{21} & a_{22} \end{bmatrix} : H \oplus \mathbb{C} \to H \oplus \mathbb{C},\ b = \begin{bmatrix} b_1 \\ b_2 \end{bmatrix} : \mathbb{C} \to H \oplus \mathbb{C},$$

$$c = \begin{bmatrix} c_1 & c_2 \end{bmatrix} : H \oplus \mathbb{C} \to \mathbb{C},\ d : \mathbb{C} \to \mathbb{C}$$

such that $\begin{bmatrix} a & b \\ c & d \end{bmatrix}$ *is unitary on $(H \oplus \mathbb{C}) \oplus \mathbb{C}$ and, for all $\lambda = (\lambda_1, \lambda_2) \in \mathbb{D}^2$,*

$$h(\lambda) = d + c\Lambda(1 - a\Lambda)^{-1} b$$

where

$$\Lambda = \operatorname{diag}\{\lambda_1 1_H, \lambda_2 1_{\mathbb{C}}\}$$

Proof. By [4, Theorem 3] there exists a 2×2 matricial function ψ in the Schur class such that

$$\psi_{11} = \frac{1}{2} s = -\psi_{22}$$

and

$$p = -\det \psi = \frac{1}{4} s^2 + \psi_{12}\psi_{21}.$$

We have $h(\lambda) = \mathcal{F}_{\psi(\lambda_1)}(\lambda_2)$. By Lemma 1.8 there exist a Hilbert space H and functions

$$f : \mathbb{D}^2 \to H, \quad g : \mathbb{D}^2 \to \mathbb{C}$$

such that, for all $\lambda, \mu \in \mathbb{D}^2$,

$$1 - \bar{h}(\mu) h(\lambda) = (1 - \bar{\mu}_1 \lambda_1) \langle f(\lambda), f(\mu) \rangle_H + (1 - \bar{\mu}_2 \lambda_2) \bar{g}(\mu) g(\lambda).$$

We may re-write this relation without using minus signs:

$$\begin{aligned} &\langle \lambda_1 f(\lambda), \mu_1 f(\mu) \rangle_H + \langle \lambda_2 g(\lambda), \mu_2 g(\mu) \rangle_{\mathbb{C}} + 1 \\ &= \langle f(\lambda), f(\mu) \rangle_H + \langle g(\lambda), g(\mu) \rangle_{\mathbb{C}} + \langle h(\lambda), h(\mu) \rangle_{\mathbb{C}} \end{aligned}$$

This equation can be interpreted as the equality of the Gramians of two families of vectors in $H \oplus \mathbb{C} \oplus \mathbb{C}$:

$$\left\langle \begin{bmatrix} \lambda_1 f(\lambda) \\ \lambda_2 g(\lambda) \\ 1 \end{bmatrix}, \begin{bmatrix} \mu_1 f(\mu) \\ \mu_2 g(\mu) \\ 1 \end{bmatrix} \right\rangle = \left\langle \begin{bmatrix} f(\lambda) \\ g(\lambda) \\ h(\lambda) \end{bmatrix}, \begin{bmatrix} f(\mu) \\ g(\mu) \\ h(\mu) \end{bmatrix} \right\rangle$$

for λ, $\mu \in \mathbb{D}^2$. Hence there exists an isometry

$$L : \operatorname{span} \left\{ \begin{bmatrix} \lambda_1 f(\lambda) \\ \lambda_2 g(\lambda) \\ 1 \end{bmatrix} : \lambda \in \mathbb{D}^2 \right\} \longrightarrow \operatorname{span} \left\{ \begin{bmatrix} f(\lambda) \\ g(\lambda) \\ h(\lambda) \end{bmatrix} : \lambda \in \mathbb{D}^2 \right\}$$

such that

$$\begin{bmatrix} f(\lambda) \\ g(\lambda) \\ h(\lambda) \end{bmatrix} = L \begin{bmatrix} \lambda_1 f(\lambda) \\ \lambda_2 g(\lambda) \\ 1 \end{bmatrix} \tag{8}$$

for all $\lambda \in \mathbb{D}^2$. We can suppose that the domain and codomain of L are both of countably infinite codimension in $H \oplus \mathbb{C} \oplus \mathbb{C}$, for if necessary we may replace H by $H \oplus l^2$ and f by $f \oplus 0$. Hence there exists a unitary U on $H \oplus \mathbb{C} \oplus \mathbb{C}$ which extends L. Write

$$U = \begin{bmatrix} a & b \\ c & d \end{bmatrix}$$

with the operators a, b, c, d as in the statement of the lemma. Let

$$x(\lambda) = \begin{bmatrix} f(\lambda) \\ g(\lambda) \end{bmatrix} \in H \oplus \mathbb{C}$$

By equation (8) we have, for all $\lambda \in \mathbb{D}^2$,

$$\begin{bmatrix} x(\lambda) \\ h(\lambda) \end{bmatrix} = \begin{bmatrix} a & b \\ c & d \end{bmatrix} \begin{bmatrix} \Lambda x(\lambda) \\ 1 \end{bmatrix},$$

and so

$$\begin{aligned} x(\lambda) &= a\Lambda x(\lambda) + b1 \\ h(\lambda) &= c\Lambda x(\lambda) + d1. \end{aligned}$$

Since $a\Lambda$ is a strict contraction, $1 - a\Lambda$ is invertible and so we may eliminate $x(\lambda)$ to obtain

$$h(\lambda) = d + c\Lambda(1 - a\Lambda)^{-1}b.$$

□

We can now show that the two notions of realization of a function $\varphi : \mathbb{D} \to \Gamma$ coincide.

Theorem 1.10. *Let $\varphi = (s,p) : \mathbb{D} \to \Gamma$ be analytic and let*

$$h(\lambda) = \frac{2\lambda_2 p(\lambda_1) - s(\lambda_1)}{2 - \lambda_2 s(\lambda_1)}$$

for $\lambda = (\lambda_1, \lambda_2) \in \mathbb{D}^2$. Let H be a Hilbert space and let $\begin{bmatrix} a & b \\ c & d \end{bmatrix}$ be a unitary operator on $(H \oplus \mathbb{C}) \oplus \mathbb{C}$ such that

$$h(\lambda) = d + c\Lambda(1 - a\Lambda)^{-1}b \tag{9}$$

for all $\lambda \in \mathbb{D}^2$ and $\Lambda = \operatorname{diag}\{\lambda_1 I_H, \lambda_2 I_{\mathbb{C}}\}$, as in Theorem 1.9. Let

$$\psi = \left[\begin{array}{c|cc} a_{11} & a_{12} & b_1 \\ \hline a_{21} & a_{22} & b_2 \\ c_1 & c_2 & d \end{array}\right].$$

Then ψ is a 2×2 matricial function in the Schur class and

$$\psi_{11} = \frac{1}{2}s = -\psi_{22} \qquad \text{and} \qquad \det \psi = -p.$$

In other words, if

$$A = a_{11}, \quad B_1 = a_{12}, \quad B_2 = b_1$$

$$C = \begin{bmatrix} a_{21} \\ c_1 \end{bmatrix}, \quad D = \begin{bmatrix} a_{22} & b_2 \\ c_2 & d \end{bmatrix} = [\, D_{ij} \,]$$

then

$$s = 2\left[\begin{array}{c|c} A & B_1 \\ \hline C_1 & D_{11} \end{array}\right] = -2\left[\begin{array}{c|c} A & B_2 \\ \hline C_2 & D_{22} \end{array}\right]$$

and

$$p = -\det\left[\begin{array}{c|c} A & B \\ \hline C & D \end{array}\right].$$

Proof. Note that

$$\psi(z) = \begin{bmatrix} a_{22} & b_2 \\ c_2 & d \end{bmatrix} + \begin{bmatrix} a_{21} \\ c_1 \end{bmatrix} z(1 - a_{11}z)^{-1} [\, a_{12} \quad b_1 \,], \tag{10}$$

and so

$$\psi_{11} = \left[\begin{array}{c|c} a_{11} & a_{12} \\ \hline a_{21} & a_{22} \end{array}\right], \quad \psi_{22} = \left[\begin{array}{c|c} a_{11} & b_1 \\ \hline c_1 & d \end{array}\right].$$

We shall first show that $s = -2\psi_{22}$. From the definition of h we have, for $\lambda_1 \in \mathbb{D}$,

$$h(\lambda_1, 0) = -\frac{1}{2}s(\lambda_1),$$

while, by the realization formula (9)

$$\begin{aligned} h(\lambda_1, 0) &= d + c\begin{bmatrix} \lambda_1 & \\ & 0 \end{bmatrix}\left(1 - a\begin{bmatrix} \lambda_1 & \\ & 0 \end{bmatrix}\right)^{-1} b \\ &= d + [\, c_1\lambda_1 \quad 0 \,]\begin{bmatrix} 1 - a_{11}\lambda_1 & 0 \\ -a_{21}\lambda_1 & 1 \end{bmatrix}^{-1}\begin{bmatrix} b_1 \\ b_2 \end{bmatrix} \\ &= d + [\, c_1\lambda_1 \quad 0 \,]\begin{bmatrix} (1 - a_{11}\lambda_1)^{-1} & 0 \\ a_{21}\lambda_1(1 - a_{11}\lambda_1)^{-1} & 1 \end{bmatrix}\begin{bmatrix} b_1 \\ b_2 \end{bmatrix} \\ &= d + c_1\lambda_1(1 - a_{11}\lambda_1)^{-1}b_1 \\ &= \left[\begin{array}{c|c} a_{11} & b_1 \\ \hline c_1 & d \end{array}\right](\lambda_1) = \psi_{22}(\lambda_1). \end{aligned}$$

Thus $s = -2\psi_{22}$ as claimed.

Fix $\lambda_1 \in \mathbb{D}$ and consider the function $h(\lambda_1, z)$ of the variable z. It is defined and analytic for $z \in \mathbb{D}$, but being a linear fractional function, it has a meromorphic extension to the Riemann sphere; its extension has a unique pole, which is at $\dfrac{2}{s(\lambda_1)}$. We can also identify the pole of $h(\lambda_1, \cdot)$ from the realization formula (9). Let us write

$$x(z) = \begin{bmatrix} x_1(z) \\ x_2(z) \end{bmatrix} = \left(1 - a\begin{bmatrix} \lambda_1 & 0 \\ 0 & z \end{bmatrix}\right)^{-1} b \in H \oplus \mathbb{C}$$

for $z \in \mathbb{D}$, where λ_1 remains fixed. Thus

$$h(\lambda_1, \lambda_2) = d + c\Lambda x(\lambda_2).$$

We have

$$\begin{bmatrix} 1-a_{11}\lambda_1 & -a_{21}z \\ -a_{12}\lambda_1 & 1-a_{22}z \end{bmatrix} \begin{bmatrix} x_1(z) \\ x_2(z) \end{bmatrix} = \begin{bmatrix} b_1 \\ b_2 \end{bmatrix}.$$

Eliminate $x_1(z)$ to obtain

$$\begin{aligned} &\left[\, 1-a_{22} \quad -a_{21}\lambda_1(1-a_{11}\lambda_1)^{-1}a_{12}z \,\right] x_2(z) \\ &= b_2 + a_{21}\lambda_1(1-a_{11}\lambda_1)^{-1}b_1 \end{aligned}$$

which relation can be re-written

$$x_2(z) = \left(1 - \left[\begin{array}{c|c} a_{11} & a_{12} \\ \hline a_{21} & a_{22} \end{array}\right](\lambda_1)z\right)^{-1} \left[\begin{array}{c|c} a_{11} & b_1 \\ \hline a_{21} & b_2 \end{array}\right](\lambda_1).$$

It is immediately apparent that $x_2(\cdot)$ extends to a linear fractional function whose pole is at

$$\left[\begin{array}{c|c} a_{11} & a_{12} \\ \hline a_{21} & a_{22} \end{array}\right](\lambda_1)^{-1} = \psi_{11}(\lambda_1)^{-1}.$$

Furthermore,

$$(1-a_{11}\lambda_1)x_1(z) - a_{21}zx_2(z) = b_1,$$

and hence

$$x_1(z) = (1-a_{11}\lambda_1)^{-1}\left(b_1 + a_{21}zx_2(z)\right).$$

It follows that the analytic function $x_1 : \mathbb{D} \to H$ also has meromorphic extension to the Riemann sphere, with a unique pole, namely $\psi_{11}(\lambda_1)^{-1}$. Since

$$h(\lambda_1, z) = d + c_1\lambda_1 x_1(z) + c_2 z x_2(z),$$

the scalar function $h(\lambda_1, \cdot)$ extends to a meromorphic function which is analytic everywhere except possibly at $\psi_{11}(\lambda_1)^{-1}$. It must therefore be that $\dfrac{2}{s(\lambda_1)} = \dfrac{1}{\psi_{11}(\lambda_1)}$, that is

$$s = 2\psi_{11}.$$

It remains to show that $p = -\det\psi$, that is

$$\begin{aligned} p &= -\psi_{11}\psi_{22} + \psi_{12}\psi_{21} \\ &\doteq \frac{s^2}{4} + \psi_{12}\psi_{21}. \end{aligned}$$

From equation (10) we have

$$\psi_{12} = \left[\begin{array}{c|c} a_{11} & b_1 \\ \hline a_{21} & b_2 \end{array}\right], \qquad \psi_{21} = \left[\begin{array}{c|c} a_{11} & a_{12} \\ \hline c_1 & c_2 \end{array}\right].$$

It will therefore suffice to prove that

$$\left[\begin{array}{c|c} a_{11} & b_1 \\ \hline a_{21} & b_2 \end{array}\right] \left[\begin{array}{c|c} a_{11} & a_{12} \\ \hline c_1 & c_2 \end{array}\right] = p - \frac{s^2}{4}. \tag{11}$$

Observe that

$$\frac{\partial h}{\partial z}(\lambda_1, z)\bigg|_{z=0} = p(\lambda_1) - \frac{s(\lambda_1)^2}{4}.$$

Let us also calculate $\frac{\partial h}{\partial z}$ from the realization formula (9). If $\Lambda = \operatorname{diag}\{\lambda_1, z\}$, and $\Lambda^\sim = \operatorname{diag}\{\lambda_1, w\}$, then

$$\begin{aligned}\frac{\partial}{\partial z}\Lambda(1-a\Lambda)^{-1} &= \lim_{w\to z}\frac{1}{z-w}\left\{\Lambda(1-a\Lambda)^{-1}-(1-\Lambda^\sim a)^{-1}\Lambda^\sim\right\}\\ &= \lim_{w\to z}(1-\Lambda^\sim a)^{-1}\frac{1}{z-w}\{(1-\Lambda^\sim a)\Lambda-\Lambda^\sim(1-a\Lambda)\}(1-a\Lambda)^{-1}\\ &= (1-\Lambda a)^{-1}\begin{bmatrix}0&0\\0&1\end{bmatrix}(1-a\Lambda)^{-1}.\end{aligned}$$

Hence, on differentiating the realization formula (9), we obtain

$$\frac{\partial h}{\partial z}(\lambda_1, z) = c(1-\Lambda a)^{-1}\begin{bmatrix}0\\1\end{bmatrix}\begin{bmatrix}0 & 1\end{bmatrix}(1-a\Lambda)^{-1}b$$

and so

$$\begin{aligned}\frac{\partial h}{\partial z}(\lambda_1, 0) &= c\begin{bmatrix}1-\lambda_1 a_{11} & -\lambda_1 a_{12}\\ 0 & 1\end{bmatrix}^{-1}\begin{bmatrix}0\\1\end{bmatrix}\begin{bmatrix}0 & 1\end{bmatrix}\begin{bmatrix}1-a_{11}\lambda_1 & 0\\ -a_{21}\lambda_1 & 1=\end{bmatrix}^{-1}b\\ &= \left[\begin{array}{c|c}a_{11} & a_{12}\\ \hline c_1 & c_2\end{array}\right](\lambda_1)\left[\begin{array}{c|c}a_{11} & b_1\\ \hline a_{21} & b_2\end{array}\right](\lambda_1).\end{aligned}$$

On equating the two expressions for $\frac{\partial h}{\partial z}(\lambda_1, 0)$ we obtain the desired relation (11). □

We can summarise this theorem by saying: every unitary realization of $\Phi(\lambda_2, \varphi(\lambda_1))$ is a realization of φ, for any analytic function $\varphi : \mathbb{D} \to \Gamma$. To introduce two notions and prove they are equivalent is not usually regarded as the height of mathematical taste, but the identity of these two realizations has real computational power. This has already been illustrated in the derivation of a realization of φ_β, and will be further demonstrated by some less simple examples in Section 3.

2. Complex geodesics in the symmetrised bidisc

In Theorems 2.1 and 2.2 below we describe two concrete classes of rational functions of low degree which map $\mathbb{D}$ analytically into Γ. Functions in these two classes furnish appropriate examples, we believe, on which to try out the calculation of realizations in the sense of Section 2. We shall now give some of the background to the selection of these particular functions. Readers who are willing to take this on trust can proceed to the description of the two classes. In fact we have already met the functions described in Theorem 2.1; the second class is described in Theorem 2.2.

The notion of a complex geodesic is defined for a wide class of complex manifolds [12, 14]. In the case of the domain G the complex geodesics were calculated explicitly in [4]. The most elementary way to think of complex geodesics of G is as extremal solutions of two-point interpolation problems. Consider any pair z_1, z_2 of distinct points of G. Pick the smallest possible $r \in (0,\ 1)$ such that there exists an analytic function $\varphi : \mathbb{D} \to G$ such that $\varphi(0) = z_1$ and $\varphi(r) = z_2$. Then φ is a complex geodesic of G through the points z_1 and z_2. The problem of finding the smallest r and an appropriate φ is called the *Kobayashi extremal problem* for G, z_1 and z_2. More generally, the definition of complex geodesics is as follows.

For any domain $\Omega \subset \mathbb{C}^n$ the *Carathéodory pseudodistance* C_Ω on Ω is defined, for any pair of points $z_1, z_2 \in \Omega$, by

$$C_\Omega(z_1, z_2) = \sup d(F(z_1), F(z_2))$$

where the supremum is taken over all analytic functions $F : \Omega \to \mathbb{D}$ and d denotes the hyperbolic distance on $\mathbb{D}$. It is easily seen that C_Ω is a pseudodistance, that is, it is symmetric and satisfies the triangle inequality. If in addition C_Ω separates the points of Ω then one says that Ω is *Carathéodory-hyperbolic*, and C_Ω is called the Carathéodory distance on Ω. An analytic function $\varphi : \mathbb{D} \to \Omega$ is called a *complex geodesic* in Ω if

$$C_\Omega(\varphi(\lambda),\ \varphi(\mu)) = d(\lambda,\ \mu)$$

for all $\lambda, \mu \in \mathbb{D}$. If φ is a complex geodesic then so is $\varphi \circ v$ for any Möbius function v; for the purposes of this paper, φ and $\varphi \circ v$ will be identified as their ranges are the same subset of Ω.

Complex geodesics play a particularly significant role in convex domains. A theorem of Lempert asserts that in a convex domain Ω the complex geodesics coincide with the solutions of the Kobayashi extremal problem for Ω. The symmetrised bidisc G is not convex, but it is nevertheless true that the complex geodesics of G coincide with the solutions of the Kobayashi extremal problem for G. This fact is also proved in [4]. Moreover, the complex geodesics of G are of one of two particularly simple forms.

Theorem 2.1. *Let $|\beta| < 1$ and let $\varphi_\beta : \mathbb{D} \to \mathbb{C}$ be defined by*

$$\varphi_\beta(\lambda) = (\beta\lambda + \bar{\beta}, \lambda), \qquad \lambda \in \mathbb{D}.$$

Then φ_β maps $\mathbb{D}$ into G, and φ_β is a complex geodesic of G.

Theorem 2.2. *Let $w = (t, q) \in G$ and let ω_0 be any value of $\omega \in \mathbb{T}$ which maximises*

$$\mathrm{Re}\ (2\omega t - \omega^2 q).$$

Let $\psi_w = (s, p)$ where, for any $\lambda \in \mathbb{D}$,

$$p(\lambda) \;=\; \frac{(4 - 2\omega_0 t + \omega_0^2 q)\lambda^2 + 2(t - \omega_0 q)\lambda + q}{4 - 2\bar{\omega_0}\bar{t} + \bar{\omega_0}^2\bar{q} + 2(\bar{t} - \bar{\omega_0}\bar{q})\lambda + \bar{q}\lambda^2} \tag{12}$$

and

$$s(\lambda) \;=\; 2\frac{\omega_0 p(\lambda) - \lambda}{1 - \omega_0\lambda}. \tag{13}$$

Then p is a Blaschke product of degree 2, ψ_w maps $\mathbb{D}$ analytically into G and ψ_w is a complex geodesic of G.

It can happen that there are two values of $\omega_0 \in \mathbb{T}$ for which Re $(2\omega t - \omega^2 q)$ is maximised, but it transpires that they give rise to (two parametrisations of) the same geodesic, so that we do not need to display the dependence of ψ_w on ω_0.

Theorem 2.3. *Let f be a complex geodesic of G. Either $f = \varphi_\beta$ for some $\beta \in \mathbb{D}$ or $f = \psi_w$ for some $w \in G$ (up to composition with a Möbius function).*

Complex geodesics of the form φ_β, $\beta \in \mathbb{D}$, will be called *affine geodesics*; those of the form ψ_w, $w \in G$, will be called *geodesics of second degree.*

In the next section we shall require a technical result about the complex number ω_0 occurring in Theorem 2.2.

Lemma 2.4. *Let $A,\ B \in \mathbb{C}$. The following statements are equivalent:*
(1) *there exist $x, y \in \mathbb{C}$ such that $x\bar{y} = A$ and $|x|^2 + |y|^2 = B$;*
(2) $2|A| \le B$.

The proof is elementary.

Lemma 2.5. *Let $(t, q) \in G$ and $\omega_0 \in \mathbb{T}$. The following statements are equivalent:*
(1) Re $(2\omega t - \omega^2 q)$ *attains its maximum over $\omega \in \mathbb{T}$ at ω_0;*
(2) $\omega_0 t - {\omega_0}^2 q \ \ge\ |q|$.

Proof. Statement (1) asserts that

$$\text{Re }(2\omega_0 t - {\omega_0}^2 q) \ge \text{Re }(2\omega t - \omega^2 q)$$

or equivalently

$$q\omega^2 - 2t\omega + 2\text{Re }(2\omega_0 t - {\omega_0}^2 q) - 2\bar{t}\bar{\omega} + \bar{q}\bar{\omega}^2 \ge 0$$

for all $\omega \in \mathbb{T}$. By the Fejér-Riesz Theorem [17, Sec. 53], the latter condition holds if and only if the left-hand side is expressible as $|\ P(\omega)\ |^2$ on $\mathbb{T}$ for some polynomial P. Since the left-hand side vanishes at $\omega_0 \in \mathbb{T}$, any such P has $\omega - \omega_0$ as a factor, and must therefore be expressible in the form

$$P(\omega) = (\omega - \omega_0)(x\omega + y)$$

for some $x,\ y \in \mathbb{C}$. Hence (1) holds if and only if there exist $x,\ y \in \mathbb{C}$ such that

$$\begin{aligned}
&q\omega^2 - 2t\omega + 2\text{Re }(2\omega_0 t - {\omega_0}^2 q) - 2\bar{t}\bar{\omega} + \bar{q}\bar{\omega}^2 \\
&= |\omega - \omega_0|^2 |x\omega + y|^2 \\
&= (2 - \bar{\omega}_0\omega - \omega_0\bar{\omega})(|x|^2 + |y|^2 + x\bar{y}\omega + \bar{x}y\bar{\omega}) \\
&= -\bar{\omega}_0 x\bar{y}\omega^2 + (2x\bar{y} - \bar{\omega}_0|x|^2 - \bar{\omega}_0|y|^2)\omega \\
&\quad +2|x|^2 + 2|y|^2 - \bar{\omega}_0\bar{x}y - \omega_0 x\bar{y} \\
&\quad +(2\bar{x}y - \omega_0|x|^2 - \omega_0|y|^2)\bar{\omega} - \omega_0\bar{x}y\bar{\omega}^2
\end{aligned} \tag{14}$$

for every $\omega \in \mathbb{T}$. On equating coefficients we deduce that (1) holds if and only if there exist $x, y \in \mathbb{C}$ such that

$$x\bar{y} = -\omega_0 q$$

and

$$\begin{aligned} |x|^2 + |y|^2 &= -\omega_0(-2t - 2x\bar{y}) \\ &= 2(\omega_0 t - {\omega_0}^2 q). \end{aligned}$$

It may be checked that this pair of equations implies the equality of the constant terms in equation (14). By Lemma 2.4, (1) holds if and only if

$$2\,|-\omega_0 q\,| \leq 2(\omega_0 t - {\omega_0}^2)q\ .$$

which is equivalent to (2). □

3. Realization of complex geodesics

In Section 1 we established the existence of realizations of analytic functions from $\mathbb{D}$ to Γ and we calculated realizations of affine geodesics φ_β, $\beta \in \mathbb{D}$ (Theorems 1.5 and 1.6). It remains a question as to whether it is feasible to calculate realizations of less special analytic functions from $\mathbb{D}$ to Γ, and so we explore this issue further by computing realizations of a class of more complicated (but still very special) functions. We believe that the complex geodesics of G constitute a natural class of functions on which to test our notion of realization. These functions are in a sense extremal analytic functions from $\mathbb{D}$ to G, and so it may be hoped that they have particularly simple realizations. This turns out to be true up to a point: they have realizations with state spaces of modest dimension, but the matrices and the computations needed to find them are complicated.

We turn to complex geodesics of G of second degree, as described in Theorem 2.2. Consider a point $w = (t,\ q)\ \in\ G$, and let ω_0 be any value of $\omega \in \mathbb{T}$ which maximises Re $(2\omega t - \omega^2 q)$. By Lemma 2.5 we have

$$|q|\ \leq\ \omega_0 t - {\omega_0}^2 q\ \in\ \mathbb{R}. \tag{15}$$

Let $\psi_w\ =\ (s,\ p)\ :\ \ \mathbb{D} \to G$, where p, s are given by equations (12) and (13) respectively. It is convenient to introduce the notation

$$v_1 = 2(t - \omega_0 q),\ \ v_2 = 4 - 2\omega_0 t + {\omega_0}^2 q, \tag{16}$$

so that

$$p(\lambda) = \frac{v_2\lambda^2 + v_1\lambda + q}{\bar{q}\lambda^2 + \bar{v}_1\lambda + \bar{v}_2} \tag{17}$$

and

$$s(\lambda) = 2\,\frac{\omega_0(v_2\lambda^2 + v_1\lambda + q) - \lambda(\bar{q}\lambda^2 + \bar{v}_1\lambda + \bar{v}_2)}{(1-\omega_0\lambda)(\bar{q}\lambda^2 + \bar{v}_1\lambda + \bar{v}_2)}.$$

By construction, the factor $1 - \omega_0\lambda$ in the latter expression cancels, and in fact one may verify that

$$s(\lambda) = 2\,\frac{\bar{\omega}_0\bar{q}\lambda^2 + \eta\lambda + \omega_0 q}{\bar{q}\lambda^2 + \bar{v}_1\lambda + \bar{v}_2}, \tag{18}$$

where

$$\eta = -4 + 2\text{Re } (2\omega_0 t - {\omega_0}^2 q) = 4 - 2\text{Re } v_2. \tag{19}$$

The following relations between the variables will be useful.

Lemma 3.1. *Let* $(t, q) \in G$ *and let* $\omega_0 \in \mathbb{T}$ *maximise* $\text{Re } (2\omega t - \omega^2 q)$ *over* $\omega \in \mathbb{T}$. *Let* v_1, v_2 *and* η *be given by equations* (16) *and* (19). *The following relations hold.*

(i) $2\ |q| \ \leq\ |v_1| \ =\ \omega_0 v_1 \ <\ 2;$

(ii) $v_2 \ =\ 4 - |v_1| - {\omega_0}^2 q;$

(iii) $v_2 - \bar{\omega_0}^2 \bar{q} \ =\ |v_1| - \eta \ =\ 4 - 2\text{Re } (\omega_0 t) \ >\ 0;$

(iv) $|v_2| \ >\ 1.$

Proof. (i) By Lemma 2.5,

$$\omega_0 v_1 \ =\ 2(\omega_0 t - {\omega_0}^2 q) \ \geq\ 2|q|,$$

so that

$$2|q| \ \leq\ |v_1| \ =\ \omega_0 v_1.$$

Suppose that $|v_1| \geq 2$. Then

$$\omega_0 t - \omega^2 q \ =\ 1 + \varepsilon$$

for some $\varepsilon \geq 0$. Let $t_0 = \omega_0 t,\ q_0 = {\omega_0}^2 q$, so that $(t_0,\ q_0) \in G$ and $t_0 = q_0 + 1 + \varepsilon$. By Theorem 1.1,

$$|t_0 - \bar{t}_0 q_0| \ <\ 1 - |q_0|^2,$$

that is,

$$|\ q_0 + 1 + \varepsilon - (\bar{q}_0 + 1 + \varepsilon) q_0\ | < \ 1 - |q_0|^2,$$

which implies that

$$|\ 1 + \frac{\varepsilon}{1 - |q_0|^2}(1 - q_0)\ | \ <\ 1.$$

However,

$$\begin{aligned} \text{Re } (1 + \frac{\varepsilon}{1 - |q_0|^2}(1 - q_0)) \ &=\ 1 + \frac{\varepsilon}{1 - |q_0|^2}(1 - \text{Re } q_0) \\ &\geq\ 1, \end{aligned}$$

a contradiction. Thus $|v_1| \ <\ 2$.

(ii) $v_2 \ =\ 4 - 2\omega_0 t + {\omega_0}^2 q \ =\ 4 - 2(\omega_0 t - {\omega_0}^2 q) - {\omega_0}^2 q \ =\ 4 - |v_1| - {\omega_0}^2 q.$

(iii) Since $\omega_0 t - {\omega_0}^2 q$ is real, we have

$$\omega_0 t - {\omega_0}^2 q \ =\ \bar{\omega_0} \bar{t} - \bar{\omega_0}^2 \bar{q}.$$

Use this relation to eliminate $\bar{\omega_0}^2 \bar{q}$:

$$\begin{aligned} v_2 - \bar{\omega_0}^2 \bar{q} \ &=\ 4 - 2\omega_0 t + {\omega_0}^2 q - \bar{\omega_0}^2 \bar{q} \\ &=\ 4 - 2\omega_0 t + {\omega_0}^2 q + (\omega_0 t - {\omega_0}^2 q - \bar{\omega}_0 \bar{t}) \\ &=\ 4 - \omega_0 t - \bar{\omega}_0 \bar{t} \ =\ 4 - 2\text{Re } (\omega_0 t). \end{aligned}$$

(iv) $|v_2| \ = |\ 4 - |v_1| - {\omega_0}^2 q\ | \geq |\ 4 - |v_1|\ | - |q| \ >\ 2 - |q| \ >\ 1.$

Since $(t,\ q) \in G$, we have $|t| < 2$ and so

$$4 - 2\text{Re }(\omega_0 t) \ > \ 4 - 2 \cdot 2 \ = \ 0.$$

Furthermore, by part (ii), $v_2 + {\omega_0}^2 q = 4 - |v_1|$ and so

$$\begin{aligned}\eta &= 4 - \bar{v}_2 - v_2 \ = \ 4 - (\bar{v}_2 - {\omega_0}^2 q) - (v_2 + {\omega_0}^2 q) \\ &= 4 - (v_2 - \bar{\omega_0}^2 \bar{q}) - (4 - |v_1|) \\ &= |v_1| - (v_2 - \bar{\omega_0}^2 \bar{q}).\end{aligned}$$

Hence $v_2 - \bar{\omega_0}^2 \bar{q} \ = \ |v_1| - \eta.$ □

In view of Lemma 15 (iii) we may define $Q > 0$ by

$$\begin{aligned}Q &= (v_2 - \bar{\omega_0}^2 \bar{q})^{1/2} \ = \ (|v_1| - \eta)^{1/2} \\ &= (4 - 2\text{Re }(\omega_0 t))^{1/2}.\end{aligned} \tag{20}$$

We shall now compute a realization of $\psi_w,\ w \in G$, by the strategy that succeeded in Section 2 for the realization of $\varphi_\beta,\ \beta \in \mathbb{D}$. We write $\lambda = (\lambda_1, \lambda_2),\ \mu = (\mu_1, \mu_2)$ and introduce the function h in the Schur class on $\mathbb{D}^2$ defined by

$$h(\lambda) = \Phi(\lambda_2, \psi_w(\lambda_1)). \tag{21}$$

We shall then seek an identity of the form

$$1 - \bar{h}(\mu)h(\lambda) = (1 - \bar{\mu}_1\lambda_1)\langle\ g_1(\lambda), g_1(\mu)\ \rangle + (1 - \bar{\mu}_2\lambda_2)\langle\ g_2(\lambda), g_2(\mu)\ \rangle$$

for suitable vector-valued functions $g_1,\ g_2$ on $\mathbb{D}^2$. From such an identity, with the aid of Theorem 1.10, we shall derive a realization of ψ_w.

From equations (17), (18) and (21) we have

$$\begin{aligned}h(\lambda) &= \Phi(\lambda_2,\ s(\lambda_1),\ p(\lambda_1)) \\ &= \frac{2\lambda_2 p(\lambda_1) - s(\lambda_1)}{2 - \lambda_2 s(\lambda_1)} \\ &= \frac{\lambda_2 f_1(\lambda_1) - f_2(\lambda_1)}{\tilde{f}_1(\lambda_1) - \lambda_2 f_2(\lambda_1)}\end{aligned} \tag{22}$$

where

$$f_1(\lambda_1) = v_2{\lambda_1}^2 + v_1\lambda_1 + q,\ \ f_2(\lambda_1) = \bar{\omega}_0\bar{q}{\lambda_1}^2 + \eta\lambda_1 + \omega_0 q \tag{23}$$

and

$$\tilde{f}_1(\lambda_1) \ = \ {\lambda_1}^2 \bar{f}_1(1/\bar{\lambda}_1) \ = \ \bar{q}{\lambda_1}^2 + \bar{v}_1\lambda_1 + \bar{v}_2.$$

Note that the polynomial $1-\bar{h}(\mu)h(\lambda)$ is congruent (in an obvious sense) to the kernel

$$\begin{aligned} X(\lambda,\mu) &= (\tilde{f}_1(\mu_1)-\mu_2 f_2(\mu_1))^-(\tilde{f}_1(\lambda_1)-\lambda_2 f_2(\lambda_1)) \\ &\quad -(\mu_2 f_1(\mu_1)-f_2(\mu_1))^-(\lambda_2 f_1(\lambda_1)-f_2(\lambda_1)) \qquad (24)\\ &= \overline{\tilde{f}_1(\mu_1)}\tilde{f}_1(\lambda_1)-\bar{f}_2(\mu_1)f_2(\lambda_1) \\ &\quad +\bar{\mu}_2(\bar{\tilde{f}}_1(\mu_1)f_2(\lambda_1)-\bar{f}_2(\mu_1)\tilde{f}_1(\lambda_1)) \\ &\quad +\lambda_2(\bar{f}_2(\mu_1)f_1(\lambda_1)-\overline{\tilde{f}_1(\mu_1)}f_2(\lambda_1)) \\ &\quad +\bar{\mu}_2\lambda_2(\bar{f}_2(\mu_1)f_2(\lambda_1)-\bar{f}_1(\mu_1)f_1(\lambda_1)). \qquad (25)\end{aligned}$$

The key identity for the realization of ψ_w is the following, in which X is defined by equation (24) and v_1 and Q are as previously, that is,

$$v_1=2(t-\omega_0 q), \qquad Q=\sqrt{4-2\mathrm{Re}\ (\omega_0 t)}$$

(cf. equation (20)).

Lemma 3.2. *Let* $w=(t,q)\in G$ *and let* $\omega_0\in\mathbb{T}$ *maximise* Re $(2\omega t-\omega^2 q)$ *over* $\omega\in\mathbb{T}$. *For any* $\lambda,\ \mu\in\mathbb{C}^2$,

$$X(\lambda,\mu)=Q^2 Y(\lambda,\mu) \qquad (26)$$

where

$$\begin{aligned} Y(\lambda,\mu) &= (1-\bar{\mu}_1\lambda_1)[(4-2|v_1|)(1+\bar{\mu}_1\bar{\mu}_2)(1+\lambda_1\lambda_2) \\ &\quad +(|v_1|-\bar{\omega}_0\bar{q}\bar{\mu}_2-\omega_0 q\lambda_2)(1+\bar{\omega}_0\bar{\mu}_1)(1+\omega_0\lambda_1)] \\ &\quad +(1-\bar{\mu}_2\lambda_2)(|v_1|\bar{\mu}_1\lambda_1+\omega_0 q\bar{\mu}_1+\bar{\omega}_0\bar{q}\lambda_1) \\ &\quad (1+\bar{\omega}_0\bar{\mu}_1)(1+\omega_0\lambda_1). \qquad (27)\end{aligned}$$

Proof. Equation (25) shows that X can be written

$$\begin{aligned} X(\lambda,\mu) &= X_{00}(\lambda_1,\mu_1)+X_{01}(\lambda_1,\mu_1)\bar{\mu}_2^0\lambda_2 \\ &\quad +X_{10}(\lambda_1,\mu_1)\bar{\mu}_2{\lambda_2}^0+X_{11}(\lambda_1,\mu_1)\bar{\mu}_2\lambda_2\ ,\end{aligned}$$

where each X_{ij} is a polynomial of degree 2 in λ_1 and $\bar{\mu}_1$. Because of the symmetry

$$X(\mu,\lambda)\ =\ X(\lambda,\mu)^-$$

it will suffice to check that X_{00}, X_{01} and X_{11} agree with the corresponding coefficients on the right-hand side of equation (27). We start with X_{00}: we must show that, for any $x,\ y\in\mathbb{C}$,

$$X_{00}(x,y)\ =\ Q^2 Y_{00}(x,y) \qquad (28)$$

where we write

$$Y(\lambda,\mu)\ =\ \Sigma_{i,j}Y_{ij}(\lambda_1,\mu_1)\bar{\mu}_2^i{\lambda_2}^j.$$

From equation (25) we have

$$\begin{aligned} X_{00}(x,y) &= \overline{\tilde{f}_1(y)}\tilde{f}_1(x) - \overline{f_2(y)}f_2(x) \\ &= (q\bar{y}^2 + v_1\bar{y} + v_2)(\bar{q}x^2 + \bar{v}_1 x + \bar{v}_2) \\ &\quad -(\omega_0 q\bar{y}^2 + \eta\bar{y} + \bar{\omega}_0\bar{q})(\bar{\omega}_0\bar{q}x^2 + \eta x + \omega_0 q) \\ &= |v_2|^2 - |q|^2 + (v_1\bar{v}_2 - \eta\omega_0 q)\bar{y} \\ &\quad +(v_2\bar{v}_1 - \bar{\omega}_0\bar{q}\eta)x + (|v_1|^2 - \eta^2)\bar{y}x \\ &\quad +(q\bar{v}_2 - {\omega_0}^2 q^2)\bar{y}^2 + (v_2\bar{q} - \bar{\omega}_0^2\bar{q}^2)x^2 \\ &\quad +(v_1\bar{q} - \eta\bar{\omega}_0\bar{q})\bar{y}x^2 + (q\bar{v}_1 - \omega_0 q\eta)\bar{y}^2 x \\ &\quad +0\bar{y}^2x^2. \end{aligned} \tag{29}$$

On the other hand, from equation (27)

$$\begin{aligned} Y_{00}(x,y) &= Y(x,0,y,0) \\ &= (1-\bar{y}x)[4 - 2|v_1| + |v_1|(1+\bar{\omega}_0\bar{y})(1+\omega_0 x)] \\ &\quad +(|v_1|\bar{y}x + \omega_0 q\bar{y} + \bar{\omega}_0\bar{q}x)(1+\bar{\omega}_0\bar{y})(1+\omega_0 x) \\ &= 4 - |v_1| + (|v_1|\bar{\omega}_0 + \omega_0 q)\bar{y} + (|v_1|\omega_0 + \bar{\omega}_0\bar{q})x \\ &\quad +(-4 + 3|v_1| + {\omega_0}^2 q + \bar{\omega}_0^2\bar{q})\bar{y}x + q\bar{y}^2 + \bar{q}x^2 \\ &\quad +\omega_0 q\bar{y}^2 x + \bar{\omega}_0\bar{q}\bar{y}x^2 + 0\bar{y}^2x^2. \end{aligned} \tag{30}$$

Thus the relation (28) is equivalent to the following set of five equations:

$$|v_2|^2 - |q|^2 = Q^2(4 - |v_1|) \tag{31}$$

$$v_1\bar{v}_2 - \eta\omega_0 q = Q^2(\bar{\omega}_0|v_1| + \omega_0 q) \tag{32}$$

$$|v_1|^2 - \eta^2 = Q^2(-4 + 3|v_1| + {\omega_0}^2 q + \bar{\omega}_0^2\bar{q}) \tag{33}$$

$$q\bar{v}_2 - {\omega_0}^2 q^2 = Q^2 q \tag{34}$$

$$q\bar{v}_1 - \omega_0 q\eta = Q^2\omega_0 q. \tag{35}$$

Begin with equation (31). In view of Lemma 15 (ii) we have

$$\begin{aligned} Q^2(4 - |v_1|) &= (\bar{v}_2 - {\omega_0}^2 q)(v_2 + {\omega_0}^2 q) \\ &= |v_2|^2 - {\omega_0}^2 q(v_2 - \bar{v}_2 + {\omega_0}^2 q). \end{aligned}$$

Since $v_2 - \bar{\omega}_0^2\bar{q}$ is real we have

$$\begin{aligned} v_2 - \bar{v}_2 + {\omega_0}^2 q &= v_2 - \bar{\omega}_0^2\bar{q} - (\bar{v}_2 - {\omega_0}^2 q) + \bar{\omega}_0^2\bar{q} \\ &= \bar{\omega}_0^2\bar{q}. \end{aligned}$$

Hence

$$\begin{aligned} Q^2(4 - |v_1|) &= |v_2|^2 - {\omega_0}^2 q \cdot \bar{\omega}_0^2\bar{q} \\ &= |v_2|^2 - |q|^2, \end{aligned}$$

and equation (31) is established. Next,

$$\begin{aligned} Q^2(\bar{\omega}_0|v_1| + \omega_0 q) &= (\bar{v}_2 - {\omega_0}^2 q)(v_1 + \omega_0 q) \\ &= v_1\bar{v}_2 + \omega_0 q(\bar{v}_2 - |v_1| - {\omega_0}^2 q). \end{aligned}$$

By Lemma 15 (ii),

$$\begin{aligned}\bar{v}_2 - |v_1| - {\omega_0}^2 q &= \bar{v}_2 + v_2 - 4\\ &= 2\text{Re } v_2 - 4\\ &= -\eta.\end{aligned}$$

Thus

$$Q^2(\bar{\omega}_0|v_1| + \omega_0 q) = v_1\bar{v}_2 - \eta\omega_0 q,$$

and so equation (32) is valid.

To verify equation (33) observe that, by Lemma 15 (ii),

$$\begin{aligned}|v_1| + \eta &= |v_1| + 4 - 2\text{Re } v_2\\ &= |v_1| + 4 - 2\text{Re } (4 - |v_1| - {\omega_0}^2 q)\\ &= -4 + 3|v_1| + {\omega_0}^2 q + \bar{\omega}_0^2\bar{q}.\end{aligned}$$

Hence, since $Q^2 = |v_1| - \eta$, we have

$$\begin{aligned}|v_1|^2 - \eta^2 &= (|v_1| - \eta)(|v_1| + \eta)\\ &= Q^2(-4 + 3|v_1| + {\omega_0}^2 q + \bar{\omega}_0^2\bar{q}),\end{aligned}$$

as required.

Equation (34) is immediate from the fact that $v_2 - \bar{\omega}_0^2\bar{q}$ is real, and furthermore

$$Q^2\omega_0 q = (|v_1| - \eta)\omega_0 q = q\bar{v}_1 - \omega_0 q\eta,$$

so that equation (35) holds. Hence the polynomial identity (28) is valid. Now we compare X_{01} and Y_{01}. We have

$$\begin{aligned}X_{01}(x,y) &= \overline{f_2(y)}f_1(x) - \overline{\tilde{f}_1(y)}f_2(x)\\ &= (\omega_0 q\bar{y}^2 + \eta\bar{y} + \bar{\omega}_0\bar{q})(v_2x^2 + v_1x + q)\\ &\quad -(q\bar{y}^2 + v_1\bar{y} + v_2)(\bar{\omega}_0\bar{q}x^2 + \eta x + \omega_0 q)\\ &= \bar{\omega}_0|q|^2 - v_2\omega_0 q + (\eta q - v_1\omega_0 q)\bar{y}\\ &\quad +(\bar{\omega}_0\bar{q}v_1 - v_2\eta)x + (\eta v_1 - v_1\eta)\bar{y}x\\ &\quad +(\omega_0 q^2 - \omega_0 q^2)\bar{y}^2 + (\bar{\omega}_0\bar{q}v_2 - v_2\bar{\omega}_0\bar{q})x^2\\ &\quad +(\eta v_2 - v_1\bar{\omega}_0\bar{q})\bar{y}x^2 + (\omega_0 q v_1 - q\eta)\bar{y}^2x\\ &\quad +(\omega_0 q v_2 - \bar{\omega}_0|q|^2)\bar{y}^2x^2. \qquad (36)\end{aligned}$$

From definition (27),

$$\begin{aligned}Y_{01}(x,y) &= (1 - \bar{y}x)[\ (4 - 2|v_1|)x - \omega_0 q(1 + \bar{\omega}_0\bar{y})(1 + \omega_0 x)\ ]\\ &= (1 - \bar{y}x)[\ -\omega_0 q - q\bar{y} + (4 - 2|v_1| - {\omega_0}^2 q)x - \omega_0 q\bar{y}x\ ].\end{aligned}$$

It is straightforward to verify that the right-hand side of equation (36) factorizes, to give

$$\begin{aligned} X_{01}(x,y) &= (1-\bar{y}x)[\,-\omega_0 q(v_2-\bar{\omega}_0^2\bar{q}) - q(|v_1|-\eta)\bar{y} \\ &\quad +(\bar{\omega}_0\bar{q}v_1 - v_2\eta)x - \omega_0 q(v_2-\bar{\omega}_0^2\bar{q})\bar{y}x\,] \\ &= (1-\bar{y}x)[\,-\omega_0 qQ^2 - qQ^2\bar{y} \\ &\quad +(\bar{\omega}_0\bar{q}v_1 - v_2\eta)x - \omega_0 qQ^2\bar{y}x\,]. \end{aligned}$$

In the formulae for X_{01} and Q^2Y_{01} all but one term are manifestly equal, and the desired relation

$$X_{01} = Q^2Y_{01} \tag{37}$$

will follow provided that the x terms agree, that is,

$$\bar{\omega}_0\bar{q}v_1 - v_2\eta = Q^2(4-2|v_1|-\omega_0^2 q). \tag{38}$$

Use the relations

$$\omega_0 v_1 = |v_1|, \quad v_2 = Q^2 + \bar{\omega}_0^2\bar{q}$$

to obtain

$$\begin{aligned} \bar{\omega}_0\bar{q}v_1 - v_2\eta &= \bar{\omega}_0^2\bar{q}|v_1| - (Q^2+\bar{\omega}_0^2\bar{q})\eta \\ &= \bar{\omega}_0^2\bar{q}(|v_1|-\eta) - Q^2\eta \\ &= Q^2(-\eta+\bar{\omega}_0^2\bar{q}). \end{aligned}$$

From the definition (19) of η we have

$$\begin{aligned} -\eta+\bar{\omega}_0^2\bar{q} &= 4-2\mathrm{Re}\,(2\omega_0 t - \omega_0^2 q) + \bar{\omega}_0^2 q \\ &= 4-2\mathrm{Re}\,(2\omega_0 t - 2\omega_0^2 q + \omega_0^2 q) + \bar{\omega}_0^2 q \\ &= 4-2|v_1| - 2\mathrm{Re}\,(\omega_0^2 q) + \bar{\omega}_0^2\bar{q} \\ &= 4-2|v_1| - {\omega_0}^2 q. \end{aligned}$$

Hence

$$\bar{\omega}_0\bar{q}v_1 - v_2\eta = Q^2(4-2|v_1|-\omega_0^2 q)$$

as required, and so the relation (38) is indeed valid.

To complete the proof that $X = Q^2Y$ we need to show that $X_{11} = Q^2Y_{11}$. This will follow by symmetry from the fact that $X_{00} = Q^2Y_{00}$. It may be checked with the aid of equation (25) that

$$\begin{aligned} X_{11}(x,y) &= -(\bar{v}_1-\omega_0\eta)q\bar{y} - (v_1-\bar{\omega}_0\eta)\bar{q}x \\ &\quad -Q^2q\bar{y}^2 - (|v_1|^2-\eta^2)\bar{y}x - Q^2\bar{q}x^2 \\ &\quad -(v_1\bar{v}_2-\omega_0 q\eta)\bar{y}^2x - (\bar{v}_1v_2-\bar{\omega}_0\bar{q}\eta)\bar{y}x^2 \\ &\quad -(|v_2|^2-|q|^2)\bar{y}^2x^2. \end{aligned}$$

By inspection, from equation (29), we have

$$X_{11}(x,y) = -\bar{y}^2x^2\bar{X}_{00}(1/\bar{x},\ 1/\bar{y})$$

(this relation can also be derived directly from equation (25)). On the other hand, from equation (27) we have

$$\begin{aligned} Y_{11}(x,y) &= (1-\bar{y}x)(4-2|v_1|)\bar{y}x \\ &\quad -(|v_1|\bar{y}x+\omega_0 q\bar{y}+\bar{\omega}_0\bar{q}x)(1+\bar{\omega}_0\bar{y})(1+\omega_0 x) \\ &= -\omega_0 q\bar{y}-\bar{\omega}_0\bar{q}x-q\bar{y}^2-\bar{q}x^2 \\ &\quad -(-4+3|v_1|+2\text{Re }(\omega_0^2 q))\bar{y}x-(\bar{\omega}_0|v_1|+\omega_0 q)\bar{y}^2x \\ &\quad -(\omega_0|v_1|+\bar{\omega}_0\bar{q})\bar{y}x^2-(4-|v_1|)\bar{y}^2x^2, \end{aligned}$$

and, on comparing with equation (30), we deduce that

$$Y_{11}(x,y) = -\bar{y}^2x^2\bar{Y}_{00}(1/\bar{x},\ 1/\bar{y}).$$

Since we know that $X_{00}=Q^2Y_{00}$ we can conclude:

$$\begin{aligned} Q^2Y_{11}(x,y) &= -Q^2\bar{y}^2x^2\bar{Y}_{00}(1/\bar{x},\ 1/\bar{y}) \\ &= -\bar{y}^2x^2\bar{X}_{00}(1/\bar{x},\ 1/\bar{y}) \\ &= X_{11}(x,y). \end{aligned}$$

Thus $X=Q^2Y$ and the key identity is established. □

The next step is to use the key identity to obtain a relation of the form

$$1-\bar{h}(\mu)h(\lambda)=(1-\bar{\mu}_1\lambda_1)\langle\, g_1(\lambda),g_1(\mu)\,\rangle+(1-\bar{\mu}_2\lambda_2)\langle\, g_2(\lambda),g_2(\mu)\,\rangle$$

for suitable vector-valued functions $g_1,\ g_2$ on $\mathbb{D}^2$. The key identity does not quite suffice for this purpose as it stands: in the expression (27) for Y, we need to subtract and add the quantity

$$M(1-\bar{\mu}_1\lambda_1)(1-\bar{\mu}_2\lambda_2)(1+\bar{\omega}_0\bar{\mu}_1)(1+\omega_0\lambda_1)$$

for a suitable positive M so that the coefficients of $1-\bar{\mu}_1\lambda_1$ and $1-\bar{\mu}_2\lambda_2$ are positive kernels on $\mathbb{D}^2$. If we do this we obtain a modified expression for Y:

$$\begin{aligned} Y(\lambda,\mu) &= (1-\bar{\mu}_1\lambda_1)[\ (4-2|v_1|)(1+\bar{\mu}_1\bar{\mu}_2)(1+\lambda_1\lambda_2) \\ &\quad +(|v_1|-M-\bar{\omega}_0\bar{q}\bar{\mu}_2-\omega_0 q\lambda_2+M\bar{\mu}_2\lambda_2)(1+\bar{\omega}_0\bar{\mu}_1)(1+\omega_0\lambda_1)\ ] \\ &\quad +(1-\bar{\mu}_2\lambda_2)\,[(|v_1|-M)\bar{\mu}_1\lambda_1+\omega_0 q\bar{\mu}_1+\bar{\omega}_0\bar{q}\lambda_1+M]\,(1+\bar{\omega}_0\bar{\mu}_1)(1+\omega_0\lambda_1). \end{aligned} \tag{39}$$

Note that if we can choose $M>0$ such that

$$|v_1|-M = \frac{|q|^2}{M} \tag{40}$$

then we have

$$\begin{aligned} &|v_1|-M-\bar{\omega}_0\bar{q}\bar{\mu}_2-\omega_0 q\lambda_2+M\bar{\mu}_2\lambda_2 \\ &= (M^{1/2}\bar{\mu}_2-M^{-1/2}\omega_0 q)(M^{1/2}\lambda_2-M^{-1/2}\bar{\omega}_0\bar{q}). \end{aligned}$$

Accordingly, let us choose

$$M=\frac{1}{2}\{\ |v_1|+\sqrt{|v_1|^2-4|q|^2}\ \}$$

as a solution of equation (40). By Lemma 3.1 (i) M is non-negative; it is zero only if $v_1 = q = 0$, that is, $t = q = 0$. Suppose for the moment that t and q are not both 0, and let

$$\xi = M^{1/2} > 0$$

The identity (39) becomes

$$\begin{aligned} Y(\lambda,\mu) &= (1-\bar{\mu}_1\lambda_1)[\,(4-2|v_1|)(1+\bar{\mu}_1\bar{\mu}_2)(1+\lambda_1\lambda_2) \\ &\quad +(\omega_0 q\xi^{-1}-\xi\bar{\mu}_2)(\bar{\omega}_0\bar{q}\xi^{-1}-\xi\lambda_2)(1+\bar{\omega}_0\bar{\mu}_1)(1+\omega_0\lambda_1)\,] \\ &\quad +(1-\bar{\mu}_2\lambda_2)(\xi+\omega_0 q\xi^{-1}\bar{\mu}_1)(\xi+\bar{\omega}_0\bar{q}\xi^{-1}\lambda_1)(1+\bar{\omega}_0\bar{\mu}_1)(1+\omega_0\lambda_1) \\ &= (1-\bar{\mu}_1\lambda_1)\langle G_1(\lambda),G_1(\mu)\rangle_{\mathbb{C}^2} + (1-\bar{\mu}_2\lambda_2)\langle G_2(\lambda),G_2(\mu)\rangle_{\mathbb{C}} \end{aligned} \tag{41}$$

where $G_1 : \mathbb{D}^2 \to \mathbb{C}^2$, $G_2 : \mathbb{D}^2 \to \mathbb{C}^2$ are defined by

$$\begin{aligned} G_1(\lambda) &= \begin{bmatrix} (4-2|v_1|)^{1/2} & 1+\lambda_1\lambda_2 \\ \bar{\omega}_0\bar{q}\xi^{-1}-\xi\lambda_2 & 1+\omega_0\lambda_1 \end{bmatrix}, \\ G_2(\lambda) &= (\xi+\bar{\omega}_0\bar{q}\xi^{-1}\lambda_1)(1+\omega_0\lambda_1). \end{aligned}$$

Note that, by virtue of Lemma 3.1 (i), $(4-2|v_1|)^{1/2} > 0$. Combine equations (26) and (41) to obtain

$$\begin{aligned} X(\lambda,\mu) &= (1-\bar{\mu}_1\lambda_1)\langle QG_1(\lambda), QG_1(\mu)\rangle \\ &\quad +(1-\bar{\mu}_2\lambda_2)\langle QG_2(\lambda), QG_2(\mu)\rangle. \end{aligned}$$

Now divide through by

$$(\tilde{f}_1(\mu_1)-\mu_2 f_2(\mu_1))^{-}(\tilde{f}_1(\lambda_1)-\lambda_2 f_2(\lambda_1))$$

to obtain the following result.

Theorem 3.3. *Let $w = (t,q) \in G$, $w \neq (0,0)$, and let*

$$h(\lambda) = \Phi(\lambda_2, \psi_w(\lambda_1)),\ \lambda = (\lambda_1,\lambda_2) \in \mathbb{D}^2.$$

For all $\lambda, \mu \in \mathbb{D}^2$,

$$1-\bar{h}(\mu)h(\lambda) = (1-\bar{\mu}_1\lambda_1)\langle g_1(\lambda),g_1(\mu)\rangle_{\mathbb{C}^2} + (1-\bar{\mu}_2\lambda_2)\langle g_2(\lambda),g_2(\mu)\rangle_{\mathbb{C}} \tag{42}$$

where $g_1 : \mathbb{D}^2 \to \mathbb{C}^2$, $g_2 : \mathbb{D}^2 \to \mathbb{C}^2$ are defined as follows. Let $\omega_0 \in \mathbb{T}$ maximise Re $(2\omega t - \omega^2 q)$ *over $\omega \in \mathbb{T}$ and let*

$$\begin{aligned} Q &= \sqrt{4-2\mathrm{Re}\ (\omega_0 t)}, \\ v_1 &= 2(t-\omega_0 q),\ v_2 = 4-2\omega_0 t+{\omega_0}^2 q, \\ \eta &= 4-2\mathrm{Re}\ \ v_2, \\ f_1(x) &= v_2 x^2 + v_1 x + q,\ f_2(x) = \bar{\omega}_0\bar{q}x^2+\eta x+\omega_0 q, \\ \tilde{f}_1(x) &= \bar{q}x^2+\bar{v}_1 x+\bar{v}_2, \\ \xi &= \frac{1}{\sqrt{2}}\sqrt{|v_1|+\sqrt{|v_1|^2-4|q|^2}}. \end{aligned}$$

Then $Q > 0$, $\xi > 0$ *and*

$$g_1(\lambda) = \frac{Q}{\tilde{f}_1(\lambda_1) - \lambda_2 f_2(\lambda_1)} \begin{bmatrix} (4-2|v_1|)^{1/2} & 1+\lambda_1\lambda_2 \\ \bar{\omega}_0\bar{q}\xi^{-1} - \xi\lambda_2 & 1+\omega_0\lambda_1 \end{bmatrix},$$

$$g_2(\lambda) = \frac{Q}{\tilde{f}_1(\lambda_1) - \lambda_2 f_2(\lambda_1)}(\xi + \bar{\omega}_0\bar{q}\xi^{-1}\lambda_1)(1+\omega_0\lambda_1).$$

Proof. The foregoing calculations establish that the identity (42) is valid for all $\lambda, \mu \in \mathbb{D}^2$ such that

$$(\tilde{f}_1(\mu_1) - \mu_2 f_2(\mu_1))^-(\tilde{f}_1(\lambda_1) - \lambda_2 f_2(\lambda_1)) \neq 0.$$

Hence, for all $\lambda \in \mathbb{D}^2$ not on the variety

$$\{\ \lambda : \tilde{f}_1(\lambda_1) = \lambda_2 f_2(\lambda_1)\ \},$$

we have

$$1 - |h(\lambda)|^2 = (1-|\lambda_1|^2)\|g_1(\lambda)\|^2 + (1-|\lambda_2|^2)\|g_2(\lambda)\|^2,$$

and hence, for λ in a dense open set in $\mathbb{D}^2$,

$$1 \geq 1 - |h(\lambda)|^2 = \frac{Q^2}{|\tilde{f}_1(\lambda_1) - \lambda_2 f_2(\lambda_1)|^2}\{\ (4-2|v_1|)|1+\lambda_1\lambda_2|^2$$
$$+|\bar{\omega}_0\bar{q}\xi^{-1} - \xi\lambda_2||1+\omega_0\lambda_1|^2 + |\bar{\omega}_0\bar{q}\xi^{-1}\lambda_1 + \xi|^2|1+\omega_0\lambda_1|^2\ \}.$$

Hence

$$|\tilde{f}_1(\lambda_1) - \lambda_2 f_2(\lambda_1)| \geq Q(4-2|v_1|)|1+\lambda_1\lambda_2|$$

on a dense open set in $\mathbb{D}^2$. By continuity the inequality remains valid on all of $\mathbb{D}^2$. The right-hand side is strictly positive (recall that $|v_1| < 2$ by Lemma 15 (i)), and so $\tilde{f}_1(\lambda_1) - \lambda_2 f_2(\lambda_1)$ does not vanish on $\mathbb{D}^2$. Hence the identity (42) holds for all $\lambda, \mu \in \mathbb{D}^2$. □

From the identity in Theorem 3.3 we derive a realization of h of Agler type. Rewrite relation (42) without minus signs:

$$1 + \langle\lambda_1 g_1(\lambda), \mu_1 g_1(\mu)\rangle + \langle\lambda_2 g_2(\lambda), \mu_2 g_2(\mu)\rangle$$
$$= \bar{h}(\mu)h(\lambda) + \langle g_1(\lambda), g_1(\mu)\rangle + \langle g_2(\lambda), g_2(\mu)\rangle.$$

This can be expressed as the equality of the Gramians of two families of vectors in $\mathbb{C}^2 \oplus \mathbb{C} \oplus \mathbb{C}$:

$$\left\langle \begin{bmatrix} \lambda_1 g_1(\lambda) \\ \lambda_2 g_2(\lambda) \\ 1 \end{bmatrix}, \begin{bmatrix} \mu_1 g_1(\mu) \\ \mu_2 g_2(\mu) \\ 1 \end{bmatrix} \right\rangle = \left\langle \begin{bmatrix} g_1(\lambda) \\ g_2(\lambda) \\ h(\lambda) \end{bmatrix}, \begin{bmatrix} g_1(\mu) \\ g_2(\mu) \\ h(\mu) \end{bmatrix} \right\rangle$$

for all $\lambda, \mu \in \mathbb{D}^2$. Hence there is an isometry

$$L : \text{span}\{ \begin{bmatrix} \lambda_1 g_1(\lambda) \\ \lambda_2 g_2(\lambda) \\ 1 \end{bmatrix} : \lambda \in \mathbb{D}^2\} \to \text{span}\{ \begin{bmatrix} g_1(\lambda) \\ g_2(\lambda) \\ h(\lambda) \end{bmatrix} : \lambda \in \mathbb{D}^2\} \tag{43}$$

such that, for all $\lambda \in \mathbb{D}^2$,

$$L \begin{bmatrix} \lambda_1 g_1(\lambda) \\ \lambda_2 g_2(\lambda) \\ 1 \end{bmatrix} = \begin{bmatrix} g_1(\lambda) \\ g_2(\lambda) \\ h(\lambda) \end{bmatrix}.$$

We leave it as an exercise to show that the two spans are in fact the whole of $\mathbb{C}^4$. Hence L is a unitary operator and is uniquely determined. Write

$$L = \begin{bmatrix} a & b \\ c & d \end{bmatrix}, \quad \Lambda = \operatorname{diag}\{\lambda_1, \lambda_1, \lambda_2\}$$

where a is 3×3 and d is 1×1. Then we have

$$a\Lambda \begin{bmatrix} g_1(\lambda) \\ g_2(\lambda) \end{bmatrix} + b1 = \begin{bmatrix} g_1(\lambda) \\ g_2(\lambda) \end{bmatrix}, \tag{44}$$

$$c\Lambda \begin{bmatrix} g_1(\lambda) \\ g_2(\lambda) \end{bmatrix} + d = h(\lambda). \tag{45}$$

It follows that, for all $\lambda \in \mathbb{D}^2$,

$$h(\lambda) = d + c\Lambda(1 - a\Lambda)^{-1}b.$$

We now calculate a, b, c and d.

$$\begin{aligned} d &= h(0,\ 0) = \Phi(0,\ s(0),\ p(0)) \\ &= -\frac{1}{2}s(0) = -\omega_0 q/\bar{v}_2, \end{aligned}$$

while, from equation (44) we have

$$b = \begin{bmatrix} g_1(0,\ 0) \\ g_2(0,\ 0) \end{bmatrix} = \frac{Q}{\bar{v}_2} \begin{bmatrix} (4 - 2|v_1|)^{1/2} \\ \bar{\omega}_0 \bar{q} \xi^{-1} \\ \xi \end{bmatrix}.$$

For brevity we shall write

$$R = (4 - 2|v_1|)^{1/2} = 2(1 - \omega_0 t + \omega_0^2 q)^{1/2}, \quad q_0 = \omega_0 q. \tag{46}$$

Let us find c ($= [c_1\ c_2\ c_3]$, say). On putting $\lambda_1 = 0$ in equation (45) we get

$$c_3 \lambda_2 g_2(0,\ \lambda_2) = h(0,\ \lambda_2) - d.$$

Here everything but c_3 is known, and the equation reduces to

$$c_3 \lambda_2 \frac{Q}{\bar{v}_2 - q_0 \lambda_2} \xi = \frac{Q^2 q \lambda_2}{(\bar{v}_2 - q_0 \lambda_2)\bar{v}_2},$$

whence

$$c_3 = \frac{Qq}{\bar{v}_2 \xi}.$$

On substituting $\lambda_2 = 0$ in equation (45) we have

$$\begin{aligned}[c_1\ c_2]xg_1(x,\ 0) &= h(x,\ 0) - d \\ &= -\frac{f_2(x)}{\tilde{f}_1(x)} + \frac{q_0}{\bar{v}_2} \\ &= \frac{-\bar{q}_0(\bar{v}_2 - \omega_0{}^2 q)x^2 - (\bar{v}_2\eta - \bar{v}_1\omega_0 q)x}{\bar{v}_2(\bar{q}x^2 + \bar{v}_1 x + \bar{v}_2)}.\end{aligned}$$

Cancelling x and bearing in mind the relation (39), we may write this equation

$$[c_1\ c_2]\frac{Q}{\bar{q}x^2 + \bar{v}_1 x + \bar{v}_2}\begin{bmatrix} R \\ \bar{q}_0\xi^{-1}(1+\omega_0 x)\end{bmatrix} = \frac{-Q^2\bar{q}_0 x + Q^2(4 - 2|v_1| - \bar{\omega}_0^2\bar{q})}{\bar{v}_2(\bar{q}x^2 + \bar{v}_1 x + \bar{v}_2)}$$

which reduces to

$$c_1 R + c_2\bar{\omega}_0\bar{q}\xi^{-1}(1+\omega_0 x) = \frac{Q}{\bar{v}_2}(-\bar{\omega}_0\bar{q}x + 4 - 2|v_1| - \bar{\omega}_0^2\bar{q}),$$

which has the unique solution for $c_1,\ c_2$,

$$c_1 = \frac{QR}{\bar{v}_2},\quad c_2 = -\frac{Q\xi\bar{\omega}_0}{\bar{v}_2}.$$

Thus

$$c = \frac{Q}{\bar{v}_2}[\ R\ \ -\bar{\omega}_0\xi\ q\xi^{-1}\ ].$$

It remains to find a. Write

$$a = [a_1\ a_2\ a_3]$$

and put $\lambda_1 = 0,\ \lambda_2 = x$ in equation (44) to get

$$a_3 x g_2(0,\ x) = \begin{bmatrix} g_1(0,\ x) \\ g_2(0,\ x)\end{bmatrix} - b,$$

that is,

$$a_3 x\frac{Q\xi}{\bar{v}_2 - q_0 x} = \frac{Q}{\bar{v}_2 - q_0 x}\begin{bmatrix} R \\ \bar{q}_0\xi^{-1} - \xi x \\ \xi\end{bmatrix} - \frac{Q}{\bar{v}_2}\begin{bmatrix} R \\ \bar{q}_0\xi^{-1} \\ \xi\end{bmatrix}.$$

This relation simplifies to

$$a_3\xi\bar{v}_2 x = \begin{bmatrix} Rq_0 \\ -\xi\bar{v}_2 + |q|^2\xi^{-1} \\ \xi q_0\end{bmatrix} x.$$

Hence

$$a_3 = \frac{1}{\xi\bar{v}_2}\begin{bmatrix} R\omega_0 q \\ -\xi\bar{v}_2 + |q|^2\xi^{-1} \\ \xi\omega_0 q\end{bmatrix}. \tag{47}$$

Now substitute $\lambda_1 = x,\ \lambda_2 = 0$ into equation (44).

$$[a_1\ a_2]xg_1(x,\ 0) = \begin{bmatrix} g_1(x,\ 0) \\ g_2(x,\ 0) \end{bmatrix} - b$$

$$= \frac{Q}{\bar{q}x^2 + \bar{v}_1 x + \bar{v}_2} \begin{bmatrix} R \\ \bar{q}_0\xi^{-1}(1+\omega_0 x) \\ (\xi + \bar{q}_0\xi^{-1}x)(1+\omega_0 x) \end{bmatrix} - \frac{Q}{\bar{v}_2}\begin{bmatrix} R \\ \bar{q}_0\xi^{-1} \\ \xi \end{bmatrix},$$

that is,

$$[a_1\ a_2]x\frac{Q}{\bar{q}x^2 + \bar{v}_1 x + \bar{v}_2}\begin{bmatrix} R \\ \bar{q}_0\xi^{-1}(1+\omega_0 x) \end{bmatrix} =$$

$$\frac{Q}{(\bar{q}x^2 + \bar{v}_1 x + \bar{v}_2)\bar{v}_2}\begin{bmatrix} -R\bar{v}_1 - R\bar{q}x \\ \bar{q}\xi^{-1}(\bar{v}_2 - \bar{\omega}_0\bar{v}_1) - \bar{\omega}_0\bar{q}^2\xi^{-1}x \\ \bar{v}_2(\bar{\omega}_0\bar{q}\xi^{-1} + \omega_0\xi) - \bar{v}_1\xi + \bar{q}(\bar{v}_2\xi^{-1} - \xi)x \end{bmatrix} x,$$

which can be written

$$Ra_1 + \bar{q}_0\xi^{-1}(1+\omega_0 x)a_2 = \frac{1}{\bar{v}_2}\begin{bmatrix} -R\bar{v}_1 - R\bar{q}x \\ \bar{q}\xi^{-1}(\bar{v}_2 - \bar{\omega}_0\bar{v}_1) - \bar{\omega}_0\bar{q}^2\xi^{-1}x \\ \bar{v}_2(\bar{\omega}_0\bar{q}\xi^{-1} + \omega_0\xi) - \bar{v}_1\xi + \bar{q}(\bar{v}_2\xi^{-1} - \xi)x \end{bmatrix}. \tag{48}$$

Equating coefficients of x we find that

$$\bar{q}\xi^{-1}a_2 = \frac{1}{\bar{v}_2}\begin{bmatrix} -R\bar{q} \\ -\bar{\omega}_0\bar{q}^2\xi^{-1} \\ \bar{q}(\bar{v}_2\xi^{-1} - \xi) \end{bmatrix},$$

and so

$$a_2 = \frac{1}{\bar{v}_2}\begin{bmatrix} -R\xi \\ -\bar{\omega}_0\bar{q} \\ \bar{v}_2 - \xi^2 \end{bmatrix}. \tag{49}$$

Substituting back into equation (48) and comparing the terms independent of x we obtain

$$\bar{v}_2 Ra_1 = -\bar{\omega}_0\begin{bmatrix} -R\bar{q} \\ -\bar{\omega}_0\bar{q}^2\xi^{-1} \\ \bar{q}(\bar{v}_2\xi^{-1} - \xi) \end{bmatrix} + \begin{bmatrix} -R\bar{v}_1 \\ \bar{q}\xi^{-1}(\bar{v}_2 - \bar{\omega}_0\bar{v}_1) \\ \bar{v}_2(\bar{\omega}_0\bar{q}\xi^{-1} + \omega_0\xi) - \bar{v}_1\xi \end{bmatrix}$$

$$= \begin{bmatrix} R(\bar{\omega}_0\bar{q} - \bar{v}_1) \\ \bar{q}\xi^{-1}(\bar{v}_2 - \bar{\omega}_0\bar{v}_1 + \bar{\omega}_0^2\bar{q}) \\ \xi(\bar{\omega}_0\bar{q} + \omega_0\bar{v}_2 - \bar{v}_1) \end{bmatrix} = \begin{bmatrix} R(-2\bar{t} + 3\bar{\omega}_0\bar{q}) \\ 4q\xi^{-1}(1 - \bar{\omega}_0\bar{t} + \bar{\omega}_0^2\bar{q}) \\ 4\omega_0\xi(1 - \bar{\omega}_0\bar{t} + \bar{\omega}_0^2\bar{q}) \end{bmatrix}. \tag{50}$$

On assembling the columns (47), (49) and (50) we obtain a, and hence a realization of h. The outcome of our calculations is as follows.

Theorem 3.4. *Let $w \in G$, $w \neq (0,\ 0)$, and let*

$$h(\lambda) = \Phi(\lambda_2,\ \psi_w(\lambda_1)).$$

Then h is in the Schur class on $\mathbb{D}^2$, and for any $\lambda \in \mathbb{D}^2$,

$$h(\lambda) = d + c\Lambda(1 - a\Lambda)^{-1}b$$

where

$$\Lambda = \operatorname{diag}\{\lambda_1,\ \lambda_1,\ \lambda_2\} \qquad \text{and} \qquad \begin{bmatrix} a & b \\ c & d \end{bmatrix}$$

is a unitary matrix defined as follows.

$$\begin{aligned}
a &= \bar{v}_2^{-1}\begin{bmatrix} -2\bar{t}+3\bar{\omega}_0\bar{q} & -\xi R & \omega_0 q\xi^{-1}R \\ q\xi^{-1}R & -\bar{\omega}_0\bar{q} & -\bar{v}_2+|q|^2\xi^{-2} \\ \omega_0\xi R & \bar{v}_2-\xi^2 & \omega_0 q \end{bmatrix}, \\
b &= Q\bar{v}_2^{-1}\begin{bmatrix} R \\ \bar{\omega}_0\bar{q}\xi^{-1} \\ \xi \end{bmatrix}, \\
c &= Q\bar{v}_2^{-1}\begin{bmatrix} R & -\bar{\omega}_0\xi & q\xi^{-1} \end{bmatrix}, \\
d &= -\bar{v}_2^{-1}\omega_0 q, \\
R &= 2(1-\omega_0 t+\omega_0^2 q)^{1/2} > 0
\end{aligned}$$

and ω_0, ξ, v_2 *and* Q *are as in the statement of Theorem* 3.3.

Note that in writing down the (2,1) and (3,1) entries of a we have used the fact that

$$4R^{-1}(1-\bar{\omega}_0\bar{t}+\bar{\omega}_0^2\bar{q}) = R.$$

This is so because, from the definition (46),

$$\begin{aligned}
R &= (4-2|v_1|)^{1/2} = (4-2\omega_0 v_1)^{1/2} = (4-4\omega_0 t+4\omega_0^2 q)^{1/2} \\
&= 2(1-\omega_0 t+\omega_0^2 q)^{1/2} = 2(1-\bar{\omega}_0\bar{t}+\bar{\omega}_0^2\bar{t})^{1/2}.
\end{aligned}$$

We have now found realizations of the functions $h = \Phi(\cdot,\ \psi(\cdot))$ for all complex geodesics ψ of G with a single exception, to wit $\psi = \psi_{(0,0)}$. In this case the parameter ξ in the foregoing theorem is zero and so the formulae in the description of the realization are undefined. For completeness we give a realization of the exceptional geodesic.

Consider $w = (t,\ q) = (0,\ 0)$. Here Re $(2\omega t - \omega^2 q)$ is identically, so $\omega_0 \in \mathbb{T}$ is arbitrary. The formulae (12), (13) give $\psi = \psi_w = (s,\ p)$ where

$$p(\lambda) = \lambda^2,\ s(\lambda) = -2\lambda,$$

and then the Schur function h on $\mathbb{D}^2$ is given by

$$h(\lambda) = \Phi(\lambda_2,\ -2\lambda_1,\ \lambda_1^2) = \lambda_1.$$

A realization of h is thus

$$h(\lambda_1,\ \lambda_2) = 0 + 1\lambda_1(1-0\lambda_1)^{-1}1,$$

corresponding to the unitary matrix $\begin{bmatrix} 0 & 1 \\ 1 & 0 \end{bmatrix}$.

We may now apply Theorem 1.10 to construct matricial realizations (of the type described in Theorem 1.2) of complex geodesics ψ_w.

Theorem 3.5. *Let $w = (t,\ q) \in G$, $w \neq (0,\ 0)$, and let $\psi_w = (s,\ p)$ be a corresponding complex geodesic of G, as described in Theorem 2.2. Let*

$$\begin{aligned} A &= \bar{v}_2^{-1}\begin{bmatrix} -2\bar{t}+3\bar{\omega}_0\bar{q} & -\xi R \\ q\xi^{-1}R & -\bar{\omega}_0\bar{q} \end{bmatrix}, \\ B &= \begin{bmatrix} B_1 & B_2 \end{bmatrix} = \bar{v}_2^{-1}\begin{bmatrix} \omega_0 q\xi^{-1}R & QR \\ -\bar{v}_2+|q|^2\xi^{-2} & \bar{\omega}_0\bar{q}Q\xi^{-1} \end{bmatrix}, \\ C &= \begin{bmatrix} C_1 \\ C_2 \end{bmatrix} = \bar{v}_2^{-1}\begin{bmatrix} \omega_0\xi R & \bar{v}_2-\xi^2 \\ QR & -\bar{\omega}_0 Q\xi \end{bmatrix}, \\ D &= \begin{bmatrix} D_{ij} \end{bmatrix} = \bar{v}_2^{-1}\begin{bmatrix} \omega_0 q & Q\xi \\ qQ\xi^{-1} & -\omega_0 q \end{bmatrix} \end{aligned}$$

where ω_0, ξ, v_2, Q and R are as in the statement of Theorem 3.4. Then

$$\begin{bmatrix} A & B \\ C & D \end{bmatrix}$$

is a unitary matrix and

$$\begin{aligned} s &= 2\left[\begin{array}{c|c} A & B_1 \\ \hline C_1 & D_{11} \end{array}\right] = -2\left[\begin{array}{c|c} A & B_2 \\ \hline C_2 & D_{22} \end{array}\right], \\ p &= -\det\left[\begin{array}{c|c} A & B \\ \hline C & D \end{array}\right]. \end{aligned}$$

The proof is simply to combine Theorems 1.10 and 3.4.
Again, for completeness, we deal with the special case

$$\psi(\lambda) = \psi_{(0,0)}(\lambda) = (-2\lambda,\ \lambda^2).$$

Here we may take

$$\begin{aligned} B &= 1,\ A = D = 0 \text{ of type } 2\times 2, \\ C &= \operatorname{diag}\{-1,\ 1\}. \end{aligned}$$

Then the $ABCD$ matrix is unitary and

$$\left[\begin{array}{c|c} A & B \\ \hline C & D \end{array}\right](\lambda) = \begin{bmatrix} -\lambda & 0 \\ 0 & \lambda \end{bmatrix},$$

so that $\psi = (s,\ p)$ where

$$s(\lambda) = 2\left[\begin{array}{c|c} A & B_1 \\ \hline C_1 & D_{11} \end{array}\right](\lambda) = -2\left[\begin{array}{c|c} A & B_2 \\ \hline C_2 & D_{22} \end{array}\right](\lambda)$$

and

$$p(\lambda) = -\det\left[\begin{array}{c|c} A & B \\ \hline C & D \end{array}\right](\lambda).$$

4. Conclusion

In an earlier paper [4] a notion of realization of an analytic function from $\mathbb{D}$ to G was obtained as a means to the solution of the spectral Nevanlinna-Pick problem for 2×2 matrices. We have shown in this paper that, at least for a certain natural class of special functions (the complex geodesics of G), realizations can be explicitly calculated. To this end we introduced a second notion of realization and proved that the two notions give rise to the same $ABCD$-matrix. The virtue of the second notion is that there is a natural way to approach its calculation, and by this means we were able to realize all complex geodesics of G. It must be admitted that the calculation is far from straightforward: one has to discover and establish an identity of a certain form (cf. equation (43)), a task which requires considerable amounts of experimentation and algebraic ingenuity. It remains to be seen to what extent the realization formula will prove practically useful, but in principle it can be used to construct analytic 2×2 matrix functions F on $\mathbb{D}$ with the property that $F(\lambda)$ is a stable matrix for every $\lambda \in \mathbb{D}$ (in the sense that the eigenvalues of $F(\lambda)$ lie in $\mathbb{D}$). Indeed, if $\begin{bmatrix} A & B \\ C & D \end{bmatrix}$ is unitary, with D of type 2×2, then we may write

$$
\begin{aligned}
s &= \left[\begin{array}{c|c} A & B_1 \\ \hline C_1 & D_{11} \end{array}\right] - \left[\begin{array}{c|c} A & B_2 \\ \hline C_2 & D_{22} \end{array}\right], \\
p &= \left(\frac{1}{4}\mathrm{tr}^2 - \det\right) \left[\begin{array}{c|c} A & B \\ \hline C & D \end{array}\right],
\end{aligned}
$$

and for any 2×2 analytic matrix function L on $\mathbb{D}$,

$$
F(\lambda) = \mathrm{e}^{-L(\lambda)} \begin{bmatrix} 0 & 1 \\ -p(\lambda) & s(\lambda) \end{bmatrix} \mathrm{e}^{L(\lambda)} \tag{51}
$$

has the stated property. The construction of such analytic matrix functions is relevant to some questions of engineering design. The formula (51) does not parametrise *all* analytic functions from $\mathbb{D}$ to the stable matrices: any F of the form (51) has the property that $F(\lambda)$ is non-derogatory for every $\lambda \in \mathbb{D}$ (in the 2×2 case, this means that $F(\lambda)$ is not a scalar multiple of the identity matrix). However, the formula does give an extensive class of analytic functions.

It is natural to ask whether the results extend to higher dimensions. Some of the results in the thesis of D. Ogle [16] are relevant to this question, but up to now we have not tried to extend the present results to dimensions greater than 2.

Other natural questions arise concerning the dynamical systems associated with the realizations which we have constructed in the form of transfer functions. Is there an illuminating input-state-output description of analytic functions from $\mathbb{D}$ to Γ? Theorem 1.10 suggests that there must be an intimate connection between a certain 2D system and a (one time-dimensional) system with two inputs and two outputs. What is that connection?

References

[1] J. Agler, On the representation of certain holomorphic functions defined on a polydisc, in *Topics in Operator Theory: Ernst D. Hellinger Memorial Volume*(Ed. L. de Branges, I. Gohberg and J. Rovnyak), Operator Theory: Advances and Applications, **48**, Birkhäuser Verlag, Basel, 47–66.

[2] J. Agler and N.J. Young, The two-point spectral Nevanlinna-Pick problem, *Integral Equations Operator Theory* **37** (2000) 375–385.

[3] J. Agler and N.J. Young, A model theory for Γ-contractions, *J. Operator Theory*, to appear.

[4] J. Agler and N.J. Young, The two-by-two spectral Nevanlinna-Pick problem, *Trans. Amer. Maths. Soc.*, to appear.

[5] J. Agler and N.J. Young, The complex geometry of the symmetrised bidisc, in preparation.

[6] B. Aupetit, Caractérisation spectrale des algèbres de Banach de dimension finie, *J. Functional Analysis* **26** (1977) 232–250.

[7] J.A. Ball, I.C. Gohberg and L. Rodman, *Interpolation of Rational Matrix Functions*, Operator Theory: Advances and Applications Vol. 45, Birkhäuser Verlag, Basel, 1990.

[8] J.A. Ball and N.J. Young, Problems on the realization of functions, *Fields Inst. Communications* **25** (2000) 179–185 (the Proceedings of a conference on applied operator theory, Winnipeg, 1998).

[9] H. Bart, I.C. Gohberg and M.A. Kaashoek, The state space method in problems of analysis, *Proc. of the First International Conference on Industrial and Applied Mathematics*, Contributions from the Netherlands, CWI, Amsterdam, [1987], 1–16.

[10] H. Bercovici, C. Foiaş and A. Tannenbaum, Spectral variants of the Nevanlinna-Pick interpolation problem, *Signal processing, scattering and operator theory, and numerical methods (Amsterdam, 1989)*, 23–45, Progr. Systems Control Theory, 5, Birkhäuser Boston, Boston, MA, 1990.

[11] H. Bercovici, C. Foiaş and A. Tannenbaum, A spectral commutant lifting theorem, *Trans. Amer. Math. Soc.* **325** (1991), 741–763.

[12] S. Dineen, *The Schwarz Lemma*, Oxford University Press, 1989.

[13] J.C. Doyle, Analysis of feedback systems with structured uncertainties, *Proc. IEE-D* **129** (1982), 242–250.

[14] S.G. Krantz, *Function theory of several complex variables.* Reprint of the 1992 edition. AMS Chelsea Publishing, Providence, RI, 2001.

[15] *Matlab μ-Analysis and Synthesis Toolbox*, The Math Works Inc., Natick, Massachusetts (http://www.mathworks.com/products/muanalysis/).

[16] D.J. Ogle, *Operator and function theory of the symmetrized polydisc* , Ph.D. Thesis, University of Newcastle upon Tyne, 1999.

[17] F. Riesz and B. Sz.-Nagy, *Functional Analysis,* Dover Publications Inc. New York, 1990.

[18] B. Sz.-Nagy and C. Foiaş, *Harmonic Analysis of Operators on Hilbert Space*, Akadémiai Kiadó, Budapest, 1970.

J. Agler
Department of Mathematics
University of California at San Diego
La Jolla
California CA 92093, USA

F.B. Yeh
Department of Mathematics
Tunghai University
Taichung
407, Taiwan

N.J. Young
School of Mathematics and Statistics
Merz Court
University of Newcastle upon Tyne
NE1 7RU, England

Operator Theory:
Advances and Applications, Vol. 143, 39–76

A Basic Interpolation Problem for Generalized Schur Functions and Coisometric Realizations

D. Alpay, T.Ya. Azizov, A. Dijksma, H. Langer, and G. Wanjala

Abstract. The basic interpolation problem for Schur functions is: Find all Schur functions $s(z)$ for which $s(0)$ has a given value. In this paper we consider the same basic interpolation problem but now for the class of generalized Schur functions with finitely many negative squares which are holomorphic at $z = 0$. In Section 3 its solutions are given by three fractional linear transformations in which the main parameter runs through a subset of the class of generalized Schur functions.

A generalized Schur function can be written as the characteristic function of a minimal coisometric colligation with a Pontryagin state space. In the second part of this paper we describe the colligation of a solution $s(z)$ of the basic interpolation problem for generalized Schur functions in terms of the colligation of the corresponding parameter function and the interpolation data. First we consider the canonical coisometric realization of $s(z)$ in which the state space is the reproducing kernel Pontryagin space with kernel $\frac{1-s(z)s(w)^*}{1-zw^*}$; see for example [6]. In the final section we follow a direct approach more in line with [2, 3].

1. Introduction

A Schur function is a holomorphic function on the open unit disk $\mathbb{D}$ which is bounded by 1 there. We denote the class of Schur functions by $\mathbf{S}_0$. Consider the following *basic interpolation problem* for Schur functions:

(P_0) Given $\sigma_0 \in \mathbb{C}$, determine all functions $s(z) \in \mathbf{S}_0$ with $s(0) = \sigma_0$.

The answer to this problem is well known: If $|\sigma_0| > 1$, then there is no solution. If $|\sigma_0| = 1$, by the maximum modulus principle there is only one solution namely

---

2000 *Mathematics Subject Classification.* Primary 47A48, 47A57, 47B32, 47B50.

Key words and phrases. Generalized Schur functions, coisometric realizations, reproducing kernel spaces.

The research for this paper was supported by the Netherlands Organization of Scientific Research NWO (grant 047-008-008), by the Russian Foundation for Basic Research RFBR (grant 02-01-00353), and by the Research Training Network HPRN-CT-2000-00116 of the European Union.

$s(z) \equiv \sigma_0$. If $|\sigma_0| < 1$, then there are infinitely many solutions, and the formula

$$s(z) = \frac{z\widetilde{s}(z) + \sigma_0}{z\sigma_0^*\widetilde{s}(z) + 1} \tag{1.1}$$

gives a one-to-one correspondence between all solutions and the functions $\widetilde{s}(z) \in \mathbf{S}_0$; this is a simple consequence of Schwarz's Lemma. The formula (1.1) is called a *parametrization formula for all solutions* and the function $\widetilde{s}(z)$ is called the *parameter for the solution* $s(z)$.

In this paper we pose the same basic question but now we look for all solutions $s(z)$ in the class $\mathbf{S}_\kappa^0$ of generalized Schur functions with κ negative squares which are holomorphic at $z = 0$. This class and some more general classes of functions used in the sequel are defined as follows. By $\mathbf{S}_\kappa$ we denote the set of all complex functions $s(z)$ which are meromorphic on $\mathbb{D}$ and have one of the following equivalent properties (see, for example, [17], [6]):

(i) $s(z)$ admits the representation

$$s(z) = \prod_{j=1}^{\kappa} \frac{1 - zz_j^*}{z - z_j} s_0(z)$$

with points $z_j \in \mathbb{D}$ and a function $s_0(z) \in \mathbf{S}_0$, $s_0(z_j) \neq 0$, $j = 1, 2, \ldots, \kappa$.

(ii) $s(z)$ has κ poles in $\mathbb{D}$ (counted according to their multiplicities) and

$$\limsup_{r \uparrow 1} |s(r\,\mathrm{e}^{i\,t})| \leq 1, \quad \text{for almost all } t \in [0, 2\pi].$$

(iii) The kernel

$$K_s(z,w) := \frac{1 - s(z)s(w)^*}{1 - zw^*}, \quad z, w \in \mathcal{D}(s),$$

has κ negative squares; here $\mathcal{D}(s)$ denotes the domain of holomorphy of $s(z)$ in $\mathbb{D}$.

The functions of the class $\mathbf{S}_\kappa$, $\kappa = 1, 2, \ldots$, are called *generalized Schur functions with* κ *negative squares.* The set of generalized Schur functions which have κ negative squares and which are holomorphic at the origin will be denoted by $\mathbf{S}_\kappa^0$. Evidently, $\mathbf{S}_0^0 = \mathbf{S}_0$ and $\mathbf{S}_\kappa^0 = \mathbf{S}_\kappa \cap \mathbf{A}^0$, where by $\mathbf{A}^0$ we denote the set of functions which are holomorphic at $z = 0$. Finally we set $\mathbf{S}^0 = \cup_{\kappa \in \mathbb{N}_0} \mathbf{S}_\kappa^0$.

Now we formulate the *basic interpolation problem* for the class $\mathbf{S}_\kappa^0$ with $\kappa \in \mathbb{N}_0$ and describe all its solutions.

(P_κ) Given $\sigma_0 \in \mathbb{C}$, determine all functions $s(z) \in \mathbf{S}_\kappa^0$ with $s(0) = \sigma_0$.

As for the problem (P_0), we consider three cases for the problem (P_κ). Case 1 corresponds to $|\sigma_0| < 1$, Case 2 to $|\sigma_0| > 1$, and Case 3 to $|\sigma_0| = 1$. In each of these cases we give a parametrization formula for the solutions. In the sequel we write the Taylor expansion of $s(z) \in \mathbf{A}^0$ at $z = 0$ as

$$s(z) = \sigma_0 + \sigma_1 z + \sigma_2 z^2 + \cdots + \sigma_n z^n + \cdots, \quad \sigma_n = s^{(n)}(0)/n!, \; n = 0, 1, \ldots.$$

In Case 1, for every $\kappa \in \mathbb{N}_0$ there are infinitely many solutions $s(z) \in \mathbf{S}_\kappa^0$ as the following theorem shows.

Theorem 1.1 ($|\sigma_0| < 1$). *The formula*

$$s(z) = \frac{z\widetilde{s}(z) + \sigma_0}{z\sigma_0^*\widetilde{s}(z) + 1}$$

gives a one-to-one correspondence between all solutions $s(z) \in \mathbf{S}_\kappa^0$ of the problem (P_κ) *and all parameters $\widetilde{s}(z) \in \mathbf{S}_\kappa^0$.*

If $s(z)$ is a solution given by the above formula with parameter $\widetilde{s}(z)$, the formula

$$s(z) - \sigma_0 = \frac{(1 - |\sigma_0|^2)z\widetilde{s}(z)}{z\sigma_0^*\widetilde{s}(z) + 1}$$

and the fact that for any integer $k \geq 1$,

$$\widetilde{s}(z) \in \mathbf{S}_\kappa^0 \iff z^k\widetilde{s}(z) \in \mathbf{S}_\kappa^0,$$

imply that the function $s(z) - \sigma_0$ has a zero of order k at $z = 0$ if and only if $\widetilde{s}(z)$ has a zero of order $k-1$ at $z = 0$. In other words, the formula

$$s(z) = \frac{z^k\widetilde{s}(z) + \sigma_0}{z^k\sigma_0^*\widetilde{s}(z) + 1}$$

gives a one-to-one correspondence between all solutions $s(z) \in \mathbf{S}_\kappa^0$ for which $\sigma_1 = \sigma_2 = \cdots = \sigma_{k-1} = 0$, $\sigma_k \neq 0$, and all parameters $\widetilde{s}(z) \in \mathbf{S}_\kappa^0$ with $\widetilde{s}(0) \neq 0$.

We now consider Case 2. As already mentioned there are no solutions in $\mathbf{S}_0$. The following theorem shows that for each $\kappa \geq 1$ there are infinitely many solutions $s(z) \in \mathbf{S}_\kappa^0$. In Section 3 we show that if $s(z)$ is a solution then $s(z) - \sigma_0$ has a zero of order k at $z = 0$ with $1 \leq k \leq \kappa$. This implies that the next theorem lists all solutions of the problem (P_κ) in Case 2.

Theorem 1.2 ($|\sigma_0| > 1$). *For each integer k with $1 \leq k \leq \kappa$, the formula*

$$s(z) = \frac{\sigma_0\widetilde{s}(z) + z^k}{\widetilde{s}(z) + \sigma_0^* z^k}$$

gives a one-to-one correspondence between all solutions $s(z) \in \mathbf{S}_\kappa^0$ of the problem (P_κ) *with $\sigma_1 = \sigma_2 = \cdots = \sigma_{k-1} = 0$ and $\sigma_k \neq 0$ and all parameters $\widetilde{s}(z) \in \mathbf{S}_{\kappa-k}^0$ with $\widetilde{s}(0) \neq 0$.*

Case 3 is more complicated than the first two cases. There is a unique solution in the class $\mathbf{S}_0$, namely $s(z) \equiv \sigma_0$. By the next theorem, for $\kappa \geq 1$, there are infinitely many solutions $s(z) \in \mathbf{S}_\kappa^0$. To describe these we use the following notation. We associate with any k complex numbers $s_0 \neq 0, s_1, \ldots, s_{k-1}$ the polynomial

$$\begin{aligned} Q(z) &= Q(z; s_0, s_1, \ldots, s_{k-1}) \\ &= c_0 + c_1 z + \cdots + c_{k-1}z^{k-1} - (c_{k-1}^* z^{k+1} + c_{k-2}^* z^{k+2} + \cdots + c_0^* z^{2k}) \end{aligned}$$

of degree $2k$, where the coefficients $c_0, c_1, \dots, c_{k-1}$ are determined by the formula

$$\begin{pmatrix} c_0 & & \\ \vdots & \ddots & \\ c_{k-1} & \cdots & c_0 \end{pmatrix}\begin{pmatrix} s_0 & & \\ \vdots & \ddots & \\ s_{k-1} & \cdots & s_0 \end{pmatrix} = \sigma_0 I_k;$$

here the matrices on the left-hand side are lower triangular. Assume $s(z)$ is a solution. If $s(z) \not\equiv \sigma_0$, then there is an integer $k \geq 1$ such that

$$s(z) - \sigma_0 = \sigma_k z^k + \cdots,$$

where $\sigma_k \neq 0$ and $\cdots$ indicates that the power series at $z = 0$ continues with summands with higher powers of z. Then with $Q(z) = Q(z; \sigma_k, \sigma_{k+1}, \dots, \sigma_{2k-1})$ we have

$$Q(z)(s(z) - \sigma_0) - \sigma_0 z^k \not\equiv 0$$

(otherwise $s(z) \notin \mathbf{S}_0$, see [2, p. 5, item (6)]), and hence there is an integer $\ell \geq 1$ and a complex number $t_\ell \neq 0$ such that

$$Q(z)(s(z) - \sigma_0) - \sigma_0 z^k = \sigma_0 t_\ell z^\ell + \cdots. \tag{1.2}$$

A straightforward calculation shows that $\ell = 2k + q$ for some integer $q \geq 0$. In Section 3 we show that necessarily $1 \leq k \leq \kappa$ and $0 \leq q \leq \kappa - k$.

Theorem 1.3 ($|\sigma_0| = 1$). *For each integer k with $1 \leq k \leq \kappa$, any choice of complex numbers $s_0 \neq 0, s_1, \dots, s_{k-1}$, and each integer q with $0 \leq q \leq \kappa - k$ the formula*

$$s(z) = \frac{(Q(z) + z^k)\widetilde{s}(z) - \sigma_0 Q(z) z^q}{\sigma_0^* Q(z)\widetilde{s}(z) - (Q(z) - z^k) z^q}$$

with $Q(z) = Q(z; s_0, s_1, \dots, s_{k-1})$ gives a one-to-one correspondence between all solutions $s(z) \in \mathbf{S}_\kappa^0$ of the problem (P$_\kappa$) *with $\sigma_1 = \sigma_2 = \cdots = \sigma_{k-1} = 0$ and*

$$\sigma_j = s_{j-k}, \quad j = k, k+1, \dots, 2k-1,$$

and all parameters $\widetilde{s}(z) \in \mathbf{S}_{\kappa-k-q}^0$ with $\widetilde{s}(0) \neq \sigma_0$ if $q = 0$ and $\widetilde{s}(0) \neq 0$ if $q > 0$.

If $s(z)$ is a solution, then, evidently, k is the order of the zero of $s(z) - \sigma_0$ at $z = 0$. Moreover, the coefficients $\sigma_{2k}, \sigma_{2k+1}, \dots, \sigma_{2k+q-1}$ of $s(z)$ are determined by $s_0, s_1, \dots, s_{k-1}$ and the coefficient σ_{2k+q} satisfies an inequality, due to the fact that, on account of (1.2), the function $Q(z)(s(z) - \sigma_0) - \sigma_0 z^k$ has a zero of order $2k + q$ at $z = 0$.

Note that if we invert formula (1.1) we get the Schur transformation

$$s(z) \mapsto \widetilde{s}(z) = \frac{1}{z}\frac{s(z) - \sigma_0}{1 - \sigma_0^* s(z)},$$

which goes back to Schur [18, 19]. If we invert the parametrization formulas in Theorems 1.1, 1.2, and 1.3 we obtain the three formulas for the *generalized Schur transform*; see (3.5). They constitute the main step in the generalized Schur algorithm which has been studied in [12], [15], [11], [13], [1], and also in [2, 3, 4]. The three theorems together are a corollary of Theorem 3.1 in Section 3, where the basic interpolation problem is treated from a slightly different point of view for functions in the class $\mathbf{S}^0$.

In the second part of this paper, Sections 4, 5, and 6, we describe the solutions of the interpolation problem as characteristic functions of minimal coisometric colligations obtained from the interpolation data and the minimal coisometric colligation of the parameter function. As an example we consider problem (P_0). If the parameter $\widetilde{s}(z)$ is given as the characteristic function $s_{\widetilde{V}}(z) = \widetilde{\gamma} + z\langle (I - z\widetilde{T})^{-1}\widetilde{u}, \widetilde{v}\rangle_{\widetilde{\mathcal{P}}}$ of a minimal coisometric colligation

$$\widetilde{V} = \begin{pmatrix} \widetilde{T} & \widetilde{u} \\ \langle\,\cdot\,, \widetilde{v}\rangle_{\mathcal{P}} & \widetilde{\gamma} \end{pmatrix} : \begin{pmatrix} \widetilde{\mathcal{P}} \\ \mathbb{C} \end{pmatrix} \to \begin{pmatrix} \widetilde{\mathcal{P}} \\ \mathbb{C} \end{pmatrix}$$

in which $\widetilde{T}$ is a bounded operator on a Pontryagin space $\widetilde{\mathcal{P}}$, $\widetilde{u}$ and $\widetilde{v}$ are elements of $\widetilde{\mathcal{P}}$ and $\widetilde{\gamma} = \widetilde{s}(0)$, then the solution $s(z)$ to the interpolation problem corresponding to this parameter coincides with the characteristic function $s_V(z)$ of the minimal coisometric colligation

$$V = \begin{pmatrix} -\widetilde{\gamma}\sigma_0^* & \frac{1}{r}\langle\,\cdot\,, \widetilde{v}\rangle_{\mathcal{P}} & \widetilde{\gamma} \\ -r\sigma_0\widetilde{u} & \widetilde{T} & r\widetilde{u} \\ r^2 & 0 & \sigma_0 \end{pmatrix} : \left(\begin{pmatrix} \mathbb{C}_{r^2} \\ \widetilde{\mathcal{P}} \end{pmatrix} \atop \mathbb{C} \right) \to \left(\begin{pmatrix} \mathbb{C}_{r^2} \\ \widetilde{\mathcal{P}} \end{pmatrix} \atop \mathbb{C} \right),$$

where $r = \sqrt{1 - |\sigma_0|^2}$ and $\mathbb{C}_{r^2}$ is the space $\mathbb{C}$ equipped with the inner product

$$\langle \alpha, \beta \rangle = r^2 \beta^* \alpha.$$

In this example, which is a special case of Theorem 6.1 below, the extending space $\mathbb{C}_{r^2}$ is a Hilbert space. In general the extending space will be a finite-dimensional Pontryagin space.

Note that the colligation for $\widetilde{s}(z)$ can be retrieved from that for $s(z)$: $\widetilde{T}$ is the compression of the main operator

$$T = \begin{pmatrix} -\widetilde{\gamma}\sigma_0^* & \frac{1}{r}\langle\,\cdot\,, \widetilde{v}\rangle \\ -r\sigma_0\widetilde{u} & \widetilde{T} \end{pmatrix} : \begin{pmatrix} \mathbb{C}_{r^2} \\ \widetilde{\mathcal{P}} \end{pmatrix} \to \begin{pmatrix} \mathbb{C}_{r^2} \\ \widetilde{\mathcal{P}} \end{pmatrix}$$

of the colligation V to the space $\widetilde{\mathcal{P}}$, $\widetilde{u}$ is the projection of the element

$$\frac{1}{r}\begin{pmatrix} \widetilde{\gamma} \\ r\widetilde{u} \end{pmatrix} \in \begin{pmatrix} \mathbb{C}_{r^2} \\ \widetilde{\mathcal{P}} \end{pmatrix}$$

onto the space $\widetilde{\mathcal{P}}$, and finally $\widetilde{v}$ is the projection of the element

$$\frac{1}{r}T^* \begin{pmatrix} r^2 \\ 0 \end{pmatrix} \in \begin{pmatrix} \mathbb{C}_{r^2} \\ \widetilde{\mathcal{P}} \end{pmatrix}$$

onto the space $\widetilde{\mathcal{P}}$. In [2, 3] we generalized this retrieving by compression and projection to the case of generalized Schur functions. These results will be obtained here too, but as consequences of the method.

Section 2 consists of three subsections. The first two contain results needed in the sequel about reproducing kernel Pontryagin spaces and minimal coisometric and canonical coisometric realizations of functions which are holomorphic near

$z = 0$ in the open unit disk $\mathbb{D}$. In the third subsection we apply these results to the generalized Schur functions. The main theorem in Section 3 is Theorem 3.1 which describes all solutions of the basic interpolation problem but now formulated for the class $\mathbf{S}^0$. Theorems 1.1, 1.2, and 1.3 stated above are immediate consequences of this result. The canonical coisometric colligation of a solution is described in Section 4 in terms of the interpolation data and the canonical coisometric colligation of the parameter. We show in Section 5 how from this description we can retrieve some of the results in [2, 3]. In Section 6 we provide a direct approach in line with the method employed in [2, 3]. Here we describe the minimal coisometric colligation of a solution in terms of the interpolation data and an abstract minimal coisometric colligation of the parameter. In the canonical colligation the state space is a reproducing kernel Pontryagin space consisting of functions which are holomorphic at $z = 0$ and the main operator is the difference quotient operator at $z = 0$ with which one can perform "calculations". In the direct approach the state space and the main operator are more abstract as in the example above.

Minimal and canonical unitary realizations will be considered elsewhere.

2. Preliminaries

1. **Reproducing kernel Pontryagin spaces.** A *Krein space* $\mathcal{K}$ is a complex linear space with an inner product $\langle \cdot\, , \cdot \rangle_{\mathcal{K}}$ such that $\mathcal{K}$ admits a fundamental decomposition $\mathcal{K} = \mathcal{K}_+ \oplus \mathcal{K}_-$ in which $\{\mathcal{K}_\pm, \pm\langle \cdot\, , \cdot\rangle_{\mathcal{K}}\}$ are Hilbert spaces and $\mathcal{K}_+ \perp \mathcal{K}_-$. The *positive/negative index* $\operatorname{ind}_\pm \mathcal{K} := \dim \mathcal{K}_\pm$ is independent of such decompositions. In this note a *Pontryagin space* is a Krein space with a finite negative index.

Let κ be a nonnegative integer. An $n \times n$ matrix-valued function $K(z, w)$ defined on some open set $\Omega \subset \mathbb{C}$ is said to have κ *positive* (*negative*, respectively) *squares*, in formula $\mathrm{sq}_\pm(K) = \kappa$, if

$$K(z, w)^* = K(w, z), \qquad z, w \in \Omega, \tag{2.1}$$

and for all choices of $\ell \in \mathbb{N}$, and $z_1, z_2, \ldots, z_\ell \in \Omega$, the Hermitian matrix

$$\left(K(z_i, z_j)\right)_{i,j=1}^{\ell} \tag{2.2}$$

has at most κ and for at least one such choice exactly κ positive (negative, respectively) eigenvalues. If (2.1) holds and for every $m \in \mathbb{N}$ there is a matrix of the form (2.2) which has m or more positive (negative, respectively) eigenvalues, then $K(z, w)$ is said to have infinitely many positive (negative, respectively) squares, in formula $\mathrm{sq}_\pm(K) = \infty$. In this context the function $K(z, w)$ is called a *kernel* or *Hermitian form.*

If $K(z, w)$ is an $n \times n$ matrix kernel with a finite number of negative squares, we denote by $\mathcal{P}(K)$ the *reproducing kernel Pontryagin space with reproducing kernel* $K(z, w)$. The space is the completion of the linear span of the functions

$z \mapsto K(z,w)\alpha$ on Ω, $w \in \Omega$, $\alpha \in \mathbb{C}^n$, equipped with the inner product

$$\langle K(\cdot,w)\alpha, K(\cdot,v)\beta\rangle_{\mathcal{P}(K)} = \beta^* K(v,w)\alpha, \quad w,v \in \Omega, \alpha,\beta \in \mathbb{C}^n.$$

It can be shown that $\mathrm{ind}_\pm \mathcal{P}(K) = \mathrm{sq}_\pm(K)$. (An equality like this means that the number on the left- and the number on the right-hand side are either both infinite or both finite and equal.) The elements of the space are n-vector functions on Ω and if $K(z,w)$ is holomorphic in $z \in \Omega$, then they are holomorphic on Ω. The key feature of the space $\mathcal{P}(K)$ is the reproducing property of the kernel $K(z,w)$:

$$\langle f, K(\cdot,w)\alpha\rangle_{\mathcal{P}(K)} = \alpha^* f(w), \quad f \in \mathcal{P}(K), w \in \Omega, \alpha \in \mathbb{C}^n.$$

In the sequel we shall use the following theorems. For more details and more general statements we refer to, for example, [6, Section 1.5].

Theorem 2.1. *Assume $K(z,w) = K_1(z,w) + K_2(z,w)$, where for $j = 1,2$, $K_j(z,w)$ is an $n \times n$ matrix kernel on Ω with a finite number of negative squares. Then*

(i) $\mathrm{sq}_-(K) \le \mathrm{sq}_-(K_1) + \mathrm{sq}_-(K_2)$.

(ii) *The following statements are equivalent:*
 (a) $\mathcal{P}(K) = \mathcal{P}(K_1) \oplus \mathcal{P}(K_2)$;
 (b) $\mathcal{P}(K_1) \subset \mathcal{P}(K)$ *and the inclusion mapping* $\mathcal{K}_1 \hookrightarrow \mathcal{K}$ *is isometric*;
 (c) $\mathcal{P}(K_1) \cap \mathcal{P}(K_2) = \{0\}$.

(iii) *If* (a), (b), *and* (c) *of* (ii) *hold then* $\mathrm{sq}_-(K) = \mathrm{sq}_-(K_1) + \mathrm{sq}_-(K_2)$.

Theorem 2.2. *Assume $K_1(z,w) = A(z)K(z,w)A(w)^*$, where $K(z,w)$ is an $n \times n$ matrix kernel on Ω with a finite number of negative squares and $A(z)$ is an $m \times n$ matrix-valued function on Ω. Then*

(i) $\mathrm{sq}_-(K_1) \le \mathrm{sq}_-(K)$,

(ii) $\mathcal{P}(K_1) = A\mathcal{P}(K)$,

(iii) *the operator*

$$f(z) \in \mathcal{P}(K) \mapsto A(z)f(z) \in \mathcal{P}(K_1)$$

of multiplication by $A(z)$ is a unitary map if and only if

$$\{f \in \mathcal{P}(K) | A(z)f(z) = 0, z \in \Omega\} = \{0\};$$

in this case $\mathrm{sq}_-(K_1) = \mathrm{sq}_-(K)$.

2. **Minimal and canonical coisometric realizations**. In the following theorem $S(z)$ is an $n \times n$ matrix function which is holomorphic in a connected neighborhood $\mathcal{D}(S)$ of the origin in $\mathbb{D}$ and $K_S(z,w)$ is the kernel

$$K_S(z,w) = \frac{J - S(z)JS(w)^*}{1 - zw^*}, \quad z,w \in \mathcal{D}(S),$$

where J is an $n \times n$ *signature matrix*, meaning that $J = J^* = J^{-1}$. The space $\mathbb{C}^n$ equipped with the inner product

$$\langle \alpha, \beta\rangle_J = \beta^* J\alpha, \quad \alpha,\beta \in \mathbb{C}^n,$$

is a Pontryagin space and will be denoted by $\mathbb{C}^n_J$. The following theorem can be proved as in, for example, [10] and [14].

Theorem 2.3. *Let $S(z)$ be an $n \times n$ matrix function holomorphic in a connected neighborhood $\mathcal{D}(S)$ of $z = 0$ in $\mathbb{D}$.*

(i) *There exist a Krein space $(\mathcal{K}, \langle \cdot, \cdot \rangle_{\mathcal{K}})$ and bounded operators $T : \mathcal{K} \to \mathcal{K}$, $F : \mathbb{C}^n \to \mathcal{K}$, and $G : \mathcal{K} \to \mathbb{C}^n$ such that the function*

$$S_V(z) = S(0) + zG^*(I - zT)^{-1}F$$

associated with the operator matrix

$$V := \begin{pmatrix} T & F \\ G & S(0) \end{pmatrix} : \begin{pmatrix} \mathcal{K} \\ \mathbb{C}^n_J \end{pmatrix} \to \begin{pmatrix} \mathcal{K} \\ \mathbb{C}^n_J \end{pmatrix}$$

coincides with $S(z)$ in a neighborhood of $z = 0$.

(ii) *In this representation V can always be chosen such that V is coisometric, that is, $VV^* = I$ in the space $\mathcal{K} \oplus \mathbb{C}^n_J$ and such that V is minimal, which means that*

$$\overline{\operatorname{span}}\,\{T^{*n}G^*\alpha \,|\, \alpha \in \mathbb{C}^n,\, n = 0, 1, \ldots\} = \mathcal{K},$$

and then

$$\operatorname{sq}_\pm(K_S) = \operatorname{ind}_\pm \mathcal{K}.$$

(iii) *If $\operatorname{sq}_-(K_S) < \infty$ then a minimal coisometric operator V in the representation for $S(z)$ is uniquely determined up to an isomorphism on the Pontryagin space $\mathcal{K}$, and $S(z) = S_V(z)$, $z \in \mathcal{D}(S)$.*

The operator V in (i) is called an *operator colligation* on the space $\mathcal{K} \oplus \mathbb{C}^n_J$ with entries T, F, G (and $S(0)$), T is called the *main operator* and $\mathcal{K}$ the *inner* or *state space* of the colligation. The function $S_V(z)$ is called the *characteristic function* of V. The identity

$$S(z) = S(0) + zG^*(I - zT)^{-1}F \quad \text{near } z = 0$$

is called the *operator representation* or *realization* of $S(z)$ induced by V. If V is (minimal) coisometric the realization is also called *(minimal) coisometric.* Instead of the term minimal, in connection with a coisometric colligation also the term *closely outer connected* is frequently used.

Now we assume that the kernel $K_S(z, w)$ has finitely many negative squares. Then $\mathcal{P}(K_S)$ is a reproducing kernel Pontryagin space. By what has been written in Subsection 2.1, the elements of $\mathcal{P}(K_S)$ are n-vector functions which are holomorphic on $\mathcal{D}(S)$; in particular they are holomorphic at the origin. A proof of the next theorem can be found in, for example, [6, Theorem 2.2.1].

Theorem 2.4. (i) *The operators*

$$\begin{aligned} R_0 &: \mathcal{P}(K_S) \to \mathcal{P}(K_S), & R_0 f(z) &= \frac{f(z) - f(0)}{z}, \\ E_0 &: \mathcal{P}(K_S) \to \mathbb{C}^n, & E_0 f &= f(0), \\ R_0 S &: \mathbb{C}^n \to \mathcal{P}(K_S), & R_0 S\alpha(z) &= \frac{S(z) - S(0)}{z}\alpha, \end{aligned}$$

are bounded operators.

(ii) *The colligation*

$$V = \begin{pmatrix} R_0 & R_0 S \\ E_0 & S(0) \end{pmatrix} : \begin{pmatrix} \mathcal{P}(K_S) \\ \mathbb{C}_J^n \end{pmatrix} \to \begin{pmatrix} \mathcal{P}(K_S) \\ \mathbb{C}_J^n \end{pmatrix}$$

is coisometric: $VV^* = I$ *on* $\mathcal{P}(K_S) \oplus \mathbb{C}_J^n$ *and minimal:*

$$\cap_{z \in \mathcal{D}(S)} \ker E_0 (I - zR_0)^{-1} = \{0\}.$$

(iii) $S(z)$ *has the corresponding minimal coisometric realization:*

$$S(z) = S(0) + zE_0(I - zR_0)^{-1} R_0 S \text{ near } z = 0. \tag{2.3}$$

The minimality conditions in Theorems 2.3 and 2.4 coincide and so by Theorem 2.3 (iii) the colligations in these theorems are the same up to an isomorphism of the state space. The colligation in Theorem 2.4 is called the *canonical coisometric colligation* for $S(z)$ and the identity (2.3) is called the *canonical coisometric realization* of $S(z)$.

We apply Theorems 2.3 and 2.4 for the case $n = 1$ in the next subsection. For the case $n = 2$ and

$$J = \begin{pmatrix} 1 & 0 \\ 0 & -1 \end{pmatrix}$$

we recall some results concerning 2×2 matrix polynomials $\Theta(z)$ which are J-unitary, that is, have the property

$$\Theta(z)^* J \Theta(z) = J, \quad z \in \mathbb{T},$$

where $\mathbb{T}$ stands for the unit circle. We denote by $\mathbf{H}_2$ the Hardy space of 2-vector functions $f(z) = \sum_{n=0}^{\infty} f_n z^n$ on $\mathbb{D}$ with coefficients $f_n \in \mathbb{C}^2$ such that $\sum_{n=0}^{\infty} f_n^* f_n < \infty$ and inner product

$$[f, g]_{\mathbf{H}_2} = \sum_{n=0}^{\infty} g_n^* f_n, \quad f(z) = \sum_{n=0}^{\infty} f_n z^n, \; g(z) = \sum_{n=0}^{\infty} g_n z^n \in \mathbf{H}_2.$$

The linear space $\mathbf{H}_2$ equipped with the inner product

$$[f, g]_{\mathbf{H}_{2,J}} = \sum_{n=0}^{\infty} g_n^* J f_n, \quad f(z) = \sum_{n=0}^{\infty} f_n z^n, \; g(z) = \sum_{n=0}^{\infty} g_n z^n \in \mathbf{H}_2,$$

is a Krein space and will be denoted by $\mathbf{H}_{2,J}$. For the following theorem we refer to [16] and [9]. Now that J is fixed, we write $\mathcal{P}(\Theta)$ instead of $\mathcal{P}(K_\Theta)$.

Theorem 2.5. *Let* $\Theta(z)$ *be a* J*-unitary matrix polynomial.*

(i) $\mathcal{P}(\Theta)$ *is a subspace of* $\mathbf{H}_{2,J}$ *and on* $\mathcal{P}(\Theta)$ *the* $\mathcal{P}(\Theta)$*-inner product coincides with the* $\mathbf{H}_{2,J}$*-inner product.*

(ii) *The McMillan degree* d *of* $\Theta(z)$ *equals the degree of* $\Theta(z)$ *as a matrix polynomial and* $d = \dim \mathcal{P}(\Theta)$.

(iii) $\mathcal{P}(\Theta) = \operatorname{span}\{R_0^k \Theta(z)\alpha \mid k = 1, 2, \ldots, d, \; \alpha \in \mathbb{C}^2\}$.

That the set on the right-hand side of the equality in part (iii) is contained in $\mathcal{P}(\Theta)$ follows from Theorem 2.4 (i). Later we shall use the following simple example. Let $\Theta_0(z)$ be a J-unitary 2×2 matrix polynomial and set

$$\Theta_q(z) = \Theta_0(z)\Psi_q(z), \quad \Psi_q(z) \begin{pmatrix} 1 & 0 \\ 0 & z^q \end{pmatrix}, \quad q = 0, 1, 2 \ldots.$$

Then $\Psi_q(z)$ and $\Theta_q(z)$ are J-unitary.

We claim

$$\mathcal{P}(\Theta_q) = \mathcal{P}(\Theta_0) \oplus \Theta_0(z)\mathcal{P}(\Psi_q).$$

To see this we use the decomposition

$$\begin{aligned} K_{\Theta_q}(z,w) &= \frac{J - \Theta(z)J\Theta(w)^*}{1 - zw^*} \\ &= \frac{J - \Theta_0(z)J\Theta_0(w)^*}{1 - zw^*} + \Theta_0(z)\frac{J - \Psi_q(z)J\Psi_q(w)^*}{1 - zw^*}\Theta_0(w)^* \\ &= K_{\Theta_0}(z,w) + \Theta_0(z)K_{\Psi_q}(z,w)\Theta(w)^*. \end{aligned}$$

From Theorem 2.5 (iii) it follows that $\mathcal{P}(\Theta_0) \subset \mathcal{P}(\Theta_q)$ and Theorem 2.5 (i) implies that the inclusion mapping $\mathcal{P}(\Theta_0) \hookrightarrow \mathcal{P}(\Theta_q)$ is isometric. Hence, by Theorem 2.1 (ii), the claim holds. Note that

$$\mathcal{P}(\Psi_q) = \{p(z)\Theta_0(z)\begin{pmatrix} 0 \\ 1 \end{pmatrix} \,|\, p(z) \text{ is a polynomial of degree } \leq q - 1\}.$$

3. **Generalized Schur functions**. By Theorem 2.3 with $n = 1$ and $J = 1$, every function $s(z) \in \mathbf{A}^0$ has a minimal coisometric realization of the form

$$s(z) = \gamma + \langle (1 - zT)^{-1}u, v\rangle_{\mathcal{K}}, \tag{2.4}$$

where $\mathcal{K}$ is a Krein space, u and v are elements of $\mathcal{K}$, $\gamma = s(0)$ and the colligation

$$V = \begin{pmatrix} T & u \\ \langle\,\cdot\,, v\rangle_{\mathcal{K}} & \gamma \end{pmatrix} \tag{2.5}$$

is coisometric and minimal. Applying the Riesz representation theorem we replaced the operator F in Theorem 2.3 by u and represented G^* by v. These "abstract" realizations are considered in Section 6.

If near $z = 0$, $s(z)$ has the Taylor expansion

$$s(z) = \sum_{n=0}^{\infty} \sigma_n z^n$$

then the operator representation (2.4) of $s(z)$ implies that

$$\sigma_0 = \gamma, \quad \sigma_{n+1} = \langle T^n u, v\rangle_{\mathcal{K}}, \; n = 0, 1, 2, \ldots.$$

The fact that the colligation V in the representation is coisometric in $\mathcal{K} \oplus \mathbb{C}$ can be characterized by its entries T, u, v, γ:

$$\begin{aligned} TT^* + \langle \cdot, u \rangle_{\mathcal{K}} u &= I, & (2.6) \\ \langle v, v \rangle_{\mathcal{K}} + |\gamma|^2 &= 1, & (2.7) \\ Tv + \gamma^* u &= 0. & (2.8) \end{aligned}$$

Finally, the minimality of V in (2.5) is equivalent to

$$\mathcal{K} = \overline{\operatorname{span}} \{ T^{*n} v \,|\, n = 0, 1, \ldots \}.$$

In particular $s(z) \in \mathbf{S}^0$ has a minimal coisometric realization of this form with a Pontryagin state space $\mathcal{K}$, and $\operatorname{ind}_- \mathcal{K} = \kappa$ if and only if $s(z) \in \mathbf{S}^0_\kappa$. To obtain the canonical coisometric realization of $s(z)$ we apply Theorem 2.4 with $n = 1$ and $J = 1$ (throughout the sequel for $s(z) \in \mathbf{S}^0$, we write $\mathcal{P}(s)$ instead of $\mathcal{P}(K_s)$):

$$V = \begin{pmatrix} R_0 & R_0 s \\ E_0 & s(0) \end{pmatrix} : \begin{pmatrix} \mathcal{P}(s) \\ \mathbb{C} \end{pmatrix} \to \begin{pmatrix} \mathcal{P}(s) \\ \mathbb{C} \end{pmatrix}$$

is the canonical coisometric colligation for $s(z)$ and

$$\begin{aligned} s(z) &= s(0) + z E_0 (I - z R_0)^{-1} R_0 S \\ &= s(0) + z \langle (I - z R_0)^{-1} R_0 s, K_s(\cdot, 0) \rangle_{\mathcal{P}(s)} \text{ near } z = 0 \end{aligned}$$

is the canonical coisometric realization. This realization of $s(z) \in \mathbf{S}^0$ coincides with the realization (2.4) if we set $\mathcal{K} = \mathcal{P}(s)$ and

$$T = R_0, \quad u = R_0 s, \quad v = K_s(\cdot, 0), \quad \gamma = s(0).$$

These canonical colligations play a central role in Sections 4 and 5.

3. Solutions of the interpolation problem

In the following theorem we describe all solutions of the interpolation problem:

(P^0) Given $\sigma_0 \in \mathbb{C}$, find all functions $s(z) \in \mathbf{S}^0$ with $s(0) = \sigma_0$.

The parametrization Theorems 1.1, 1.2, and 1.3 are simple consequences. Recall that we write the Taylor expansion of $s(z) \in \mathbf{S}^0$ at $z = 0$ as

$$s(z) = \sigma_0 + \sigma_1 z + \sigma_2 z^2 + \cdots + \sigma_n z^n + \cdots, \quad \sigma_n = s^{(n)}(0)/n!, \; n = 0, 1, \ldots.$$

Theorem 3.1. *Let k, q be integers with $k \geq 1$ and $q \geq 0$, let $s_0 \neq 0, s_1, \ldots, s_{k-1}$ be k complex numbers, and let $Q(z) = Q(z, s_0, s_1, \ldots, s_{k-1})$.*

(i) *The formula*

$$s(z) = \begin{cases} \dfrac{z^k\widetilde{s}(z) + \sigma_0}{\sigma_0^* z^k \widetilde{s}(z) + 1} & \text{in Case 1,} \\ \dfrac{\sigma_0\widetilde{s}(z) + z^k}{\widetilde{s}(z) + \sigma_0^* z^k} & \text{in Case 2,} \\ \dfrac{(Q(z) + z^k)\widetilde{s}(z) - \sigma_0 Q(z) z^q}{\sigma_0^* Q(z)\widetilde{s}(z) - (Q(z) - z^k) z^q} & \text{in Case 3,} \end{cases} \tag{3.1}$$

establishes a one-to-one correspondence between all solutions $s(z) \in \mathbf{S}^0$ of the problem (P^0) *with the property that in all three cases $\sigma_1 = \sigma_2 = \cdots = \sigma_{k-1} = 0$ and $\sigma_k \neq 0$, and in* Case 3 *with the additional property that*

$$\sigma_j = s_{j-k}, \quad j = k, k+1, \ldots, 2k-1,$$

and all parameters $\widetilde{s}(z) \in \mathbf{S}^0$ with

$$\widetilde{s}(0) \neq \begin{cases} 0 & \text{in Cases 1, 2, and 3 with } q > 0, \\ \sigma_0 & \text{in Case 3 with } q = 0. \end{cases} \tag{3.2}$$

(ii) *If $\widetilde{s}(z) \in \mathbf{S}^0_{\widetilde{\kappa}}$ then $s(z)$ defined by* (3.1) *belongs to the class $\mathbf{S}^0_\kappa$ with κ given by*

$$\kappa = \begin{cases} \widetilde{\kappa} & \text{in Case 1,} \\ \widetilde{\kappa} + k & \text{in Case 2,} \\ \widetilde{\kappa} + k + q & \text{in Case 3.} \end{cases} \tag{3.3}$$

In Case 3 the solution $s(z)$ in (3.1) has the property that $Q(z)(s(z)-\sigma_0)-\sigma_0 z^k$ has a zero of order $2k+q$ at $z=0$. This will be shown in the proof of the theorem below.

The expressions on the right-hand side of (3.1) are fractional linear transformations of the form

$$s(z) = T_{\Theta(z)}\widetilde{s}(z) = \frac{a(z)\widetilde{s}(z) + b(z)}{c(z)\widetilde{s}(z) + d(z)}, \tag{3.4}$$

where

$$\Theta(z) = \begin{pmatrix} a(z) & b(z) \\ c(z) & d(z) \end{pmatrix}$$

can be chosen as follows. In Case 1:

$$\Theta(z) = \Theta_1(z) := \frac{1}{\sqrt{1-|\sigma_0|^2}} \begin{pmatrix} 1 & \sigma_0 \\ \sigma_0^* & 1 \end{pmatrix} \begin{pmatrix} z^k & 0 \\ 0 & 1 \end{pmatrix},$$

in Case 2:

$$\Theta(z) = \Theta_2(z) := \frac{1}{\sqrt{|\sigma_0|^2 - 1}} \begin{pmatrix} \sigma_0 & 1 \\ 1 & \sigma_0^* \end{pmatrix} \begin{pmatrix} 1 & 0 \\ 0 & z^k \end{pmatrix},$$

and in Case 3:

$$\Theta(z) = \Theta_3(z) := \begin{pmatrix} Q(z) + z^k & \sigma_0 Q(z) \\ \sigma_0^* Q(z) & Q(z) - z^k \end{pmatrix} J \begin{pmatrix} 1 & 0 \\ 0 & z^q \end{pmatrix},$$

where J is the signature matrix

$$J = \begin{pmatrix} 1 & 0 \\ 0 & -1 \end{pmatrix}.$$

These Θ's are precisely the elementary J-unitary 2×2 matrix polynomials studied in [4]. With $\Theta(z) = \Theta_j(z)$, $j = 1, 2$, or 3, we associate the kernel

$$K_\Theta(z, w) = \frac{J - \Theta(z) J \Theta(w)^*}{1 - zw^*}$$

and denote the corresponding reproducing kernel space by $\mathcal{P}(\Theta)$ as in Section 2. In the theorem that follows we collect some facts related to this space needed in the proofs of Theorems 3.1 and 3.3. By C we denote the lower triangular $k \times k$ matrix

$$C = \begin{pmatrix} c_0 & & \\ \vdots & \ddots & \\ c_{k-1} & \cdots & c_0 \end{pmatrix},$$

and by $Z_k(z)$ the $1 \times k$ matrix polynomial

$$Z_k(z) = \begin{pmatrix} 1 & z & \cdots & z^{k-1} \end{pmatrix};$$

Evidently, every polynomial $p(z)$ of degree $\leq k-1$ can be written in the form $p(z) = Z_k(z)\alpha$ for some unique $\alpha \in \mathbb{C}^k$.

Theorem 3.2.

(i) *The 2×2 matrix $\Theta(z) = \Theta_j(z)$ is J-unitary, $j = 1, 2, 3$.*

(ii) *The space $\mathcal{P}(\Theta_j)$, $j = 1, 2$, is spanned by the columns of the matrix polynomial $\begin{pmatrix} 1 \\ \sigma_0^* \end{pmatrix} Z_k(z)$ and the Gram matrix for this basis is given by $(1 - |\sigma_0|^2) I_k$.*

(iii) *The space $\mathcal{P}(\Theta_3)$ is spanned by the columns of the three matrix polynomials*

$$\begin{pmatrix} 1 \\ \sigma_0^* \end{pmatrix} Z_k(z), \quad \begin{pmatrix} 0 \\ 1 \end{pmatrix} Z_k(z) C^* + \begin{pmatrix} 1 \\ \sigma_0^* \end{pmatrix} z^k Z_k(z), \quad \begin{pmatrix} \sigma_0 Q(z) \\ Q(z) - z^k \end{pmatrix} Z_q(z)$$

and the corresponding Gram matrix is given by

$$\begin{pmatrix} 0 & -\sigma_0^* C^* & 0 \\ -\sigma_0 C & -CC^* & 0 \\ 0 & 0 & -I_q \end{pmatrix}.$$

(iv) *We have*

$$\dim \mathcal{P}(\Theta_j) = \begin{cases} k & \text{if } j = 1, \\ k & \text{if } j = 2, \\ 2k + q & \text{if } j = 3, \end{cases} \quad \textit{and} \quad \operatorname{ind}_- \mathcal{P}(\Theta_j) = \begin{cases} 0 & \text{if } j = 1, \\ k & \text{if } j = 2, \\ k + q & \text{if } j = 3. \end{cases}$$

Item (i) can be verified by direct calculation and using the fact that for $z \in \mathbb{T}$ the function $z^k Q(z)^*$ assumes purely imaginary values. Items (ii) and (iii) follow from Theorem 2.5 and the example following it. Item (iv) is a consequence of the two preceding items.

Proof of Theorem 3.1. First we assume $s(z)$ is a solution and show that it must be of the form (3.1) for some $\widetilde{s}(z)$. Define $\widetilde{s}(z)$ by

$$\widetilde{s}(z) = \begin{cases} \dfrac{1}{z^k}\dfrac{s(z)-\sigma_0}{1-\sigma_0^* s(z)} & \text{in Case 1,} \\ z^k \dfrac{1-\sigma_0^* s(z)}{s(z)-\sigma_0} & \text{in Case 2,} \\ z^q \dfrac{(Q(z)-z^k)s(z)-\sigma_0 Q(z)}{\sigma_0^* Q(z)s(z)-(Q(z)+z^k)} & \text{in Case 3.} \end{cases} \tag{3.5}$$

We show that $\widetilde{s}(z)$ is holomorphic at $z=0$ by calculating $\widetilde{s}(0)$, in this way we also show that the parameter has the properties mentioned in the theorem. We begin with Case 1: We have

$$\widetilde{s}(z) = \frac{1}{z^k}\frac{\sigma_k z^k + \cdots}{(1-|\sigma_0|^2) - \sigma_0^*\sigma_k z^k + \cdots}$$

and hence

$$\widetilde{s}(0) = \frac{\sigma_k}{1-|\sigma_0|^2} \neq 0.$$

Now we consider Case 2:

$$\widetilde{s}(z) = z^k \frac{(1-|\sigma_0|^2) - \sigma_0^*\sigma_k z^k + \cdots}{\sigma_k z^k + \cdots}.$$

Since $s(z)$ is a solution we have $\sigma_k \neq 0$ and hence

$$\widetilde{s}(0) = \frac{1-|\sigma_0|^2}{\sigma_k} \neq 0.$$

Finally in Case 3, since the denominator of the fractional linear transform cannot be identically equal to zero, we have for some integer $q \geq 0$ and some complex number $t_{2k+q} \neq 0$,

$$\begin{aligned} &\sigma_0^* Q(z)s(z) - (Q(z)+z^k) \\ &\quad = \sigma_0^*[Q(z)(s(z)-\sigma_0) - \sigma_0 z^k] \\ &\quad = t_{2k+q} z^{2k+q} + \cdots . \end{aligned} \tag{3.6}$$

The numerator has the expansion

$$\begin{aligned} &(Q(z)-z^k)s(z) - \sigma_0 Q(z) \\ &\quad = Q(z)(s(z)-\sigma_0) - z^k s(z) \\ &\quad = -z^k(s(z)-\sigma_0) + [Q(z)(s(z)-\sigma_0) - \sigma_0 z^k] \\ &\quad = (-\sigma_k z^{2k} + \cdots) + [\sigma_0 t_{2k+q} z^{2k+q} + \cdots] \end{aligned}$$

where $\sigma_k \neq 0$, because $s(z)$ has a zero of order k. It follows that if $q > 0$, then

$$\widetilde{s}(z) = -z^q \frac{\sigma_k z^{2k} + \cdots}{t_{2k+q} z^{2k+q} + \cdots} = -\frac{\sigma_k}{t_{2k+q}} + \cdots ,$$

and the constant term on the right-hand side is nonzero. If $q = 0$ we have

$$\widetilde{s}(z) = \frac{(\sigma_0 t_{2k} - \sigma_k)z^{2k} + \cdots}{t_{2k}z^{2k} + \cdots} = (\sigma_0 - \frac{\sigma_k}{t_{2k}}) + \cdots,$$

and the constant term on the right-hand side is not equal to σ_0. So $\widetilde{s}(z)$ is holomorphic at $z = 0$ and (3.2) holds.

We now show that $\widetilde{s}(z)$ has a finite number of negative squares. The equality (3.5) can be written in the form

$$\widetilde{s}(z) = \frac{d(z)s(z) - b(z)}{-c(z)s(z) + a(z)} = T_{\widetilde{\Theta}(z)}s(z), \quad \widetilde{\Theta}(z) = \begin{pmatrix} d(z) & -b(z) \\ -c(z) & a(z) \end{pmatrix}, \tag{3.7}$$

with in Case 1:

$$\widetilde{\Theta}(z) = \widetilde{\Theta}_1(z) := \frac{1}{\sqrt{1-|\sigma_0|^2}} \begin{pmatrix} 1 & 0 \\ 0 & z^k \end{pmatrix} \begin{pmatrix} 1 & -\sigma_0 \\ -\sigma_0^* & 1 \end{pmatrix},$$

in Case 2:

$$\widetilde{\Theta}(z) = \widetilde{\Theta}_2(z) := \frac{1}{\sqrt{|\sigma_0|^2 - 1}} \begin{pmatrix} z^k & 0 \\ 0 & 1 \end{pmatrix} \begin{pmatrix} \sigma_0^* & -1 \\ -1 & \sigma_0 \end{pmatrix},$$

and in Case 3:

$$\widetilde{\Theta}(z) = \widetilde{\Theta}_3(z) := \begin{pmatrix} z^q & 0 \\ 0 & 1 \end{pmatrix} \begin{pmatrix} -(Q(z) - z^k) & \sigma_0 Q(z) \\ -\sigma_0^* Q(z) & Q(z) + z^k \end{pmatrix}.$$

The matrix $\widetilde{\Theta}(z) = \widetilde{\Theta}_j(z)$, $j = 1, 2, 3$, being the inverse of $\Theta_j(z)$ apart from the factor $1/\det \Theta_j(z)$ which is of modulus 1 if $|z| = 1$, is a J-unitary 2×2 matrix polynomial and hence the kernel

$$K_{\widetilde{\Theta}}(z,w) = \frac{J - \widetilde{\Theta}(z) J \widetilde{\Theta}(w)^*}{1 - zw^*}$$

has finitely many negative squares (see Theorem 2.5). We check that $d(z) + c(z)\widetilde{s}(z) \not\equiv 0$. We use that $k \geq 1$. In Case 1:

$$d(z) + c(z)\widetilde{s}(z) = \frac{1}{\sqrt{1-|\sigma_0|^2}}(1 + \sigma_0^* z^k \widetilde{s}(z)) = \frac{1}{\sqrt{1-|\sigma_0|^2}}(1 + \cdots).$$

In Case 2:

$$d(z) + c(z)\widetilde{s}(z) = \frac{1}{\sqrt{|\sigma_0|^2 - 1}}(\sigma_0^* z^k + \widetilde{s}(z)) = \frac{1}{\sqrt{|\sigma_0|^2 - 1}}(\widetilde{s}(0) + \cdots).$$

Finally in Case 3:

$$\begin{aligned} d(z) + c(z)\widetilde{s}(z) &= -(Q(z) - z^k)z^q + \sigma_0^* Q(z)\widetilde{s}(z) \\ &= \sigma_0^* Q(z)(\widetilde{s}(z) - \sigma_0 z^q) + z^{k+q} \\ &= \begin{cases} \sigma_0^* c_0 \widetilde{s}(0) + \cdots & \text{if } q > 0, \\ -\dfrac{\sigma_k}{t_{2k}} + \cdots & \text{if } q = 0. \end{cases} \end{aligned}$$

In these three formulas the constant term on the right-hand side is not equal to zero, and so the left-hand side is not identically equal to zero. From (3.7) and (3.4) we now have

$$\begin{aligned}\begin{pmatrix}1 & -\widetilde{s}(z)\end{pmatrix}\widetilde{\Theta}(z) &= \begin{pmatrix}d(z)+c(z)\widetilde{s}(z) & -b(z)-a(z)\widetilde{s}(z)\end{pmatrix}\\ &= \big(d(z)+c(z)\widetilde{s}(z)\big)\begin{pmatrix}1 & -s(z)\end{pmatrix},\end{aligned}$$

which implies

$$\begin{aligned}K_{\widetilde{s}}(z,w) &= \begin{pmatrix}1 & -\widetilde{s}(z)\end{pmatrix}\frac{J}{1-zw^*}\begin{pmatrix}1 & -\widetilde{s}(w)\end{pmatrix}^*\\ &\ \begin{pmatrix}1 & -\widetilde{s}(z)\end{pmatrix}\frac{J-\widetilde{\Theta}(z)J\widetilde{\Theta}(w)^*+\widetilde{\Theta}(z)J\widetilde{\Theta}(w)^*}{1-zw^*}\begin{pmatrix}1 & -\widetilde{s}(w)\end{pmatrix}^*\\ &= \begin{pmatrix}1 & -\widetilde{s}(z)\end{pmatrix}\frac{J-\widetilde{\Theta}(z)J\widetilde{\Theta}(w)^*}{1-zw^*}\begin{pmatrix}1 & -\widetilde{s}(w)\end{pmatrix}^*\\ &\quad+\big(d(z)+c(z)\widetilde{s}(z)\big)\frac{1-s(z)s(w)^*}{1-zw^*}\big(d(w)+c(w)\widetilde{s}(w)\big)^*\\ &= \begin{pmatrix}1 & -\widetilde{s}(z)\end{pmatrix}K_{\widetilde{\Theta}}(z,w)\begin{pmatrix}1 & -\widetilde{s}(w)\end{pmatrix}^*\\ &\quad+\big(d(z)+c(z)\widetilde{s}(z)\big)K_s(z,w)\big(d(w)+c(w)\widetilde{s}(w)\big)^*.\end{aligned}$$

This formula shows that $\widetilde{s}(z)$ has finitely many negative squares (see Theorem 2.1 (i)) and we conclude that $\widetilde{s}(z)$ belongs to the class $\mathbf{S}^0$.

Now we show that formula (3.1) defines a solution. In Case 1 and, because of the restrictions on the values of $\widetilde{s}(0)$ in Cases 2 and 3, also in these cases we have that $s(z)\to\sigma_0$ as $z\to 0$. It follows that $s(z)$ is holomorphic at $z=0$ and $s(0)=\sigma_0$. In Case 1 we have

$$s(z)-\sigma_0=\frac{(1-|\sigma_0|^2)\widetilde{s}(0)z^k+\cdots}{1+\cdots}$$

and hence $s(z)$ has a zero of order k at $z=0$. In case 2 we have since $k\geq 1$

$$s(z)-\sigma_0=\frac{(1-|\sigma_0|^2)z^k+\cdots}{\widetilde{s}(0)+\cdots},$$

which shows that $s(z)$ has a zero of order k at $z=0$. In Case 3 again because $k\geq 1$ we have

$$s(z)-\sigma_0=\frac{z^k(\widetilde{s}(z)-\sigma_0z^q)}{\sigma_0^*Q(z)(\widetilde{s}(z)-\sigma_0z^q)+z^{k+q}}=\begin{cases}\dfrac{z^k\widetilde{s}(0)+\cdots}{\sigma_0^*c_0\widetilde{s}(0)+\cdots} & \text{if } q>0,\\[2ex] \dfrac{z^k(\widetilde{s}(0)-\sigma_0)+\cdots}{\sigma_0^*c_0(\widetilde{s}(0)-\sigma_0)+\cdots} & \text{if } q=0.\end{cases}$$

Since if $q>0$, $\widetilde{s}(0)\neq 0$ and if $q=0$, $\widetilde{s}(0)\neq\sigma_0$, we have that also in Case 3 $s(z)-\sigma_0$ has a zero of order k at $z=0$. We continue with this case and find

$$Q(z)(s(z)-\sigma_0)-\sigma_0 z^k = -\frac{\sigma_0 z^{2k+q}}{\sigma_0^* Q(z)(\widetilde{s}(z)-\sigma_0 z^q)+z^{k+q}}$$

$$=\begin{cases} -\dfrac{\sigma_0 z^{2k+q}}{\sigma_0^* c_0 \widetilde{s}(0)+\cdots} & \text{if } q>0,\\ -\dfrac{\sigma_0 z^{2k+q}}{\sigma_0^* c_0(\widetilde{s}(0)-\sigma_0)+\cdots} & \text{if } q=0.\end{cases}$$

which shows that the function on the left-hand side has a zero of order $2k+q$ at $z=0$. This proves the remark directly after Theorem 3.1. By writing out the function on the left-hand side explicitly in terms of the coefficients of $s(z)$ we find that $\sigma_j=s_{j-k}$, $j=k,k+1,\dots,2k-1$.

In the same way as for $\widetilde{s}(z)$ above we will now prove that $s(z)\in\mathbf{S}^0$. We claim

$$a(z)-c(z)s(z)=tz^{\deg\Theta}+\cdots,\quad t\neq 0. \tag{3.8}$$

(For a more general result see [4].) Indeed, in Case 1 we have

$$a(z)-c(z)s(z)=\frac{1}{\sqrt{1-|\sigma_0|^2}}(1-\sigma_0 s(z))z^k=\sqrt{1-|\sigma_0|^2}z^k+\cdots,$$

in Case 2:

$$a(z)-c(z)s(z)=\frac{1}{\sqrt{|\sigma_0|^2-1}}(\sigma_0-s(z))=-\frac{\sigma_k}{\sqrt{|\sigma_0|^2-1}}z^k+\cdots,$$

and in Case 3:

$$\begin{aligned} a(z)-c(z)s(z)&=Q(z)+z^k-\sigma_0^*Q(z)s(z)=\\ &=-\sigma_0^*(Q(z)(s(z)-\sigma_0)-\sigma_0 z^k)=-t_{2k+q}z^{2k+q}+\cdots.\end{aligned}$$

From (3.4) we have

$$\begin{aligned}\begin{pmatrix}1 & -s(z)\end{pmatrix}\Theta(z)&=\begin{pmatrix}a(z)-c(z)s(z) & b(z)-d(z)s(z)\end{pmatrix}\\ &=(a(z)-c(z)s(z))\begin{pmatrix}1 & -\widetilde{s}(z)\end{pmatrix},\end{aligned}$$

which implies

$$\begin{aligned} K_s(z,w)&=\begin{pmatrix}1 & -s(z)\end{pmatrix}\frac{J}{1-zw^*}\begin{pmatrix}1 & -s(w)\end{pmatrix}^*\\ &\begin{pmatrix}1 & -s(z)\end{pmatrix}\frac{J-\Theta(z)J\Theta(w)^*+\Theta(z)J\Theta(w)^*}{1-zw^*}\begin{pmatrix}1 & -s(w)\end{pmatrix}^*\\ &=\begin{pmatrix}1 & -s(z)\end{pmatrix}\frac{J-\Theta(z)J\Theta(w)^*}{1-zw^*}\begin{pmatrix}1 & -s(w)\end{pmatrix}^*\\ &\quad+(a(z)-c(z)s(z))\frac{1-\widetilde{s}(z)\widetilde{s}(w)^*}{1-zw^*}(a(w)-c(w)s(w))^*\\ &=\begin{pmatrix}1 & -s(z)\end{pmatrix}K_\Theta(z,w)\begin{pmatrix}1 & -s(w)\end{pmatrix}^*\\ &\quad+(a(z)-c(z)s(z))K_{\widetilde{s}}(z,w)(a(w)-c(w)s(w))^*.\end{aligned}$$

By Theorem 2.1 (i), $s(z)$ has finitely many negative squares and so $s(z) \in \mathbf{S}^0$. Formula (3.3) follows directly from (3.10) in the next theorem and Theorem 3.2 (iv). □

Note that in Theorem 3.1 we require that $s(z) - \sigma_0$ has a zero of finite order at $z = 0$; hence the solution $s(z)$ with $s(z) \equiv \sigma_0$ with $|\sigma_0| = 1$ is excluded in the parametrization. The map $s(z) \mapsto \widetilde{s}(z)$ defined by (3.5) for all functions $s(z) \in \mathbf{S}_0$ with $s(z) \not\equiv \sigma_0$ if $|\sigma_0| = 1$ is the *generalized Schur transform* studied in [12], [15], [11], [13], [1], and also in [2, 3, 4].

Theorem 3.3. *Assume $s(z) \in \mathbf{S}^0$ is a solution of the interpolation problem and $\widetilde{s}(z)$ is the corresponding parameter : $s(z) = T_{\Theta(z)}\widetilde{s}(z)$ with $\Theta(z) = \Theta_j$ corresponding to* Case *j, $j = 1, 2, 3$.*

(i) *The space $\mathcal{P}(s)$ can be decomposed as the orthogonal direct sum*

$$\mathcal{P}(s) = (1 \quad - s(z))\mathcal{P}(\Theta) \oplus (a(z) - c(z)s(z))\mathcal{P}(\widetilde{s}) \tag{3.9}$$

and

$$\operatorname{ind}_-\mathcal{P}(s) = \operatorname{ind}_-\mathcal{P}(\Theta) + \operatorname{ind}_-\mathcal{P}(\widetilde{s}). \tag{3.10}$$

(ii) *The map*

$$U : \mathcal{P}(s) \ni f \mapsto \begin{pmatrix} \mathbf{f} \\ g \end{pmatrix} \in \begin{pmatrix} \mathcal{P}(\Theta) \\ \mathcal{P}(\widetilde{s}) \end{pmatrix}$$

determined by the decomposition (3.9)*:*

$$f(z) = \begin{pmatrix} 1 & -s(z) \end{pmatrix} \mathbf{f}(z) + (a(z) - c(z)s(z))g(z) \tag{3.11}$$

is unitary.

Proof. One way to prove (3.9) is to use the equality

$$\begin{aligned} K_s(z,w) = {} & \begin{pmatrix} 1 & -s(z) \end{pmatrix} K_\Theta(z,w) \begin{pmatrix} 1 & -s(w) \end{pmatrix}^* \\ & + (a(z) - c(z)s(z))K_{\widetilde{s}}(z,w)(a(w) - c(w)s(w))^*, \end{aligned}$$

proved at the end of the proof of Theorem 3.1 and to invoke Theorem 2.1 (ii) by showing that the map

$$\begin{pmatrix} 1 & -s(z) \end{pmatrix} : \mathcal{P}(\Theta) \to \mathcal{P}(s)$$

is a well-defined isometry; see [4, Theorem 6.6] or more generally, [7, 8]. But we also want to prove that the mappings

$$\begin{pmatrix} 1 & -s(z) \end{pmatrix} : \mathcal{P}(\Theta) \to \begin{pmatrix} 1 & -s(z) \end{pmatrix} \mathcal{P}(\Theta)$$

and

$$a(z) - c(z)s(z) : \mathcal{P}(\widetilde{s}) \to (a(z) - c(z)s(z))\mathcal{P}(\widetilde{s})$$

are injective. For then we can apply Theorem 2.2 and conclude that the map U in the theorem is unitary. The injectivity of the two maps and the orthogonal decomposition (3.9) can also be proved in one stroke. On account of Theorems 2.1 (ii) (c) and 2.2 (ii) they follow immediately from the implication

$$\begin{pmatrix} 1 & -s(z) \end{pmatrix} \mathbf{f}(\mathbf{z}) = (a(z) - c(z)s(z))g(z),\ \mathbf{f} \in \mathcal{P}(\Theta),\ g \in \mathcal{P}(\widetilde{s}) \Longrightarrow \mathbf{f} \equiv 0,\ g \equiv 0.$$

To prove the implication, assume that the equality on the left-hand side holds for some $\mathbf{f} \in \mathcal{P}(\Theta)$ and $g \in \mathcal{P}(\widetilde{s})$. Then by (3.8),

$$(a(z) - c(z)s(z))g(z) \text{ has a zero of order } \geq \deg\Theta \text{ at } z = 0.$$

By Theorem 3.2 (ii), in Cases 1 and 2

$$\mathbf{f} = \begin{pmatrix} 1 \\ \sigma_0^* \end{pmatrix} p(z)$$

for some polynomial $p(z)$ of degree $\leq k-1$ and so

$$\begin{pmatrix} 1 & -s(z) \end{pmatrix} \mathbf{f}(\mathbf{z}) = ((1 - |\sigma_0|^2) - \sigma_0^* \sigma_k z^k + \cdots)p(z),$$

which does not have a zero at $z = 0$ unless $p(z) \equiv 0$ and so the implication holds. We turn to Case 3: Now $\mathbf{f} \in \mathcal{P}(\Theta_q)$ has the form

$$\mathbf{f} = \begin{pmatrix} 1 \\ \sigma_0^* \end{pmatrix} p(z) + \begin{pmatrix} 0 \\ 1 \end{pmatrix} q_1(z) + \begin{pmatrix} 1 \\ \sigma_0^* \end{pmatrix} q_2(z) + \begin{pmatrix} \sigma_0 Q(z) \\ Q(z) - z^k \end{pmatrix} r(z),$$

where $p(z)$, $q_1(z)$, $q_2(z)$ are polynomials of degree $\leq k-1$, $q_2(z)$ is related to $q_1(z)$ in the sense that if $q_1(z) \equiv 0$ then $q_2(z) \equiv 0$, and $r(z)$ is a polynomial of degree $\leq q-1$. We have in the notation of the proof of Theorem 3.1,

$$\begin{aligned}
\begin{pmatrix} 1 & -s(z) \end{pmatrix} \mathbf{f}(\mathbf{z}) &= (1 - \sigma_0^* s(z))p(z) - s(z)q_1(z) + (1 - \sigma_0^* s(z))q_2(z) \\
&\quad + (\sigma_0 Q(z) - s(z)(Q(z) - z^k))r(z) \\
&= (-\sigma_0^* \sigma_k z^k + \cdots)p(z) + (-\sigma_0 - \sigma_k z^k + \cdots)q_1(z) + (-\sigma_0^* \sigma_k z^{2k} + \cdots)q_2(z) \\
&\quad + ((-Q(z)(s(z) - \sigma_0) - \sigma_0 z^k) + z^k(s(z) - \sigma_0))r(z) \\
&= (-\sigma_0^* \sigma_k z^k + \cdots)p(z) + (-\sigma_0 - \sigma_k z^k + \cdots)q_1(z) + (-\sigma_0^* \sigma_k z^{2k} + \cdots)q_2(z) \\
&\quad + ((\sigma_{2k} z^{2k} + \cdots) - (\sigma_0 t_{2k+q} z^{2k+q} + \cdots))r(z).
\end{aligned}$$

For the right-hand side to have at $z = 0$ a zero of order at least $2k$ we see that necessarily $q_1(z) \equiv 0$ and hence $q_2(z) \equiv 0$, and then also (as $k \geq 1$) $p(z) \equiv 0$. If $q > 0$ then to have at $z = 0$ a zero of order $2k + q$ necessarily $r(z) \equiv 0$; if $q = 0$, then $r(z) \equiv 0$ being a polynomial of degree $\leq q-1$. So in Case 3 the implication also holds. □

4. Solutions in terms of colligations

In this section $s(z) \in \mathbf{S}^0$ is a solution of the basic interpolation problem (P^0) given by (3.1) with parameter $\widetilde{s}(z)$: $s(z) = T_{\Theta(z)}\widetilde{s}(z)$, where $\Theta(z) = \Theta_j(z)$ corresponding to Case j, $j = 1, 2, 3$, as in (3.4) and thereafter.

Theorem 4.1. *Under the unitary map*

$$\widehat{U} = \begin{pmatrix} U & 0 \\ 0 & 1 \end{pmatrix} : \begin{pmatrix} \mathcal{P}(s) \\ \mathbb{C} \end{pmatrix} \to \begin{pmatrix} \begin{pmatrix} \mathcal{P}(\Theta) \\ \mathcal{P}(\widetilde{s}) \end{pmatrix} \\ \mathbb{C} \end{pmatrix}$$

the canonical coisometric colligation

$$V = \begin{pmatrix} R_0 & R_0 s \\ E_0 & s(0) \end{pmatrix} : \begin{pmatrix} \mathcal{P}(s) \\ \mathbb{C} \end{pmatrix} \to \begin{pmatrix} \mathcal{P}(s) \\ \mathbb{C} \end{pmatrix}$$

for $s(z)$ is transformed into the colligation

$$\widehat{V} = \widehat{U} V \widehat{U}^{-1} = \begin{pmatrix} T & u \\ \langle\,\cdot\,, v\rangle_{\mathcal{P}(\Theta)\oplus\mathcal{P}(\widetilde{s})} & s(0) \end{pmatrix} : \begin{pmatrix} \begin{pmatrix} \mathcal{P}(\Theta) \\ \mathcal{P}(\widetilde{s}) \end{pmatrix} \\ \mathbb{C} \end{pmatrix} \to \begin{pmatrix} \begin{pmatrix} \mathcal{P}(\Theta) \\ \mathcal{P}(\widetilde{s}) \end{pmatrix} \\ \mathbb{C} \end{pmatrix} \tag{4.1}$$

with

$$T = U R_0 U^{-1} = \begin{pmatrix} R_0 - R_0\Theta \begin{pmatrix} \widetilde{s}(0) \\ 1 \end{pmatrix} \frac{\begin{pmatrix} 0 & 1 \end{pmatrix}}{n(0)} E_0 & \frac{1}{n(0)} R_0 \Theta \begin{pmatrix} d(0) \\ -c(0) \end{pmatrix} E_0 \\ R_0 \widetilde{s} \frac{\begin{pmatrix} 0 & 1 \end{pmatrix}}{n(0)} E_0 & R_0 - R_0 \widetilde{s} \frac{c(0)}{n(0)} E_0 \end{pmatrix},$$

and

$$u = \frac{1}{n(0)} \begin{pmatrix} R_0 \Theta \begin{pmatrix} \widetilde{s}(0) \\ 1 \end{pmatrix} \\ R_0 \widetilde{s} \end{pmatrix}, \quad v = \begin{pmatrix} K_\Theta(\,\cdot\,,0) \begin{pmatrix} 1 \\ -\sigma_0^* \end{pmatrix} \\ 0 \end{pmatrix} \in \begin{pmatrix} \mathcal{P}(\Theta) \\ \mathcal{P}(\widetilde{s}) \end{pmatrix},$$

where

$$n(0) = d(0) + c(0)\widetilde{s}(0) = \begin{cases} 1/\sqrt{1-|\sigma_0|^2} & \text{in Case 1}, \\ \widetilde{s}(0)/\sqrt{|\sigma_0|^2 - 1} & \text{in Case 2}, \\ (\widetilde{s}(0) - \sigma_0)/\sigma_k & \text{in Case 3 with } q = 0. \end{cases}$$

The theorem shows that the 3×3 matrix operator $\widehat{V}$ and hence the canonical coisometric colligation V can be constructed from the canonical coisometric colligation

$$\widetilde{V} = \begin{pmatrix} R_0 & R_0 \widetilde{s} \\ E_0 & \widetilde{s}(0) \end{pmatrix} : \begin{pmatrix} \mathcal{P}(\widetilde{s}) \\ \mathbb{C} \end{pmatrix} \to \begin{pmatrix} \mathcal{P}(\widetilde{s}) \\ \mathbb{C} \end{pmatrix}$$

of the parameter $\widetilde{s}(z)$ and the data of the interpolation problem, that is, $\Theta(z)$. Here and elsewhere in the sequel the same symbols R_0 and E_0 will be used for the difference quotient operator and evaluation at the origin acting on and to different spaces. For example, in the $1,1$ and $2,2$ entry of the operator T in the theorem R_0 acts on the space $\mathcal{P}(\Theta)$ and $\mathcal{P}(\widetilde{s})$, respectively. But we shall also write $R_0 f(z)$ for $\frac{f(z)-f(0)}{z}$ and $E_0 f(z)$ for $f(0)$ for any matrix or operator function $f(z)$ holomorphic at the origin.

In the notation of the theorem

$$\langle\,\cdot\,, v\rangle_{\mathcal{P}(\Theta)\oplus\mathcal{P}(\widetilde{s})} = \begin{pmatrix} \begin{pmatrix} 1 & -s(0) \end{pmatrix} E_0 & 0 \end{pmatrix} : \begin{pmatrix} \mathcal{P}(\Theta) \\ \mathcal{P}(\widetilde{s}) \end{pmatrix} \to \mathbb{C}.$$

Proof of Theorem 4.1. In the following calculations we use that for matrix functions $h(z)$ and $k(z)$, which are holomorphic near $z = 0$ and for which the product $h(z)k(z)$ is well defined,

$$R_0(hk) = hR_0k + (R_0h)k(0).$$

Applying R_0 to both sides of (3.11) we obtain

$$R_0 f = \begin{pmatrix}1 & -s\end{pmatrix}\left\{R_0\mathbf{f} + R_0\Theta\begin{pmatrix}1\\0\end{pmatrix}g(0)\right\}$$
$$+(a-sc)R_0 g - R_0 s\left\{\begin{pmatrix}0 & 1\end{pmatrix}\mathbf{f}(0) + c(0)g(0)\right\}.$$

From (3.4) we derive the matrix equality

$$\begin{pmatrix}1 & -s\end{pmatrix}\Theta = \begin{pmatrix}1 & -s\end{pmatrix}\begin{pmatrix}a & b\\ c & d\end{pmatrix} = (a-sc)\begin{pmatrix}1 & -\widetilde{s}\end{pmatrix}$$

and hence

$$\begin{pmatrix}1 & -s\end{pmatrix}R_0\Theta + \begin{pmatrix}0 & -R_0 s\end{pmatrix}\Theta(0)$$
$$= (a-sc)\begin{pmatrix}0 & -R_0\widetilde{s}\end{pmatrix} + \left\{\begin{pmatrix}1 & -s\end{pmatrix}R_0\Theta\begin{pmatrix}1\\0\end{pmatrix} - (R_0 s)c(0)\right\}\begin{pmatrix}1 & -\widetilde{s}(0)\end{pmatrix}.$$

If we multiply both sides from the right by $\begin{pmatrix}0\\1\end{pmatrix}$ and rearrange terms we get

$$R_0 s = \frac{1}{n(0)}\left\{\begin{pmatrix}1 & -s\end{pmatrix}R_0\Theta\begin{pmatrix}\widetilde{s}(0)\\1\end{pmatrix} + (a-sc)R_0\widetilde{s}\right\}.$$

If we substitute this result in the right-hand side of the formula for $R_0 f$ we obtain

$$R_0 f = \begin{pmatrix}\begin{pmatrix}1 & -s\end{pmatrix} & (a-sc)\end{pmatrix}T\begin{pmatrix}\mathbf{f}\\ g\end{pmatrix}.$$

The remainder of the proof can be worked out in the same manner and is omitted. We only mention that for v we obtain the equality

$$v = \begin{pmatrix}K_\Theta(\,\cdot\,,0)\begin{pmatrix}1\\ -s(0)^*\end{pmatrix}\\ (a(0)-s(0)c(0))^* K_{\widetilde{s}}(\,\cdot\,,0)\end{pmatrix}$$

which yields v as described in the theorem, since in the three cases considered $a(0) - s(0)c(0) = 0$; see (3.8). Also the proof of the formula for $n(0)$ is straightforward and therefore omitted. □

In Case 3 with $q > 0$ Theorem 4.1 can be sharpened. For that we write $\Theta_3(z)$ as

$$\Theta_3 = \Theta_0\Psi_q, \quad \Theta_0(z) = \begin{pmatrix}Q(z)+z^k & \sigma_0 Q(z)\\ \sigma_0^* Q(z) & Q(z)-z^k\end{pmatrix}J, \ \Psi_q = \begin{pmatrix}1 & 0\\ 0 & z^q\end{pmatrix}.$$

Then (see the example after Theorem 2.5)

$$K_{\Theta_3}(z,w) = K_{\Theta_0}(z,w) + \Theta_0(z)K_{\Psi_q}(z,w)\Theta_0(w)^*,$$

which implies that the space $\mathcal{P}(\Theta_3)$ admits the orthogonal decomposition

$$\mathcal{P}(\Theta_3) = \mathcal{P}(\Theta_0) \oplus \Theta_0(z)\mathcal{P}(\Psi_q)$$

and the map

$$W : \mathcal{P}(\Theta_3) \ni \mathbf{f} + \Theta_0 \mathbf{g} \mapsto \begin{pmatrix} \mathbf{f} \\ \mathbf{g} \end{pmatrix} \in \begin{pmatrix} \mathcal{P}(\Theta_0) \\ \mathcal{P}(\Psi_q) \end{pmatrix}$$

is a unitary map. By $\widehat{U}_q$ we denote the unitary composition of $\widehat{U}$ and W:

$$\widehat{U}_q = \begin{pmatrix} W & 0 & 0 \\ 0 & I_{\mathcal{P}(\widetilde{s})} & 0 \\ 0 & 0 & 1 \end{pmatrix} \widehat{U} : \begin{pmatrix} \mathcal{P}(s) \\ \mathbb{C} \end{pmatrix} \rightarrow \left(\begin{pmatrix} \mathcal{P}(\Theta_0) \\ \mathcal{P}(\Psi_q) \\ \mathcal{P}(\widetilde{s}) \\ \mathbb{C} \end{pmatrix} \right).$$

Using $R_0\Theta_3 = (R_0\Theta_0)\Psi_q(0) + \Theta_0(R_0\Psi_q)$, in which

$$\Psi_q(0) = \begin{pmatrix} 1 & 0 \\ 0 & 0 \end{pmatrix}, \quad \Theta_0(0) = c_0 \begin{pmatrix} 1 & \sigma_0 \\ \sigma_0^* & 1 \end{pmatrix} J,$$

one obtains from Theorem 4.1 after some straightforward manipulations the following decompositions.

Theorem 4.2. *In* Case 3 *with $q > 0$ under the unitary map $\widehat{U}_q$ the canonical coisometric colligation*

$$V = \begin{pmatrix} R_0 & R_0 s \\ E_0 & s(0) \end{pmatrix} : \begin{pmatrix} \mathcal{P}(s) \\ \mathbb{C}^n \end{pmatrix} \rightarrow \begin{pmatrix} \mathcal{P}(s) \\ \mathbb{C}^n \end{pmatrix}$$

is transformed into the colligation

$$\widehat{V}_3 = \widehat{U}_q V \widehat{U}_q^{-1} = \begin{pmatrix} T_q & u_q \\ \langle\,\cdot\,, v_q\rangle_{\mathcal{P}(\Theta_0)\oplus\mathcal{P}(\Psi_q)\oplus\mathcal{P}(\widetilde{s})} & s(0) \end{pmatrix} : \left(\begin{pmatrix} \mathcal{P}(\Theta_0) \\ \mathcal{P}(\Psi_q) \\ \mathcal{P}(\widetilde{s}) \\ \mathbb{C} \end{pmatrix} \right) \rightarrow \left(\begin{pmatrix} \mathcal{P}(\Theta_0) \\ \mathcal{P}(\Psi_q) \\ \mathcal{P}(\widetilde{s}) \\ \mathbb{C} \end{pmatrix} \right)$$

with main operator T_q equal to

$$\begin{pmatrix} R_0 - R_0\Theta_0 \begin{pmatrix} \sigma_k \\ 0 \end{pmatrix} (0 \;\; 1) E_0 & R_0\Theta_0 \begin{pmatrix} \sigma_0 \\ 1 \end{pmatrix} (0 \;\; 1)\, E_0 & 0 \\ -R_0\Psi_q \begin{pmatrix} \sigma_k \\ \frac{1}{n(0)} \end{pmatrix} (0 \;\; 1)\, E_0 & R_0 - R_0\Psi_q \begin{pmatrix} 1 \\ \frac{1}{\sigma_k n(0)} \end{pmatrix} (1 \;\; -\sigma_0) E_0 & -\frac{1}{\sigma_k n(0)} R_0\Psi_q \begin{pmatrix} 0 \\ 1 \end{pmatrix} E_0 \\ \frac{1}{n(0)} R_0\widetilde{s}\, (0 \;\; 1)\, E_0 & \frac{1}{\sigma_k n(0)} R_0\widetilde{s}\, (1 \;\; -\sigma_0)\, E_0 & R_0 - \frac{1}{\sigma_k n(0)} R_0\widetilde{s} E_0 \end{pmatrix}$$

and elements

$$u_q = \frac{1}{n(0)} \begin{pmatrix} R_0\Theta_0 \begin{pmatrix} \widetilde{s}(0) \\ 0 \end{pmatrix} \\ R_0\Psi_q \begin{pmatrix} \widetilde{s}(0) \\ 1 \end{pmatrix} \\ R_0\widetilde{s} \end{pmatrix}, \quad v_q = \begin{pmatrix} K_{\Theta_0}(\,\cdot\,, 0) \begin{pmatrix} 1 \\ -\sigma_0^* \end{pmatrix} \\ 0 \\ 0 \end{pmatrix} \in \begin{pmatrix} \mathcal{P}(\Theta_0) \\ \mathcal{P}(\Psi_q) \\ \mathcal{P}(\widetilde{s}) \end{pmatrix},$$

where $n(0) = \widetilde{s}(0)/\sigma_k$.

5. The generalized Schur transform revisited

Let T, u, and v (T_q, u_q, and v_q in Case 3 if $q>0$) be as in the previous section. If P is the projection in $\mathcal{P}(\Theta)\oplus\mathcal{P}(\widetilde{s})$ ($\mathcal{P}(\Theta_0)\oplus\mathcal{P}(\Psi_q)\oplus\mathcal{P}(\widetilde{s})$) onto the second (third) component $\mathcal{P}(\widetilde{s})$, then R_0 as an operator on $\mathcal{P}(\widetilde{s})$ can be written in the form

$$R_0=\begin{cases} PT|_{\mathcal{P}(\widetilde{s})}+\frac{c(0)}{n(0)}R_0\widetilde{s}E_0 & \text{in Cases 1, 2, and Case 3 with } q=0,\\ PT_q|_{\mathcal{P}(\widetilde{s})}+\frac{1}{\sigma_k n(0)}R_0\widetilde{s}E_0 & \text{in Case 3 with } q>0,\end{cases}$$

which shows that R_0 is the compression of T to $\mathcal{P}(\widetilde{s})$ plus a one-dimensional perturbation (that is, if $c(0)\neq 0$). Moreover, we have

$$R_0\widetilde{s}=\begin{cases} n(0)Pu & \text{in Cases 1, 2, and Case 3 with } q=0,\\ R_0\widetilde{s}=n(0)Pu_q & \text{in Case 3 with } q>0.\end{cases}$$

These formulas and the next theorem show that the canonical coisometric colligation of the parameter $\widetilde{s}(z)$ of the solution $s(z)\in\mathbf{S}^0$ can be recovered from the canonical coisometric colligation of $s(z)$. This was done in [2, 3]; there $\widetilde{s}(z)$ was interpreted as the Schur transform of $s(z)$.

Theorem 5.1. *We have*

$$K_{\widetilde{s}}(\,\cdot\,,0)=\begin{cases} n(0)^*PT^{*k}v & \text{in Cases 1 and 2,}\\[2ex] n(0)^*PT^{*(2k)}v & \text{in Case 3 with } q=0,\\[2ex] n(0)^*PT_q^{*(2k+q)}v_q & \text{in Case 3 with } q>0,\end{cases}\tag{5.1}$$

where

$$n(0)=d(0)+c(0)\widetilde{s}(0)=\begin{cases} 1/\sqrt{1-|\sigma_0|^2} & \text{in Case 1 },\\[1ex] \widetilde{s}(0)/\sqrt{|\sigma_0|^2-1} & \text{in Case 2 },\\[1ex] (\widetilde{s}(0)-\sigma_0)/\sigma_k & \text{in Case 3 with } q=0,\\[1ex] \widetilde{s}(0)/\sigma_k & \text{in Case 3 with } q>0.\end{cases}$$

Proof. Case 1. We set $r=\sqrt{1-|\sigma_0|^2}$. Then

$$\Theta=\Theta_1(z)=\frac{1}{r}\begin{pmatrix}1 & \sigma_0\\ \sigma_0^* & 1\end{pmatrix}\begin{pmatrix}z^k & 0\\ 0 & 1\end{pmatrix},$$

and the entries in (4.1) take the form

$$T=\begin{pmatrix} R_0-z^{k-1}\widetilde{s}(0)\begin{pmatrix}1\\ \sigma_0^*\end{pmatrix}\begin{pmatrix}0 & 1\end{pmatrix}E_0 & \dfrac{z^{k-1}}{r}\begin{pmatrix}1\\ \sigma_0^*\end{pmatrix}E_0\\ rR_0\widetilde{s}\begin{pmatrix}0 & 1\end{pmatrix}E_0 & R_0\end{pmatrix}:\begin{pmatrix}\mathcal{P}(\Theta)\\ \mathcal{P}(\widetilde{s})\end{pmatrix}\to\begin{pmatrix}\mathcal{P}(\Theta)\\ \mathcal{P}(\widetilde{s})\end{pmatrix}$$

$$u=\begin{pmatrix} z^{k-1}\widetilde{s}(0)\begin{pmatrix}1\\ \sigma_0^*\end{pmatrix}\\ rR_0\widetilde{s}\end{pmatrix},\quad v=\begin{pmatrix} K_\Theta(\,\cdot\,,0)\begin{pmatrix}1\\ -\sigma_0^*\end{pmatrix}\\ 0\end{pmatrix}\in\begin{pmatrix}\mathcal{P}(\Theta)\\ \mathcal{P}(\widetilde{s})\end{pmatrix},$$

and

$$T^* = \begin{pmatrix} R_0^* - \langle \cdot, z^{k-1}\widetilde{s}(0)\begin{pmatrix} 1 \\ \sigma_0^* \end{pmatrix}\rangle_{\mathcal{P}(\Theta)} K_\Theta(\cdot,0)\begin{pmatrix} 0 \\ 1 \end{pmatrix} & \langle \cdot, rR_0\widetilde{s}\rangle_{\mathcal{P}(\widetilde{s})} K_\Theta(\cdot,0)\begin{pmatrix} 0 \\ 1 \end{pmatrix} \\ \langle \cdot, \frac{z^{k-1}}{r}\begin{pmatrix} 1 \\ \sigma_0^* \end{pmatrix}\rangle_{\mathcal{P}(\Theta)} K_{\widetilde{s}}(\cdot,0) & R_0^* \end{pmatrix}.$$

Hence, by induction,

$$T^{*j}v = \begin{pmatrix} R_0^{*j} K_\Theta(\,\cdot\,,0)\begin{pmatrix} 1 \\ -\sigma_0^* \end{pmatrix} \\ 0 \end{pmatrix}, \quad j = 0,1,\dots,k-1,$$

and

$$T^{*k}v = \begin{pmatrix} * \\ rK_{\widetilde{s}}(\,\cdot\,,0) \end{pmatrix},$$

where the entry denoted by $*$ need not be specified. We conclude that

$$PT^{*k}v = rK_{\widetilde{s}}(\,\cdot\,,0) = \frac{1}{n(0)^*}K_{\widetilde{s}}(\,\cdot\,,0),$$

and thus the formula for Case 1 holds.

Case 2. Now we set $r = \sqrt{|\sigma_0|^2 - 1}$. Then

$$\Theta = \Theta_2(z) = \frac{1}{r}\begin{pmatrix} \sigma_0 & 1 \\ 1 & \sigma_0^* \end{pmatrix}\begin{pmatrix} 1 & 0 \\ 0 & z^k \end{pmatrix},$$

and $\widehat{V}$ in (4.1) has entries

$$T = \begin{pmatrix} R_0 - \dfrac{z^{k-1}}{\widetilde{s}(0)}\begin{pmatrix} 1 \\ \sigma_0^* \end{pmatrix}\begin{pmatrix} 0 & 1 \end{pmatrix} E_0 & -\dfrac{z^{k-1}}{r\widetilde{s}(0)}\begin{pmatrix} 1 \\ \sigma_0^* \end{pmatrix} E_0 \\ \dfrac{r}{\widetilde{s}(0)} R_0\widetilde{s}\begin{pmatrix} 0 & 1 \end{pmatrix} E_0 & R_0 - \dfrac{1}{\widetilde{s}(0)}R_0\widetilde{s}E_0 \end{pmatrix} : \begin{pmatrix} \mathcal{P}(\Theta) \\ \mathcal{P}(\widetilde{s}) \end{pmatrix} \to \begin{pmatrix} \mathcal{P}(\Theta) \\ \mathcal{P}(\widetilde{s}) \end{pmatrix}$$

$$u = \begin{pmatrix} \frac{z^{k-1}}{\widetilde{s}(0)}\begin{pmatrix} 1 \\ \sigma_0^* \end{pmatrix} \\ \dfrac{r}{\widetilde{s}(0)}R_0\widetilde{s} \end{pmatrix}, \quad v = \begin{pmatrix} K_\Theta(\,\cdot\,,0)\begin{pmatrix} 1 \\ -\sigma_0^* \end{pmatrix} \\ 0 \end{pmatrix} \in \begin{pmatrix} \mathcal{P}(\Theta) \\ \mathcal{P}(\widetilde{s}) \end{pmatrix}.$$

Now

$$T^* = \begin{pmatrix} R_0^* - \langle \cdot, \dfrac{z^{k-1}}{\widetilde{s}(0)}\begin{pmatrix} 1 \\ \sigma_0^* \end{pmatrix}\rangle_{\mathcal{P}(\Theta)} K_\Theta(\,\cdot\,,0)\begin{pmatrix} 0 \\ 1 \end{pmatrix} & \langle \cdot, \dfrac{r}{\widetilde{s}(0)}R_0\widetilde{s}\rangle_{\mathcal{P}(\widetilde{s})} K_\Theta(\,\cdot\,,0)\begin{pmatrix} 0 \\ 1 \end{pmatrix} \\ -\langle \cdot, \dfrac{z^{k-1}}{r\widetilde{s}(0)}\begin{pmatrix} 1 \\ \sigma_0^* \end{pmatrix}\rangle_{\mathcal{P}(\Theta)} K_{\widetilde{s}}(\,\cdot\,,0) & R_0^* - \langle \cdot, \dfrac{1}{\widetilde{s}(0)}R_0\widetilde{s}\rangle_{\mathcal{P}(\widetilde{s})} K_{\widetilde{s}}(\,\cdot\,,0) \end{pmatrix}$$

and so, by induction,

$$T^{*j}v = \begin{pmatrix} R_0^{*j} K_\Theta(\,\cdot\,,0)\begin{pmatrix} 1 \\ -\sigma_0^* \end{pmatrix} \\ 0 \end{pmatrix}, \quad j = 0,1,\dots,k-1,$$

and

$$T^{*k}v = \begin{pmatrix} * \\ \dfrac{r}{\widetilde{s}(0)^*}K_{\widetilde{s}}(\,\cdot\,,0) \end{pmatrix},$$

where the entry denoted by $*$ need not be specified. We conclude that

$$PT^{*k}v = \frac{r}{\widetilde{s}(0)^*}K_{\widetilde{s}}(\,\cdot\,,0) = \frac{1}{n(0)^*}K_{\widetilde{s}}(\,\cdot\,,0).$$

This proves the formula for Case 2.

Case 3 with $q=0$. Now we have

$$\Theta(z) = \Theta_3(z) = z^k I_2 + Q(z)\begin{pmatrix} 1 & -\sigma_0 \\ \sigma_0^* & -1 \end{pmatrix}, \quad I_2 = \begin{pmatrix} 1 & 0 \\ 0 & 1 \end{pmatrix},$$

and we find that the entries of $\widehat{V}$ in (4.1) are given by

$$T = \begin{pmatrix} R_0 + t_{2k}R_0\Theta\begin{pmatrix} \widetilde{s}(0) \\ 1 \end{pmatrix}\begin{pmatrix} 0 & 1 \end{pmatrix}E_0 & t_{2k}c_0R_0\Theta\begin{pmatrix} 1 \\ \sigma_0^* \end{pmatrix}E_0 \\ -t_{2k}R_0\widetilde{s}\begin{pmatrix} 0 & 1 \end{pmatrix}E_0 & R_0 + \sigma_0^*c_0t_{2k}R_0\widetilde{s}E_0 \end{pmatrix},$$

$$u = -t_{2k}\begin{pmatrix} R_0\Theta\begin{pmatrix} \widetilde{s}(0) \\ 1 \end{pmatrix} \\ R_0\widetilde{s} \end{pmatrix}, \quad v = \begin{pmatrix} K_\Theta(\,\cdot\,,0)\begin{pmatrix} 1 \\ -\sigma_0^* \end{pmatrix} \\ 0 \end{pmatrix} \in \begin{pmatrix} \mathcal{P}(\Theta) \\ \mathcal{P}(\widetilde{s}) \end{pmatrix},$$

where

$$t_{2k} = -\frac{1}{n(0)} = -\frac{\sigma_k}{\widetilde{s}(0)-\sigma_0}.$$

The calculations for $T^{*j}v$ are a bit more involved. The matrix of T^* is

$$\begin{pmatrix} R_0^* + t_{2k}^*\langle\,\cdot\,,R_0\Theta\begin{pmatrix} \widetilde{s}(0) \\ 1 \end{pmatrix}\rangle_{\mathcal{P}(\Theta)}K_\Theta(\,\cdot\,,0)\begin{pmatrix} 0 \\ 1 \end{pmatrix} & -t_{2k}^*\langle\,\cdot\,,R_0\widetilde{s}\rangle_{\mathcal{P}(\widetilde{s})}K_\Theta(\,\cdot\,,0)\begin{pmatrix} 0 \\ 1 \end{pmatrix} \\ t_{2k}^*c_0^*\langle\,\cdot\,,R_0\Theta\begin{pmatrix} 1 \\ \sigma_0^* \end{pmatrix}\rangle_{\mathcal{P}(\Theta)}K_{\widetilde{s}}(\,\cdot\,,0) & R_0^* + \sigma_0c_0^*t_{2k}^*\langle\,\cdot\,,R_0\widetilde{s}\rangle_{\mathcal{P}(\widetilde{s})}K_{\widetilde{s}}(\,\cdot\,,0) \end{pmatrix},$$

and we get by induction

$$T^{*j}v = \begin{pmatrix} R_0^{*j}K_\Theta(\,\cdot\,,0)\begin{pmatrix} 1 \\ -\sigma_0^* \end{pmatrix} \\ 0 \end{pmatrix}, \quad j = 0,1,\ldots,k-1,$$

$$T^{*(k+j)}v = \begin{pmatrix} R_0^{*(k+j)}K_\Theta(\cdot,0)\begin{pmatrix} 1 \\ -\sigma_0^* \end{pmatrix} - p_j(R_0^*)K_\Theta(\cdot,0)\begin{pmatrix} 1 \\ 0 \end{pmatrix} \\ 0 \end{pmatrix}, \quad j=0,1,\ldots,k-1,$$

where $p_j(z)$ is a polynomial of degree j and with leading coefficient σ_k^*. Finally we obtain

$$PT^{*2k}v = -t_{2k}^*K_{\widetilde{s}}(\,\cdot\,,0) = \frac{1}{n(0)^*}K_{\widetilde{s}}(\,\cdot\,,0),$$

which completes the proof for Case 3 with $q=0$.

Case 3 with $q > 0$. To prove the formula for this case we use the decomposition in Theorem 4.2. With the help of the formula

$$R_0^j\Theta_0(z) = \begin{cases} z^{k-j}I_2 + R_0^jQ(z)\begin{pmatrix} 1 & -\sigma_0 \\ \sigma_0^* & -1 \end{pmatrix}, & j = 0,1,\dots,k, \\ R_0^jQ(z)\begin{pmatrix} 1 & -\sigma_0 \\ \sigma_0^* & -1 \end{pmatrix}, & j = k+1, k+2, \dots, \end{cases}$$

we find for $j = 0, 1, \dots, k-1$:

$$T_q^{*j}v_q = \begin{pmatrix} R_0^{*j}K_{\Theta_0}(\cdot\,,0)\begin{pmatrix} 1 \\ -\sigma_0^* \end{pmatrix} \\ 0 \\ 0 \end{pmatrix},$$

$$T_q^{*(k+j)}v_q = \begin{pmatrix} R_0^{*(k+j)}K_{\Theta_0}(\cdot\,,0)\begin{pmatrix} 1 \\ -\sigma_0^* \end{pmatrix} - p_j(R_0^*)K_{\Theta_0}(\cdot\,,0)\begin{pmatrix} 0 \\ 1 \end{pmatrix} \\ 0 \\ 0 \end{pmatrix},$$

where, here and also below, $p_j(z)$ is some polynomial of degree j with leading coefficient σ_k^*, and

$$T_q^{*2k}v_q = \begin{pmatrix} R_0^{*2k}K_{\Theta_0}(\cdot\,,0)\begin{pmatrix} 1 \\ -\sigma_0^* \end{pmatrix} - p_k(R_0^*)K_{\Theta_0}(\cdot\,,0)\begin{pmatrix} 0 \\ 1 \end{pmatrix} \\ -\sigma_k^*K_{\Psi_q}(\cdot\,,0)\begin{pmatrix} 0 \\ 1 \end{pmatrix} \\ 0 \end{pmatrix}.$$

From the last formula we get for $j = 1, 2, \dots, q-1$,

$$T_q^{*(2k+j)}v_q = \begin{pmatrix} R_0^{*(2k+j)}K_{\theta_0}(\cdot\,,0)\begin{pmatrix} 1 \\ -\sigma_0^* \end{pmatrix} - p_{k+j}(R_0^*)K_{\Theta_0}(\cdot\,,0)\begin{pmatrix} 0 \\ 1 \end{pmatrix} \\ -\widetilde{p}_j(R_0)K_{\Psi_q}(\cdot\,,0)\begin{pmatrix} 0 \\ 1 \end{pmatrix} \\ 0 \end{pmatrix},$$

where again $\widetilde{p}_j(z)$ is a polynomial of degree j with leading coefficient σ_0^*. This finally leads to desired result:

$$T_q^{*(2k+q)}v_q = \begin{pmatrix} * \\ * \\ \frac{1}{n(0)^*}K_{\widetilde{s}}(\cdot\,,0) \end{pmatrix},$$

where the starred entries need not be specified any further. □

6. Direct approach

In this section we consider the three fractional linear transformations in (3.1), which arise from the basic interpolation problem (P^0) and correspond to the Cases 1, 2, and 3. We shall assume that the parameter $\widetilde{s}(z)$ belongs to the class $\mathbf{A}^0$ and that it is the characteristic function of a minimal coisometric colligation $\widetilde{V}$ with state space $\mathcal{K}$. Then the transformed function $s(z)$ also belongs to the class $\mathbf{A}^0$ and is the characteristic function of a minimal coisometric colligation V. The purpose of this section is to express the entries of V in terms of those of $\widetilde{V}$ and the data of the interpolation problem (P^0) corresponding to the three cases.

We shall use the following notation. Let ℓ be a positive integer. By $\mathbb{C}^\ell$ we mean the ℓ-dimensional complex vector space of ℓ-vectors α, β, ... with inner product $\langle\alpha,\beta\rangle = \beta^*\alpha$. The vectors $\varepsilon_0, \varepsilon_1, \ldots, \varepsilon_{\ell-1}$ form the standard orthonormal basis in $\mathbb{C}^\ell$. If $\mathcal{G}$ is an invertible, Hermitian $\ell\times\ell$ matrix, $\mathbb{C}^\ell_{\mathcal{G}}$ stands for the vector space $\mathbb{C}^\ell$ equipped with the inner product

$$\langle\alpha,\beta\rangle_{\mathcal{G}} = \beta^*\mathcal{G}\alpha.$$

By S we denote the shift matrix

$$\mathrm{S} = \begin{pmatrix} 0 & 1 & 0 & \cdots & 0 & 0 \\ 0 & 0 & 1 & \cdots & 0 & 0 \\ \vdots & \vdots & \ddots & \ddots & \vdots & \vdots \\ 0 & 0 & 0 & \ddots & 1 & 0 \\ 0 & 0 & 0 & \cdots & 0 & 1 \\ 0 & 0 & 0 & \cdots & 0 & 0 \end{pmatrix},$$

the order of which depends on the context in which it will be used.
Let $\mathcal{K}$ be a Krein space and consider an operator matrix V of the form

$$V := \begin{pmatrix} A & \alpha\langle\,\cdot\,,a\rangle_{\mathcal{K}} & \gamma \\ b\langle\,\cdot\,,\beta\rangle_{\mathcal{G}} & B & c \\ \langle\,\cdot\,,\delta\rangle_{\mathcal{G}} & \langle\,\cdot\,,d\rangle_{\mathcal{K}} & \sigma \end{pmatrix} : \begin{pmatrix} \mathbb{C}^l_{\mathcal{G}} \\ \mathcal{K} \\ \mathbb{C} \end{pmatrix} \to \begin{pmatrix} \mathbb{C}^l_{\mathcal{G}} \\ \mathcal{K} \\ \mathbb{C} \end{pmatrix}, \tag{6.1}$$

where A is an operator on $\mathbb{C}^l_{\mathcal{G}}$, α, β, γ and δ are vectors in $\mathbb{C}^l_{\mathcal{G}}$, a, b, c, and d are elements in the space $\mathcal{K}$, B is an operator on $\mathcal{K}$, and σ is a complex number. Then

$$V^* = \begin{pmatrix} \mathcal{G}^{-1}A^*\mathcal{G} & \beta\langle\,\cdot\,,b\rangle_{\mathcal{K}} & \delta \\ a\langle\,\cdot\,,\alpha\rangle_{\mathcal{G}} & B^* & d \\ \langle\,\cdot\,,\gamma\rangle_{\mathcal{G}} & \langle\,\cdot\,,c\rangle_{\mathcal{K}} & \sigma^* \end{pmatrix} : \begin{pmatrix} \mathbb{C}^l_{\mathcal{G}} \\ \mathcal{K} \\ \mathbb{C} \end{pmatrix} \to \begin{pmatrix} \mathbb{C}^l_{\mathcal{G}} \\ \mathcal{K} \\ \mathbb{C} \end{pmatrix}.$$

It turns out that the minimal coisometric colligation V for $s(z)$ alluded to above has this form. In the following three theorems we shall refer to the matrix V and specify its entries corresponding to Cases 1, 2, and 3 with $q=0$. For the remaining case (Case 3 with $q>0$) we shall consider a different V. Proofs of these theorems can be given much along the lines of the proofs in [2, 3]; they are straightforward and direct, but the calculations are cumbersome and lenghty. We will sketch some of the main steps. The complete details can be found in [5]. As before we write the Taylor expansion of $s(z) \in \mathbf{A}^0$ at $z = 0$ as

$$s(z) = \sigma_0 + \sigma_1 z + \sigma_2 z^2 + \cdots + \sigma_n z^n + \cdots, \quad \sigma_n = s^{(n)}(0)/n!, \ n = 0, 1, \ldots.$$

Theorem 6.1. *Let σ_0 be a complex number with $|\sigma_0| < 1$ and let k be an integer ≥ 1. For $\widetilde{s}(z) \in \mathbf{A}^0$, consider the function*

$$s(z) = \frac{z^k \widetilde{s}(z) + \sigma_0}{\sigma_0^* z^k \widetilde{s}(z) + 1}.$$

Then $s(z) \in \mathbf{A}^0$, $s(0) = \sigma_0$, $\sigma_1 = \sigma_2 = \cdots = \sigma_{k-1} = 0$, and $\sigma_k \neq 0$ if and only if $\widetilde{s}(0) \neq 0$. Let

$$\widetilde{V} = \begin{pmatrix} \widetilde{T} & \widetilde{u} \\ \langle \cdot, \widetilde{v} \rangle_{\widetilde{\mathcal{K}}} & \widetilde{\gamma} \end{pmatrix} : \begin{pmatrix} \widetilde{\mathcal{K}} \\ \mathbb{C} \end{pmatrix} \to \begin{pmatrix} \widetilde{\mathcal{K}} \\ \mathbb{C} \end{pmatrix}$$

be a minimal coisometric colligation such that $\widetilde{s}(z) = s_{\widetilde{V}}(z)$ near $z = 0$. Then the colligation V in (6.1) with $\ell = k$, $\mathcal{G} = r^2 I_k$, $r = \sqrt{1 - |\sigma_0|^2}$, $\mathcal{K} = \widetilde{\mathcal{K}}$ and with entries

$$\begin{array}{lll} A = \mathrm{S} - \widetilde{\gamma}\sigma_0^* \varepsilon_{k-1}\varepsilon_0^*, & \alpha = \frac{1}{r}\varepsilon_{k-1},\ a = \widetilde{v}, & \gamma = \widetilde{\gamma}\varepsilon_{k-1}, \\ b = \widetilde{u},\ \beta = -\frac{\sigma_0}{r}\varepsilon_0, & B = \widetilde{T}, & c = r\widetilde{u}, \\ \delta = \varepsilon_0, & d = 0, & \sigma = \sigma_0 \end{array}$$

is minimal coisometric in the space $\mathbb{C}^k_{\mathcal{G}} \oplus \widetilde{\mathcal{K}} \oplus \mathbb{C}$ and $s(z) = s_V(z)$ near $z = 0$.

The arguments for the proofs of the first statements of this and the next theorems are the same as in the proof of Theorem 3.1, where we show that the formula (3.1) defines a solution.

To show that V in Theorem 6.1 is coisometric we need to prove 9 identities like (2.6)–(2.8). We only show

$$A\mathcal{G}^{-1}A^*\mathcal{G} + \alpha\langle a, a\rangle_{\widetilde{\mathcal{K}}}\langle\,\cdot\,, \alpha\rangle_{\mathcal{G}} + \gamma\langle\,\cdot\,, \gamma\rangle_{\mathcal{G}} = I_k.$$

The other identities can be proved similarly. Using $\langle \widetilde{v}, \widetilde{v}\rangle_{\widetilde{\mathcal{K}}} = 1 - |\widetilde{\gamma}|^2$ we find that

$$\begin{aligned} &A\mathcal{G}^{-1}A^*\mathcal{G} + \alpha\langle a, a\rangle_{\widetilde{\mathcal{K}}}\langle\,\cdot\,, \alpha\rangle_{\mathcal{G}} + \gamma\langle\,\cdot\,, \gamma\rangle_{\mathcal{G}} \\ &\quad= A\mathcal{G}^{-1}A^*\mathcal{G} + \langle \widetilde{v}, \widetilde{v}\rangle_{\widetilde{\mathcal{K}}}\langle \mathcal{G}\,\cdot\,, \frac{1}{r}\varepsilon_{k-1}\rangle_{\mathbb{C}^k}\frac{1}{r}\varepsilon_{k-1} + \langle \mathcal{G}\,\cdot\,, \widetilde{\gamma}\varepsilon_{k-1}\rangle_{\mathbb{C}^k}\widetilde{\gamma}\varepsilon_{k-1} \\ &\quad= A\mathcal{G}^{-1}A^*\mathcal{G} + \langle \widetilde{v}, \widetilde{v}\rangle_{\widetilde{\mathcal{K}}}\frac{1}{r^2}\varepsilon_{k-1}\varepsilon_{k-1}^*\mathcal{G} + |\widetilde{\gamma}|^2\varepsilon_{k-1}\varepsilon_{k-1}^*\mathcal{G} \\ &\quad= AA^* + \left((1 - |\widetilde{\gamma}|^2) + |\widetilde{\gamma}|^2 r^2\right)\varepsilon_{k-1}\varepsilon_{k-1}^* \end{aligned}$$

$$\begin{aligned}
&= \left(\mathrm{S}-\widetilde{\gamma}\sigma_0^*\varepsilon_{k-1}\varepsilon_0^*\right)\left(\mathrm{S}^*-\widetilde{\gamma}^*\sigma_0\varepsilon_0\varepsilon_{k-1}^*\right)\\
&\quad+\left((1-|\widetilde{\gamma}|^2)+|\widetilde{\gamma}|^2r^2\right)\varepsilon_{k-1}\varepsilon_{k-1}^*\\
&= \mathrm{SS}^*-\widetilde{\gamma}\sigma_0^*\varepsilon_{k-1}\varepsilon_0^*\mathrm{S}^*-\mathrm{S}\widetilde{\gamma}^*\sigma_0\varepsilon_0\varepsilon_{k-1}^*+|\widetilde{\gamma}|^2|\sigma_0|^2\varepsilon_{k-1}\varepsilon_0^*\varepsilon_0\varepsilon_{k-1}^*\\
&\quad+\left((1-|\widetilde{\gamma}|^2+|\widetilde{\gamma}|^2r^2\right)\varepsilon_{k-1}\varepsilon_{k-1}^*\\
&= \mathrm{SS}^*+0+0|\widetilde{\gamma}|^2|\sigma_0|^2\varepsilon_{k-1}\varepsilon_{k-1}^*+\left((1-|\widetilde{\gamma}|^2+|\widetilde{\gamma}|^2(1-|\sigma_0|^2)\right)\varepsilon_{k-1}\varepsilon_{k-1}^*\\
&= \mathrm{SS}^*+\varepsilon_{k-1}\varepsilon_{k-1}^*\\
&= I_k.
\end{aligned}$$

To show that V is minimal we first note that

$$v=\begin{pmatrix}\delta\\ d\end{pmatrix}=\begin{pmatrix}\varepsilon_0\\ 0\end{pmatrix},$$

and

$$T^*=\begin{pmatrix}\mathcal{G}^{-1}A^*\mathcal{G} & -\frac{\sigma_0}{r}\varepsilon_0\langle\,\cdot\,,\widetilde{u}\rangle a_{\widetilde{\mathcal{K}}}\\ \frac{1}{r}\widetilde{v}\langle\,\cdot\,,\varepsilon_{k-1}\rangle_{\mathcal{G}} & \widetilde{T}^*\end{pmatrix}=\begin{pmatrix}A^* & -\frac{\sigma_0}{r}\varepsilon_0\langle\,\cdot\,,\widetilde{u}\rangle_{\widetilde{\mathcal{K}}}\\ \frac{1}{r}\widetilde{v}\langle\,\cdot\,,\varepsilon_{k-1}\rangle_{\mathcal{G}} & \widetilde{T}^*\end{pmatrix}.$$

Then

$$T^*v=\begin{pmatrix}A^* & -\frac{\sigma_0}{r}\varepsilon_0\langle\,\cdot\,,\widetilde{u}\rangle_{\widetilde{\mathcal{K}}}\\ \frac{1}{r}\widetilde{v}\langle\,\cdot\,,\varepsilon_{k-1}\rangle_{\mathcal{G}} & \widetilde{T}^*\end{pmatrix}\begin{pmatrix}\varepsilon_0\\ 0\end{pmatrix}=\begin{pmatrix}\varepsilon_1\\ 0\end{pmatrix}$$

and, by induction,

$$T^{*j}v=T^*\begin{pmatrix}\varepsilon_{j-1}\\ 0\end{pmatrix}=\begin{pmatrix}\varepsilon_j\\ 0\end{pmatrix},\quad j=2,\dots,k-1.$$

Hence

$$\operatorname{span}\left\{v,T^*v,\dots,T^{*k-1}v\right\}=\begin{pmatrix}\mathbb{C}^k\\ 0\end{pmatrix}.$$

Also by induction it can be shown that for each $n\geq 0$ there exist numbers a_j (depending on n) with $a_{k+n}\neq 0$ such that

$$T^{*(k+n)}v=a_0\begin{pmatrix}\varepsilon_0\\ 0\end{pmatrix}+\cdots+a_{k-1}\begin{pmatrix}\varepsilon_{k-1}\\ 0\end{pmatrix}+a_k\begin{pmatrix}0\\ \widetilde{v}\end{pmatrix}+\cdots+a_{k+n}\begin{pmatrix}0\\ \widetilde{T}^{*n}\widetilde{v}\end{pmatrix}$$

and thus the minimality of V follows from that of $\widetilde{V}$. That $s_V(z)=s(z)$ near $z=0$ can be proved in a similar way as in the proof of the next theorem.

Theorem 6.2. *Let σ_0 be a complex number with $|\sigma_0|>1$ and let k be an integer ≥ 1. For $\widetilde{s}(z)\in\mathbf{A}^0$ with $\widetilde{s}(0)\neq 0$, consider the function*

$$s(z)=\frac{\sigma_0\widetilde{s}(z)+z^k}{\widetilde{s}(z)+\sigma_0^*z^k}.$$

Then $s(z)\in\mathbf{A}^0$, $s(z)=\sigma_0$, $\sigma_1=\sigma_2=\cdots=\sigma_{k-1}=0$ and $\sigma_k\neq 0$. Let

$$\widetilde{V}=\begin{pmatrix}\widetilde{T} & \widetilde{u}\\ \langle\,\cdot\,,\widetilde{v}\rangle_{\widetilde{\mathcal{K}}} & \widetilde{\gamma}\end{pmatrix}:\begin{pmatrix}\widetilde{\mathcal{K}}\\ \mathbb{C}\end{pmatrix}\to\begin{pmatrix}\widetilde{\mathcal{K}}\\ \mathbb{C}\end{pmatrix}$$

be a minimal coisometric colligation such that $\widetilde{s}(z) = s_{\widetilde{V}}(z)$ *near* $z = 0$. *Then the colligation* V *in* (6.1) *with* $\ell = k$, $\mathcal{G} = -r^2 I_k$, $r = \sqrt{|\sigma_0|^2 - 1}$, $\mathcal{K} = \widetilde{\mathcal{K}}$ *and with entries*

$$A = \mathrm{S} - \tfrac{\sigma_0^*}{\widetilde{\gamma}}\varepsilon_{k-1}\varepsilon_0^*, \quad \alpha = \tfrac{1}{r\widetilde{\gamma}}\varepsilon_{k-1}, \ a = \widetilde{v}, \qquad \gamma = \tfrac{1}{\widetilde{\gamma}}\varepsilon_{k-1},$$

$$b = \widetilde{u}, \ \beta = -\tfrac{\sigma_0}{\widetilde{\gamma}^* r}\varepsilon_0, \quad B = \widetilde{T} - \tfrac{1}{\widetilde{\gamma}}\langle\,\cdot\,,\widetilde{v}\rangle_{\widetilde{\mathcal{K}}}\widetilde{u}, \qquad c = -\tfrac{r}{\widetilde{\gamma}}\widetilde{u},$$

$$\delta = \varepsilon_0, \qquad d = 0, \qquad \sigma = \sigma_0$$

is minimal coisometric in the space $\mathbb{C}^k_{\mathcal{G}} \oplus \widetilde{\mathcal{K}} \oplus \mathbb{C}$ *and* $s(z) = s_V(z)$ *near* $z = 0$.

That V is minimal coisometric can be proved as for Theorem 6.1. For this case we show that the characteristic function for V coincides with $s(z)$. We have that

$$\begin{aligned} s_V(z) &= \gamma + z\left\langle (I - zT)^{-1}u, v\right\rangle_{\mathbb{C}^k_{\mathcal{G}}\oplus\widetilde{\mathcal{K}}} \\ &= \gamma + z\left\langle \left[I - z\begin{pmatrix} A & \alpha\langle\,\cdot\,,a\rangle_{\widetilde{\mathcal{K}}} \\ b\langle\,\cdot\,,\beta\rangle_{\mathcal{G}} & B\end{pmatrix}\right]^{-1}\begin{pmatrix}\gamma\\ c\end{pmatrix}, \begin{pmatrix}\delta\\ d\end{pmatrix}\right\rangle_{\mathbb{C}^k_{\mathcal{G}}\oplus\widetilde{\mathcal{K}}} \end{aligned}$$

Let

$$\left[I - z\begin{pmatrix} A & \alpha\langle\,\cdot\,,a\rangle_{\widetilde{\mathcal{K}}} \\ b\langle\,\cdot\,,\beta\rangle_{\mathcal{G}} & B\end{pmatrix}\right]^{-1}\begin{pmatrix}\gamma\\ c\end{pmatrix} = \begin{pmatrix}\tau\\ t\end{pmatrix} \in \begin{pmatrix}\mathbb{C}^k_{\mathcal{G}}\\ \widetilde{\mathcal{K}}\end{pmatrix}. \tag{6.2}$$

Then

$$\begin{aligned} s_V(z) &= \gamma + z\left\langle \begin{pmatrix}\tau\\ t\end{pmatrix}, \begin{pmatrix}\delta\\ d\end{pmatrix}\right\rangle_{\mathbb{C}^k_{\mathcal{G}}\oplus\widetilde{\mathcal{K}}} = \sigma_0 + z\langle\tau,\delta\rangle_{\mathcal{G}} \\ &= \sigma_0 + z\,\langle \mathcal{G}\tau,\delta\rangle_{\mathbb{C}^k} = \sigma_0 + z\,\langle \mathcal{G}\tau,\varepsilon_0\rangle_{\mathbb{C}^k} = \sigma_0 - zr^2\tau_0. \end{aligned}$$

From (6.2) we get

$$\begin{aligned} \begin{pmatrix}\gamma\\ c\end{pmatrix} &= \left[I - z\begin{pmatrix} A & \alpha\langle\,\cdot\,,a\rangle_{\mathcal{K}} \\ b\langle\,\cdot\,,\beta\rangle_{\mathcal{G}} & B\end{pmatrix}\right]\begin{pmatrix}\tau\\ t\end{pmatrix} \\ &= \begin{pmatrix} I_k - zA & -z\alpha\langle\,\cdot\,,a\rangle_{\mathcal{K}} \\ -zb\langle\,\cdot\,,\beta\rangle_{\mathcal{G}} & I - zB\end{pmatrix}\begin{pmatrix}\tau\\ t\end{pmatrix} \end{aligned}$$

and that

$$\tau - z\left(A\tau + \alpha\langle t, a\rangle_{\widetilde{\mathcal{K}}}\right) = \gamma, \tag{6.3}$$

and

$$t - z\left(b\langle\tau,\beta\rangle_{\mathcal{G}} + Bt\right) = c. \tag{6.4}$$

From (6.4) we get

$$(I - zB)\,t = c + zb\langle\tau,\beta\rangle_{\mathcal{G}}.$$

Hence

$$\left[I - z(\widetilde{T} - \frac{1}{\widetilde{\gamma}}\langle\,\cdot\,,\widetilde{v}\rangle_{\widetilde{\mathcal{K}}}\widetilde{u})\right]t = \frac{-r}{\widetilde{\gamma}}\widetilde{u} - \frac{-\sigma_0^*}{r\widetilde{\gamma}}z\widetilde{u}\langle\mathcal{G}\tau,\varepsilon_0\rangle_{\mathbb{C}^k},$$

from which we see that

$$(I - z\widetilde{T})t = \frac{-r}{\widetilde{\widetilde{\gamma}}}\widetilde{u} + \frac{\sigma_0^* r}{\widetilde{\widetilde{\gamma}}} z\tau_0\widetilde{u} - \frac{1}{\widetilde{\widetilde{\gamma}}} z\langle t, \widetilde{v}\rangle_{\widetilde{\mathcal{K}}}\widetilde{u}.$$

Hence we have that

$$t = \left[-\frac{r}{\widetilde{\widetilde{\gamma}}} + \frac{\sigma_0^* r}{\widetilde{\widetilde{\gamma}}} z\tau_0 - \frac{1}{\widetilde{\widetilde{\gamma}}} z\langle t, \widetilde{v}\rangle_{\widetilde{\mathcal{K}}}\right]\left(I - z\widetilde{T}\right)^{-1}\widetilde{u}$$

and taking inner product with $\widetilde{v}$ we obtain

$$\begin{aligned}\langle t, \widetilde{v}\rangle_{\widetilde{\mathcal{K}}} &= \left[\frac{-r}{\widetilde{\widetilde{\gamma}}} + \frac{\sigma_0^* r}{\widetilde{\widetilde{\gamma}}} z\tau_0 - \frac{1}{\widetilde{\widetilde{\gamma}}} z\langle t, \widetilde{v}\rangle_{\widetilde{\mathcal{K}}}\right]\left\langle \left(I - z\widetilde{T}\right)^{-1}\widetilde{u}, \widetilde{v}\right\rangle_{\widetilde{\mathcal{K}}} \\ &= \left[\frac{-r}{\widetilde{\widetilde{\gamma}}} + \frac{\sigma_0^* r}{\widetilde{\widetilde{\gamma}}} z\tau_0 - \frac{1}{\widetilde{\widetilde{\gamma}}} z\langle t, \widetilde{v}\rangle_{\widetilde{\mathcal{K}}}\right]\frac{\widetilde{s}(z) - \widetilde{\gamma}}{z}.\end{aligned}$$

So

$$\left[1 + \frac{1}{\widetilde{\widetilde{\gamma}}} z\frac{\widetilde{s}(z) - \widetilde{\gamma}}{z}\right]\langle t, \widetilde{v}\rangle_{\widetilde{\mathcal{K}}} = \frac{-r}{\widetilde{\widetilde{\gamma}}}\left(1 - z\sigma_0^*\tau_0\right)\frac{\widetilde{s}(z) - \widetilde{\gamma}}{z}$$

and

$$\widetilde{s}(z)\langle t, \widetilde{v}\rangle_{\widetilde{\mathcal{K}}} = -r\left(1 - z\sigma_0^*\tau_0\right)\frac{\widetilde{s}(z) - \widetilde{\gamma}}{z}.$$

We now get

$$\langle t, \widetilde{v}\rangle_{\widetilde{\mathcal{K}}} = -r\left(1 - z\sigma_0^*\tau_0\right)\frac{\widetilde{s}(z) - \widetilde{\gamma}}{z\widetilde{s}(z)}.$$

But $a = \widetilde{v}$ and so from (6.3) we obtain that

$$\begin{aligned}(I_k - zA)\tau &= \gamma + z\alpha\langle t, \widetilde{v}\rangle_{\widetilde{\mathcal{K}}} = \frac{1}{\widetilde{\widetilde{\gamma}}}\varepsilon_{k-1} - \frac{r}{r\widetilde{\widetilde{\gamma}}} z\left(1 - z\sigma_0^*\tau_0\right)\frac{\widetilde{s}(z) - \widetilde{\gamma}}{z\widetilde{s}(z)}\varepsilon_{k-1} \\ &= \left[\frac{1}{\widetilde{\widetilde{\gamma}}} - \frac{1}{\widetilde{\widetilde{\gamma}}}(1 - z\sigma_0^*\tau_0)\frac{\widetilde{s}(z) - \widetilde{\gamma}}{\widetilde{s}(z)}\right]\varepsilon_{k-1} = \frac{1}{\widetilde{\widetilde{\gamma}}}\left[1 - (1 - z\sigma_0^*\tau_0)\frac{\widetilde{s}(z) - \widetilde{\gamma}}{\widetilde{s}(z)}\right]\varepsilon_{k-1}.\end{aligned}$$

Hence we have

$$\begin{pmatrix} 1 & -z & 0 & \cdots & 0 \\ 0 & 1 & -z & \cdots & 0 \\ \vdots & \vdots & \ddots & \ddots & \vdots \\ 0 & 0 & 0 & \ddots & -z \\ z\frac{\sigma_0^*}{\widetilde{\widetilde{\gamma}}} & 0 & 0 & \cdots & 1 \end{pmatrix}\tau = \frac{1}{\widetilde{\widetilde{\gamma}}}\left[1 - (1 - z\sigma_0^*\tau_0)\frac{\widetilde{s}(z) - \widetilde{\gamma}}{\widetilde{s}(z)}\right]\varepsilon_{k-1}.$$

This is equivalent to the following system of equations:

$$\begin{cases} \tau_0 - z\tau_1 = 0 \\ \tau_1 - z\tau_2 = 0 \\ \vdots \\ \tau_{k-2} - z\tau_{k-1} = 0 \end{cases}$$

and

$$\tau_0 \frac{\sigma_0^*}{\widetilde{\gamma}} z + \tau_{k-1} = \frac{1}{\widetilde{\gamma}} \left[1 - (1 - z\sigma_0^* \tau_0) \frac{\widetilde{s}(z) - \widetilde{\gamma}}{\widetilde{s}(z)} \right]. \tag{6.5}$$

Hence

$$\tau_0 = z\tau_1 = z^2\tau_2 = \cdots = z^{k-1}\tau_{k-1},$$

so that

$$\tau_{k-1} = \frac{\tau_0}{z^{k-1}}.$$

Equation (6.5) gives

$$\tau_0 = \frac{1 - \widetilde{s}(z)\tau_{k-1}}{z\sigma_0^*} = \frac{z^k - 1 - \widetilde{s}(z)\tau_0}{z^k \sigma_0^*},$$

from which we get

$$\tau_0 = \frac{z^{k-1}}{z^k \sigma_0^* + \widetilde{s}(z)}.$$

But

$$\begin{aligned} s_V(z) &= \sigma_0 - zr^2\tau_0 = \sigma_0 - \frac{z^k r^2}{z^k\sigma_0^* + \widetilde{s}(z)} = \frac{\sigma_0 \left(z^k \sigma_0^* + \widetilde{s}(z) \right) - z^k r^2}{z^k \sigma_0^* + \widetilde{s}(z)} \\ &= \frac{|\sigma_0|^2 z^k + \sigma_0 \widetilde{s}(z) - z^k(|\sigma_0|^2 - 1)}{z^k\sigma_0^* + \widetilde{s}(z)} = \frac{z^k + \sigma_0\widetilde{s}(z)}{\widetilde{s}(z) + \sigma_0^* z^k}. \end{aligned}$$

Hence $s(z) = s_V(z)$.

In the next two theorems we consider Case 3 with $q = 0$ and $q > 0$, respectively. We recall from the Introduction the definition of the polynomial $Q(z)$ of degree $2k$ associated with k complex numbers $s_0 \neq 0, s_1, \ldots, s_{k-1}$:

$$\begin{aligned} Q(z) &= Q(z; s_0, s_1, \ldots, s_{k-1}) \\ &= c_0 + c_1 z + \cdots + c_{k-1}z^{k-1} - (c_{k-1}^* z^{k+1} + c_{k-2}^* z^{k+2} + \cdots + c_0^* z^{2k}), \end{aligned}$$

where the coefficients $c_0, c_1, \ldots, c_{k-1}$ are determined by the formula

$$\begin{pmatrix} c_0 & & \\ \vdots & \ddots & \\ c_{k-1} & \cdots & c_0 \end{pmatrix} \begin{pmatrix} s_0 & & \\ \vdots & \ddots & \\ s_{k-1} & \cdots & s_0 \end{pmatrix} = \sigma_0 I_k.$$

Theorem 6.3. *Let σ_0 and for some integer $k \geq 1$, $s_0, s_1, \ldots, s_{k-1}$ be complex numbers such that $|\sigma_0| = 1$ and $s_0 \neq 0$. For $\widetilde{s}(z) \in \mathbf{A}^0$ with $\widetilde{s}(0) \neq \sigma_0$, consider the function*

$$s(z) = \frac{(Q(z) + z^k)\widetilde{s}(z) - \sigma_0 Q(z)}{\sigma_0^* Q(z)\widetilde{s}(z) - (Q(z) - z^k)},$$

where $Q(z) = Q(z; s_0, s_1, \ldots, s_{k-1})$. Then $s(z) \in \mathbf{A}^0$,

$$s(0) = \sigma_0, \ \sigma_1 = \sigma_2 = \cdots = \sigma_{k-1} = 0, \ \sigma_k = s_0, \sigma_{k+1} = s_1, \ldots, \sigma_{2k-1} = s_{k-1},$$

and

$$\sigma_{2k} = \frac{1}{c_0}(\sigma_0 t_{2k} - c_1 s_{k-1} - c_2 s_{k-2} - \cdots - c_{k-1}s_1), \quad t_{2k} = -\frac{s_0}{\widetilde{s}(0) - \sigma_0}. \tag{6.6}$$

Consider the $k \times k$ matrix

$$\Delta = \begin{pmatrix} \sigma_k & & \\ \vdots & \ddots & \\ \sigma_{2k-1} & \cdots & \sigma_k \end{pmatrix}$$

and the $2k \times 2k$ matrix

$$\mathrm{R} = \frac{1}{c_0} \begin{pmatrix} 0 & 0 & 0 & 0 & 0 & \cdots & 0 \\ 0 & 0 & 0 & 0 & 0 & \cdots & 0 \\ \vdots & \vdots & \vdots & \vdots & \vdots & \cdots & \vdots \\ 0 & 0 & 0 & 0 & 0 & \cdots & 0 \\ 0 & 0 & 0 & 0 & 0 & \cdots & 0 \\ c_0^* & \cdots & c_{k-1}^* & \frac{t_{2k}}{\sigma_0^* \sigma_k} & -c_{k-1} & \cdots & -c_1 \end{pmatrix}.$$

Let

$$\widetilde{V} = \begin{pmatrix} \widetilde{T} & \widetilde{u} \\ \langle \cdot, \widetilde{v} \rangle_{\widetilde{\mathcal{K}}} & \widetilde{\gamma} \end{pmatrix} : \begin{pmatrix} \widetilde{\mathcal{K}} \\ \mathbb{C} \end{pmatrix} \to \begin{pmatrix} \widetilde{\mathcal{K}} \\ \mathbb{C} \end{pmatrix}$$

be a minimal coisometric colligation such that $\widetilde{s}(z) = s_{\widetilde{V}}(z)$ near $z = 0$. Then the colligation V in (6.1) with $\ell = 2k$,

$$\mathcal{G} = \begin{pmatrix} 0 & -\sigma_0 \Delta^* \\ -\sigma_0^* \Delta & -\Delta \Delta^* \end{pmatrix},$$

$\mathcal{K} = \widetilde{\mathcal{K}}$ and with entries

$$A = \mathcal{G}^{-1}(\mathrm{S} + \mathrm{R})\mathcal{G}, \quad \alpha = t_{2k}\mathcal{G}^{-1}\varepsilon_{2k-1}, \quad a = \widetilde{v}, \quad \gamma = \mathcal{G}^{-1} \begin{pmatrix} \sigma_1 \\ \vdots \\ \sigma_{2k} \end{pmatrix},$$

$$b = \widetilde{u}, \quad \beta = -t_{2k}^* \mathcal{G}^{-1} \begin{pmatrix} \sigma_0 \\ \vdots \\ \sigma_{2k-1} \end{pmatrix}, \quad B = \widetilde{T} + \frac{t_{2k}}{\sigma_k} \langle \cdot, \widetilde{v} \rangle_{\widetilde{\mathcal{K}}} \widetilde{u}, \quad c = t_{2k}\widetilde{u},$$

$$\delta = \varepsilon_0, \qquad d = 0, \qquad \sigma = \sigma_0$$

is minimal coisometric in the space $\mathbb{C}_{\mathcal{G}}^{2k} \oplus \widetilde{\mathcal{K}} \oplus \mathbb{C}$ and $s(z) = s_V(z)$ near $z = 0$.

The proof of this theorem is similar to the proof of the preceding ones but more involved; as mentioned before, complete details can be found [5].

Now we consider Case 3 with $q > 0$. In this case the colligation V will take a different form:

$$V := \begin{pmatrix} A & \alpha_{(2k)} \langle \cdot, \alpha_{(q)} \rangle_{\mathcal{G}_2} & \beta_{(2k)} \langle \cdot, a \rangle_{\mathcal{K}} & \gamma_{(2k)} \\ \delta_{(q)} \langle \cdot, \delta_{(2k)} \rangle_{\mathcal{G}_1} & B & \rho_{(q)} \langle \cdot, b \rangle_{\mathcal{K}} & \mu_q \\ c \langle \cdot, \lambda_{(2k)} \rangle_{\mathcal{G}_1} & d \langle \cdot, \tau_{(q)} \rangle_{\mathcal{G}_2} & C & e \\ \eta \langle \cdot, \xi_{(2k)} \rangle_{\mathcal{G}_1} & \alpha \langle \cdot, \chi_{(q)} \rangle_{\mathcal{G}_2} & \beta \langle \cdot, f \rangle_{\mathcal{K}} & \gamma \end{pmatrix}, \tag{6.7}$$

mapping

$$\begin{pmatrix} \mathbb{C}^{2k}_{\mathcal{G}_1} \\ \mathbb{C}^{q}_{\mathcal{G}_2} \\ \mathcal{K} \\ \mathbb{C} \end{pmatrix} \quad \text{into} \quad \begin{pmatrix} \mathbb{C}^{2k}_{\mathcal{G}_1} \\ \mathbb{C}^{q}_{\mathcal{G}_2} \\ \mathcal{K} \\ \mathbb{C} \end{pmatrix}$$

Here, for example, the notations $\alpha_{(2k)}$ and $\alpha_{(q)}$ stand for vectors in $\mathbb{C}^{2k}$ and $\mathbb{C}^q$. We write $\varepsilon_{\ell,0},\ldots,\varepsilon_{\ell,\ell-1}$ for the standard orthonormal basis in $\mathbb{C}^\ell$.

Theorem 6.4. *Let σ_0 and for some integer $k \geq 1$, $s_0, s_1, \ldots, s_{k-1}$ be complex numbers such that $|\sigma_0| = 1$ and $s_0 \neq 0$. For $\widetilde{s}(z) \in \mathbf{A}^0$ with $\widetilde{s}(0) \neq 0$, consider the function*

$$s(z) = \frac{(Q(z)+z^k)\widetilde{s}(z) - \sigma_0 Q(z) z^q}{\sigma_0^* Q(z)\widetilde{s}(z) - (Q(z)-z^k)z^q},$$

where $Q(z) = Q(z; s_0, s_1, \ldots, s_{k-1})$ and q is an integer > 0. Then $s(z) \in \mathbf{A}^0$,

$$s(0) = \sigma_0,\ \sigma_1 = \sigma_2 = \cdots = \sigma_{k-1} = 0,\ \sigma_k = s_0, \sigma_{k+1} = s_1, \ldots, \sigma_{2k-1} = s_{k-1},$$

and

$$\sigma_{2k} = -\frac{1}{c_0}(c_1 s_{k-1} + c_2 s_{k-2} + \cdots + c_{k-1} s_1).$$

Consider the $k \times k$ matrix

$$\Delta = \begin{pmatrix} \sigma_k & & \\ \vdots & \ddots & \\ \sigma_{2k-1} & \cdots & \sigma_k \end{pmatrix},$$

and the $2k \times 2k$ matrix

$$\mathrm{R}_1 = \frac{1}{c_0}\begin{pmatrix} 0 & 0 & 0 & 0 & 0 & \cdots & 0 \\ 0 & 0 & 0 & 0 & 0 & \cdots & 0 \\ \vdots & \vdots & \vdots & \vdots & \vdots & \cdots & \vdots \\ 0 & 0 & 0 & 0 & 0 & \cdots & 0 \\ 0 & 0 & 0 & 0 & 0 & \cdots & 0 \\ c_0^* & \cdots & c_{k-1}^* & -1 & -c_{k-1} & \cdots & -c_1 \end{pmatrix}.$$

Let

$$\widetilde{V} = \begin{pmatrix} \widetilde{T} & \widetilde{u} \\ \langle \cdot, \widetilde{v}\rangle_{\widetilde{\mathcal{K}}} & \widetilde{\gamma} \end{pmatrix} : \begin{pmatrix} \widetilde{\mathcal{K}} \\ \mathbb{C} \end{pmatrix} \to \begin{pmatrix} \widetilde{\mathcal{K}} \\ \mathbb{C} \end{pmatrix}$$

be a minimal coisometric colligation such that $\widetilde{s}(z) = s_{\widetilde{V}}(z)$ near $z = 0$. Then the colligation V in (6.7) with

$$\mathcal{G}_1 = \begin{pmatrix} 0 & -\sigma_0\Delta^* \\ -\sigma_0^*\Delta & \Delta\Delta^* \end{pmatrix},\quad \mathcal{G}_2 = -I_q,\quad \mathcal{K} = \widetilde{\mathcal{K}}$$

and with entries on the first row of V*:*

$$A = \mathcal{G}_1^{-1}(\mathrm{S}+\mathrm{R}_1)\mathcal{G} - \sigma_0^*\mathcal{G}_1^{-1}\begin{pmatrix}\sigma_1\\ \vdots\\ \sigma_{2k-1}\\ \sigma_{2k}-\sigma_0^*\sigma_k^2\end{pmatrix}\langle\cdot,\varepsilon_{2k,0}\rangle_{\mathcal{G}_1}, \quad \alpha_{(2k)} = -\varepsilon_{2k,k-1},$$

$$\alpha_{(q)} = \varepsilon_{q,0}, \quad \beta_{(2k)} = 0, \quad a = 0, \quad \gamma_{(2k)} = \sigma_0^*\mathcal{G}_1^{-1}\begin{pmatrix}\sigma_1\\ \vdots\\ \sigma_{2k-1}\\ \sigma_{2k}-\sigma_0^*\sigma_k^2\end{pmatrix},$$

on the second row

$$\delta_{(q)} = \frac{1}{r\widetilde{\gamma}}\varepsilon_{q,q-1}, \quad \delta_{(2k)} = \sigma_0^*\varepsilon_{2k,k} + \sigma_k^*\varepsilon_{2k,0}, \quad B = \begin{pmatrix} 0 & 1 & & \\ & \ddots & \ddots & \\ & & \ddots & 1\\ \frac{\sigma_0}{\widetilde{\gamma}} & 0 & & 0\end{pmatrix},$$

$$\rho_{(q)} = \frac{1}{\widetilde{\gamma}}\varepsilon_{q,q-1}, \quad b = \widetilde{v}, \quad \mu_{(q)} = -\frac{\sigma_0^*\sigma_k}{\widetilde{\gamma}}\varepsilon_{q,q-1},$$

on the third row

$$c = \frac{r}{\widetilde{\gamma}}\widetilde{u}, \quad \lambda_{(2k)} = \sigma_0^*\varepsilon_{2k,k} + \sigma_k^*\varepsilon_{2k,0}, \quad d = \frac{r\sigma_0}{\widetilde{\gamma}}\widetilde{u}, \quad \tau_{(q)} = \varepsilon_{q,0},$$

$$C = \widetilde{T} - \frac{1}{\widetilde{\gamma}}\langle\cdot,\widetilde{v}\rangle_{\mathcal{K}}\widetilde{u}, \quad e = -\frac{r\sigma_0^*\sigma_k}{\widetilde{\gamma}}\widetilde{u},$$

and finally on the fourth row

$$\eta = \frac{r}{\widetilde{\gamma}}\widetilde{u}, \quad \xi_{2k} = \varepsilon_{2k,0}, \quad \alpha = 0, \quad \chi = 0, \quad \beta = 0, \quad f = 0, \quad \gamma = \sigma_0$$

is minimal coisometric in the space $\mathbb{C}^{2k}_{\mathcal{G}_1} \oplus \mathbb{C}^q_{\mathcal{G}_2} \oplus \widetilde{\mathcal{K}} \oplus \mathbb{C}$ *and* $s(z) = s_V(z)$ *near* $z = 0$.

In the same way as the coefficient σ_{2k} of $s(z)$ in the theorem, the coefficients $\sigma_{2k+1}, \ldots, \sigma_{2k+q-1}$ can be expressed explicitly in terms of the data $s_0, s_1, \ldots, s_{k-1}$ by writing out the equality (3.6):

$$\begin{aligned}\sigma_0^*Q(z)s(z) - (Q(z)+z^k)\\ &= \sigma_0^*[Q(z)(s(z)-\sigma_0) - \sigma_0 z^k]\\ &= t_{2k+q}z^{2k+q} + \cdots .\end{aligned}$$

The coefficient σ_{2k+q} however satisfies an inequality which corresponds to the coefficient t_{2k+q} being nonzero. If $q = 0$, then

$$t_{2k} = \sigma_0^*(c_0\sigma_{2k} + c_1\sigma_{2k-1} + \ldots + c_{k-1}\sigma_{k+1}),$$

which leads to the formula (6.6) for σ_{2k} in Theorem 6.3. The second formula in (6.6) shows that $t_{2k} \neq 0$.

In the proof of Theorem 6.4 we make use of the following lemma (see [2]). If $\mathcal{K}$ is a Krein space with inner product $\langle \cdot, \cdot \rangle_{\mathcal{K}}$, then by $\mathcal{K}'$ we denote the anti-Krein space, which is the same linear space $\mathcal{K}$ but now equipped with inner product $\langle x, y\rangle_{\mathcal{K}'} = -\langle x, y\rangle_{\mathcal{K}}$, $x, y \in \mathcal{K}$.

Lemma 6.5. *Let $s(z)$ be the characteristic function of the minimal coisometric colligation*

$$V = \begin{pmatrix} T & u \\ \langle \cdot, v\rangle_{\mathcal{K}} & \gamma \end{pmatrix} : \begin{pmatrix} \mathcal{K} \\ \mathbb{C} \end{pmatrix} \to \begin{pmatrix} \mathcal{K} \\ \mathbb{C} \end{pmatrix},$$

that is,

$$s(z) = \gamma + z\left\langle (I - zT)^{-1}u, v\right\rangle_{\mathcal{K}},$$

where $\mathcal{K}$ is a Krein space and assume $\gamma = s(0) \neq 0$. Then the function $s_i(z) := s(z)^{-1}$ is the characteristic function of the minimal coisometric colligation

$$V_i = \begin{pmatrix} T_i & u_i \\ \langle \cdot, v_i\rangle_{\mathcal{K}'} & \gamma_i \end{pmatrix} : \begin{pmatrix} \mathcal{K}' \\ \mathbb{C} \end{pmatrix} \to \begin{pmatrix} \mathcal{K}' \\ \mathbb{C} \end{pmatrix},$$

where $\mathcal{K}'$ is the anti-space of the Krein space $\mathcal{K}$, with entries

$$T_i := T - \frac{\langle \cdot, v\rangle_{\mathcal{K}}}{\gamma}u, \quad u_i = \frac{u}{\gamma},$$
$$v_i = \frac{v}{\gamma^*}, \qquad \gamma_i = \frac{1}{\gamma},$$

that is,

$$s_i(z) = \gamma_i + z\langle (I - zT_i)^{-1}u_i, v_i\rangle_{\mathcal{K}'}.$$

To prove Theorem 6.4 we apply Lemma 6.5 and construct the colligation for $\widetilde{s}_i(z) := 1/\widetilde{s}(z)$ from that of $\widetilde{s}(z)$; here we use that $\widetilde{s}(0) \neq 0$. Using Theorem 6.1 with $k = q$ we construct the colligation $\widehat{V}_i$ for $\widehat{s}_i(z) := z^q\widetilde{s}_i(z)$. Given

$$s(z) = \frac{(Q(z) + z^k)\widetilde{s}(z) - \sigma_0 Q(z)z^q}{\sigma_0^* Q(z)\widetilde{s}(z) - (Q(z) - z^k)z^q},$$

let $s_i(z) := 1/s(z)$. We associate σ_{ij}, c_{ij}, $Q_i(z)$, k_i, etc. with $s_i(z)$ in the same way as σ_j, c_j, $Q(z)$, k, etc. are associated with $s(z)$. Straightforward calculations give

$$\sigma_{ij} = \begin{cases} \sigma_0^*, & j = 0, \\ 0, & j = 1, \dots, k-1, \\ -\sigma_0^{*2}\sigma_j, & j = k, \dots, 2k-1, \\ -\sigma_0^{*2}(\sigma_{2k} - \sigma_0^*\sigma_k^2), & j = 2k, \end{cases}$$

and hence $k_i = k$, $c_{ij} = -c_j$, $Q_i(z) = -Q(z)$, and

$$t_{2k_i} = -\sigma_0^*\sigma_k + t_{2k} = -\sigma_0^*\sigma_k \neq 0,$$

which implies $q_i = 0$. Using $s(0) = s_i(0)^*$ we find that

$$s_i(z) = \frac{(Q_i(z) + z^k)\widehat{s}_i(z) - \sigma_{i0} Q_i(z)}{\sigma_{i0}^* Q_i(z)\widehat{s}_i(z) - (Q_i(z) - z^k)}.$$

Now we can apply Theorem 6.3 with $s_i(z)$ and $\widehat{s}_i(z)$ in place of $s(z)$ and $\widetilde{s}(z)$ and construct the colligation for $s_i(z)$ from the colligation $\widehat{V}_i$. It remains to use Lemma 6.5 again to obtain the colligation for $s(z)$ as in the theorem.

References

[1] R. Ackner, H. Lev-Ari, and T. Kailath, *The Schur algorithm for matrix-valued meremorphic functions*, Siam J. Matrix Anal. Appl. **15**(1) (1994), 140–150.

[2] D. Alpay, T. Azizov, A. Dijksma, and H. Langer, *The Schur algorithm for generalized Schur functions, I: Coisometric realizations*, Operator Theory: Adv. Appl., **129**, Birkhäuser Verlag, Basel, 2001, 1–36.

[3] D. Alpay, T. Azizov, A. Dijksma, and H. Langer, *The Schur algorithm for generalized Schur functions, II: Jordan chains and transformations of characteristic functions*, to appear in Monatsheft für Mathematik.

[4] D. Alpay, T. Azizov, A. Dijksma, and H. Langer, *The Schur algorithm for generalized Schur functions, III: J-unitary matrix polynomials on the circle*, submitted for publication.

[5] D. Alpay, T. Azizov, A. Dijksma, H. Langer, and G. Wanjala, *Characteristic functions of coisometric colligations and fractional linear transformations arising from a basic interpolation problem*, Department of Mathematics, University of Groningen, Groningen, The Netherlands, Research report IWI 2002.

[6] D. Alpay, A. Dijksma, J. Rovnyak, and H. de Snoo, *Schur functions, operator colligations, and reproducing kernel Pontryagin spaces*, Operator Theory: Adv. Appl., **96**, Birkhäuser Verlag, Basel, 1997.

[7] D. Alpay and H. Dym, *On applications of reproducing kernel spaces to the Schur algorithm and rational J-unitary factorization*, Operator Theory: Adv. Appl., **18**, Birkhäuser Verlag, Basel, 1986, 89–159.

[8] D. Alpay and H. Dym, *On reproducing kernel spaces, the Schur algorithm, and interpolation in a general class of domains*, Operator Theory: Adv. Appl., **59**, Birkhäuser Verlag, Basel, 1992, 30–77.

[9] D. Alpay and I. Gohberg, *Unitary rational matrix functions*, Operator Theory: Adv. Appl., **33**, Birkhäuser Verlag, Basel, 1988, 175–222.

[10] T.Ya. Azizov, I.S. Iokhvidov, *Linear operators in spaces with an indefinite metric*, Wiley, New York, 1989.

[11] M.J. Bertin, A. Decomps-Guilloux, M. Grandet-Hugot, M. Pathiaux-Delfosse, and J.P. Schreiber, *Pisot and Salem numbers*, Birkhäuser Verlag, Basel, 1992.

[12] C. Chamfy, *Fonctions méromorphes sur le cercle unité et leurs séries de Taylor*, Ann. Inst. Fourier **8** (1958), 211–251.

[13] P. Delsarte, Y. Genin, and Y. Kamp, *Pseudo-Carathéodory functions and hermitian Toeplitz matrices*, Philips J. Res. **41**(1) (1986), 1–54.

[14] A. Dijksma, H. Langer, and H.S.V. de Snoo, *Unitary colligations in Krein spaces and their role in the extension theory of isometries and symmetric linear relations in Hilbert spaces*, Functional Analysis II, Proceedings Dubrovnik 1985, Lecture Notes in Mathematics 1242, Springer–Verlag, Berlin, 1987, 1–42.

[15] J. Dufresnoy, *Sur le problème des coefficients par certaines fonctions dans le cercle unité*, Ann. Acad. Sc. Fenn. Ser. A.I, **250**(9) (1958), 1–7.

[16] H. Dym, *J contractive matrix functions, reproducing kernel Hilbert spaces and interpolation*, CBMS Regional Conference Series, Amer. Math. Soc. **71**, Providence, R.I., 1989.

[17] M.G. Krein and H. Langer, *Über einige Fortsetzungsprobleme, die eng mit der Theorie hermitescher Operatoren in Raume Π_κ zusammenhängen, Teil I: Einige Funktionenklassen und ihre Darstellungen*, Math. Nachr. **77** (1977), 187–236.

[18] I. Schur, *Über die Potenzreihen, die im Innern des Einheitskreises beschränkt sind*, J. Reine Angew. Math. **147** (1917), 205–232; English translation in: *I. Schur methods in operator theory and signal processing*, Operator Theory: Adv. Appl., **18**, Birkhäuser Verlag, Basel, 1989.

[19] I. Schur, *Über die Potenzreihen, die im Innern des Einheitskreises beschränkt sind; Fortsetzung*, J. Reine Angew. Math. **148** (1918), 122–145.

D. Alpay
Department of Mathematics, Ben-Gurion University of the Negev
P.O. Box 653, 84105 Beer-Sheva, Israel
e-mail: dany@math.bgu.ac.il

T.Ya. Azizov
Department of Mathematics, Voronezh State University
394693 Voronezh, Russia
e-mail: azizov@tom.vsu.ru

A. Dijksma
Department of Mathematics, University of Groningen
P.O. Box 800, 9700 AV Groningen, The Netherlands
e-mail: dijksma@math.rug.nl

H. Langer
Institute of Analysis and Technical Mathematics, Vienna University of Technology, Wiedner Hauptstrasse 8–10, A-1040 Vienna, Austria
e-mail: hlanger@mail.zserv.tuwien.ac.at

G. Wanjala
Department of Mathematics, University of Groningen
P.O. Box 800, 9700 AV Groningen, The Netherlands
e-mail: gerald@math.rug.nl

Operator Theory:
Advances and Applications, Vol. 143, 77–134

Formal Reproducing Kernel Hilbert Spaces: The Commutative and Noncommutative Settings

Joseph A. Ball and Victor Vinnikov

Abstract. The notion of the reproducing kernel Hilbert space has played a central role in operator theory and applications since its introduction by Aronszajn in 1950. We explore here the basic ideas and some applications of an extension of this idea to the setting of Hilbert spaces whose elements are formal power series in possibly noncommuting indeterminates.

1. Introduction

Suppose that $\mathcal{H}$ is a Hilbert space whose elements $z \mapsto f(z)$ are $\mathcal{E}$-valued functions (where $\mathcal{E}$ is some coefficient (or auxiliary) Hilbert space) defined on some set Ω (so $f\colon \Omega \mapsto \mathcal{E}$). We say that $\mathcal{H}$ is a *reproducing kernel Hilbert space* if the evaluation map $\Phi_{z'}$ defined by $\Phi(z')\colon f \mapsto f(z')$ is a bounded operator from $\mathcal{H}$ into $\mathcal{E}$ for each $z' \in \Omega$. The adjoint map $\Phi(z')^*$ is then a bounded operator from $\mathcal{E}$ into $\mathcal{H}$, and, for each $z, z' \in \Omega$, the expression $K(z,z') := \Phi(z)\Phi(z')^*$ defines a bounded operator on $\mathcal{E}$. The function $(z,z') \mapsto K(z,z')$ from $\Omega \times \Omega$ into $\mathcal{L}(\mathcal{E})$ is then said to be the *reproducing kernel* for the reproducing kernel Hilbert space $\mathcal{H}$. Note that, from the definitions, we see that $K_{z'}(\cdot)e := K(\cdot, z')e = \Phi(z')^*e \in \mathcal{H}$ for each $e \in \mathcal{E}$ and that

$$\langle f, K_{z'}e\rangle_{\mathcal{H}} = \langle f, \Phi(z')^*e\rangle_{\mathcal{H}} = \langle \Phi(z')f, e\rangle_{\mathcal{E}} = \langle f(z'), e\rangle_{\mathcal{E}},$$

i.e., $K(z,z')$ has the *reproducing kernel property: for each* $z' \in \Omega$ *and* $e \in \mathcal{E}$, $K_{z'}(\cdot)e := K(\cdot, z')e \in \mathcal{H}$ *and* $\langle f, K_{z'}e\rangle_{\mathcal{H}} = \langle f(z'), e\rangle_{\mathcal{E}}$. Moreover, if $K\colon \Omega \times \Omega \mapsto \mathcal{E}$ is the reproducing kernel for a reproducing kernel Hilbert space of $\mathcal{E}$-valued functions on Ω, then, for any finite collection $z_1, \dots, z_N$ of points in Ω and $e_1, \dots, e_N$

2000 *Mathematics Subject Classification.* Primary: 46E22; Secondary: 47A48, 93C35.

The first author is supported by NSF grant DMS-9987636; both authors are supported by a grant from the US-Israel Binational Science Foundation.

of vectors in $\mathcal{E}$, we have

$$\begin{aligned}\sum_{i,j=1}^{N} \langle K(z_i,z_j)e_j, e_i\rangle_{\mathcal{E}} &= \sum_{i,j=1}^{N} \langle \Phi(z_i)\Phi(z_j)^* e_j, e_i\rangle_{\mathcal{E}} \\ &= \sum_{i,j=1}^{N} \langle \Phi(z_j)^* e_j, \Phi(z_i)^* e_i\rangle_{\mathcal{H}} = \|\sum_{j=1}^{N} K_{z_j} e_j\|^2_{\mathcal{H}} \geq 0,\end{aligned}$$

i.e., the operator-valued function $K\colon \Omega\times\Omega \mapsto \mathcal{L}(\mathcal{E})$ is a *positive kernel* in the sense that

$$\sum_{i,j=1}^{N} \langle K(z_i,z_j)e_j, e_i\rangle_{\mathcal{E}} \geq 0 \tag{1.1}$$

for all $z_1,\dots,z_N \in \Omega$ and $e_1,\dots,e_n \in \mathcal{E}$ for all $N = 1,2,3,\dots$. Conversely, if $K\colon \Omega\times\Omega \mapsto \mathcal{E}$ is a positive kernel, one can define a pre-Hilbert space inner product on

$$\mathcal{H}_0 = \operatorname{span}\{K(\cdot,z')e\colon z'\in\Omega,\, e\in\mathcal{E}\}$$

by

$$\left\langle \sum_{j=1}^{N} K(\cdot,z_j)e_j, \sum_{i=1}^{M} K(\cdot,z_i')e_i'\right\rangle_{\mathcal{H}_0} = \sum_{j=1}^{N}\sum_{i=1}^{M} \langle K(z_i',z_j)e_j, e_i'\rangle_{\mathcal{E}}.$$

The assumption that K is a positive kernel implies that this inner product is positive definite and that the Hilbert-space completion $\mathcal{H}$ of $\mathcal{H}_0$ can be identified with a space of $\mathcal{E}$-valued functions for which $K(\cdot,\cdot)$ serves as its reproducing kernel. It then follows that $K(\cdot,\cdot)$ has a factorization $K(z,z') = \Phi(z)\Phi(z')^*$ where $\Phi(z')\colon \mathcal{H}\mapsto\mathcal{E}$ is the evaluation-at-z' operator on $\mathcal{H}$. If on the other hand, $K\colon \Omega\times\Omega\mapsto\mathcal{L}(\mathcal{E})$ is any operator-valued function which has a factorization $K(z,z') = H(z)H(z')^*$ for some operator-valued function $H\colon \Omega\mapsto\mathcal{L}(\mathcal{H},\mathcal{E})$ for some auxiliary Hilbert space $\mathcal{H}$, then

$$\sum_{i,j=1}^{N} \langle K(z_i,z_j)e_j, e_i\rangle_{\mathcal{E}} = \|\sum_{j=1}^{N} H(z_j)^* e_j\|^2 \geq 0$$

and hence K is a positive kernel. We thus have the following characterizations of operator-valued functions K arising as the reproducing kernel for a reproducing kernel Hilbert space:

Theorem 1.1. *Let $K\colon \Omega\times\Omega\mapsto\mathcal{L}(\mathcal{E})$ be an operator-valued function on $\Omega\times\Omega$. Then the following are equivalen*

1. *The operator-valued function K is the reproducing kernel function for a uniquely determined reproducing kernel Hilbert space $\mathcal{H}(K)$ of $\mathcal{E}$-valued functions defined on Ω.*

2. *There is an auxiliary Hilbert space $\mathcal{H}$ and an operator-valued function $H\colon \Omega \mapsto \mathcal{L}(\mathcal{H},\mathcal{E})$ for which the factorization*
$$K(z,z') = H(z)H(z')^*$$
holds for all $z, w \in \Omega$.

3. *K is a positive kernel, i.e.,* (1.1) *holds for all $z_1,\dots,z_N \in \Omega$, $e_1,\dots,e_N \in \mathcal{E}$ and $N = 1,2,3,\dots$.*

Moreover in this case the reproducing kernel Hilbert space $\mathcal{H}(K)$ can be defined directly in terms of the function $H(z)$ appearing in condition (2) *by*
$$\mathcal{H}(K) = \{H(z)h\colon h \in \mathcal{H}\}$$
with norm taken to be the lifted norm
$$\|H(z)h\|_{\mathcal{H}(K)} = \|Qh\|$$
where Q is the orthogonal projection of $\mathcal{H}$ onto the orthogonal complement of the kernel of the multiplication operator $M_H\colon \mathcal{H} \mapsto \mathcal{H}(K)$ given by $h \mapsto H(z)h$.

The scalar case ($\mathcal{E} = \mathbb{C}$) of Theorem 1.1 goes back to [2]. The vectorial version presented here is also now well known and has been noticed and used in a variety of contexts and applications; see, e.g., [1, 23].

If one assumes that Ω is a domain in $\mathbb{C}^d$ and that the elements f of $\mathcal{H}(K)$ are holomorphic functions on Ω, then in particular $K(\cdot,w)e = \Phi(w)^*e \in \mathcal{H}(K)$ is holomorphic on Ω for each $e \in \mathcal{E}$, and hence $K(\cdot,z')$ is a holomorphic $\mathcal{L}(\mathcal{E})$-valued function for each fixed $z' \in \Omega$. Moreover, since $K(z,z') = \Phi(z)\Phi(z')^* = (\Phi(z')\Phi(z)^*)^* = K(z',z)^*$, we see that $K(z,\cdot)$ is conjugate holomorphic on Ω for each fixed $z \in \Omega$, i.e., K is a *sesquiholomorphic kernel* on $\Omega \times \Omega$. As an application of the Cauchy integral formula, it is not hard to see that the operators $\Phi^k(z')\colon \mathcal{H}(K) \mapsto \mathcal{E}$ defined by
$$\Phi^k(z')\colon f \mapsto \frac{\partial^k f}{\partial z'^k}$$
are bounded as well, with adjoints given by
$$\Phi^k(z')^*\colon e \mapsto \frac{\partial^k K}{\partial \overline{z}'^k}(\cdot,z') \in \mathcal{H}(K)$$
where $\frac{\partial^k f}{\partial z^k}$ is the higher order partial derivative of f
$$\frac{\partial^k f}{\partial z^k} = \frac{\partial^{k_1+\dots+k_d} f}{\partial^{k_1} z_1 \cdots \partial^{k_d} z_d} \text{ if } k = (k_1,\dots,k_d) \text{ and } z = (z_1,\dots,z_d).$$

Our purpose is to extend these ideas to the setting where the Hilbert space consists of formal power series
$$f(z) = \sum_{n \in \mathbb{Z}^d_+} f_n z^n = \sum_{n_1,\dots,n_d \geq 0} f_{n_1,\dots,n_d} z_1^{n_1} \cdots z_d^{n_d} \text{ with } f_n \in \mathcal{E}$$

where the radius of convergence of an element f may be as small as zero, but evaluation of power-series coefficients $f \mapsto f_n$ is bounded from $\mathcal{H}$ to $\mathcal{E}$ and to spaces of Laurent series

$$f(z) = \sum_{n \in \mathbb{Z}^d} f_n z^n = \sum_{n_1, \dots, n_d = -\infty}^{\infty} f_{n_1, \dots, n_d} z_1^{n_1} \cdots z_d^{n_d},$$

where the Laurent series may not be convergent for any $z \in \mathbb{C}^d$ but evaluation of Laurent coefficients $f \mapsto f_n$ is bounded from $\mathcal{H}$ to $\mathcal{E}$ for each $n \in \mathbb{Z}^d$; in particular, as we shall see, one can view the familiar L^2-space of vector-valued functions $f(z) = \sum_{n=-\infty}^{\infty} f_n z^n$ (here we take $d = 1$ for simplicity) with $\|f\|_2^2 = \sum_{n=-\infty}^{\infty} \|f_n\|^2$ as a *formal reproducing kernel Hilbert space* (FRKHS). More generally, if the power series representation $K(z, z') = \sum_{n,n'} W_{n,n'} z^n z'^{-n'}$ has the Toeplitz form

$$W_{n,n'} = \widehat{W}_{n-n'} \text{ for some } \widehat{W}_j \in \mathcal{L}(\mathcal{E}), \ j \in \mathbb{Z}^d,$$

one obtains a model for a commuting unitary d-tuple of operators $\mathcal{U} = (\mathcal{U}_1, \dots, \mathcal{U}_d)$ on a Hilbert space $\mathcal{H}$ having also specification of an operator V mapping a coefficient space $\mathcal{E}$ into $\mathcal{H}$ (a *scale*, or a *control/observation* operator) – i.e., a model for a d-variable *unitary scattering system* in the terminology of [14] and [17]; the associated spaces $\mathcal{H}(K)$ – also denoted as $\mathcal{L}_W$ where W is the block Toeplitz matrix $W = [W_{n-n'}]_{n,n' \in \mathbb{Z}^d}$ – give an alternative approach (extended to the d-variable case) to the Hellinger space model L^σ used in [17] associated with a positive measure on the unit circle $\mathbb{T}$ as well as to the de Branges-Rovnyak space model $\mathcal{E}(\varphi)$ associated with a Herglotz function φ on the unit disk.

One can also consider power series in a d-tuple $z = (z_1, \dots, z_d)$ of noncommuting indeterminates:

$$f(z) = \sum_{w \in \mathcal{F}_d} f_w z^w$$

where $\mathcal{F}_d$ denotes the free semigroup generated by the alphabet $(g_1, \dots, g_d)$, the elements of which are words w of the form $w = g_{i_n} \cdots g_{i_1}$ with each index $i_j \in \{1, \dots, d\}$ and where then z^w represents the monomial $z^w = z_{i_n} \cdots z_{i_1}$ in the noncommuting indeterminates $z = (z_1, \dots, z_d)$. Spaces of this sort give rise to models for a row-isometric and row-unitary operator d-tuple $\mathcal{U} = (\mathcal{U}_1, \dots, \mathcal{U}_d)$, i.e., a d-tuple $\mathcal{U}_1, \dots, \mathcal{U}_d$ of operators on a Hilbert space $\mathcal{H}$ for which the operator row matrix

$$[\mathcal{U}_1 \ \ \dots \ \ \mathcal{U}_d] : \oplus_{j=1}^d \mathcal{H} \mapsto \mathcal{H}$$

is isometric (respectively, unitary), together with specification of a *scale operator* $V \colon \mathcal{E} \mapsto \mathcal{H}$ (a *row-isometric* (respectively, *row-unitary*) *scattering system*). This object was introduced in [5] and [6] without explicit identification of the NFRKHS structure. Here we set forth the basic properties of these spaces from the point of view of reproducing kernel Hilbert spaces. In particular, we give a new proof based on reproducing kernel techniques of the fact that a formal power series in the noncommutative d-variable Schur class can be realized as the transfer function of a noncommutative, conservative, discrete-time linear system. Here the noncommutativity of the linear system refers to the fact that the inputs, state and outputs evolve on a free semigroup rather than the usual discrete-time time axis $\mathbb{Z}$.

In Section 2 we present the commutative case, i.e., formal reproducing kernel Hilbert spaces whose elements are power (or Laurent) series in a collection of commuting variables. In Section 3 we treat the noncommutative situation.

2. Commutative formal reproducing kernel Hilbert spaces

2.1. General FRKHSs

Suppose that $\mathcal{H}$ is a Hilbert space consisting of formal Laurent series in d commuting variables $z = (z_1, \dots, z_d)$

$$f(z) = \sum_{n \in \mathbb{Z}^d} f_n z^n = \sum_{n_1, \dots, n_d = -\infty}^{\infty} f_{n_1, \dots, n_d} z_1^{n_1} \cdots z_d^{n_d}$$

with coefficients f_n in the auxiliary Hilbert space $\mathcal{E}$. We denote the space of all such formal power series as $\mathcal{E}[[z, z^{-1}]]$ (where $z = (z_1, \dots, z_d)$). We then say that $\mathcal{H}$ is a *formal reproducing kernel Hilbert space* (FRKHS) if the coefficient-evaluation functional $\Phi_n \colon f \mapsto f_n$ is a bounded operator from $\mathcal{H}$ to $\mathcal{E}$ for each $n \in \mathbb{Z}^d$.

Note that elements of $\mathcal{E}[[z, z^{-1}]]$ can be regarded as $\mathcal{E}$-valued functions on $\mathbb{Z}^d$. If $\mathcal{H}$ is a FRKHS with coefficients in $\mathcal{E}$ in the commuting variables $z = (z_1, \dots, z_d)$, we may then view $\mathcal{H}$ as a standard reproducing kernel Hilbert space over the set $\Omega = \mathbb{Z}^d$. Thus there is a reproducing kernel $K \colon \mathbb{Z}^d \times \mathbb{Z}^d \mapsto \mathcal{L}(\mathcal{E})$; in this context, given $n, n' \in \mathbb{Z}^d$ we denote the value of K at (n, n') by

$$(n, n') \mapsto K_{n,n'} \in \mathcal{L}(\mathcal{E}) \tag{2.1}$$

rather than by $K(n, n')$. Note that elements $f \in \mathcal{H}$ are viewed as power series $f(z) = \sum_{n \in \mathbb{Z}^d} f_n z^n$ rather than as functions $n \mapsto f_n$ from $\mathbb{Z}^d$ to $\mathcal{E}$. Hence, for $e \in \mathcal{E}$ and $n' \in \mathbb{Z}^d$, we write the element $\Phi_{n'}^* e \in \mathcal{H}$ as $\Phi_{n'}^* e = K_{n'}(\cdot) e$ where

$$K_{n'}(z) = \sum_{n \in \mathbb{Z}^d} K_{n,n'} z^n \in \mathcal{L}(\mathcal{E})[[z, z^{-1}]]$$

and $K_{n,n'}$ is as in (2.1). Then the reproducing kernel property takes the form

$$\langle f, K_{n'}(\cdot) e \rangle_{\mathcal{H}} = \langle f_{n'}, e \rangle_{\mathcal{E}} \text{ for all } f \in \mathcal{H} \text{ and } e \in \mathcal{E}. \tag{2.2}$$

We can make the notation more suggestive of the case of a standard reproducing kernel Hilbert space of analytic functions with convergent power series representations as follows. In general, given a coefficient Hilbert space $\mathcal{C}$, we define a pairing

$$\langle \cdot, \cdot \rangle_{\mathcal{C} \times \mathcal{C}[[z', z'^{-1}]]} \colon \mathcal{C} \times \mathcal{C}[[z', z'^{-1}]] \mapsto \mathbb{C}[[z', z'^{-1}]]$$

by

$$\langle c, \sum_{n' \in \mathbb{Z}^d} f_{n'} z'^{-n'} \rangle_{\mathcal{C} \times \mathcal{C}[[z', z'^{-1}]]} = \sum_{n' \in \mathbb{Z}^d} \langle c, f_{n'} \rangle_{\mathcal{C}} z'^{n'}.$$

When convenient we on occasion take the pairing in the reverse order:

$$\langle \sum_{m' \in \mathbb{Z}^d} f_{m'} z'^{m'}, c \rangle_{\mathcal{C}[[z', z'^{-1}]] \times \mathcal{C}} = \sum_{m' \in \mathbb{Z}^d} \langle f_{m'}, c \rangle_{\mathcal{C}} z'^{m'}.$$

These are actually special cases of a more general partially defined pairing on $\mathcal{C}[[z', z'^{-1}]]$ with values in $\mathbb{C}[[z', z'^{-1}]]$ given by

$$\left\langle \sum_{m' \in \mathbb{Z}^d} f_{m'} z'^{m'}, \sum_{n' \in \mathbb{Z}^d} g_{n'} z'^{n'} \right\rangle_{\mathcal{C}[[z', z'^{-1}]] \times \mathcal{C}[[z', z'^{-1}]]}$$

$$= \sum_{m' \in \mathbb{Z}^d} \left[\sum_{n' \in \mathbb{Z}^d} \langle f_{m'-n'}, g_{n'} \rangle_{\mathcal{C}} \right] z'^{m'}$$

where the sum required to compute the coefficient of $z^{m'}$ is finite under the assumption that at least one of f or g is a polynomial in $z_1, \dots, z_d$ and $z_1^{-1}, \dots, z_d^{-1}$.

Then the reproducing kernel property (2.2) for the FRKHS $\mathcal{H}$ can be written more suggestively as

$$\langle f, K(\cdot, z')e \rangle_{\mathcal{H} \times \mathcal{H}[[z', z'^{-1}]]} = \langle f(z'), e \rangle_{\mathcal{E}[[z', z'^{-1}]] \times \mathcal{E}}$$

where we have set

$$K(z, z') = \sum_{n, n' \in \mathbb{Z}^d} K_{n,n'} z^n z'^{-n'} \in \mathcal{L}(\mathcal{E})[[z, z^{-1}, z', z'^{-1}]].$$

We then say that $K(z, z')$ is the *reproducing kernel* for the FRKHS $\mathcal{H}$. As $\mathcal{H}$ is uniquely determined by K, we then write $\mathcal{H} = \mathcal{H}(K)$.

The following theorem gives a characterization for reproducing kernels for FRKHSs.

Theorem 2.1. *Let* $K(z, z') = \sum_{n, n' \in \mathbb{Z}^d} K_{n,n'} z^n z'^{-n'} \in \mathcal{L}(\mathcal{E})[[z, z^{-1}, z', z'^{-1}]]$. *Then the following are equivalent:*

1. *K is the reproducing kernel for a uniquely determined FRKHS $\mathcal{H}(K)$ of formal Laurent series in the commuting variables $z = (z_1, \dots, z_d)$ with coefficients in $\mathcal{E}$.*
2. *There is an auxiliary Hilbert space $\mathcal{H}$ and a formal power series $H(z) \in \mathcal{L}(\mathcal{H}, \mathcal{E})[[z, z^{-1}]]$ so that*
$$K(z, z') = H(z) H(z')^*$$
where we use the convention
$$(z'^n)^* = z'^{-n}.$$
3. *$K(z, z')$ is positive definite in the sense that*
$$\sum_{n, n' \in \mathbb{Z}^d} \langle K_{n,n'} e_{n'}, e_n \rangle_{\mathcal{E}} \geq 0$$
for all finitely supported $\mathcal{E}$-valued functions $n \mapsto e_n$ on $\mathbb{Z}^d$.

Moreover, in this case the FRKHS $\mathcal{H}(K)$ can be defined directly in terms of the formal power series $H(z)$ appearing in condition (2) by

$$\mathcal{H}(K) = \{H(z)h \colon h \in \mathcal{H}\}$$

with norm taken to be the "lifted norm"

$$\|H(z)h\|_{\mathcal{H}(K)} = \|Qh\|_{\mathcal{H}}$$

where Q is the orthogonal projection of $\mathcal{H}$ onto the orthogonal complement of the kernel of the map $M_H \colon \mathcal{H} \mapsto \mathcal{E}[[z, z^{-1}]]$ given by

$$M_H \colon h \mapsto H(z)h.$$

Proof. As explained in the preceding discussion, the reproducing kernel $K(z, z')$ for a FRKHS $\mathcal{H}$ of d-variable formal Laurent series may be viewed as a standard reproducing kernel Hilbert space on $\mathbb{Z}^d$; when this is done, then the reproducing kernel for the associated standard RKHS is $(n, n') \mapsto K_{n,n'}$, where the $K(z, z') = \sum_{n,n' \in \mathbb{Z}^d} K_{n,n'} z^n z'^{-n}$ is the formal reproducing kernel for the FRKHS. Thus $K(z, z') = \sum_{n,n' \in \mathbb{Z}^d} K_{n,n'} z^n z'^{-n}$ is the reproducing kernel for a FRKHS if and only if $(n, n') \mapsto K_{n,n'}$ is the reproducing kernel for a standard RKHS of $\mathcal{E}$-valued functions defined on $\mathbb{Z}^d$. By Theorem 1.1, this in turn is equivalent to the positivity of $K_{n,n'}$ as a function on $\mathbb{Z}^d \times \mathbb{Z}^d$; in this way we see the equivalence of (1) and (3) in Theorem 2.1. Also by Theorem 1.1, we know that (3) is equivalent to the existence of a factorization

$$K_{n,n'} = H_n H_{n'}^*$$

for some function $n \mapsto H_n$ defined on $\mathbb{Z}^d$ with values in $\mathcal{L}(\mathcal{H}, \mathcal{E})$ for some auxiliary Hilbert space $\mathcal{H}$. In terms of the formal power series $K(z, z')$, this latter condition is equivalent to

$$K(z, z') = H(z)H(z')^*$$

where we have set

$$H(z) = \sum_{n \in \mathbb{Z}^d} H_n z^n.$$

In this way we see the equivalence of (2) and (3). The description of $\mathcal{H}(K)$ in terms of the factor $H(z)$ appearing in condition (2) can be obtained by viewing $\mathcal{H}(K)$ as a standard reproducing kernel Hilbert space over $\mathbb{Z}^d$ and using the corresponding result for the standard case as stated in Theorem 1.1. This completes the proof of Theorem 2.1. □

We now present an equivalent operator-theoretic presentation of a FRKHS having reproducing kernel $K(z, z') = \sum_{n,n' \in \mathbb{Z}^d} W_{n,n'} z^n z'^{-n}$ directly in terms of the Fourier coefficients $W_{n,n'}$ $(n, n' \in \mathbb{Z}^d)$ of the kernel. Assume therefore that $W_{n,n'}$ is a given collection of operators satisfying the positivity condition in part (3) of Theorem 2.1 (with $W_{n,n'}$ in place of $K_{n,n'}$). In addition to the space $\mathcal{E}[[z, z^{-1}]]$ of infinite power series in the variables $z_1, \ldots, z_d, z_1^{-1}, \ldots, z_d^{-1}$, we introduce the space

$\mathcal{E}[z, z^{-1}]$ of polynomials in $z_1, \dots, z_d, z_1^{-1}, \dots, z_d^{-1}$. Note that the usual scalar-valued L^2-pairing between power series with square-summable coefficients can also be used to define a pairing between $\mathcal{E}[[z, z^{-1}]]$ and $\mathcal{E}[z, z^{-1}]$:

$$\langle \sum_{n \in \mathbb{Z}^d} f_n z^n, \sum_{n \in \mathbb{Z}^d} p_n z^n \rangle_{L^2} = \sum_{n \in \mathbb{Z}^d} \langle f_n, p_n \rangle_{\mathcal{E}}. \tag{2.3}$$

(Note that the sum on the right-hand side of (2.3) is finite under the assumption that p is a polynomial in z and z^{-1}.) Informally, we refer to this inner product as simply the *L^2-inner product* (extended to $\mathcal{E}[[z, z^{-1}]] \times \mathcal{E}[z, z^{-1}]$).

Let now $W = [W_{n,n'}]_{n,n' \in \mathbb{Z}^d}$ be a matrix with rows and columns indexed by $\mathbb{Z}^d$ and with matrix entries $W_{n,n'} \in \mathcal{L}(\mathcal{E})$ equal to operators on $\mathcal{E}$. Then W defines an operator, also denoted W, from $\mathcal{E}[z, z^{-1}]$ into $\mathcal{E}[[z, z^{-1}]]$ in the usual way:

$$W \colon \sum_{n \in \mathbb{Z}^d} p_n z^n \mapsto \sum_{n, n' \in \mathbb{Z}^d} W_{n,n'} p_{n'} z^n.$$

Note that the sum over n' is finite and hence convergent since p is a polynomial while the sum over n can be infinite but is taken only in a formal sense to define an element of $\mathcal{E}[[z, z^{-1}]]$. Given such a block matrix W, we say that W is a *positive semidefinite* ($\mathcal{L}(\mathcal{E})$-valued) *weight* on $\mathbb{Z}^d$ if

$$\langle Wp, p \rangle_{L^2} \geq 0 \text{ for all } p \in \mathcal{E}[z, z^{-1}].$$

Equivalently, the collection of operators $W_{n,n'}$ (for $n, n' \in \mathbb{Z}^d$) satisfies the condition in part (3) of Theorem 2.1 (with $W_{n,n'}$ in place of $K_{n,n'}$). Given such a positive semidefinite weight W, we define an inner product $\langle \cdot, \cdot \rangle_{\mathcal{L}_W}$ on $W\mathcal{E}[z, z^{-1}]$ via

$$\langle Wp, Wp' \rangle_{\mathcal{L}_W} = \langle Wp, p' \rangle_{L^2}. \tag{2.4}$$

Under the assumption that W is a positive semidefinite weight, the inner product $\langle \cdot, \cdot \rangle_{\mathcal{L}_W}$ is positive semidefinite on $W\mathcal{E}[z, z^{-1}]$; if an element $Wp \in W\mathcal{E}[z, z^{-1}]$ had zero-norm in the $\mathcal{L}_W$-inner product, then, by the Cauchy-Schwarz inequality it would follow that

$$0 = \langle Wp, W(z^n e) \rangle_{\mathcal{L}_W} = \langle Wp, z^n e \rangle_{L^2} = \langle [Wp]_n, e \rangle_{\mathcal{E}}$$

for all $e \in \mathcal{E}$, where $[Wp]_n$ is the Laurent coefficient associated with the formal power series $Wp \in \mathcal{E}[[z, z^{-1}]]$ for a given $n \in \mathbb{Z}^d$, from which it would follow that in fact $Wp = 0$ as an element of $\mathcal{E}[[z, z^{-1}]]$. Thus we see that the $\mathcal{L}_W$-inner product (2.4) is in fact positive definite. The above computation also shows that the Laurent coefficient $[Wp]_n$ of an element $Wp \in W\mathcal{E}[z, z^{-1}]$ for a given $n \in \mathbb{Z}^d$ is given by

$$\langle [Wp]_n, e \rangle_{\mathcal{E}} = \langle Wp, W(z^n e) \rangle_{\mathcal{L}_W} \text{ for all } e \in \mathcal{E}. \tag{2.5}$$

By a standard construction, we may complete the pre-Hilbert space $W\mathcal{E}[z, z^{-1}]$ to a Hilbert space, denoted by $\mathcal{L}_W$ – so far we are merely reproducing the proof of (3) $\Longrightarrow$ (1) in Theorem 1.1 (where $\Omega = \mathbb{Z}^d$ in this context). Given an element $f \in \mathcal{L}_W$ of the completion, we may use the formula (2.5) to define Fourier coefficients for f

$$\langle f_n, e \rangle_{\mathcal{E}} = \langle f, W(z^n e) \rangle_{\mathcal{L}_W}$$

and identify f with an element $f(z) = \sum_{n\in\mathbb{Z}^d} f_n z^n$ of $\mathcal{E}[[z, z^{-1}]]$. In this way we have arrived at a Hilbert space $\mathcal{L}_W$ of formal power series in $\mathcal{E}[[z, z^{-1}]]$ for which the maps $\Phi_{n'}: f \mapsto f_{n'}$ are given by

$$\langle f_{n'}, e\rangle_{\mathcal{E}} = \langle f, W(z^{n'}e)\rangle_{\mathcal{L}_W} = \langle f, \sum_{n\in\mathbb{Z}^d} W_{n,n'} e z^n\rangle_{\mathcal{L}_W}.$$

As was observed in the proof of Theorem 2.1, the kernel $K(z, z')$ for the FRKHS is recovered from the positive kernel $(n, n') \mapsto W_{n,n'}$ on $\mathbb{Z}^d \times \mathbb{Z}^d$ via $K(z, z') = \sum_{n,n'\in\mathbb{Z}^d} W_{n,n'} z^n z'^{-n'}$. We now have arrived at the following result.

Proposition 2.2. *Let W be an $\mathcal{L}(\mathcal{E})$-valued positive semidefinite weight over $\mathbb{Z}^d$ as described above. Then there is a FRKHS $\mathcal{L}_W \subset \mathcal{E}[[z, z^{-1}]]$ with dense subset $W\mathcal{E}[z, z^{-1}]$ with inner product densely given by (2.4) and with reproducing kernel $K_W(z, z')$ given by*

$$K_W(z, z') = \sum_{n,n'\in\mathbb{Z}^d} W_{n,n'} z^n z'^{-n'}.$$

Conversely, if $\mathcal{H}$ is any FRKHS of formal power series in the variable $z_1, \dots, z_d$, $z_1^{-1}, \dots, z_d^{-1}$ with coefficients in $\mathcal{E}$ and with reproducing kernel

$$K(z, z') = \sum_{n,n'\in\mathbb{Z}^d} K_{n,n'} z^n z'^{-n'},$$

then the formula $W_{n,n'} = K_{n,n'}$ gives rise to a positive semidefinite weight

$$W = [W_{n,n'}]_{n,n'\in\mathbb{Z}^d}: \mathcal{E}[z, z^{-1}] \mapsto \mathcal{E}[[z, z^{-1}]]$$

such that $\mathcal{H} = \mathcal{L}_W$.

As a first example we consider the space $L^2(\mathbb{T}^d, \mathcal{E})$ of $\mathcal{E}$-valued, norm-square integrable, measurable functions $z \mapsto f(z)$ on the d-torus $\mathbb{T} = \{z = (z_1, \dots, z_d): |z_j| = 1 \text{ for } j = 1, \dots, d\}$, where $\mathcal{E}$ is an auxiliary Hilbert space. As is well known, elements f of $L^2(\mathbb{T}^d, \mathcal{E})$ alternatively can be characterized via their d-variable Fourier series:

$$f \cong \sum_{n\in\mathbb{Z}^d} f_n z^n \tag{2.6}$$

where $f_n = \int_{\mathbb{T}^d} f(z) z^{-n}\, d\lambda(z)$ (with λ is normalized Lebesgue measure on $\mathbb{T}^d$) are the *Fourier coefficients* of f and we have

$$\|f\|_2^2 = \sum_{n\in\mathbb{Z}^d} \|f_n\|_{\mathcal{E}}^2$$

by the Parseval theorem. While the analysis of pointwise convergence of the Fourier series (2.6) to f is rather delicate, as $\{z^n \otimes e_j: n \in \mathbb{Z}^d,\ j \in \mathbf{N}\}$ is a complete orthonormal set for $L^2(\mathbb{T}^d, \mathcal{E})$ whenever $\{e_j: j \in \mathbf{N}\}$ is a complete orthonormal set for $\mathcal{E}$, it follows from the elementary theory of generalized Fourier series in a Hilbert space that the partial sums of the series on the right-hand side of (2.6) converges to f in L^2-norm. However, as point evaluation is not continuous in L^2-norm for any point $z \in \mathbb{T}^d$, the space $L^2(\mathbb{T}^d, \mathcal{E})$ is not a reproducing kernel Hilbert space in the traditional sense. However, the evaluation of Fourier coefficients $\Phi_n \colon f \mapsto f_n$ does define a continuous map $\Phi_n \colon L^2(\mathbb{T}^d, \mathcal{E}) \mapsto \mathcal{E}$, and hence we may consider $L^2(\mathbb{T}^d, \mathcal{E})$ as a FRKHS of formal Laurent series in the sense explained above. A natural question then is to compute the formal reproducing kernel. This we do in the following proposition.

Proposition 2.3. *The space $L^2(\mathbb{T}^d, \mathcal{E})$, viewed as a FRKHS of formal Laurent series, has formal reproducing kernel $k_{\mathbb{Z}^d}(z, z')$ (*two-sided Szegö kernel*) given by*

$$k_{\mathbb{Z}^d}(z, z') = \sum_{n \in \mathbb{Z}^d} z^n z'^{-n}. \tag{2.7}$$

Proof. Note that, for $n \in \mathbb{Z}^d$, the nth Fourier coefficient of a given $f \in L^2(\mathbb{T}^d, \mathcal{E})$ is given by

$$\langle f_n, e \rangle_{\mathcal{E}} = \langle f, z^n e \rangle_{L^2(\mathbb{T}^d, \mathcal{E})}.$$

By the general considerations above, it follows that $k_{\mathbb{Z}^d}(z, z') = \sum_{n \in \mathbb{Z}^d} z^n z'^{-n}$ is the associated reproducing kernel as asserted. □

We remark that the subspace $L^2_{\Xi}(\mathbb{T}^d, \mathcal{E})$ consisting of functions whose Fourier coefficients are supported on a given subset $\Xi \subseteq \mathbb{Z}^d$ has a formal reproducing kernel

$$k_{\Xi}(z, z') = \sum_{n \in \Xi} z^n z'^{-n}. \tag{2.8}$$

Note that $\mathcal{H}(k_{\mathbb{Z}^d}) \cong L^2(\mathbb{T}^d, \mathcal{E})$ has no bounded point evaluations, i.e., there is no $z \in \mathbb{C}^d$ such that evaluation of the formal power series $f(z) = \sum_{n \in \mathbb{Z}^d} f_n z^n$ at z gives rise to a bounded linear transformation from $\mathcal{H}(k_{\mathbb{Z}^d})$ to $\mathcal{E}$. Elements f of $\mathcal{H}(k_{\mathbb{Z}^d})$ however do represent functions on $\mathbb{T}^d$ in the sense of measure theory, i.e., elements are equivalence classes of functions well defined on $\mathbb{T}^d$ up to sets of Lebesgue measure zero on $\mathbb{T}^d$. Alternatively, we can arrive at a reproducing kernel Hilbert space in the standard sense by considering harmonic extensions of these functions $f \in L^2(\mathbb{T}^d, \mathcal{E})$. More precisely, for the multivariable setting discussed here, we are interested in *separately harmonic* extensions, i.e., functions $\xi \mapsto h(\xi)$ of d variables $\xi = (\xi_1, \dots, \xi_d)$ defined on the polydisk $\mathbb{D}^d$ which are harmonic in each variable separately (for each $k = 1, \dots, d$, $h(\xi_1, \dots, \xi_k, \dots, \xi_d)$ is harmonic in ξ_k for each fixed value of ξ_j with $|\xi_j| < 1$ with $j \neq k$). Indeed the harmonic extension of $f \in L^2(\mathbb{T}^d, \mathcal{E})$ is given by

$$f^{\wedge h}(\xi) = \sum_{n \in \mathbb{Z}^d} f_n \cdot (z^n)^{\wedge h}(\xi) \text{ if } f(z) = \sum_{n \in \mathbb{Z}^d} f_n z^n \tag{2.9}$$

where

$$(z^n)^{\wedge h}(\xi) = (z_1^{n_1})^{\wedge h}(\xi_1) \cdots (z_d^{n_d})^{\wedge h}(\xi) \text{ if } n = (n_1, \dots, n_d)$$

where

$$(z_k^{n_k})^{\wedge h}(\xi) = \begin{cases} \xi_k^{n_k} & \text{if } n_k \geq 0, \\ (\overline{\xi_k})^{-n_k} & \text{if } n_k < 0. \end{cases} \quad \text{for } k \in \{1, \dots, d\},\ n_k \in \mathbb{Z}.$$

One can easily check that, for each fixed $\xi' \in \mathbb{D}^d$,

$$\widetilde{k}^h_{\mathbb{Z}^d}(z,\xi') := \sum_{n\in\mathbb{Z}^d} z^n (z^{-n})^{\wedge h}(\xi')$$

has the property that $\widetilde{k}^h_{\mathbb{Z}^d}(z,\xi')e \in \mathcal{H}(k_{\mathbb{Z}^d})$ for each $e \in \mathcal{E}$ (as a formal power series in z) and that

$$\langle f, \widetilde{k}^h_{\mathbb{Z}^d}(\cdot,\xi')e\rangle_{\mathcal{H}(k_{\mathbb{Z}^d})} = \langle f^{\wedge h}(\xi'), e\rangle_{\mathcal{E}}.$$

If we let $\mathcal{H}(k_{\mathbb{Z}^d})^{\wedge h}$ be the image of $\mathcal{H}(k_{\mathbb{Z}^d})$ under the harmonic extension operator $f \mapsto f^{\wedge h}$ and assign a norm on $\mathcal{H}(k_{\mathbb{Z}^d})^{\wedge h}$ to make the harmonic extension operator $f \mapsto f^{\wedge h}$ unitary from $\mathcal{H}(k_{\mathbb{Z}^d})$ to $\mathcal{H}(k_{\mathbb{Z}^d})^{\wedge h}$, then $\mathcal{H}(k_{\mathbb{Z}^d})^{\wedge h}$ becomes a standard reproducing kernel Hilbert space consisting of separately harmonic functions on the unit polydisk $\mathbb{D}^d$ with reproducing kernel $k^h_{\mathbb{Z}^d}(\xi,\xi') = \sum_{n\in\mathbb{Z}^d}\left((z^n)^{\wedge h}(\xi)\right)\left((z^{-n})^{\wedge h}(\xi')\right)$. The next proposition gives a sufficient condition for separately-harmonic extension giving rise to a standard reproducing kernel Hilbert space on the unit polydisk $\mathbb{D}^d$ for a general FRKHS $\mathcal{H}(K) = \mathcal{L}_W$.

Proposition 2.4. *Let $W = [W_{n,n'}]_{n,n'\in\mathbb{Z}^d}$ be a positive semidefinite weight with matrix entries in $\mathcal{L}(\mathcal{E})$ giving rise to the FRKHS $\mathcal{L}_W = \mathcal{H}(K)$ (where $K(z,z') = \sum_{n,n'\in\mathbb{Z}^d} W_{n,n'} z^n z'^{-n'}$). Then a sufficient condition for the separately-harmonic extension operator $f \mapsto f^{\wedge h}(\xi')$ (see (2.9)) to give rise to a bounded linear transformation from $\mathcal{L}_W$ to $\mathcal{E}$ for every $\xi' \in \mathbb{D}$ is that*

$$M := \sup_{n\in\mathbb{Z}^d} \|W_{n,n}\|_{\mathcal{L}(\mathcal{E})} < \infty. \tag{2.10}$$

In this case

$$\widetilde{K}^h(z,\xi') := \sum_{n,n'\in\mathbb{Z}^d} W_{n,n'} z^n \cdot (z^{-n'})^{\wedge h}(\xi') \tag{2.11}$$

is such that $\widetilde{K}^h(z,\xi')e$ is in $\mathcal{L}_W$ for each $\xi' \in \mathbb{D}^d$ and $e \in \mathcal{E}$ as a formal power series in z and we have the reproducing property

$$\langle f, \widetilde{K}^h(\cdot,\xi')e\rangle_{\mathcal{L}_W} = \langle f^{\wedge h}(\xi'), e\rangle_{\mathcal{E}}.$$

If we define $\mathcal{L}_W^{\wedge h}$ to be the image of $\mathcal{L}_W$ under the separately-harmonic extension operator $f \mapsto f^{\wedge h}$ with norm defined so that $f \mapsto f^{\wedge h}$ is unitary from $\mathcal{L}_W$ onto $\mathcal{L}_W^{\wedge h}$, then $\mathcal{L}_W^{\wedge h}$ is a standard reproducing kernel Hilbert space of separately harmonic functions

$$\langle f, K^h(\cdot,\xi')\rangle_{\mathcal{L}_W^{\wedge h}} = \langle f(\xi'), e\rangle_{\mathcal{E}} \text{ for } f \in \mathcal{L}_W^{\wedge h} \text{ and } e \in \mathcal{E}$$

with reproducing kernel given by

$$K^h(\xi,\xi') := \sum_{n,n'\in\mathbb{Z}^d} W_{n,n'}\left((z^n)^{\wedge h}(\xi)\right)\left((z^{-n'})^{\wedge h}(\xi')\right).$$

Proof. The remaining assertions are straightforward once we verify that $\widetilde{K}^h(z,\xi')$ given by (2.11) has the property that $\widetilde{K}^h(z,\xi')e \in \mathcal{L}_W$ as a formal power series in z for each $\xi' \in \mathbb{D}^d$ and $e \in \mathcal{E}$. For each fixed $n' \in \mathbb{Z}^d$ and $e \in \mathcal{E}$, note that

$$\sum_{n\in\mathbb{Z}^d} W_{n,n'} z^n e = W[z^{n'}e]$$

has norm determined by

$$\|W[z^{n'}e]\|^2_{\mathcal{L}_W} = \langle W[z^{n'}e], z^{n'}e\rangle_{L^2} = \langle W_{n',n'}e, e\rangle_{\mathcal{E}} \le \|W_{n',n'}\|\|e\|^2 \le M\|e\|^2$$

where M given by (2.10) is finite by hypothesis. Moreover, for $\xi' \in \mathbb{D}^d$ we have

$$|(z^{-n'})^{\wedge h}(\xi')| \le r^{|n|}$$

where $r = \max_{j=1,\dots,d} |\xi'_j| < 1$ and $|n| = |n_1| + \cdots + |n_d|$ if $n = (n_1,\dots,n_d) \in \mathbb{Z}^d$. We conclude that

$$\sum_{n'\in\mathbb{Z}^d} \| \sum_{n\in\mathbb{Z}^d} W_{n,n'} z^n e\|_{\mathcal{L}_W} \cdot |(z^{-n'})^{\wedge h}(\xi')| \le \sqrt{M}\|e\| \cdot \sum_{n'\in\mathbb{Z}^d} r^{|n'|} < \infty$$

and hence the series defining $\widetilde{K}^h(\cdot,\xi')e$ is absolutely summable in $\mathcal{L}_W$-norm. We conclude that this series converges in $\mathcal{L}_W$-norm to a well-defined element $\widetilde{K}^h(\cdot,\xi')e$ of $\mathcal{L}_W$ as wanted. □

Just as in the classical case (see [2, Sections I.5–I.7]), we have the following general decomposition properties for FRKHSs.

Theorem 2.5.

1. *Suppose that*

$$K^1(z,z') = \sum_{n,n'\in\mathbb{Z}^d} W^1_{n,n'} z^n z'^{-n'} \quad \textit{and} \quad K^2(z,z') = \sum_{n,n'\in\mathbb{Z}^d} W^2_{n,n'} z^n z'^{-n'}$$

are the reproducing kernels for respective FRKHSs $\mathcal{H}(K^1) = \mathcal{L}_{W^1}$ and $\mathcal{H}(K^2) = \mathcal{L}_{W^2}$, where the coefficients $W^1_{n,n'} \in \mathcal{L}(\mathcal{E})$ and $W^2_{n,n'} \in \mathcal{L}(\mathcal{E})$ are linear operators on the same Hilbert space $\mathcal{E}$. Then

$$K(z,z') := K^1(z,z') + K^2(z,z') = \sum_{n,n'\in\mathbb{Z}^d} [W^1_{n,n'} + W^2_{n,n'}] z^n z'^{-n'}$$

is the reproducing kernel for a FRKHS $\mathcal{H}(K) = \mathcal{H}(K^1) + \mathcal{H}(K^2)$ with norm given by

$$\|f\|^2_{\mathcal{H}(K^1+K^2)} = \min\{\|f^1\|^2_{\mathcal{H}(K^1)} + \|f^2\|^2_{\mathcal{H}(K^2)} : f = f^1 + f^2 \text{ with } f^1 \in \mathcal{H}(K^1) \text{ and } f^2 \in \mathcal{H}(K^2).\} \tag{2.12}$$

2. *If $K(z,z') = \sum_{n,n'\in\mathbb{Z}^d} W_{n,n'} z^n z'^{-n'}$ and $K^1(z,z') = \sum_{n,n'\in\mathbb{Z}^d} W^1_{n,n'} z^n z'^{-n'}$ are reproducing kernels for respective FRKHSs $\mathcal{H}(K)$ and $\mathcal{H}(K^1)$ with coefficients $W_{n,n'}$ and $W^1_{n,n'}$ in $\mathcal{L}(\mathcal{E})$ for the same Hilbert space $\mathcal{E}$, and if $K^2(z,z') := K(z,z') - K^1(z,z')$ is a positive kernel (i.e., $(n,n') \mapsto W_{n,n'} -$*

$W^1_{n,n'}$ is a positive kernel on $\mathbb{Z}^d \times \mathbb{Z}^d$), then the associated FRKHS $\mathcal{H}(K^1)$ is contractively included in $\mathcal{H}(K)$:

$$\mathcal{H}(K^1) \subset \mathcal{H}(K) \text{ with } \|f\|^2_{\mathcal{H}(K)} \le \|f\|^2_{\mathcal{H}(K^1)} \text{ for } f \in \mathcal{H}(K^1).$$

In this case

$$\mathcal{H}(K) = \mathcal{H}(K^1) + \mathcal{H}(K^2) \tag{2.13}$$

where we have set $K^2 = K - K^1$, and $\|f\|^2_{\mathcal{H}(K)}$ can be computed as in (2.12) for any $f \in \mathcal{H}(K) = \mathcal{H}(K^1) + \mathcal{H}(K^2)$. Moreover, for a given $f \in \mathcal{H}(K)$, there is a unique $\tau_1(f) \in \mathcal{H}(K^1)$ and $\tau_2(f) \in \mathcal{H}(K^2)$ so that $f = \tau_1(f) + \tau_2(f)$ (called the minimal decomposition *of a given $f \in \mathcal{H}(K)$ with respect to $\mathcal{H}(K^1)$ and $\mathcal{H}(K^2)$) such that $\|f\|^2_{\mathcal{H}(K)} = \|\tau_1(f)\|^2_{\mathcal{H}(K^1)} + \|\tau_2(f)\|^2_{\mathcal{H}(K^2)}$. The decomposition (2.13) forms a Hilbert-space orthogonal decomposition of $\mathcal{H}(K)$ if and only if the* overlapping space *$\mathcal{H}(K^1) \cap \mathcal{H}(K^2)$ is equal to $\{0\}$.*

Conversely, if $\mathcal{H}_1$ is an linear submanifold of $\mathcal{H}(K)$ which is a Hilbert space in a norm $\|\cdot\|_1$ which gives a contractive inclusion of $\mathcal{H}_1$ into $\mathcal{H}(K)$

$$\|f\|^2_{\mathcal{H}(K)} \le \|f\|^2_1 \text{ for } f \in \mathcal{H}_1,$$

then there is a positive kernel $K^1 \in \mathcal{E}[[z, z^{-1}, z', z'^{-1}]]$ such that $K - K^1$ is a positive kernel and so that $\mathcal{H}_1$ is equal isometrically to the FRKHS $\mathcal{H}(K^1)$.

Proof. The results follow from the corresponding results for the classical case (see [2]) via viewing these FRKHSs as standard reproducing kernel Hilbert spaces of functions defined over $\Omega = \mathbb{Z}^d$. □

In the setting of Theorem 2.5, 2, the contractively included subspace $\mathcal{H}(K^2)$ of $\mathcal{H}(K)$ is the *complementary subspace* (or the *generalized orthogonal complement*) of a contractively included subspace $\mathcal{H}(K^1)$. This concept appears in the original paper of Aronszajn [2], has been a systematic part of the work of de Branges and Rovnyak [11, 12, 9] and now has been extended to the setting of reproducing kernel Kreĭn spaces (see [10]).

Again as in the classical case, we can construct a FRKHS using a multiplier.

Theorem 2.6. *Suppose that $\mathcal{H}(K)$ is a FRKHS with elements having coefficients in the Hilbert spaces $\mathcal{E}$, and suppose that $H(z) \in \mathcal{L}(\mathcal{E}, \mathcal{E}_*)[[z, z^{-1}]]$ (where $\mathcal{E}_*$ is another Hilbert space). Then*

$$H(z) \cdot f(z) = \sum_{n \in \mathbb{Z}^d} \left(\sum_{n', n'' \colon n' + n'' = n} H_{n'} f_{n''} \right) z^n$$

is well defined for every $f \in \mathcal{H}(K)$ (i.e., the series giving the coefficient of z^n above converges in $\mathcal{E}_*$ for every n) if and only if $H(z) \cdot K(z, z') \cdot H(z')^*$ is well defined in the strong operator topology on $\mathcal{L}(\mathcal{E}_*)$ (i.e., the series giving every coefficient in $H(z) \cdot K(z, z') \cdot H(z')^*$ converges in strong operator topology). In this case, define a subspace $\mathcal{M}(H) \subset \mathcal{E}_*[[z, z^{-1}]]$ by

$$\mathcal{M}(H) = H \cdot \mathcal{H}(K) = \{H(z) \cdot f(z) \colon f \in \mathcal{H}(K)\},$$

with norm

$$\|H \cdot f\|_{\mathcal{M}(H)} = \|Qf\|_{\mathcal{H}(K)}$$

where Q is the orthogonal projection of $\mathcal{H}(K)$ onto the orthogonal complement of the kernel of the linear operator $M_H \colon f \mapsto H \cdot f$; then $\mathcal{M}(H)$ is a FRKHS with reproducing kernel $K^{\mathcal{M}(H)}$ given by

$$K^{\mathcal{M}(H)}(z, z') = H(z) \cdot K(z, z') \cdot H(z')^*.$$

Proof. The result can be reduced to the classical case by viewing a FRKHS as a standard reproducing kernel Hilbert space of vector-valued functions on $\mathbb{Z}^d$. The classical case is well known, but for the sake of completeness we sketch the proof. Note that, for $z' \in \Omega$, $f \in \mathcal{H}(K)$ and $e \in \mathcal{E}_*$, we have

$$\begin{aligned}\langle H(z')f(z'), e_* \rangle_{\mathcal{E}_*} &= \langle f(z'), H(z')^* e_* \rangle_{\mathcal{E}} = \langle f, K_{z'} H(z')^* e_* \rangle_{\mathcal{H}(K)} \\ &= \langle H \cdot f, H K_{z'} H(z') e_* \rangle_{\mathcal{M}(H)}.\end{aligned}$$

□

Remark 2.7. We note that the equivalence of parts (1) and (2) in Theorem 2.1 amounts to the special case of Theorem 2.6 where $K(z, z') = I_{\mathcal{H}}$.

2.2. Toeplitz weights and models for commuting tuples of unitary operators

In this section we consider the case of a commutative FRKHS $\mathcal{H}(K) = \mathcal{L}_W$ where the weight matrix $W = [W_{n,n'}]_{n,n' \in \mathbb{Z}^d}$ is a Toeplitz matrix, i.e., the operator W is multiplication by a symbol $\widehat{W}(z) = \sum_{n \in \mathbb{Z}^d} \widehat{W}_n z^n \in \mathcal{L}(\mathcal{E}[[z, z^{-1}]]$:

$$\begin{aligned}W \colon p(z) \mapsto \widehat{W}(z) \cdot p(z) &:= \left(\sum_{n \in \mathbb{Z}^d} \widehat{W}_n z^n \right) \cdot \left(\sum_{n' \in \mathbb{Z}^d} p_{n'} z^{n'} \right) \\ &= \sum_{n \in \mathbb{Z}^d} \left(\sum_{n' \in \mathbb{Z}^d} \widehat{W}_{n-n'} p_{n'} \right) z^n.\end{aligned}$$

Thus we see that a necessary and sufficient condition for a weight

$$W = [W_{n,n'}]_{n,n' \in \mathbb{Z}^d}$$

to have the form of a multiplication operator $p(z) \mapsto \widehat{W}(z)p(z)$ is that W has the Toeplitz structure

$$W_{n,n'} = \widehat{W}_{n-n'} \text{ for some collection of operators } \widehat{W}_n \in \mathcal{L}(\mathcal{E}),\ n \in \mathbb{Z}^d.$$

We have already seen a particular instance of this where W is the identity block matrix $W = [\delta_{n,n'} I_{\mathcal{E}}]_{n,n' \in \mathbb{Z}^d}$ and the associated FRKHS $\mathcal{L}_W = \mathcal{H}(k_{\mathbb{Z}^d})$ is the FRKHS associated with $L^2(\mathbb{T}^d, \mathcal{E})$ by identifying an element f in L^2 with its Fourier series $f(z) \equiv \sum_{n \in \mathbb{Z}^d} f_n z^n$ with $f_n = \int_{\mathbb{T}^d} f(t) t^{-n} \, d\lambda(t)$ with λ equal to the Lebesgue measure on the torus $\mathbb{T}^d$. In this case the associated kernel $K(z, z') = \sum_{n,n' \in \mathbb{Z}^d} W_{n,n'} z^n z'^{-n'}$ turns out to be the two-sided Szegö kernel $k_{\mathbb{Z}^d}(z, z')$ given by (2.7).

For the case of a general Toeplitz weight $W = [\widehat{W}_{n-n'}]_{n,n' \in \mathbb{Z}^d}$, the associated kernel $K(z, z')$ is given by

$$\begin{aligned} K(z,z') &= \sum_{n,n' \in \mathbb{Z}^d} \widehat{W}_{n-n'} z^n z'^{-n'} \\ &= \sum_{n'',n' \in \mathbb{Z}^d} \widehat{W}_{n''} z^{n''+n'} z'^{-n'} = \widehat{W}(z) \cdot (k_{\mathbb{Z}^d}(z,z') \otimes I_{\mathcal{E}}). \end{aligned}$$

We thus have arrived at the following result.

Proposition 2.8. *Suppose that the element $\widehat{W}(z) = \sum_{n \in \mathbb{Z}^d} \widehat{W}_n z^n$ generates a positive semidefinite Toeplitz weight*

$$W = [W_{n,n'}]_{n,n' \in \mathbb{Z}^d} \text{ with } W_{n,n'} = \widehat{W}_{n-n'}.$$

Then $\mathcal{L}_W$ is a FRKHS with reproducing kernel $K(z, z')$ given by

$$K(z,z') = \widehat{W}(z) k_{\mathbb{Z}^d}(z,z'). \tag{2.14}$$

Remark 2.9. As we have seen the kernel $K(z, z')$ for a FRKHS must be selfadjoint in the sense that

$$K(z,z') = (K(z',z))^*.$$

It is not obvious from the form of K in (2.14) that the kernel for a Toeplitz FRKHS has this selfadjoint property. To see that it does, we note the identity

$$(z^n - z'^n) \cdot k_{\mathbb{Z}^d}(z,z') = 0.$$

For the case of a Toeplitz weight $W = [\widehat{W}_{n-n'}]_{n,n' \in \mathbb{Z}^d}$ we see that $W_{n',n'} = \widehat{W}_0$ and hence

$$\sup_{n' \in \mathbb{Z}^d} \|W_{n',n'}\| = \|\widehat{W}_0\| < \infty.$$

Hence the sufficient condition for the separately-harmonic extension map to be bounded from $\mathcal{L}_W$ to $\mathcal{E}$ for each $\xi' \in \mathbb{D}^d$ given in Proposition 2.4 is automatically satisfied in the case of a Toeplitz weight, and we arrive at the following corollary.

Corollary 2.10. *Let $W = [\widehat{W}_{n-n'}]_{n,n' \in \mathbb{Z}^d}$ be a positive semidefinite Toeplitz weight. Then the separately-harmonic extension map $f \mapsto f^{\wedge h}(\xi')$ defines a bounded linear transformation from $\mathcal{L}_W$ into $\mathcal{E}$ for each $\xi' \in \mathbb{D}^d$. The image $\mathcal{L}_W^{\wedge h}$ of $\mathcal{L}_W$ under the separately-harmonic extension maps, with norm chosen to make $f \mapsto f^{\wedge h}$ unitary from $\mathcal{L}_W$ onto $\mathcal{L}_W^{\wedge h}$, is a standard reproducing kernel Hilbert space of separately harmonic functions on $\mathbb{D}^d$ with reproducing kernel $K^h(\xi, \xi')$ given by*

$$K^h(\xi,\xi') = \sum_{n,n' \in \mathbb{Z}^d} \widehat{W}_{n-n'} (z^n)^{\wedge h}(\xi) \cdot (z^{-n})^{\wedge h}(\xi').$$

In particular, the kernel function for evaluation at $0 \in \mathbb{D}^d$, *namely*

$$K^h(\xi, 0) = \sum_{n \in \mathbb{Z}^d} \widehat{W}_n(z^n)^{\wedge h}(\xi) = (\widehat{W})^{\wedge h}(\xi)$$

is the harmonic extension of the symbol $\widehat{W}(z)$ *of the Toeplitz weight*

$$W = [\widehat{W}_{n-n'}]_{n,n' \in \mathbb{Z}^d}.$$

For the case of a Toeplitz weight $W = [\widehat{W}_{n-n'}]_{n,n' \in \mathbb{Z}^d}$ the spaces $\mathcal{L}_W$ are closely connected with models for commuting d-tuples of unitary operators on a Hilbert space $\mathcal{H}$ with prescribed cyclic subspace as we now explain. Suppose that $\mathcal{U} = (\mathcal{U}_1, \dots, \mathcal{U}_d)$ is a d-tuple of commuting unitary operators on a Hilbert space $\mathcal{H}$ and that $V \colon \mathcal{E} \mapsto \mathcal{H}$ is a linear map from a coefficient space into $\mathcal{H}$. Define a (generalized) Fourier representation operator $\Phi \colon \mathcal{H} \mapsto \mathcal{E}[[z, z^{-1}]]$ by

$$\Phi \colon h \mapsto \sum_{n \in \mathbb{Z}^d} (\Phi_n h) z^n$$

where $\Phi_n \colon \mathcal{H} \mapsto \mathcal{E}$ is given by

$$\Phi_n \colon h \mapsto V^* \mathcal{U}^{*n} h$$

where we use the standard multivariable notation

$$\mathcal{U}^{*n} = \mathcal{U}_1^{*n_1} \cdots \mathcal{U}_d^{*n_d} \text{ if } n = (n_1, \dots, n_d) \in \mathbb{Z}^d.$$

If we set $\Phi(z) = \sum_{n \in \mathbb{Z}^d} \Phi_n z^n \in \mathcal{L}(\mathcal{H}, \mathcal{E})[[z, z^{-1}]]$, then we see that the operator Φ amounts to multiplication by $\Phi(z)$. We define a norm $\| \cdot \|_0$ on $\operatorname{im} \Phi$ as the lifted norm

$$\|\Phi(z) h\|_0 = \|Qh\|_{\mathcal{H}}$$

where Q is the orthogonal projection from $\mathcal{H}$ to the kernel of the operator Φ. From the characterization of spaces $\mathcal{H}(K)$ in Theorem 2.1, we see that $(\operatorname{im} \Phi, \| \cdot \|_0)$ is a FRKHS $\mathcal{H}(K)$ with kernel $K(z, z')$ given by $K(z, z') = \Phi(z)\Phi(z')^*$, or equivalently, $(\operatorname{im} \Phi, \| \cdot \|_0)$ is a space $\mathcal{L}_W$ with the weight $W = [W_{n,n'}]_{n,n' \in \mathbb{Z}^d}$ given by

$$W_{n,n'} = \Phi_n \Phi_{n'}^* = V^* \mathcal{U}^{*n} \mathcal{U}^{n'} V.$$

Since by assumption $\mathcal{U}$ is a commuting d-tuple of unitary operators, it follows that $\mathcal{U}^{*n} \mathcal{U}^{n'} = \mathcal{U}^{*n-n'}$, and hence the weight W has the Toeplitz property

$$W = [\widehat{W}_{n-n'}]_{n,n' \in \mathbb{Z}^d} \text{ where } \widehat{W}_n = V^* \mathcal{U}^{*n} V.$$

Moreover, we have

$$\begin{aligned} \Phi \mathcal{U}_j h &= \sum_{n \in \mathbb{Z}^d} (V^* \mathcal{U}^{*n} \mathcal{U}_j h) z^n \\ &= z_j \cdot \left(\sum_{n \in \mathbb{Z}^d} (V^* \mathcal{U}^{*n} \mathcal{U}_j h) z^n z_j^{-1} \right) \\ &= z_j (\Phi h) \end{aligned}$$

and hence we have the intertwining property

$$\Phi\mathcal{U}_j = \mathcal{U}_{W,j}\Phi$$

where we write $\mathcal{U}_{W,j}$ for the operator of multiplication by the independent formal variable z_j on the space $\mathcal{L}_W$. We note that Φ by definition is a coisometry from $\mathcal{H}$ onto $\mathcal{L}_W$ with initial space equal to the orthogonal complement of the kernel of Φ. From the definition of Φ, this initial space can be identified as the smallest subspace of $\mathcal{H}$ reducing for each of the operators $\mathcal{U}_1, \dots, \mathcal{U}_d$ and containing the image space $\operatorname{im} V = V\mathcal{E}$ of V; in particular, the initial space of Φ is all of $\mathcal{H}$ (i.e., Φ is unitary) exactly when $V\mathcal{E}$ is a $*$-cyclic subspace for the d-tuple $\mathcal{U}$.

Conversely, let $W = [\widehat{W}_{n-n'}]_{n,n'\in\mathbb{Z}^d}$ be any positive semidefinite Toeplitz weight determining a FRKHS $\mathcal{L}_W$. Then we may define a d-tuple of operators $\mathcal{U}_W = (\mathcal{U}_{W,1}, \dots, \mathcal{U}_{W,d})$ on $\mathcal{L}_W$ by

$$\mathcal{U}_{W,j} \colon f(z) \mapsto z_j f(z) \text{ for } f \in \mathcal{L}_W \text{ and } j = 1, \dots, d. \tag{2.15}$$

Using the Toeplitz property of W, it is not hard to verify that $\mathcal{U}_W$ is a commuting d-tuple of unitary operators on $\mathcal{L}_W$; indeed, as a sample computation, observe that for $f = Wp \in \mathcal{L}_W$ where $p \in \mathcal{E}[z, z^{-1}]$ is a polynomial (a dense subset of $\mathcal{L}_W$) we have $z_j \cdot Wp = z_j \cdot \widehat{W}(z)p(z) = \widehat{W}(z) \cdot z_j p(z) \in \mathcal{L}_W$ and

$$\begin{aligned}\|z_j \cdot Wp\|^2_{\mathcal{L}_W} &= \|\widehat{W}(z) \cdot z_j p(z)\|^2_{\mathcal{L}(W)} = \langle z_j \widehat{W}(z)p(z), z_j p(z)\rangle_{L^2} \\ &= \langle \widehat{W}(z)p(z), p(z)\rangle_{L^2} = \|Wp\|^2_{\mathcal{L}_W}.\end{aligned}$$

Moreover, if we let $V \colon \mathcal{E} \mapsto \mathcal{L}_W$ be the map given by

$$V \colon e \mapsto We \in \mathcal{L}_W,$$

then we compute, for $e, e' \in \mathcal{E}$,

$$\langle \mathcal{U}^{*n} We, We'\rangle_{\mathcal{L}_W} = \langle \widehat{W} z^{-n} e, e'\rangle_{L^2} = \langle W_{0,-n} e, e'\rangle_{\mathcal{E}} = \langle \widehat{W}_n e, e'\rangle_{\mathcal{E}}$$

and hence we recover $W = [\widehat{W}_{n-n'}]_{n,n'\in\mathbb{Z}^d}$ from $\widehat{W}_n = V^* \mathcal{U}^* V$.

Following the terminology in [14] and [17], we define a *d-variable unitary scattering system* $\mathfrak{S}$ to be the collection of operators and spaces

$$\mathfrak{S} = (\mathcal{U} = (\mathcal{U}_1, \dots, \mathcal{U}_d), V; \mathcal{H}, \mathcal{E}) \tag{2.16}$$

where $\mathcal{H}$ and $\mathcal{E}$ are Hilbert spaces, $\mathcal{U}$ is a d-tuple of commuting unitary operators on $\mathcal{H}$ and $V \colon \mathcal{E} \mapsto \mathcal{H}$ is a linear operator (called a *scale* in [17]; it could be also called a *control/observation* operator). Given a positive semidefinite Toeplitz weight $W = [\widehat{W}_{n-n'}]_{n,n'\in\mathbb{Z}^d}$, there is an associated *model d-variable unitary scattering system* $\mathfrak{S}_W$

$$\mathfrak{S}_W = (\mathcal{U}_W = (\mathcal{U}_{W,1}, \dots, \mathcal{U}_{W,d}), V_W; \mathcal{L}_W, \mathcal{E}) \tag{2.17}$$

where $\mathcal{U}_W$ on $\mathcal{L}_W$ is given by (2.15) and where $V_W \colon \mathcal{E} \mapsto \mathcal{L}_W$ is defined by

$$V_W \colon e \mapsto We.$$

We summarize this discussion as follows.

Theorem 2.11. *Let $W = [\widehat{W}_{n-n'}]_{n,n' \in \mathbb{Z}^d}$ be a semidefinite Toeplitz weight with matrix entries $\widehat{W}_{n-n'} \in \mathcal{L}(\mathcal{E})$. Then $\mathfrak{S}_W$ defined by (2.17) is a d-variable unitary scattering system with the additional property that $\operatorname{im} V_W$ is a $*$-cyclic subspace for $\mathcal{U}_W$. Moreover, we recover $W = [\widehat{W}_{n-n'}]_{n,n' \in \mathbb{Z}^d}$ from $\mathfrak{S}_W$ via*

$$\widehat{W}_n = V_W^* \mathcal{U}_W^{*n} V_W \text{ for } n \in \mathbb{Z}^d.$$

Conversely, if $\mathfrak{S}$ as in (2.16) is any d-variable unitary scattering system, then the map $\Phi \colon \mathcal{H} \mapsto \mathcal{L}_W$ given by

$$\Phi \colon h \mapsto \Phi(z)h \text{ where } \Phi(z) = \sum_{n \in \mathbb{Z}^d} (V^* \mathcal{U}^{*n} h) z^n$$

is a coisometry from $\mathcal{H}$ onto $\mathcal{L}_W$ (with initial space equal to the smallest subspace reducing for $\mathcal{U}$ and containing $\operatorname{im} V$) which intertwines $\mathcal{U}$ with $\mathcal{U}_W$ and V with V_W:

$$\mathcal{U}_{W,j} \Phi = \Phi \mathcal{U}_j \text{ for } j = 1, \dots, d, \qquad V_W = \Phi V.$$

Here $W = [\widehat{W}_{n-n'}]_{n,n' \in \mathbb{Z}^d}$ is the positive semidefinite Toeplitz weight with symbol $\widehat{W}(z) = \sum_{n \in \mathbb{Z}^d} \widehat{W}_n z^n$ determined by

$$\widehat{W}_n = V^* \mathcal{U}^{*n} V \text{ for } n \in \mathbb{Z}^d.$$

Remark 2.12. Let us say that two d-variable unitary scattering systems $\mathfrak{S} = (\mathcal{U}, V; \mathcal{H}, \mathcal{E})$ and $\mathfrak{S}' = (\mathcal{U}', V'; \mathcal{H}', \mathcal{E})$ with the same coefficient space $\mathcal{E}$ are *unitarily equivalent* if there is a unitary transformation $\Phi \colon \mathcal{H} \mapsto \mathcal{H}'$ so that

$$\Phi \mathcal{U}_j = \mathcal{U}_j' \Phi \text{ for } j = 1, \dots, d, \qquad \Phi V = V'.$$

Let us also say that a d-variable unitary scattering system $\mathfrak{S}$ as in (2.16) is *minimal* if $\operatorname{im} V$ is $*$-cyclic for $\mathcal{U}$. Part of the content of Theorem 2.11 then is that *a minimal d-variable unitary scattering system $\mathfrak{S} = (\mathcal{U}, V; \mathcal{H}, \mathcal{E})$ is unitarily equivalent to the model d-variable unitary scattering system $\mathfrak{S}_W$, where $W = [\widehat{W}_{n-n'}]_{n,n' \in \mathbb{Z}^d}$ with $\widehat{W}_n = V^* \mathcal{U}^{n*} V$ for $n \in \mathbb{Z}^d$.*

The Toeplitz analogue of Theorem 2.5 is as follows.

Theorem 2.13.

1. *Suppose that $W^1 = [\widehat{W}^1_{n-n'}]_{n,n' \in \mathbb{Z}^d}$ and $W^2 = [\widehat{W}^2_{n-n'}]_{n,n' \in \mathbb{Z}^d}$ are two positive semidefinite Toeplitz weights with Toeplitz symbols $\widehat{W}^1(z) = \sum_{n \in \mathbb{Z}^d} \widehat{W}^1_n z^n$ and $\widehat{W}^2(z) = \sum_{n \in \mathbb{Z}^d} \widehat{W}^2_n z^n$ having coefficients $\widehat{W}^1_n \in \mathcal{L}(\mathcal{E})$ and $\widehat{W}^2_n \in \mathcal{L}(\mathcal{E})$ equal to operators on the same Hilbert space $\mathcal{E}$.*

   Then $W = [\widehat{W}^1_{n-n'} + \widehat{W}^2_{n-n'}]_{n,n' \in \mathbb{Z}^d}$ is a positive semidefinite Toeplitz weight with associated FRKHS $\mathcal{L}_W$ equal to the sum $\mathcal{L}_{W^1} + \mathcal{L}_{W^2}$ of $\mathcal{L}_{W^1}$ and $\mathcal{L}_{W^2}$ and with norm given by

$$\|f\|^2_{\mathcal{L}_W} = \min\{\|f^1\|^2_{\mathcal{L}_{W^1}} + \|f^2\|^2_{\mathcal{L}_{W^2}} : f = f^1 + f^2 \text{ with } f_j \in \mathcal{L}_{W^j} \text{ for } j = 1, 2\}. \tag{2.18}$$

2. *If* $W = [\widehat{W}_{n-n'}]_{n,n' \in \mathbb{Z}^d}$ *and* $W^1 = [\widehat{W}^1_{n-n'}]_{n,n' \in \mathbb{Z}^d}$ *are two positive semidefinite Toeplitz weights with Toeplitz symbols* $\widehat{W}(z) = \sum_{n \in \mathbb{Z}^d} \widehat{W}_n z^n$ *and* $\widehat{W}^1(z) = \sum_{n \in \mathbb{Z}^d} \widehat{W}^1_n z^n$ *having coefficients* $\widehat{W}_n \in \mathcal{L}(\mathcal{E})$ *and* $\widehat{W}^1_n \in \mathcal{L}(\mathcal{E})$ *equal to operators on the same Hilbert space* $\mathcal{E}$, *and if* $W^2 := [\widehat{W}_{n-n'} - \widehat{W}^1_{n-n'}]_{n,n' \in \mathbb{Z}^d}$ *is also a positive semidefinite Toeplitz weight, then both the associated FRKHSs* $\mathcal{L}_{W^1}$ *and* $\mathcal{L}_{W^2}$ *are contractively included in* $\mathcal{L}_W$:

$$\mathcal{L}_{W^j} \subset \mathcal{L}_W \text{ with } \|f\|^2_{\mathcal{L}_W} \le \|f\|^2_{\mathcal{L}_{W^j}} \text{ for } f \in \mathcal{L}_{W^j} \text{ for } j = 1, 2.$$

In this case

$$\mathcal{L}_W = \mathcal{L}_{W^1} + \mathcal{L}_{W^2} \tag{2.19}$$

and $\|f\|^2_{\mathcal{L}_W}$ *can be computed as in* (2.18) *for any* $f \in \mathcal{L}_W = \mathcal{L}_{W^1} + \mathcal{L}_{W^2}$. *Moreover, for a given* $f \in \mathcal{L}_W$, *there is a unique* $\tau_1(f) \in \mathcal{L}_{W^1}$ *and* $\tau_2(f) \in \mathcal{L}_{W^2}$ *so that* $f = \tau_1(f) + \tau_2(f)$ *(called the* minimal decomposition *of a given* $f \in \mathcal{L}_W$ *with respect to* $\mathcal{L}_{W^1}$ *and* $\mathcal{L}_{W^2}$*) such that* $\|f\|^2_{\mathcal{L}_W} = \|\tau_1(f)\|^2_{\mathcal{L}_{W^1}} + \|\tau_2(f)\|^2_{\mathcal{L}_{W^2}}$. *The decomposition* (2.19) *forms a Hilbert-space orthogonal decomposition of* $\mathcal{L}_W$ *if and only if the* overlapping space $\mathcal{L}_{W^1} \cap \mathcal{L}_{W^2}$ *is equal to* $\{0\}$.

Conversely, if $\mathcal{L}_1$ *is an linear submanifold of* $\mathcal{L}_W$ *which is a Hilbert space in a norm* $\|\cdot\|_1$ *which gives a contractive inclusion of* $\mathcal{L}_1$ *into* $\mathcal{L}_W$

$$\|f\|^2_{\mathcal{L}_W} \le \|f\|^2_1 \text{ for } f \in \mathcal{L}_1$$

and on which the operator

$$\mathcal{U}_{1;j} \colon f(z) \mapsto z_j \cdot f(z)$$

with adjoint

$$\mathcal{U}^*_{1;j} \colon f(z) \mapsto z_j^{-1} \cdot f(z)$$

is unitary for $j = 1, \ldots, d$, *then there is a positive semidefinite Toeplitz weight* $W^1 = [\widehat{W}^1_{n-n'}]_{n,n' \in \mathbb{Z}^d}$ *with coefficients* $\widehat{W}^1_{n-n'} \in \mathcal{L}(\mathcal{E})$ *so that* $W^2 := [\widehat{W}_{n-n'} - \widehat{W}^1_{n-n'}]_{n,n' \in \mathbb{Z}^d}$ *is a positive semidefinite Toeplitz weight and so that* $\mathcal{L}_1$ *is equal isometrically to the FRKHS* $\mathcal{L}_{W^1}$.

Proof. As most of the points in the theorem are covered by the discussion preceding the theorem, we discuss here only the converse statement in part (2) in the statement of Theorem 2.13. Let therefore $\mathcal{L}_1$ be a linear submanifold of $\mathcal{L}_W$ with a Hilbert space norm $\|\cdot\|_1$ with respect to which $\mathcal{L}_1$ is contractively included in $\mathcal{L}_W$. By the general theory of minimal decompositions and generalized orthogonal complements presented in Theorem 2.5 we know that $(\mathcal{L}_1, \|\cdot\|_1)$ is a FRKHS $\mathcal{L}_{W^1}$ for some positive semidefinite weight $W^1 = [W^1_{n,n'}]_{n,n' \in \mathbb{Z}^d}$. The remaining point is to note that the assumption that the operators $\mathcal{U}_{1,j} \colon f(z) \mapsto f(z) \cdot z_j$ are all unitary on $(\mathcal{L}_1, \|\cdot\|_1)$ then forces the weight W^1 to have the Toeplitz structure $W^1_{n,n'} = \widehat{W}^1_{n-n'}$ where $\widehat{W}^1_n = W^1_{n,0}$. This concludes the proof of Theorem 2.13. □

2.3. Model spaces in terms of an operator-valued positive measure on $\mathbb{T}^d$

The *d-variable trigonometric moment problem* is as follows: *given a subset Σ of $\mathbb{Z}^d$ and given operators $\widehat{W}_n \in \mathcal{L}(\mathcal{E})$ for each $n \in \Sigma$, find a positive measure σ on the d-torus $\mathbb{T}^d$ so that*

$$\widehat{W}_n = \int_{\mathbb{T}^d} t^{-n} \, d\sigma(t) \text{ for all } n \in \Sigma.$$

While the problem is rather intractable for a proper subset Σ of $\mathbb{Z}^d$ in the multivariable ($d > 1$) case, it is straightforward for the full case ($\Sigma = \mathbb{Z}^d$) (see, e.g., [7, 25]); here we simply state the solution of the full d-variable trigonometric moment problem.

Proposition 2.14. *Suppose that we are given operators $\widehat{W}_n \in \mathcal{L}(\mathcal{E})$ for each $n \in \mathbb{Z}^d$. Then a necessary and sufficient condition for the existence of a positive, finite $\mathcal{L}(\mathcal{E})$-valued measure on $\mathbb{T}^d$ for which $\widehat{W}_n = \int_{\mathbb{T}^d} t^{-n} \, d\sigma(t)$ for all $n \in \mathbb{Z}^d$ is that the associated Toeplitz matrix $W = [W_{n-n'}]_{n,n' \in \mathbb{Z}^d}$ be a positive semidefinite weight on $\mathbb{Z}^d$. Moreover, in this case the $\mathcal{L}(\mathcal{E})$-valued measure σ is uniquely determined by the moments $\widehat{W}_n$ ($n \in \mathbb{Z}^d$).*

A consequence of Proposition 2.14 is that there is a one-to-one correspondence between positive semidefinite Toeplitz weights $W = [\widehat{W}_{n-n'}]_{n,n' \in \mathbb{Z}^d}$ (with matrix entries in $\mathcal{L}(\mathcal{E})$) and finite, positive $\mathcal{L}(\mathcal{E})$-valued measures on $\mathbb{T}^d$. Alternatively, given a $\mathcal{L}(\mathcal{E})$-valued measure σ with associated positive semidefinite Toeplitz weight $W = [\widehat{W}_{n-n'}]_{n,n' \in \mathbb{Z}^d}$ (with $\widehat{W}_n = \int_{\mathbb{T}^d} t^{-n} \, d\sigma(t)$), we may associate a nonnegative-valued separately-harmonic function on $\mathbb{D}^d$ given by

$$\begin{aligned}(\widehat{W})^{\wedge h}(\xi) &= \sum_{n \in \mathbb{Z}^d} \widehat{W}_n (z^n)^{\wedge h}(\xi) \\ &= \int_{\mathbb{T}^d} \mathcal{P}^h(t, \xi) \, d\sigma(t)\end{aligned}$$

where we have set $\mathcal{P}^h(t, \xi)$ equal to the d-variable Poisson kernel defined for $t \in \mathbb{T}^d$ and $\xi \in \mathbb{D}^d$ with values in $\mathbb{C}$:

$$\mathcal{P}^h(t, \xi) = \sum_{n \in \mathbb{Z}^d} t^{-n} (z^n)^{\wedge h}(\xi) = \frac{1 - |\xi_1 \bar{t}_1|^2}{|1 - \xi_1 \bar{t}_1|^2} \cdots \frac{1 - |\xi_d \bar{t}_d|^2}{|1 - \xi_d \bar{t}_d|^2}.$$

Conversely, if $\widehat{W}(\xi)$ is a nonnegative-valued, separately harmonic function on $\mathbb{D}^d$, a standard application of the Helly selection principle produces a finite, positive operator-valued measure σ on $\mathbb{T}^d$ for which $\widehat{W}(\xi) = \int_{\mathbb{T}^d} \mathcal{P}^h(t, \xi) \, d\sigma(t)$.

Our next proposition gives the formulas expressing the Toeplitz symbol and the reproducing kernels $K(z, z')$ and $K^h(z, z')$ for the associated FRKHS $\mathcal{L}_W$ and its separately-harmonic extension $\mathcal{L}_W^{\wedge h}$ directly in terms of the measure σ rather than in terms of the associated positive semidefinite Toeplitz weight $W = [\widehat{W}_{n-n'}]_{n,n' \in \mathbb{Z}^d}$.

Proposition 2.15. *Let σ be a finite, positive $\mathcal{L}(\mathcal{E})$-valued measure on $\mathbb{T}^d$ with associated positive semidefinite Toeplitz weight $W = [\widehat{W}_{n-n'}]_{n,n' \in \mathbb{Z}^d}$ where $W_n = \int_{\mathbb{T}^d} t^{-n}\, d\sigma(t)$. Then the kernel function $K(z,z')$ for the FRKHS $\mathcal{L}_W$ associated with the Toeplitz weight W can be expressed directly in terms of σ as*

$$\widehat{W}(z) = \int_{\mathbb{T}^d} \mathcal{P}(t,z)\, d\sigma(t)$$

$$K(z,z') = \widehat{W}(z) \cdot k_{\mathbb{Z}^d}(z,z') = \int_{\mathbb{T}^d} \mathcal{P}(t,z)\, d\sigma(t) \cdot k_{\mathbb{Z}^d}(z,z')$$

where we have set $\mathcal{P}(t,z)$ equal to the formal d-variable Poisson kernel

$$\mathcal{P}(t,z) = \sum_{n \in \mathbb{Z}^d} t^{-n} z^n$$

defined as an element of $\mathbb{C}[[z,z^{-1}]]$ for each $t \in \mathbb{T}^d$.

The nonnegative-valued separately harmonic function $(\widehat{W})^{\wedge h}(\xi)$ associated with σ and the kernel function $K^h(\xi,\xi')$ for the standard RKHS $\mathcal{L}_W^{\wedge h}$ obtained as the image of the separately-harmonic extension $f \mapsto f^{\wedge h}$ of elements of $\mathcal{L}_W$ to the polydisk $\mathbb{D}^d$ (see Proposition 2.4) is then given explicitly in terms of σ as

$$(\widehat{W})^{\wedge h}(\xi) = \int_{\mathbb{T}^d} \mathcal{P}^h(t,\xi)\, d\sigma(t),$$

$$K^h(\xi,\xi') = \int_{\mathbb{T}^d} \mathcal{P}^h(t,\xi)\, d\sigma(t) \cdot k^h_{\mathbb{Z}^d}(\xi,\xi')$$

where $\mathcal{P}^h(t,\xi)$ is the harmonic d-variable Poisson kernel

$$\mathcal{P}^h(t,\xi) = \sum_{n \in \mathbb{Z}^d} t^{-n} (z^n)^{\wedge h}(\xi)$$

(defined as a $\mathbb{C}$-valued function on $\mathbb{T}^d \times \mathbb{D}^d$ which is separately harmonic in the ξ-variables), and where $k^h_{\mathbb{Z}^d}$ is the harmonic d-variable Szegö kernel

$$k^h_{\mathbb{Z}^d}(\xi,\xi') = \sum_{n \in \mathbb{Z}^d} (z^n)^{\wedge h}(\xi) \cdot (z^{-n})^{\wedge h}(\xi').$$

The space $\mathcal{L}_W$ can be given an alternate form expressed directly in terms of the measure σ, namely the *Hellinger space L^σ* associated with the operator-valued measure σ; for a detailed discussion of the $d = 1$ case, we refer to [17] and the references there. Given a finite, positive $\mathcal{L}(\mathcal{E})$-valued measure defined on $\mathbb{T}^d$, let $\sigma \cdot \mathcal{E}[z,z^{-1}]$ be the space of vector measures ν on $\mathbb{T}^d$ (called "charts" rather than "vector measures" in [17]) such that

$$\nu(\Delta) = \int_\Delta d\sigma(t) p(t) \text{ for } \Delta \text{ a Borel subset of } \mathbb{T}^d \tag{2.20}$$

where $p(z) \in \mathcal{E}[z,z^{-1}]$ is a polynomial in $z_1, \ldots, z_d, z_1^{-1}, \ldots, z_d^{-1}$ with coefficients in $\mathcal{E}$; informally, we may write simply $\nu = \sigma \cdot p$. For $p, q \in \mathcal{E}[z,z^{-1}]$ equal to two such polynomials, define the L^σ inner product by

$$\langle \sigma \cdot p, \sigma \cdot q \rangle_{L^\sigma} = \int_{\mathbb{T}^d} \langle d\sigma(t) p(t), q(t) \rangle_{\mathcal{E}}. \tag{2.21}$$

The completion of $\sigma \cdot \mathcal{E}[z, z^{-1}]$ in the L^σ-norm can be identified concretely as the space of $\mathcal{E}$-valued measures ν, denoted by L^σ, such that there exists a scalar measure μ on $\mathbb{T}^d$ for which the operator

$$\begin{bmatrix} \mu(\Delta) & \nu(\Delta)^* \\ \nu(\Delta) & \sigma(\Delta) \end{bmatrix} \in \mathcal{L}(\mathbb{C} \oplus \mathcal{E}) \tag{2.22}$$

is positive semidefinite for each Borel subset $\Delta \subset \mathbb{T}^d$; in fact, this is how the space L^σ is defined in [17]. It can be shown (see [17]) that there is a minimal such scalar measure μ_ν in the sense that, if μ is any scalar measure for which (2.22) is positive semidefinite for all Borel $\Delta \subset \mathbb{T}^d$, then necessarily

$$\mu(\Delta) \geq \mu_\nu(\Delta) \text{ for all } \Delta \subset \mathbb{T}^d.$$

We refer to this unique measure μ_ν as the *Hellinger measure* of the vector measure $\nu \in L^\sigma$ (with respect to σ). It turns out that the norm of $\nu \in L^\sigma$ can be given explicitly in terms of the associated Hellinger measure μ_ν as

$$\|\nu\|_{L^\sigma} = \mu_\nu(\mathbb{T}^d). \tag{2.23}$$

It is not difficult to see that the measure $t_j \cdot \nu$ is in L^σ whenever $\nu \in L^\sigma$ and $\|t_j \cdot \nu\|_{L^\sigma} = \|\nu\|_{L^\sigma}$ for each $j = 1, \dots, d$, where $t_j \cdot \nu$ is the vector measure given by

$$(t_j \cdot \nu)(\Delta) = \int_\Delta t_j \, d\nu(t) \text{ for } \Delta \subset \mathbb{T}^d;$$

indeed it is easily seen that

$$\mu_{t_j \cdot \nu} = \mu_\nu \text{ for } j = 1, \dots, d.$$

Similar observations hold for $t_j^{-1} \cdot \nu$ for each $\nu \in L^\sigma$ and $j = 1, \dots, d$. It follows that the d-tuple of operators

$$\mathcal{U}_\sigma = (\mathcal{U}_{\sigma,1}, \dots, \mathcal{U}_{\sigma,d})$$

is a d-tuple of unitary operators on L^σ, where each $\mathcal{U}_{\sigma,j} \colon L^\sigma \mapsto L^\sigma$ is given by

$$\mathcal{U}_{\sigma,j} \colon \nu \mapsto t_j \cdot \nu.$$

If we next define a map $V_\sigma \colon \mathcal{E} \mapsto L^\sigma$ by

$$V_\sigma \colon e \mapsto \sigma \cdot e$$

(where $\sigma \cdot e(\Delta) = \sigma(\Delta)e = \int_\Delta d\sigma(t)e$), then we see that $\mathfrak{S}_\sigma$ given by

$$\mathfrak{S}_\sigma = (\mathcal{U}_\sigma = (\mathcal{U}_{\sigma,1}, \dots, \mathcal{U}_{\sigma,d}), V_\sigma; L^\sigma, \mathcal{E}) \tag{2.24}$$

is a d-variable, unitary scattering system. In fact this is an alternative model for a general unitary scattering system $\mathfrak{S}$ as in (2.16) for which $\operatorname{im} V$ is $*$-cyclic for $\mathcal{U}$, just as is the case for the model $\mathfrak{S}_W$. We lay this out in the following theorem; the one-variable case is essentially contained in Theorem 4.1 of [17], so we omit the proof.

Theorem 2.16. *Let σ be a finite, positive $\mathcal{L}(\mathcal{E})$-valued measure. Then $\mathfrak{S}_\sigma$ defined as in (2.24) is a d-variable unitary scattering system with the additional property that* im V_σ *is $*$-cyclic for $\mathcal{U}_\sigma$. Moreover, we recover the positive operator-valued measure σ from the associated unitary scattering system $\mathfrak{S}_\sigma$ via*

$$\sigma(\Delta) = V_\sigma^* E_{\mathcal{U}_\sigma}(\Delta) V_\sigma \text{ for } \Delta \text{ Borel subset of } \mathbb{T}^d \tag{2.25}$$

where $\Delta \mapsto E_{\mathcal{U}_\sigma}(\Delta)$ is the spectral measure on $\mathbb{T}^d$ for the commuting, unitary d-tuple $\mathcal{U}_\sigma$.

Conversely, if $\mathfrak{S}$ as in (2.16) is any d-variable unitary scattering system, define a finite, positive $\mathcal{L}(\mathcal{E})$-valued Borel measure σ on $\mathbb{T}^d$ by

$$\sigma(\Delta) = V^* E_{\mathcal{U}}(\Delta) V$$

where $\Delta \mapsto E_{\mathcal{U}}(\Delta)$ is the spectral measure on $\mathbb{T}^d$ for $\mathcal{U}$. Then the map Φ defined by $\Phi\colon h \mapsto \nu_h$ where

$$\nu_h(\Delta) = V^* E_{\mathcal{U}}(\Delta) h \text{ for } \Delta \text{ Borel subset of } \mathbb{T}^d$$

is a coisometry from $\mathcal{H}$ onto L^σ (with initial space equal to the smallest reducing subspace for $\mathcal{U}$ containing im V*) which intertwines $\mathcal{U}$ with $\mathcal{U}_\sigma$ and V with V_σ:*

$$\mathcal{U}_{\sigma,j}\Phi = \Phi\mathcal{U}_j \text{ for } j = 1, \dots, d, \qquad L_\sigma = \Phi L.$$

Remark 2.17. Just as was the case for the $\mathcal{L}_W$ model space in Remark 2.12, part of the content of Theorem 2.16 is that *a minimal d-variable unitary scattering system is unitarily equivalent to the model d-variable unitary scattering system $\mathfrak{S}_\sigma$ where the measure σ on $\mathbb{T}^d$ is given by* (2.25).

As is discussed in [17], the space L^σ can also be realized as a space of vector-valued separately-harmonic functions on $\mathbb{D}^d$. Indeed, any vector measure $\nu \in L^\sigma$ admits a separately harmonic continuation to the polydisk $\xi \mapsto \nu^{\wedge h}(\xi)$ given by the Poisson integral:

$$\nu^{\wedge h}(\xi) = \int_{\mathbb{T}^d} \mathcal{P}^h(t, \xi)\, d\nu(t).$$

We may thus define the space $L^{\sigma,h}$ to be the set of all such $\mathcal{E}$-valued harmonic functions on $\mathbb{D}^d$ arising from a vector measure $\nu \in L^\sigma$ with

$$\|\nu^{\wedge h}\|_{L^{\sigma,h}} = \|\nu\|_{L^\sigma}.$$

From the intrinsic definition (2.23) of the norm for the space L^σ one can see that an intrinsic characterization of the norm on $L^{\sigma,h}$ is given by

$$\|f\|_{L^{\sigma,h}} = h_f(0)$$

where $\xi \mapsto h_f(\xi)$ is the minimal scalar separately harmonic function h on $\mathbb{D}^d$ such that the $\mathcal{L}(\mathbb{C} \oplus \mathcal{E})$-valued separately harmonic function

$$\xi \mapsto \begin{bmatrix} h(\xi) & f(\xi)^* \\ f(\xi) & \sigma^{\wedge h}(\xi) \end{bmatrix}$$

has positive semidefinite values on $\mathbb{D}^d$. Both these incarnations are closely related to the FRKHS $\mathcal{L}_W$ associated with a positive semidefinite Toeplitz weight $W =$

$[W_{n-n'}]_{n,n'\in\mathbb{Z}^d}$ discussed in Section 2.2, where $W = [W_{n-n'}]_{n,n'\in\mathbb{Z}^d}$ is the positive semidefinite Toeplitz weight with symbol $\widehat{W}(z)$ having harmonic extension $\widehat{W}^{\wedge h}(\xi)$ equal to the harmonic extension $\sigma^{\wedge h}(\xi)$ of the measure σ. We state the precise connections as follows.

Theorem 2.18. *Let σ be a finite, positive $\mathcal{L}(\mathcal{E})$-valued measure on $\mathbb{T}^d$. Define a positive semidefinite Toeplitz weight $W = [W_{n-n'}]_{n,n'\in\mathbb{T}^d}$ by*

$$\widehat{W}_n = \int_{\mathbb{T}^d} t^{-n}\, d\sigma(t). \tag{2.26}$$

Then the operator $U_{\sigma,W}$ given by

$$U_{\sigma,W} : \nu \mapsto \sum_{n\in\mathbb{Z}^d} [\nu]_n z^n \tag{2.27}$$

where $[\nu]_n$ are the moments of the vector measure ν

$$[\nu]_n = \int_{\mathbb{T}^d} t^{-n}\, d\nu(t)$$

is unitary from L^σ onto $\mathcal{L}_W$ and intertwines $\mathcal{U}_\sigma$ with $\mathcal{U}_W$:

$$U_{\sigma,W}\mathcal{U}_{\sigma,j} = \mathcal{U}_{W,j}U_{\sigma,W}.$$

Moreover, for any $\nu \in L^\sigma$, the separately harmonic extension $\nu^{\wedge h}$ of the measure ν agrees with the separately harmonic extension $(U_{\sigma,W}\nu)^{\wedge h}$ of the formal power series $U_{\sigma,W}\nu \in \mathcal{L}_W$:

$$\nu^{\wedge h}(\xi) = (U_{\sigma,W}\nu)^{\wedge h}(\xi) \text{ for all } \xi \in \mathbb{D}^d.$$

Conversely, suppose that $W = [\widehat{W}_{n-n'}]_{n,n'\in\mathbb{Z}^d}$ is a positive semidefinite Toeplitz weight with matrix entries in $\mathcal{L}(\mathcal{E})$. Let σ be the unique positive $\mathcal{L}(\mathcal{E})$-valued measure which solves the moment problem (2.26) (see Proposition 2.14). Then the operator $U_{W,\sigma}$ given by

$$U_{W,\sigma} : f(z) \mapsto \nu_f \tag{2.28}$$

where ν_f is the measure defined as the weak-$*$ limit as $r \to 1$ of the measures $\nu_{f,r}$ $(0 < r < 1)$ given by $\Delta \mapsto \nu_{f,r}(\Delta) = \int_\Delta f^{\wedge h}(rt)d\lambda(t)$ (where λ is normalized Lebesgue measure on $\mathbb{T}^d$) is a unitary transformation of $\mathcal{L}_W$ onto L^σ, and intertwines $\mathcal{U}_W$ with $\mathcal{U}_\sigma$:

$$U_{W,\sigma}\mathcal{U}_{W,j} = \mathcal{U}_{\sigma,j}U_{W,\sigma} \text{ for } j = 1, \dots, d.$$

Moreover, for any $f \in \mathcal{L}_W$, the separately harmonic extension $f^{\wedge h}$ of the formal power series f agrees with the separately harmonic extension $(U_{W,\sigma}f)^{\wedge h}$ of the vector measure $U_{W,\sigma}f$:

$$f^{\wedge h}(\xi) = (U_{W,\sigma}f)^{\wedge h}(\xi) \text{ for all } \xi \in \mathbb{D}^d,$$

and the maps $U_{\sigma,W}$ and $U_{W,\sigma}$ are inverses of each other:

$$U_{\sigma,W} = U_{W,\sigma}^{-1} = U_{W,\sigma}^{*},$$
$$U_{W,\sigma} = U_{\sigma,W}^{-1} = U_{\sigma,W}^{*}.$$

Proof. The first part of the theorem essentially follows by applying Theorem 2.11 with the concrete unitary scattering system $\mathfrak{S}_\sigma$ in place of the abstract unitary scattering system $\mathfrak{S}$. Similarly, the second part of the theorem essentially follows by applying 2.16 with the concrete unitary scattering system $\mathfrak{S}_W$ in place of the abstract $\mathfrak{S}$. This completes our discussion of the proof. □

As a corollary of all these connections between FRKHSs and Hellinger spaces, we obtain the following intrinsic characterization as to when a formal power series $f \in \mathcal{E}[[z, z^{-1}]]$ is in a space $\mathcal{L}_W$. As the proof follows along the same lines as is done in [17] for the one-variable case, we omit the proof.

Corollary 2.19. *Suppose that $W = [\widehat{W}_{n-n'}]_{n,n'}$ is a positive semidefinite Toeplitz weight with matrix entries $W_{n-n'} \in \mathcal{L}(\mathcal{E})$ with associated symbol*

$$\widehat{W}(z) = \sum_{n\in\mathbb{Z}^d} \widehat{W}_n z^n \in \mathcal{L}(\mathcal{E})[[z, z^{-1}]],$$

and let $\mathcal{L}_W$ be the associated FRKHS as in Theorem 2.1. Then a given

$$f = \sum_{n\in\mathbb{Z}^d} f_n z^n \in \mathcal{E}[[z, z^{-1}]]$$

is in $\mathcal{L}_W$ if and only if there exists a

$$g(z) = \sum_{n\in\mathbb{Z}^d} g_n z^n \in \mathbb{C}[[z, z^{-1}]]$$

such that the formal power series

$$\begin{bmatrix} g(z) & f(z)^* \\ f(z) & \widehat{W}(z) \end{bmatrix} \tag{2.29}$$

is positive *in the sense that its separately harmonic extension has positive semidefinite values on $\mathbb{D}^d$, or equivalently, if and only if there is a $g(z) = \sum_{n\in\mathbb{Z}^d} g_n z^n \in \mathbb{C}[[z, z^{-1}]]$ such that*

$$(n, n') \mapsto \begin{bmatrix} g_{n-n'} & f_{n-n'}^* \\ f_{n-n'} & \widehat{W}_{n-n'} \end{bmatrix} \tag{2.30}$$

is a positive kernel on $\mathbb{Z}^d \times \mathbb{Z}^d$. Moreover, there is a $g^f(z) \in \mathbb{C}[[z, z^{-1}]]$ which is minimal among formal power series g satisfying (2.29) (or (2.30)) in the sense that $(n, n') \mapsto g_{n,n'} - g_{n-n'}^f$ is a positive kernel on $\mathbb{Z}^d \times \mathbb{Z}^d$ for any $g(z) = \sum_{n\in\mathbb{Z}^d} g_n z^n$ satisfying (2.30), and the norm of $f \in \mathcal{L}_W$ is then given by

$$\|f\|_{\mathcal{L}_W} = g_0^f.$$

2.4. Models associated with Herglotz functions

So far we have seen bijective correspondences between the following objects:

1. positive semidefinite Toeplitz weights $W = [\widehat{W}_{n-n'}]_{n,n' \in \mathbb{Z}^d}$,
2. positive, separately harmonic functions on $\mathbb{D}^d$, and
3. finite, positive semidefinite, operator-valued Borel measures on $\mathbb{T}^d$.

In the $d = 1$ case, separately harmonic is the same is pluriharmonic (see, e.g., [18]), and it follows that there is yet a fourth point of view: *any positive harmonic function $h(\xi)$ on $\mathbb{D}$ can also be viewed as twice the real part of an analytic function φ on $\mathbb{D}$* having positive real part: $h(\xi) = \varphi(\xi) + \varphi(\xi)^*$ where $\varphi(\xi) = \sum_{n=0}^{\infty} \varphi_n \xi^n$ is analytic (where in general $\varphi_n \in \mathcal{L}(\mathcal{E})$ for some Hilbert space $\mathcal{E}$). The connection with a positive measure σ is then given by the Herglotz representation:

$$\varphi(\xi) = i\Im\varphi(0) + \int_{\mathbb{T}} \mathcal{H}(t,\xi)\, d\sigma(t)$$

where $\mathcal{H}(t,\xi)$ is the *Herglotz kernel*

$$\mathcal{H}(t,\xi) = 1 + 2\sum_{n=1}^{\infty} t^{-n}\xi^n = \frac{1+t^{-1}\xi}{1-t^{-1}\xi}.$$

Associated with any such Herglotz function is the de Branges-Rovnyak model space $\mathcal{E}(\varphi)$ which is a reproducing kernel Hilbert space consisting of $\mathcal{E} \oplus \mathcal{E}$-valued analytic functions $\begin{bmatrix} f(\xi) \\ g(\xi) \end{bmatrix}$ on the unit disk $\mathbb{D}$ with reproducing kernel $K^{\varphi}(z,z')$ given by

$$K^{\varphi}(\xi,\xi') = \frac{1}{2}\begin{bmatrix} \frac{\varphi(\xi)+\varphi(\xi')^*}{1-\xi\overline{\xi}'} & \frac{\varphi(\xi)-\varphi(\overline{\xi}')}{\xi-\overline{\xi}'} \\ \frac{\widetilde{\varphi}(\xi)-\widetilde{\varphi}(\overline{\xi}')}{\xi-\overline{\xi}'} & \frac{\widetilde{\varphi}(\xi)+\widetilde{\varphi}(\xi')^*}{1-\xi\overline{\xi}'} \end{bmatrix} \tag{2.31}$$

where we have set

$$\widetilde{\varphi}(\xi) = \varphi(\overline{\xi})^*.$$

These spaces have also been used as models for (1-variable) unitary scattering systems. The following theorem summarizes the situation.

Theorem 2.20 (See [3, 24]). *Let $\xi \mapsto \varphi(\xi)$ be an $\mathcal{L}(\mathcal{E})$-valued Herglotz function on the unit disk $\mathbb{D}$ with associated reproducing kernel Hilbert space $\mathcal{E}(\varphi)$ with reproducing kernel $K^{\varphi}(\xi,\xi')$ (2.31) as above. Define an operator U_{φ} by*

$$\mathcal{U}_{\varphi}\colon \begin{bmatrix} f(\xi) \\ g(\xi) \end{bmatrix} \mapsto \begin{bmatrix} \xi f(\xi) + g(0) \\ (g(\xi)-g(0))/\xi \end{bmatrix} \quad \text{for} \quad \begin{bmatrix} f(\xi) \\ g(\xi) \end{bmatrix} \in \mathcal{E}(\varphi). \tag{2.32}$$

Then $\mathcal{U}_{\varphi}$ is unitary on $\mathcal{E}(\varphi)$ with adjoint given by

$$\mathcal{U}_{\varphi}^*\colon \begin{bmatrix} f(\xi) \\ g(\xi) \end{bmatrix} \mapsto \begin{bmatrix} (f(\xi)-f(0))/\xi \\ \xi g(\xi) + f(0) \end{bmatrix}. \tag{2.33}$$

If we define $V_{\varphi}\colon \mathcal{E} \mapsto \mathcal{E}(\varphi)$ by

$$V_{\varphi}\colon e \mapsto K^{\varphi}(\xi,0)e \in \mathcal{E}(\varphi),$$

then

$$\mathfrak{S}_\varphi = (\mathcal{U}_\varphi, V_\varphi; \mathcal{E}(\varphi), \mathcal{E})$$

is a 1*-variable unitary scattering system.*

Conversely, if

$$\mathfrak{S} = (\mathcal{U}, V; \mathcal{H}, \mathcal{E})$$

is any 1*-variable unitary scattering system and we set*

$$\varphi(\xi) = V^*(I + \xi U^*)(I - \xi U^*)^{-1}V,$$

then $\xi \mapsto \varphi(\xi)$ *is an* $\mathcal{L}(\mathcal{E})$*-valued Herglotz function on* $\mathbb{D}$ *and the map*

$$\Phi\colon h \mapsto \begin{bmatrix} V^*(I - \xi U^*)^{-1}h \\ V^*(U^* - \xi I)^{-1}h \end{bmatrix}$$

is a coisometry of $\mathcal{H}$ *onto* $\mathcal{E}(\varphi)$ *(with initial space equal to the smallest reducing subspace for* $\mathcal{U}$ *containing* $\operatorname{im} V$*) which intertwines* $\mathcal{U}$ *with* $\mathcal{U}_\varphi$ *and* V *with* V_φ*:*

$$\Phi\mathcal{U} = \mathcal{U}_\varphi\Phi, \qquad V_\varphi = \Phi V.$$

Thus, in case $\mathfrak{S}$ *is a minimal unitary scattering system,* Φ *implements a unitary equivalence between the* 1*-variable unitary scattering system* $\mathfrak{S}$ *and the model* 1*-variable unitary scattering system* $\mathfrak{S}_\varphi$.

As $\mathcal{L}_W$, L^σ and $\mathcal{E}(\varphi)$ all serve as models for the minimal unitary scattering system in the 1-variable case, necessarily there must be a direct unitary equivalence between various pairs of these spaces for an appropriately related choice of parameters W, σ and φ. Our next result lays out these correspondences which involve the space $\mathcal{E}(\varphi)$ and parallels Theorem 2.18. The proof is a routine exercise tracing through the various maps.

Theorem 2.21. *Let* $\varphi(\xi) = \sum_{n=0}^\infty \varphi_n \xi^n$ *be an* $\mathcal{L}(\mathcal{E})$*-valued Herglotz function on the unit disk* $\mathbb{D}$ *and set* $W = [\widehat{W}_{n-n'}]_{n,n' \in \mathbb{Z}}$ *where*

$$\widehat{W}_n = \begin{cases} \frac{1}{2}(\varphi_0 + \varphi_0^*) & \text{if } n = 0, \\ \frac{1}{2}\varphi_n & \text{if } n \geq 1, \\ \frac{1}{2}\varphi_{-n}^* & \text{if } n \leq -1. \end{cases}$$

Then W *is a positive semidefinite Toeplitz weight and the map*

$$U_{\varphi,W}\colon \begin{bmatrix} f(\xi) \\ g(\xi) \end{bmatrix} = \begin{bmatrix} \sum_{n=0}^\infty f_n z^n \\ \sum_{n=0}^\infty g_n z^n \end{bmatrix} \mapsto \sum_{n=0}^\infty f_n \xi^n + \xi^{-1} \sum_{n=0}^\infty g_n \xi^{-n}$$

is a unitary transformation of $\mathcal{E}(\varphi)$ *onto* $\mathcal{L}_W$ *which implements a unitary equivalence between the* 1*-variable unitary scattering system* $\mathfrak{S}_\varphi$ *and the* 1*-variable unitary scattering system* $\mathfrak{S}_W$*:*

$$U_{\varphi,W}\mathcal{U}_\varphi = \mathcal{U}_W U_{\varphi,W}, \qquad V_W = U_{\varphi,W} V_\varphi.$$

Similarly, if we choose the positive $\mathcal{L}(\mathcal{E})$-valued measure σ on $\mathbb{T}^d$ so that φ has the Herglotz representation

$$\varphi(\xi) = i\Im\varphi(0) + \int_{\mathbb{T}} \frac{1+t^{-1}\xi}{1-t^{-1}\xi}\, d\sigma(t), \tag{2.34}$$

then the linear transformation $U_{\varphi,\sigma}$ defined by

$$U_{\varphi,\sigma} \colon \begin{bmatrix} f(\xi) \\ g(\xi) \end{bmatrix} \mapsto \nu_{f,g}$$

where $\nu_{f,g}$ is the Borel $\mathcal{E}$-valued measure on $\mathbb{T}$ with harmonic extension $(\nu_{f,g})^{\wedge h}$ given by

$$(\nu_{f,g})^{\wedge h}(\xi) = f(\xi) + \overline{\xi}g(\overline{\xi})$$

is unitary from $\mathcal{E}(\varphi)$ onto L^σ and implements a unitary equivalence between the 1-variable unitary scattering systems $\mathfrak{S}_\varphi$ and $\mathfrak{S}_\sigma$:

$$U_{\varphi,\sigma}\mathcal{U}_\varphi = \mathcal{U}_\sigma U_{\varphi,\sigma}, \qquad U_{\varphi,\sigma}L_\varphi = L_\sigma.$$

Conversely, suppose that $W = [\widehat{W}_{n-n'}]_{n,n'\in\mathbb{Z}}$ is a (1-variable) positive semidefinite Toeplitz weight, and we set $\varphi(z) = \widehat{W}_0 + 2\sum_{n=1}^{\infty} \widehat{W}_n z^n$. Then $\xi \mapsto \varphi(\xi)$ is a $\mathcal{L}(\mathcal{E})$-valued Herglotz function on the unit disk $\mathbb{D}$ and the map

$$U_{W,\varphi} \colon \sum_{n=-\infty}^{\infty} f_n z^n \mapsto \begin{bmatrix} \sum_{n=0}^{\infty} f_n \xi^n \\ \sum_{n=0}^{\infty} f_{-n-1}\xi^n \end{bmatrix}$$

is a unitary transformation from $\mathcal{L}_W$ onto $\mathcal{E}(\varphi)$ which implements a unitary equivalence between the 1-variable unitary scattering systems $\mathfrak{S}_W$ and $\mathfrak{S}_\varphi$:

$$U_{W,\varphi}\mathcal{U}_W = \mathcal{U}_\varphi U_{W,\varphi}, \qquad V_\varphi = U_{W,\varphi}V_W.$$

Similarly, if σ is a positive $\mathcal{E}$-valued Borel measure on $\mathbb{T}$ and we define a $\mathcal{E}$-valued Herglotz function $\xi \mapsto \varphi(\xi)$ on $\mathbb{D}$ via the Herglotz representation (2.34) induced by σ (with $\Im\varphi(0)$ chosen arbitrarily), then the operator $U_{\sigma,\varphi}$ given by

$$U_{\sigma,\varphi} \colon \nu \mapsto \begin{bmatrix} \int_{\mathbb{T}} (1-\xi t^{-1})^{-1}\, d\nu(t) \\ \int_{\mathbb{T}} (t^{-1}-\xi)^{-1}\, d\nu(t) \end{bmatrix}$$

is unitary from L^σ onto $\mathcal{E}(\varphi)$ which implements a unitary equivalence between the 1-variable unitary scattering systems $\mathfrak{S}_\sigma$ and $\mathfrak{S}_\varphi$:

$$U_{\sigma,\varphi}\mathcal{U}_\sigma = \mathcal{U}_\varphi U_{\sigma,\varphi}, \qquad V_\varphi = U_{\sigma,\varphi}V_\sigma.$$

Moreover the maps $U_{W,\varphi}$ and $U_{\varphi,W}$ are inverses of each other

$$U_{W,\varphi} = U_{\varphi,W}^{-1} = U_{\varphi,W}^*$$
$$U_{\varphi,W} = U_{W,\varphi}^{-1} = U_{W,\varphi}^*.$$

and the maps $U_{\sigma,\varphi}$ and $U_{\varphi,\sigma}$ are inverses of each other:

$$U_{\sigma,\varphi} = U_{\varphi,\sigma}^{-1} = U_{\varphi,\sigma}^*,$$
$$U_{\varphi,\sigma} = U_{\sigma,\varphi}^{-1} = U_{\sigma,\varphi}^*.$$

3. Noncommutative formal reproducing kernel Hilbert spaces

3.1. General NFRKHSs

In this section we study Hilbert spaces $\mathcal{H}$ whose elements are formal power series

$$f(z) = \sum_{v \in \mathcal{F}_d} f_v z^v \tag{3.1}$$

in a collection of noncommuting indeterminates $z = (z_1, \dots, z_d)$. Here $\mathcal{F}_d$ denotes the free semigroup on d generators $g_1, \dots, g_d$ with identity. A generic element of $\mathcal{F}_d$ (apart from the unit element) has the form of a word $w = g_{i_n} \cdots g_{i_1}$, i.e., a string of symbols $\alpha_n \cdots \alpha_1$ of finite length n with each symbol α_k belonging to the alphabet $\{g_1, \dots, g_d\}$. We shall write $|w|$ for the length n of the word $w = \alpha_n \cdots \alpha_1$. If $w = \alpha_n \cdots \alpha_1$ and $v = \beta_m \cdots \beta_1$ are words, then the product vw of v and w is the new word formed by the juxtapositions of v and w:

$$vw = \beta_m \cdots \beta_1 \alpha_n \cdots \alpha_1.$$

We define the *transpose* $w^\top$ of the word $w = g_{i_n} \cdots g_{i_1}$ by $w^\top = g_{i_1} \cdots g_{i_n}$. We denote the unit element of $\mathcal{F}_d$ by $\emptyset$ (corresponding to the empty word). In particular, if g_k is a word of unit length, we write $g_k w$ for $g_k \alpha_n \cdots \alpha_1$ if $w = \alpha_n \cdots \alpha_1$. Although $\mathcal{F}_d$ is a semigroup, we will on occasion work with expressions involving inverses of words in $\mathcal{F}_d$; the meaning is as follows: if w and v are words in $\mathcal{F}_d$, the expression wv^{-1} means w' if there is a $w' \in \mathcal{F}_d$ for which $w = w'v$; otherwise we say that wv^{-1} is undefined. An analogous interpretation applies for expressions of the form $w^{-1}v$. This convention requires some care as associativity can fail: in general it is not the case that $(wv^{-1}) \cdot w' = w \cdot (v^{-1}w')$. Our multivariable notation for the noncommutative case is:

$$z^w = z_{i_n} \cdots z_{i_1} \text{ if } w = g_{i_n} \cdots g_{i_1} \in \mathcal{F}_d.$$

Hence $z_j = z^{g_j}$ and we set $z^\emptyset = 1$. If $v, w \in \mathcal{F}_d$ are such that uv^{-1} is undefined in $\mathcal{F}_d$, we set $z^{uv^{-1}} = 0$.

In general, let us say that a Hilbert space $\mathcal{H}$ consisting of formal power series $f(z) = \sum_{v \in \mathcal{F}_d} f_v z^v \in \mathcal{E}[[z]]$ in the noncommuting indeterminates $z = (z_1, \dots, z_d)$ with coefficients f_v in the Hilbert space $\mathcal{E}$ is a NFRKHS (*noncommutative formal reproducing kernel Hilbert space*) if, for each $\alpha \in \mathcal{F}_d$, the linear operator $\Phi_\alpha \colon \mathcal{H} \mapsto \mathcal{E}$ defined by $f(z) = \sum_v f_v z^v \to f_\alpha$ is continuous. As any such power series is completely determined by the list of its coefficients $v \mapsto f_v$ for $v \in \mathcal{F}_d$, we can equivalently view the elements $f(z)$ as functions $v \mapsto f_v$ on $\mathcal{F}_d$. As $f \mapsto f_v$ is continuous by assumption, when viewed in this way such a Hilbert space $\mathcal{H}$ is a standard reproducing kernel Hilbert space of functions on $\mathcal{F}_d$. Hence, by Theorem 1.1 there is a reproducing kernel $K \colon \mathcal{F}_d \mapsto \mathcal{L}(\mathcal{E})$; in this context we denote the value of K at (v, v') by

$$(v, v') \mapsto K_{v,v'} \in \mathcal{L}(\mathcal{E}) \tag{3.2}$$

rather than by $(v, v') \mapsto K(v, v')$. Since we view elements f of $\mathcal{H}$ as formal power series $\sum_{v \in \mathcal{F}_d} f_v z^v$ rather than as functions $v \mapsto f_v$ on $\mathcal{F}_d$, we write, for a given

$v' \in \mathcal{F}_d$ and $e \in \mathcal{E}$, the element $\Phi_{v'}^* e \in \mathcal{H}$ as $\Phi_{v'}^* e = K_{v'}(\cdot)e$ where

$$K_{v'}(z) = \sum_{v \in \mathcal{F}_d} K_{v,v'} z^v$$

where $K_{v,v'}$ is as in (3.2). Then the reproducing kernel property can be written as

$$\langle f, K_{v'}(\cdot)e \rangle_{\mathcal{H}} = \langle f_v, e \rangle_{\mathcal{E}}. \tag{3.3}$$

As was done for the commutative case in Section 2, we can make the notation more suggestive of the classical case as follows. In general, given a coefficient Hilbert space $\mathcal{C}$, we can use the $\mathcal{C}$-inner product to define a pairing

$$\langle \cdot, \cdot \rangle_{\mathcal{C} \times \mathcal{C}[[z']]} \mapsto \mathbb{C}[[z']]$$

by

$$\langle c, \sum_{v' \in \mathcal{F}_d} f_{v'} z'^{v'} \rangle_{\mathcal{C} \times \mathcal{C}[[z']]} = \sum_{v' \in \mathcal{F}_d} \langle c, f_{v'} \rangle_{\mathcal{C}} z'^{v'^\top} \in \mathbb{C}[[z']].$$

When convenient we can also use the pairing in the reverse order:

$$\langle \sum_{w' \in \mathcal{F}_d} f_{w'} z'^{w'}, c \rangle_{\mathcal{C}[[z']] \times \mathcal{C}} = \sum_{w' \in \mathcal{F}_d} \langle f_{w'}, c \rangle_{\mathcal{C}} z'^{w'} \in \mathbb{C}[[z']].$$

These pairings can be seen as special cases of the more general pairing

$$\left\langle \sum_{w' \in \mathcal{F}_d} f_{w'} z'^{w'}, \sum_{v' \in \mathcal{F}_d} g_{v'} z'^{v'} \right\rangle_{\mathcal{C}[[z']] \times \mathcal{C}[[z']]} = \sum_{w' \in \mathcal{F}_d} \left[\sum_{v', v'' : w' = v''(v')^\top} \langle f_{v''}, g_{v'} \rangle_{\mathcal{C}} \right] z'^{w'}.$$

Here $z' = (z'_1, \dots, z'_d)$ is a second d-tuple of noncommuting indeterminates, each of which is assumed however to commute with each of the indeterminates $z = (z_1, \dots, z_d)$. Then the reproducing kernel property (3.3) can be written more suggestively as

$$\langle f, K(\cdot, z')e \rangle_{\mathcal{H} \times \mathcal{H}[[z']]} = \langle f(z'), e \rangle_{\mathcal{E}[[z']] \times \mathcal{E}}. \tag{3.4}$$

where we have set

$$K(z, z') = \sum_{v, v' \in \mathcal{F}_d} K_{v,v'} z^v z'^{v'^\top} \in \mathcal{H}[[z, z']].$$

The following theorem gives the characterization of NFRKHSs parallel to that given for the commutative case in Theorem 2.1.

Theorem 3.1. *Let* $K(z,z') = \sum_{v,v' \in \mathcal{F}_d} K_{v,v'} z^v z'^{v'^\top}$ *be a given element of* $\mathcal{L}(\mathcal{E})[[z,z']]$ *where* $z = (z_1, \dots, z_d)$ *and* $z' = (z'_1, \dots, z'_d)$ *are* d*-tuples of noncommuting indeterminates which do commute with each other* ($z_i z'_j = z'_j z_i$ *for all* $i, j = 1, \dots, d$). *Then the following conditions are equivalent:*

1. *K is the reproducing kernel for a uniquely determined NFRKHS $\mathcal{H}(K)$ of formal power series in the set of noncommuting indeterminates $z = (z_1, \dots, z_d)$.*

2. *There is an auxiliary Hilbert space $\mathcal{H}$ and a noncommutative formal power series $H(z) \in \mathcal{L}(\mathcal{H}, \mathcal{E})[[z]]$ such that*

$$K(z, z') = H(z)H(z')^*$$

where we use the convention

$$(z'^v)^* = z'^{v^\top}.$$

3. $K(z, z')$ *is a* positive kernel *in the sense that*

$$\sum_{v,v' \in \mathcal{F}_d} \langle K_{v,v'} e_{v'}, e_v \rangle_{\mathcal{E}} \geq 0$$

for all finitely supported $\mathcal{E}$-valued functions $v \mapsto e_v$ on $\mathcal{F}_d$.

Moreover, in this case the NFRKHS $\mathcal{H}(K)$ can be defined directly in terms of the formal power series $H(z)$ appearing in condition (2) by

$$\mathcal{H}(K) = \{H(z)h \colon h \in \mathcal{H}\}$$

with norm taken to be the "lifted norm"

$$\|H(z)h\|_{\mathcal{H}(K)} = \|Qh\|_{\mathcal{H}}$$

where Q is the orthogonal projection of $\mathcal{H}$ onto the orthogonal complement of the kernel of the map $M_H \colon \mathcal{H} \mapsto \mathcal{E}[[z]]$ given by $M_H \colon h \mapsto H(z)h$.

Proof. The proof follows in the same spirit as the proof of Theorem 2.1 by tracing through the identification of $\mathcal{H}(K)$ with a standard reproducing kernel Hilbert space over $\mathcal{F}_d \times \mathcal{F}_d$ and using the results in Theorem 1.1 for the standard case. □

Just as was done for the commutative case in Proposition 2.2, there is an equivalent operator-theoretic presentation for a NFRKHS as follows. Assume that $W_{v,v'}$ is a collection of operators in $\mathcal{L}(\mathcal{E})$ satisfying the positivity condition in Theorem 3.1. The usual scalar-valued L^2-pairing between formal power series with norm-squared summable coefficients can be used to define a pairing between $\mathcal{E}[[z]]$ and $\mathcal{E}[z]$:

$$\langle \sum_{v \in \mathcal{F}_d} f_v z^v, \sum_{v \in \mathcal{F}_d} p_{v'} z^v \rangle_{L^2} = \sum_{v \in \mathcal{F}_d} \langle f_v, p_v \rangle_{\mathcal{E}}.$$

(Note that the sum on the right is finite by the assumption that $p(z)$ is a polynomial.) Informally we refer to this inner product simply as the L^2-inner product (extended to $\mathcal{E}[[z]] \times \mathcal{E}[z]$).

Let now $W = [W_{v,v'}]_{v,v' \in \mathcal{F}_d}$ be a matrix with rows and columns indexed by $\mathcal{F}_d$ with matrix entries $W_{v,v'} \in \mathcal{L}(\mathcal{E})$. Then W defines an operator, also denoted as W, from $\mathcal{E}[z]$ into $\mathcal{E}[[z]]$ in the usual way:

$$W \colon \sum_{v \in \mathcal{F}_d} p_v z^v \mapsto \sum_{v,v' \in \mathcal{F}_d} W_{v,v'} p_{v'} z^v.$$

Note that the sum over v' is finite since $p(z)$ is a polynomial while the sum over v can be infinite but is taken only in a formal sense to define an element of $\mathcal{E}[[z]]$.

Given such a block matrix W, we say that W is a *positive semidefinite $\mathcal{L}(\mathcal{E})$-valued weight* on $\mathcal{F}_d$ if

$$\langle Wp, p\rangle_{L^2} \geq 0 \text{ for all } p \in \mathcal{E}[z].$$

Equivalently, the collection of operators $W_{v,v'}$ (for $v, v' \in \mathcal{F}_d$) satisfies the condition in part (3) of Theorem 3.1 (with $W_{v,v'}$ in place of $K_{v,v'}$). Given such a positive semidefinite weight W, we define an inner product $\langle \cdot, \cdot \rangle_{\mathcal{L}_W}$ on $W\mathcal{E}[z]$ via

$$\langle Wp, Wp' \rangle_{\mathcal{L}_W} = \langle Wp, p' \rangle_{L^2}. \tag{3.5}$$

Under the assumption that W is a positive semidefinite weight, this inner product is positive semidefinite on $W\mathcal{E}[z]$; as in the commutative case, one can see that elements of zero norm are actually equal to the zero noncommutative formal power series, and that the power series coefficient $[Wp]_{v'}$ of an element $Wp \in W\mathcal{E}[z]$ for a given $v' \in \mathcal{F}_d$ is given by

$$\langle [Wp]_{v'}, e\rangle_{\mathcal{E}} = \langle Wp, W(z^{v'}e)\rangle_{\mathcal{L}_W} \tag{3.6}$$

Then one may complete the pre-Hilbert space $W\mathcal{E}[z]$ to a Hilbert space, denoted by $\mathcal{L}_W$ (where here W is a weight on $\mathcal{F}_d$ rather than on $\mathbb{Z}^d$ as in Section 2.1). Given an element f of the completion $\mathcal{L}_W$, we may use the formula (3.6) to define coefficients

$$\langle f_{v'}, e\rangle_{\mathcal{E}} = \langle f, W(z^{v'}e)\rangle_{\mathcal{L}_W} \tag{3.7}$$

and identify $f \in \mathcal{L}_W$ with a noncommutative formal power series

$$f(z) = \sum_{v \in \mathcal{F}_d} f_v z^v.$$

From (3.7) we see that the coefficient evaluation maps $\Phi_{v'} \colon f \mapsto f_{v'}$ are continuous from $\mathcal{L}_W$ into $\mathcal{E}$ for each $v' \in \mathcal{F}_d$, and hence $\mathcal{L}_W$ is a NFRKHS. We have already observed in Theorem 3.1 that the reproducing kernel $K(z, z')$ is recovered from the positive kernel $(v, v') \mapsto W_{v,v'}$ on $(\mathcal{F}_d \times \mathcal{F}_d)$ via $K(z, z') = \sum_{v,v' \in \mathcal{F}_d} W_{v,v'} z^v z'^{v'^\top}$. We have now arrived at the following noncommutative analogue of Proposition 2.2.

Proposition 3.2. *Let W be a positive semidefinite $\mathcal{L}(\mathcal{E})$-weight over $\mathcal{F}_d$ as described above. Then there is a NFRKHS $\mathcal{L}_W \subset \mathcal{E}[[z]]$ with dense subset $W\mathcal{E}[z]$ with inner product given by* (3.5) *and with reproducing kernel $K_W(z, z')$ given by*

$$K_W(z, z') = \sum_{v,v' \in \mathcal{F}_d} W_{v,v'} z^v z'^{v'^\top}.$$

Conversely, if $\mathcal{H}$ is any NFRKHS of formal power series in the noncommuting indeterminates $z = (z_1, \dots, z_d)$ with coefficients in $\mathcal{E}$ and with reproducing kernel

$$K(z, z') = \sum_{v,v' \in \mathcal{F}_d} K_{v,v'} z^v z'^{v'^\top},$$

then the formula $W_{v,w} = K_{v,w}$ gives rise to a positive semidefinite weight

$$W = [K_{v,v'}]_{v,v' \in \mathcal{F}_d} \colon \mathcal{E}[z] \mapsto \mathcal{E}[[z]]$$

such that $\mathcal{H} = \mathcal{L}_W$.

As a first example of a NFRKHS consider the space $L^2(\mathcal{F}_d, \mathcal{E})$ of all formal power series $f(z) = \sum_{v \in \mathcal{F}_d} f_v z^v$ with coefficients in the Hilbert space $\mathcal{E}$ with norm-squared summable coefficients:

$$\|f\|^2_{L^2(\mathcal{F}_d, \mathcal{E})} = \sum_{v \in \mathcal{F}_d} \|f_v\|^2_{\mathcal{E}}.$$

Then clearly the coefficient-evaluation maps $\Phi_{v'} \colon f(z) \mapsto f_{v'}$ are bounded from $L^2(\mathcal{F}_d, \mathcal{E})$ into $\mathcal{E}$ for each $v' \in \mathcal{F}_d$. The content of the next proposition is the computation of the associated reproducing kernel function $K(z, z') \in \mathcal{L}(\mathcal{E}[[z, z']]$.

Proposition 3.3. *The space $L^2(\mathcal{F}_d, \mathcal{E})$ consisting of formal power series $f(z) = \sum_{v \in \mathcal{F}_d} f_v z^v$ with coefficients $f_v \in \mathcal{E}$ satisfying $\sum_{v \in \mathcal{F}_d} \|f_v\|^2 < \infty$ is a NFRKHS with reproducing kernel equal to*

$$k_{\mathcal{F}_d}(z, z') = \sum_{v \in \mathcal{F}_d} z^v z'^{v^\top}. \tag{3.8}$$

Proof. One can see that $k_{\mathcal{F}_d}(z, z')$ is the kernel function for $L^2(\mathcal{F}_d, \mathcal{E})$ by direct verification. Alternatively, note that $L^2(\mathcal{F}_d, \mathcal{E})$ is equal to the lifted norm space $\mathcal{L}_W$ with weight $W = [\delta_{v,v'} I_{\mathcal{E}}]_{v,v' \in \mathcal{F}_d}$ where $\delta_{v,v} = 1$ and $\delta_{v,v'} = 0$ if $v \neq v'$ in $\mathcal{F}_d$. From the general connection

$$K(z, z') = \sum_{v,v' \in \mathcal{F}_d} W_{v,v'} z^v z'^{v'^\top}$$

between the weight W and the kernel function $K(z, z')$ when $\mathcal{H}(K) = \mathcal{L}_W$ explained in Proposition 3.2, we see that the kernel function for $L^2(\mathcal{F}_d, \mathcal{E})$ is equal to $k_{\mathcal{F}_d}(z, z')$ as in (3.8) as wanted. □

The analogue of Theorem 2.5 holds in this setting as well.

Theorem 3.4.

1. *Suppose that*

$$K^1(z, z') = \sum_{v,v' \in \mathcal{F}^d} W^1_{v,v'} z^v z'^{v'^\top} \quad \text{and} \quad K^2(z, z') = \sum_{v,v' \in \mathcal{F}^d} W^2_{v,v'} z^v z'^{v'^\top}$$

are the reproducing kernels for respective NFRKHSs $\mathcal{H}(K_1) = \mathcal{L}_{W^1}$ and $\mathcal{H}(K_2) = \mathcal{L}_{W^2}$, where the coefficients $W^1_{v,v'} \in \mathcal{L}(\mathcal{E})$ and $W^2_{v,v'} \in \mathcal{L}(\mathcal{E})$ are linear operators on the same Hilbert space $\mathcal{E}$. Then

$$K(z, z') := K^1(z, z') + K^2(z, z') = \sum_{v,v' \in \mathbb{Z}^d} [W^1_{v,v'} + W^2_{v,v'}] z^v z'^{v'^\top}$$

is the reproducing kernel for a NFRKHS $\mathcal{H}(K) = \mathcal{H}(K^1) + \mathcal{H}(K^2)$ with norm given by

$$\begin{aligned} \|f\|^2_{\mathcal{H}(K^1+K^2)} = \min\{ &\|f^1\|^2_{\mathcal{H}(K^1)} + \|f^2\|^2_{\mathcal{H}(K^2)} \colon f = f^1 + f^2 \\ &\text{with } f^1 \in \mathcal{H}(K^1) \text{ and } f^2 \in \mathcal{H}(K^2).\} \end{aligned} \tag{3.9}$$

2. *If* $K(z,z') = \sum_{v,v' \in \mathcal{F}_d} W_{v,v'} z^v z'^{v'^\top}$ *and* $K^1(z,z') = \sum_{n,n' \in \mathbb{Z}^d} W^1_{v,v'} z^v z'^{v'^\top}$ *are reproducing kernels for respective NFRKHSs* $\mathcal{H}(K)$ *and* $\mathcal{H}(K^1)$ *with coefficients* $W_{v,v'}$ *and* $W^1_{v,v'}$ *in* $\mathcal{L}(\mathcal{E})$ *for the same Hilbert space* $\mathcal{E}$, *and if* $K^2(z,z') := K(z,z') - K^1(z,z')$ *is a positive kernel (i.e.,* $(v,v') \mapsto W_{v,v'} - W^1_{v,v'}$ *is a positive kernel on* $\mathbb{Z}^d \times \mathbb{Z}^d$*), then the associated NFRKHS* $\mathcal{H}(K^1)$ *is contractively included in* $\mathcal{H}(K)$:
$$\mathcal{H}(K^1) \subset \mathcal{H}(K) \text{ with } \|f\|^2_{\mathcal{H}(K)} \le \|f\|^2_{\mathcal{H}(K^1)} \text{ for } f \in \mathcal{H}(K^1).$$
In this case
$$\mathcal{H}(K) = \mathcal{H}(K^1) + \mathcal{H}(K^2) \tag{3.10}$$
where we have set $K^2 = K - K^1$, *and* $\|f\|^2_{\mathcal{H}(K)}$ *can be computed as in* (3.9) *for any* $f \in \mathcal{H}(K) = \mathcal{H}(K^1) + \mathcal{H}(K^2)$. *Moreover, for a given* $f \in \mathcal{H}(K)$, *there is a unique* $\tau_1(f) \in \mathcal{H}(K^1)$ *and* $\tau_2(f) \in \mathcal{H}(K^2)$ *so that* $f = \tau_1(f) + \tau_2(f)$ *(called the* minimal decomposition *of a given* $f \in \mathcal{H}(K)$ *with respect to* $\mathcal{H}(K^1)$ *and* $\mathcal{H}(K^2)$*) such that* $\|f\|^2_{\mathcal{H}(K)} = \|\tau_1(f)\|^2_{\mathcal{H}(K^1)} + \|\tau_2(f)\|^2_{\mathcal{H}(K^2)}$. *The decomposition* (3.10) *forms a Hilbert space orthogonal decomposition of* $\mathcal{H}(K)$ *if and only if the* overlapping space $\mathcal{H}(K^1) \cap \mathcal{H}(K^2)$ *is equal to* $\{0\}$.

Conversely, if $\mathcal{H}_1$ *is a linear submanifold of* $\mathcal{H}(K)$ *which is a Hilbert space in a norm* $\|\cdot\|_1$ *which gives a contractive inclusion of* $\mathcal{H}_1$ *into* $\mathcal{H}(K)$
$$\|f\|^2_{\mathcal{H}(K)} \le \|f\|^2_1 \text{ for } f \in \mathcal{H}_1,$$
then there is a positive kernel $K^1 \in \mathcal{E}[[z,z']]$ *such that* $K - K^1$ *is a positive kernel and so that* $\mathcal{H}_1$ *is equal to the NFRKHS* $\mathcal{H}(K^1)$ *isometrically.*

We now present a second method of constructing a NFRKHS which uses a multiplier rather than via a positive semidefinite weight $W = [W_{v,v'}]_{v,v' \in \mathcal{F}_d}$; the precise result is the following analogue of Theorem 2.6.

Theorem 3.5. *Suppose that* $\mathcal{H}(K)$ *is a NFRKHS with elements having coefficients in the Hilbert spaces* $\mathcal{E}$, *and suppose that* $H(z) \in \mathcal{L}(\mathcal{E}, \mathcal{E}_*)[[z]]$ *(where* $\mathcal{E}_*$ *is another Hilbert space). Define a subspace* $\mathcal{M}(H) \subset \mathcal{E}_*[[z]]$ *by*
$$\mathcal{M}(H) = H \cdot \mathcal{H}(K) = \{H(z) \cdot f(z) \colon f \in \mathcal{H}(K)\},$$
where $H(z) \cdot f(z)$ *denotes the multiplication of formal power series*
$$H(z) \cdot f(z) = \sum_{v \in \mathcal{F}_d} \left(\sum_{v',v'' \colon v'v''=v} H_{v'} f_{v''} \right) z^v,$$
with norm
$$\|H \cdot f\|_{\mathcal{M}(H)} = \|Qf\|_{\mathcal{H}(K)}$$
where Q *is the orthogonal projection of* $\mathcal{H}(K)$ *onto the orthogonal complement of the kernel of the linear operator* $M_H \colon f \mapsto H \cdot f$. *Then* $\mathcal{M}(H)$ *is a NFRKHS with reproducing kernel* $K^{\mathcal{M}(H)}$ *given by*
$$K^{\mathcal{M}(H)}(z,z') = H(z) \cdot K(z,z') \cdot H(z')^*.$$

Remark 3.6. As with Theorem 2.6, we note that the equivalence of parts (1) and (2) in Theorem 3.1 amounts to the special case of Theorem 3.5 where $K(z,z') = I_{\mathcal{H}}$.

For the study of row-unitary operators (as opposed to row-isometries), one has to consider also Hilbert spaces whose elements are formal power series

$$f(z,\zeta) = \sum_{v,w\in\mathcal{F}_d} f_{v,w} z^v \zeta^w$$

in two d-tuples $z = (z_1,\dots,z_d)$ and $\zeta = (\zeta_1,\dots,\zeta_d)$ of noncommuting indeterminates which do pairwise commute with each other:

$$z_i\zeta_j = \zeta_j z_i \text{ for all } i,j = 1,\dots,d.$$

We denote the set of all such formal power series f with coefficients in the space $\mathcal{X}$ by $\mathcal{X}[[z,\zeta]]$. For $\mathcal{E}$ a Hilbert space, a subspace $\mathcal{H}$ of $\mathcal{E}[[z,\zeta]]$ is said to be a noncommutative formal reproducing kernel Hilbert space (NFRKHS) if the coefficient-evaluation maps $\Phi_{v',w'} \colon \mathcal{H} \mapsto \mathcal{E}$

$$\Phi_{v'w'} \colon f(z,\zeta) = \sum_{v,w\in\mathcal{F}_d} f_{v,w} z^v \zeta^w \mapsto f_{v',w'}$$

is bounded for each $v', w' \in \mathcal{F}_d$. We then obtain analogues of Theorems 3.1, Proposition 3.2 and Theorem 3.4 for this hybrid situation; we leave the statements to the reader – see Section 3.3.

3.2. Noncommutative Toeplitz weights and models for row isometries

In this section we consider the case of a NFRKHS $\mathcal{H}(K) = \mathcal{L}_W$ where the positive semidefinite weight matrix $W = [W_{v,v'}]_{v,v'\in\mathcal{F}_d}$ has the noncommutative Toeplitz property

$$W_{vg_j,v'g_k} = \delta_{j,k} W_{v,v'} \tag{3.11}$$

(where $\delta_{j,k}$ is the Kronecker delta). Note that any selfadjoint block matrix W having the noncommutative Toeplitz property is characterized by

$$W_{v,v'} = \begin{cases} \widehat{W}_{vv'^{-1}} & \text{if } |v| \geq |v'|, \\ (\widehat{W}_{v'v^{-1}})^* & \text{if } |v'| \geq |v| \end{cases} \tag{3.12}$$

where $\widehat{W}_v = W_{v,\emptyset}$. Spaces $\mathcal{L}_W$ with W as in (3.12) are closely connected with models for a *row isometry*, i.e., a d-tuple $\mathcal{S} = (\mathcal{S}_1,\dots,\mathcal{S}_d)$ of operators on a Hilbert space $\mathcal{H}$ for which the block operator row matrix $\mathcal{S} = \begin{bmatrix} \mathcal{S}_1 & \dots & \mathcal{S}_d \end{bmatrix} \colon \mathcal{H}\otimes\mathbb{C}^d \mapsto \mathcal{H}$ is an isometry. In terms of the individual operators $\mathcal{S}_1,\dots,\mathcal{S}_d$, an equivalent condition is that each $\mathcal{S}_j$ is an isometry and that the associated image spaces $\operatorname{im}\mathcal{S}_j$ are pairwise orthogonal in $\mathcal{H}$. Such operators and the operator algebras, representation theory and noncommutative function theory associated with them have appeared in a number of studies of late (see [19, 20, 21, 22, 15, 16, 5, 6]). Here we give a reproducing kernel interpretation for the model theory developed in [5] and [6].

Suppose now that $\mathcal{S} = (\mathcal{S}_1, \dots, \mathcal{S}_d)$ is any row isometry on a Hilbert space $\mathcal{H}$ and that we are also given an operator (a *scale*, or a *control/observation* operator) $V \colon \mathcal{E} \mapsto \mathcal{H}$. Let $z = (z_1, \dots, z_d)$ be a d-tuple of noncommuting indeterminates and define a Fourier representation operator $\Phi \colon \mathcal{H} \mapsto \mathcal{E}[[z]]$ by

$$\Phi \colon h \mapsto \sum_{v \in \mathcal{F}_d} (\Phi_v h) z^v$$

where we have set $\Phi_v \colon \mathcal{H} \mapsto \mathcal{E}$ equal to

$$\Phi_v \colon h \mapsto V^* \mathcal{S}^{*v} e$$

for $v \in \mathcal{F}_d$, where we have used the noncommutative multivariable notation

$$\mathcal{S}^{*v} = \mathcal{S}_{i_n}^* \cdots \mathcal{S}_{i_1}^* \text{ if } v = g_{i_n} \cdots g_{i_1}.$$

If we set $\Phi(z) = \sum_{v \in \mathcal{F}_d} \Phi_v z^v \in \mathcal{L}(\mathcal{H}, \mathcal{E})[[z]]$, then we see that the operator $\Phi \colon h \mapsto \Phi h$ amounts to multiplication of the vector h by the formal power series $\Phi(z)$. Define a norm $\| \cdot \|_0$ on $\operatorname{im} \Phi$ as the lifted norm

$$\|\Phi(z) \cdot h\|_0 = \|Qh\|_{\mathcal{H}}$$

where Q is the orthogonal projection from $\mathcal{H}$ to the kernel of the operator Φ. From the general considerations of Theorem 3.1 we see that $\operatorname{im} \Phi$ is a NFRKHS $\mathcal{H}(K)$ with kernel $K(z, z')$ given by $K(z, z') = \Phi(z) \cdot \Phi(z')^*$, or equivalently, $(\operatorname{im} \Phi, \| \cdot \|_0)$ is a space $\mathcal{L}_W$ with the weight $W = [W_{v,v'}]_{v,v' \in \mathcal{F}_d}$ given by

$$W_{v,v'} = \Phi_v \Phi_{v'}^* = V^* \mathcal{S}^{*v} \mathcal{S}^{v'^\top} V. \tag{3.13}$$

The row isometry property of $\mathcal{S}$ implies that any $W = [W_{v,v'}]_{v,v' \in \mathcal{F}_d}$ of the form (3.13) has the noncommutative Toeplitz property (3.11). Moreover, if $h = \Phi^* p = \sum_{v' \in \mathcal{F}_d} \Phi_{v'}^* p_{v'}$ for some polynomial $p(z) = \sum_{v' \in \mathcal{F}_d} p_{v'} z^{v'} \in \mathcal{E}[z]$, then

$$\mathcal{S}_j h = \mathcal{S}_j \Phi^* p = \mathcal{S}_j \left(\sum_{v' \in \mathcal{F}_d} \Phi_{v'}^* p_{v'} \right) = \sum_{v' \in \mathcal{F}_d} \mathcal{S}_j \mathcal{S}^{v'^\top} V p_{v'} = \sum_{v' \in \mathcal{F}_d} \mathcal{S}^{(v' g_j)^\top} V p_{v'}$$

and hence

$$\begin{aligned} \Phi \mathcal{S}_j \Phi^* p &= \sum_{v, v' \in \mathcal{F}_d} V^* \mathcal{S}^{*v} \mathcal{S}^{(v' g_j)^\top} V p_{v'} z^v = \sum_{v, v' \in \mathcal{F}_d} W_{v, v' g_j} p_{v'} z^v \\ &= \sum_{v, v'' \in \mathcal{F}_d} W_{v, v''} p_{v'' g_j^{-1}} z^v = W(p(z) \cdot z_j). \end{aligned}$$

Since we also have $\Phi \Phi^* p = W p$, we conclude that $\Phi \mathcal{S}_j = \mathcal{S}_{W,j} \Phi$ on the dense subset $\Phi^* \mathcal{E}[z]$ of the cokernel of Φ, where $\mathcal{S}_{W,j}$ is defined on the dense subset $W \mathcal{E}[z]$ of $\mathcal{L}_W$ by

$$\mathcal{S}_{W,j} \colon W p \mapsto W(p(z) \cdot z_j). \tag{3.14}$$

Another computation shows that

$$\Phi \mathcal{S}_j^* h = \sum_{v \in \mathcal{F}_d} (V \mathcal{S}^{*v} \mathcal{S}_j^* h) z^v = \sum_{v \in \mathcal{F}_d} (V \mathcal{S}^{*v g_j} h) z^v = \sum_{v' \in \mathcal{F}_d} (V \mathcal{S}^{*v'} h) z^{v'} \cdot z_j^{-1}$$

from which we conclude that

$$\Phi \mathcal{S}_j^* = (S_j^{R+})^{[*]}\Phi$$

where the operator $(S_j^{R+})^{[*]}$ given by

$$(S_j^{R+})^{[*]}\colon f(z) \mapsto f(z) \cdot z_j^{-1}$$

is the L^2-adjoint of the right shift operator

$$S_j^{R+}\colon f(z) \mapsto f(z) \cdot z_j$$

acting on $\mathcal{E}[[z]]$.

Conversely, if W is a positive semidefinite noncommutative Toeplitz weight $W = [W_{v,v'}]_{v,v' \in \mathcal{F}_d}$ (determined from $\widehat{W}_v$ for $v \in \mathcal{F}_d$ as in (3.12)), then we can verify directly that $\mathcal{S}_W = (\mathcal{S}_{W,1}, \dots, \mathcal{S}_{W,d})$ with $\mathcal{S}_{W,j}$ defined densely by (3.14) is a row isometry:

$$\begin{aligned}\langle \mathcal{S}_{W,j} W(z^v e), \mathcal{S}_{W,k} W(z^{v'} e')\rangle_{\mathcal{L}_W} &= \langle W(z^{vg_j} e), W(z^{v' g_k})e'\rangle_{\mathcal{L}_W} \\ &= \langle W(z^{vg_j} e), z^{v' g_k} e'\rangle_{L^2} \\ &= \delta_{k,j}\langle W_{v',v} e, e'\rangle_{L^2} \\ &= \delta_{k,j}\langle W(z^v e), W(z^{v'} e)\rangle_{\mathcal{L}_W}.\end{aligned}$$

We have essentially arrived at the following result. For purposes of stating the result, it is convenient to define the collection of spaces and operators

$$\mathfrak{S} = (\mathcal{S} = (\mathcal{S}_1, \dots, \mathcal{S}_d), V; \mathcal{H}, \mathcal{E})$$

to be a *row-isometric scattering system* if $\mathcal{S}$ is a row-isometric d-tuple of operators on the Hilbert space $\mathcal{H}$ and $V\colon \mathcal{E} \mapsto \mathcal{H}$ is a linear operator from the coefficient Hilbert space $\mathcal{E}$ into $\mathcal{H}$. The notion of two such row-isometric scattering systems being *unitarily equivalent* parallels that for the unitary equivalence of two d-variable unitary scattering systems (see Remark 2.12).

Theorem 3.7.

1. *Let* $W = [W_{v,v'}]_{v,v' \in \mathcal{F}_d}$ *be a positive semidefinite noncommutative Toeplitz weight (so* $W_{v,v'}$ *has the form* (3.12) *for a one-parameter family* $v \mapsto \widehat{W}_v$ *of operators in* $\mathcal{L}(\mathcal{E})$*). Define operators* $\mathcal{S}_W = (\mathcal{S}_{W,1}, \dots, \mathcal{S}_{W,d})$ *as in* (3.14) *and define an operator* $V_W\colon \mathcal{E} \mapsto \mathcal{L}_W$ *by*

$$V_W\colon e \mapsto We.$$

Then the collection $\mathfrak{S}_W$ *given by*

$$\mathfrak{S}_W = (\mathcal{S}_W = (\mathcal{S}_{W,1}, \dots, \mathcal{S}_{W,d}), V_W; \mathcal{L}_W, \mathcal{E})$$

is a row-isometric scattering system. The adjoint $\mathcal{S}_{W,j}^*$ *of* $\mathcal{S}_{W,j}$ *is the restriction of* $(S_j^{R+})^{[*]}$ *to* $\mathcal{L}_W$*:*

$$\mathcal{S}_{W,j}^* f = (S_j^{R+})^{[*]} f \text{ for } f \in \mathcal{L}_W.$$

Moreover, we recover W via

$$W_{v,v'} = V_W^* \mathcal{S}_W^{v*} \mathcal{S}_W^{v^\top} V. \tag{3.15}$$

2. *Conversely, if*

$$\mathfrak{S} = (\mathcal{S} = (\mathcal{S}_1, \ldots, \mathcal{S}_d), V; \mathcal{H}, \mathcal{E})$$

is any row-isometric scattering system, define $W = [W_{v,v'}]_{v,v' \in \mathcal{F}_d}$ *by*

$$W_{v,v'} = V^* \mathcal{S}^{*v} \mathcal{S}^{v'^\top} V.$$

Then W is a positive semidefinite noncommutative Toeplitz weight, and the map $\Phi \colon \mathcal{H} \mapsto \mathcal{L}_W$ *given by*

$$\Phi \colon h \mapsto \Phi(z)h \text{ where } \Phi(z) = \sum_{v \in \mathcal{F}_d} (V^* \mathcal{S}^{*v} h) z^v$$

is a coisometry from $\mathcal{H}$ *onto* $\mathcal{L}_W$ *(with initial space equal to the smallest subspace invariant for* $\mathcal{S}$ *containing* $\operatorname{im} V$*) which intertwines* $\mathcal{S}$ *with* $\mathcal{S}_W$ *and V with* V_W*:*

$$\mathcal{S}_{W,j} \Phi = \Phi \mathcal{S}_j \text{ for } j = 1, \ldots, d; \qquad V_W = \Phi V.$$

Hence the abstract row-isometric scattering system $\mathfrak{S}$ *is unitarily equivalent to the model row-isometric scattering system* $\mathfrak{S}_W$ *with* $W = [W_{v,v'}]_{v,v' \in \mathcal{F}_d}$ *as in* (3.15) *in case* $\operatorname{im} V$ *is a cyclic subspace for* $\mathcal{S}$*.*

Proof. As most points were covered in the discussion preceding the theorem, here we only verify the formula for $\mathcal{S}_{W,j}^*$. Let $W = [W_{v,v'}]_{v,v' \in \mathcal{F}_d}$ be a positive semidefinite noncommutative Toeplitz weight. From the identity (3.15) we may view W as of the form $W_{v,v'} = \Phi_v \Phi_{v'}^*$ where $\Phi_v = V^* \mathcal{S}^{*v} \colon \mathcal{H} \mapsto \mathcal{E}$ with $\mathcal{S} = (\mathcal{S}_1, \ldots, \mathcal{S}_d)$ equal to a row isometry on a Hilbert space $\mathcal{H}$ and where V is a scale operator $V \colon \mathcal{E} \mapsto \mathcal{H}$. Then elements $f(z)$ of $\mathcal{L}_W$ are characterized as those elements $f \in \mathcal{E}[[z]]$ which have a representation

$$f(z) = \Phi h = \sum_{v \in \mathcal{F}_d} (V^* \mathcal{S}^{*v} h) z^v$$

for some $h \in \mathcal{H}$. We then compute

$$\begin{aligned}
(S_j^{R+})^{[*]} \Phi h &= (S_j^{R+})^* \left(\sum_{v \in \mathcal{F}_d} (V^* \mathcal{S}^{*v} h) z^v \right) = \sum_{v \in \mathcal{F}_d} (V^* \mathcal{S}_W^{*v} h) z^{v g_j^{-1}} \\
&= \sum_{v' \in \mathcal{F}_d} (V^* \mathcal{S}^{*v' g_j} h) z^{v'} = \sum_{v' \in \mathcal{F}_d} (V^* \mathcal{S}^{*v'} \mathcal{S}_j^* h) z^{v'} \\
&= \Phi(\mathcal{S}_j^* h)
\end{aligned}$$

from which we conclude that $\mathcal{L}_W$ is invariant under the backward shift operators $(S_j^{R+})^{[*]}$ for $j = 1, \ldots, d$. Once this fact is established, from the computation (for any polynomials $p, q \in \mathcal{E}[z]$)

$$\begin{aligned}\langle \mathcal{S}_{W,j} Wp, Wq\rangle_{\mathcal{L}_W} &= \langle W(p(z)z_j), W(q(z))\rangle_{\mathcal{L}_W} = \langle S_j^{R+}p, Wq\rangle_{L^2} \\ &= \langle p, (S_j^{R+})^{[*]}Wq\rangle_{L^2} = \langle Wp, (S_j^{R+})^{[*]}Wq\rangle_{\mathcal{L}_W},\end{aligned}$$

we see that $\mathcal{S}_{W,j}^* = (S_j^{R+})^{[*]}|_{\mathcal{L}_W}$ as asserted. □

We also have a theory of minimal decomposition and generalized orthogonal complements for contractively-included subspaces as follows.

Theorem 3.8. *Theorem* 2.13 *holds verbatim with "positive semidefinite Toeplitz weight" replaced by "positive semidefinite noncommutative Toeplitz weight". Explicitly, the noncommutative version of the converse statement in part* (2) *of Theorem* 2.13 *is as follows: Suppose that W is a positive semidefinite noncommutative Toeplitz weight and that $\mathcal{L}_1$ is a linear submanifold of $\mathcal{L}_W$ such that*

1. *$\mathcal{L}_1$ is a Hilbert space in the norm $\|\cdot\|_1$ and is contractively included in $\mathcal{L}_W$:*

$$\|f\|_{\mathcal{L}_W}^2 \leq \|f\|_1^2 \text{ for all } f \in \mathcal{L}_1.$$

2. *$\mathcal{L}_1$ is invariant under each of the backward shift operators $(S_j^{R+})^{[*]}$*

$$(S_j^{R+})^{[*]}\colon f(z) \mapsto f(z)\cdot z_j^{-1}$$

for $j = 1, \ldots, d$ and the d-tuple of adjoints

$$\mathcal{S}_1 = \left(\left((S_1^{R+})^{[*]}|_{\mathcal{L}_1}\right)^*, \ldots, \left((S_d^{R+})^{[*]}|_{\mathcal{L}_1}\right)^*\right) \tag{3.16}$$

defines a row isometry on $\mathcal{L}_1$.

Then there is a positive semidefinite noncommutative Toeplitz weight $[W_{v,v'}^1]_{v,v' \in \mathcal{F}_d}$ such that $W^2 := [W_{v,v'} - W_{v,v'}^1]_{v,v' \in \mathcal{F}_d}$ is also positive semidefinite for which $(\mathcal{L}_1, \|\cdot\|_1)$ is equal to $\mathcal{L}_{W^1}$ isometrically.

Proof. We only remark that the proof of the last statement relies on the general theory of generalized orthogonal complements for NFRKHSs as presented in Theorem 3.4 combined with the following fact: *given a NFRKHS $\mathcal{L}_W$ generated by a positive semidefinite weight $W^1 = [W_{v,v'}^1]_{v,v' \in \mathcal{F}_d}$, then the weight has the noncommutative Toeplitz structure* (3.12) *if and only if $\mathcal{L}_{W^1}$ is invariant under each of the backward shift operators $(S_j^{R+})^{[*]}$ for $j = 1, \ldots, d$, and the d-tuple of adjoints $\mathcal{S}_1$ as in* (3.16) *forms a row isometry on $\mathcal{L}_W$*. To see this fact, first verify that, under the assumption that $\mathcal{L}_{W^1}$ is invariant under $(S_j^{R+}))^{[*]}$, then the adjoint $\mathcal{S}_{1,j} := \left((S_j^{R+})^{[*]}|_{\mathcal{L}_{W^1}}\right)^*$ is given by $\mathcal{S}_{1,j}\colon W^1(p) \mapsto W^1(S_j^{R+}p)$: indeed, for all $p, q \in \mathcal{E}[z]$ we have

$$\begin{aligned}\langle (S_j^{R+})^{[*]}W^1p, W^1q\rangle_{\mathcal{L}_{W^1}} &= \langle (S_j^{R+})^{[*]}W^1p, q\rangle_{L^2} = \langle W^1p, S_j^{R+}p\rangle_{L^2} \\ &= \langle W^1p, W^1S_j^{R+}q\rangle_{\mathcal{L}_{W^1}}.\end{aligned}$$

Next, note that the row isometry property of $\mathcal{S}_1$ is then equivalent to

$$\begin{aligned}
\delta_{k,j}\langle W^1_{v,v'}e', e\rangle_{\mathcal{E}} &= \delta_{k,j}\langle W^1(z^{v'}e'), z^v e\rangle_{L^2} \\
&= \delta_{k,j}\langle W^1(z^{v'}e'), W^1(z^v e)\rangle_{\mathcal{L}_{W^1}} \\
&= \langle \mathcal{S}_{1,k}W^1(z^{v'}e'), \mathcal{S}_{1,j}W^1(z^v e)\rangle_{\mathcal{L}_{W^1}} \\
&= \langle W^1(z^{v'g_k}e'), W^1(z^{vg_j}e)\rangle_{\mathcal{L}_{W^1}} \\
&= \langle W^1(z^{v'g_k}e), z^{vg_j}e'\rangle_{L^2} \\
&= \langle W^1_{vg_j,v'g_k}e, e'\rangle)\mathcal{E}
\end{aligned}$$

for all $e, e' \in \mathcal{E}$ and $j, k = 1, \ldots, d$. In this way we see that the row isometry property of $\mathcal{S}_1$ is equivalent to the Toeplitz property (3.11) for W^1. This concludes the proof of Theorem 3.8. □

Remark 3.9. The spaces $\mathcal{L}_W$ with W a positive semidefinite noncommutative Toeplitz weight ($W_{v,v'}$ as in (3.12)) can be viewed as noncommutative generalizations of the de Branges-Rovnyak spaces $\mathcal{L}(\varphi)$ associated with a Herglotz function φ on the unit disk. Such spaces $\mathcal{L}(\varphi)$ have reproducing kernel

$$K^{\varphi}_{11}(\xi, \xi') = \frac{1}{2}\frac{\varphi(\xi) + \varphi(\xi')^*}{1 - \xi\overline{\xi}'}$$

and amount to restricting the elements $\begin{bmatrix} f(\xi) \\ g(\xi) \end{bmatrix}$ of a space $\mathcal{E}(\varphi)$ discussed in Section 2.4 to the first component $f(\xi)$. We do not know an analogue of the harmonic extensions and associated positive measure on the unit circle for this noncommutative setting.

3.3. Noncommutative Haplitz weights and models for row unitaries

As announced at the end of Section 3.1, we consider now Hilbert spaces $\mathcal{H}$ whose elements are formal power series

$$f(z, \zeta) = \sum_{v,w \in \mathcal{F}_d} f_{v,w} z^v \zeta^w \tag{3.17}$$

in a collection of two sets of indeterminates $z = (z_1, \ldots, z_d)$ and $\zeta = (\zeta_1, \ldots, \zeta_d)$. Here it is assumed that the z_i's do not commute with each other, the ζ_i's do not commute with each other for $i = 1, \ldots, d$, but it is the case that $z_i\zeta_j = \zeta_j z_i$ for all $i, j = 1, \ldots, d$. We say that a Hilbert space whose elements consist of such power series f with coefficients $f_{v,w}$ in the Hilbert space $\mathcal{E}$ (i.e., $f \in \mathcal{E}[[z, \zeta]]$) is a *noncommutative formal reproducing kernel Hilbert space* (NFRKHS) if the coefficient evaluation map $\Phi_{v',w'}\colon f(z) \mapsto f_{v',w'}$ is a bounded linear operator from $\mathcal{H}$ to $\mathcal{E}$ for each $(v, w) \in \mathcal{F}_d \times \mathcal{F}_d$. All the results of Section 3.1 carry over into this context with the obvious modifications. In particular, any such NFRKHS has a kernel function

$$K(z, \zeta; z', \zeta') \in \mathcal{L}(\mathcal{E})[[z, \zeta; z', \zeta']]$$

(where $z' = (z'_1, \dots, z'_d)$ and $\zeta' = (\zeta'_1, \dots, \zeta'_d)$ is a second pair of noncommuting d-tuples of indeterminates commuting with each other and with the z and ζ indeterminates) such that

$$K(z, \zeta; z', \zeta')e \in \mathcal{H}[[z', \zeta']]$$

for each $e \in \mathcal{E}$ and

$$\langle f(z, \zeta), K(z, \zeta; z', \zeta')e\rangle_{\mathcal{H}\times\mathcal{H}[[z',\zeta']]} = \langle f(z', \zeta'), e\rangle_{\mathcal{E}[[z',\zeta']]\times\mathcal{E}}$$

for all $e \in \mathcal{E}$. As the kernel function K then uniquely determines $\mathcal{H}$, we then write $\mathcal{H} = \mathcal{H}(K)$. As in Theorem 3.1 elements $K(z, \zeta; z', \zeta') \in \mathcal{L}(\mathcal{E})[[z, \zeta; z', \zeta']]$ arising in this way are characterized by the existence of a factorization

$$K(z, \zeta; z', \zeta') = H(z, \zeta)H(z', \zeta')^*$$

for some $H(z, \zeta) \in \mathcal{L}(\mathcal{H}, \mathcal{E})[[z, \zeta]]$ for some auxiliary Hilbert space $\mathcal{H}$, and by the positivity property

$$\sum_{v,w,v',w' \in \mathcal{F}_d} \langle K_{v,w;v',w'} e_{v',w'}, e_{v,w}\rangle_{\mathcal{E}} \geq 0$$

for all finitely supported $\mathcal{E}$-valued functions $(v, w) \mapsto e_{v,w}$ on $\mathcal{F}_d \times \mathcal{F}_d$. Alternatively, if we let $W_{v,w;v',w'}$ to be defined by

$$K(z, \zeta; z', \zeta') = \sum_{v,w,v',w' \in \mathcal{F}_d} W_{v,w;v',w'} z^v \zeta^w z'^{v'^{\top}} \zeta'^{w'^{\top}},$$

then we may view $W = [W_{v,w;v',w'}]_{(v,w),(v',w') \in \mathcal{F}_d \times \mathcal{F}_d}$ as a block matrix with rows and columns indexed by $\mathcal{F}_d \times \mathcal{F}_d$ and define an operator $W \colon \mathcal{E}[z, \zeta] \mapsto \mathcal{E}[[z, \zeta]]$ by

$$W(\sum_{v,w \in \mathcal{F}_d} p_{v,w} z^v \zeta^w) = \sum_{v,w \in \mathcal{F}_d} \left(\sum_{v',w' \in \mathcal{F}_d} W_{v,w;v',w'} p_{v',w'} \right) z^v \zeta^w.$$

Note that the assumption that p is a polynomial ($p \in \mathcal{E}[z, \zeta]$) implies that the inside sum is finite and hence well defined, while the outside sum is a formal sum defining an element of $\mathcal{E}[[z, \zeta]]$. We then say that W *is a positive semidefinite weight* if

$$\langle Wp, p\rangle_{L^2} \geq 0 \text{ for all } p \in \mathcal{E}[z, \zeta].$$

Here the inner product is simply the usual L^2-inner product on the coefficients adapted to give a pairing between $\mathcal{E}[[z, \zeta]]$ and $\mathcal{E}[z, \zeta]$:

$$\langle \sum_{v,w \in \mathcal{F}_d} f_{v,w} z^v, \zeta^w, \sum_{v,w \in \mathcal{F}_d} p_{v,w} z^v \zeta^w\rangle_{L^2} = \sum_{v,w \in \mathcal{F}_d} \langle f_{v,w}, p_{v,w}\rangle_{\mathcal{E}}$$

where the sum is really finite since $p_{v,w} = 0$ for all but finitely many $(v, w) \in \mathcal{F}_d \times \mathcal{F}_d$. We then define the $\mathcal{L}_W$ inner product on $W\mathcal{E}[z, \zeta]$ by

$$\langle Wp, Wq\rangle_{\mathcal{L}_W} = \langle Wp, q\rangle_{L^2}$$

and define $\mathcal{L}_W$ to be the completion of $W\mathcal{E}[z,\zeta]$ with respect to the $\mathcal{L}_W$-norm, identified as a linear submanifold of $\mathcal{E}[[z,\zeta]]$. For $f = \sum_{v,w} f_{v,w} z^v \zeta^w \in \mathcal{L}_W \subset \mathcal{E}[[z,\zeta]]$, we have

$$\langle f_{v',w'}, e\rangle_{\mathcal{E}} = \langle f, W(z^{v'}\zeta^{w'}e)\rangle_{\mathcal{L}_W}$$

and hence we see that $\mathcal{L}_W$ is a (double-variable) NFRKHS with reproducing kernel

$$K(z,\zeta;z',\zeta') = \sum_{v,w,v',w'\in\mathcal{F}_d} W_{v,w;v',w'} z^v \zeta^w z'^{v'^\top} \zeta'^{w'^\top}$$

and we have a complete correspondence between double-variable NFRKHSs represented as $\mathcal{H}(K)$ and $\mathcal{L}_W$ as in the single d-tuple variable case, with K and $W = [W_{v,w;v',w'}]_{(v,w),(v',w')\in\mathcal{F}_d\times\mathcal{F}_d}$ related according to

$$K(z,\zeta;z',\zeta') = \sum_{v,w,v',w'\in\mathcal{F}_d} W_{v,w;v',w'} z^v \zeta^w z'^{v'^\top} \zeta'^{w'^\top}.$$

Here we shall be primarily interested in positive semidefinite weights $W = [W_{v,w;v',w'}]_{v,w,v',w'\in\mathcal{F}_d}$ on $\mathcal{F}_d \times \mathcal{F}_d$ (with values in $\mathcal{L}(\mathcal{E})$ for a Hilbert space $\mathcal{E}$) which have the three general properties

$$W_{\emptyset,w;v'g_j,w'} = W_{\emptyset,wg_j;v',w'} \tag{3.18}$$

$$W_{vg_k,w;v'g_j,v'} = \delta_{k,j} W_{v,w;v',w'} \tag{3.19}$$

$$W_{\emptyset,w;\emptyset,w'} = \sum_{j=1}^{d} W_{\emptyset,wg_j;\emptyset,w'g_j} \tag{3.20}$$

Following [5] and [6], we refer to (3.18) as the *Hankel* property, to (3.19) as the *Toeplitz* property; an operator satisfying these two together we then call a *Haplitz* operator. We refer to (3.20) as the Cuntz property, due to its connection with representations of the Cuntz algebra which we shall explain shortly (see [5]). We refer to a positive semidefinite weight W satisfying all three properties (3.18)–(3.20) simply as a positive semidefinite *Cuntz weight*.

Just as in the noncommutative Toeplitz case discussed in Section 3.2, a Cuntz weight $W = [W_{v,w;v',w'}]_{v,w,v',w'\in\mathcal{F}_d}$ can be recovered from its symbol

$$\widehat{W}(z,\zeta) = \sum_{v,w\in\mathcal{F}_d} \widehat{W}_{v,w} z^v \zeta^w$$

defined by

$$\widehat{W}(z,\zeta)e = W(e) \text{ for } e \in \mathcal{E},$$

or, in other words,

$$\widehat{W}_{v,w} = W_{v,w;\emptyset,\emptyset}.$$

As verified in [5], a Cuntz weight W is completely determined by its symbol $\widehat{W}(z,\zeta) = \sum_{v,w\in\mathcal{F}_d} \widehat{W}_{v,w} z^v \zeta^w$ according to the formula

$$W_{v,w;v',w'} = \begin{cases} \widehat{W}_{(vv'^{-1})w'^\top,w} & \text{if } |v| \geq |v'|, \\ \widehat{W}_{w'^\top,w(v'v^{-1})^\top} & \text{if } |v| \leq |v'|. \end{cases} \tag{3.21}$$

Furthermore, given an element $\widehat{W}(z,\zeta) \in \mathcal{L}(\mathcal{E})[[z,\zeta]]$, one can always use (3.21) to define an operator $W\colon \mathcal{E}[z,\zeta] \mapsto \mathcal{E}[[z,\zeta]]$. This operator W then is a positive semidefinite Cuntz weight if and only if $\widehat{W}(z,\zeta)$ has a factorization of the form

$$\widehat{W}(z,\zeta) = Y(\zeta)Y(z)^* \tag{3.22}$$

for some $Y(\zeta) \in \mathcal{L}(\widetilde{\mathcal{H}},\mathcal{E})[[\zeta]]$ for some auxiliary Hilbert space $\widetilde{\mathcal{H}}$ (i.e., if and only if $(\zeta,\zeta') \mapsto W(\zeta',\zeta)$ is a positive kernel in $\mathcal{L}(\mathcal{E})[[\zeta,\zeta']]$), and in addition the so-called *Cuntz defect* is zero:

$$\widehat{W}(z,\zeta) - \sum_{j=1}^{d} z_j^{-1}\widehat{W}(z,\zeta)\zeta_j^{-1} = 0. \tag{3.23}$$

The Haplitz property (3.18)–(3.19) can also be characterized by the intertwining condition

$$WS_j^R = U_j^R W \text{ on } \mathcal{E}[z,\zeta] \text{ for } j = 1,\dots,d \tag{3.24}$$

where $S_j^R\colon \mathcal{E}[[z,\zeta]] \mapsto \mathcal{E}[[z,\zeta]]$ (a type of right unilateral shift operator) and $U_j^R\colon \mathcal{E}[[z,\zeta]] \mapsto \mathcal{E}[[z,\zeta]]$ (a type of right bilateral shift operator if one views ζ_j roughly as z_j^{-1}) are given by

$$S_j^R\colon f(z,\zeta) \mapsto f(z,\zeta)\cdot z_j \tag{3.25}$$

$$U_j^R\colon f(z,\zeta) \mapsto f(0,\zeta)\cdot \zeta_j^{-1} + f(z,\zeta)\cdot z_j \tag{3.26}$$

where we have set

$$f(0,\zeta) = \sum_{w\in\mathcal{F}_d} f_{\emptyset,w} z^{\emptyset}\zeta^w \text{ if } f(z,\zeta) = \sum_{v,w\in\mathcal{F}_d} f_{v,w} z^v \zeta^w.$$

Then the L^2 formal adjoints of S_j^R and U_j^R are given by

$$S_j^{R[*]}\colon f(z,\zeta) \mapsto f(z,\zeta)\cdot z_j^{-1} \tag{3.27}$$

$$U_j^{R[*]}\colon f(z,\zeta) \mapsto f(0,\zeta)\cdot \zeta_j + f(z,\zeta)\cdot z_j^{-1} \tag{3.28}$$

for $j = 1,\dots,d$, and, under the assumption that $W = W^{[*]}$ is a selfadjoint Haplitz operator, the intertwining relations (3.24) imply the additional intertwining relations

$$WU_j^{R[*]} = S_j^{R[*]}W \text{ on } \mathcal{E}[z,\zeta] \text{ for } j = 1,\dots,d. \tag{3.29}$$

As a Cuntz weight (or, more generally, a Haplitz operator)

$$W = [W_{v,w;v',w'}]_{v,w,v',w'\in\mathcal{F}_d}$$

is uniquely determined by its corresponding symbol, in principle any property of or object associated with a Cuntz weight can be calculated in terms of its symbol, just as in the commutative case in Section 2.2. However, in general the symbol calculus for Haplitz operators gets more complicated than that for the commutative case, as illustrated by the following formula for the kernel function for the NFRKHS $\mathcal{L}_W$ in terms of the symbol $\widehat{W}(z,\zeta)$ of W analogous to formula (2.14) for the commutative case.

Proposition 3.10. *Suppose that* $W = [W_{v,w;v'w'}]_{v,w,v'w'\in\mathcal{F}_d}$ *is a positive semidefinite Haplitz weight with symbol* $\widehat{W}(z,\zeta) = \sum_{v,w} \widehat{W}_{v,w} z^v \zeta^w$ *(where* $\widehat{W}_{v,w} = W_{v,w;\emptyset,\emptyset}$*). Then the associated kernel function*

$$K(z,\zeta;z',\zeta') = \sum_{v,w,v',w'\in\mathcal{F}_d} W_{v,w;v',w'} z^v \zeta^w z'^{v'^\top} \zeta'^{w'^\top}$$

can be given directly in terms of the symbol as

$$K(z,\zeta;z',\zeta') = \left(\widehat{W}(z,\zeta)\left(\sum_{\beta\in\mathcal{F}_d} (z^{-1})^\beta \zeta'^{\beta^\top}\right)\right) \cdot k_{\mathcal{F}_d}(z,z') + \widehat{W}(\zeta',\zeta)\cdot k_{\mathcal{F}_d}(z,z')\cdot \sum_{\gamma\colon \gamma\neq\emptyset} (\zeta^{-1})^{\gamma^\top} z'^{\gamma^\top} \tag{3.30}$$

where $k_{\mathcal{F}_d}(z,z') = \sum_{v\in\mathcal{F}_d} z^v z'^{v^\top}$ *is the noncommutative Szegö kernel.*

Proof. We compute

$$\begin{aligned}
K(z,\zeta;z',\zeta') &= \sum_{v,w,\alpha,\beta\in\mathcal{F}_d} W_{v,w;\alpha,\beta} z^v \zeta^w z'^{\alpha^\top} \zeta'^{\beta^\top} \\
&= \sum_{v,w,\alpha,\beta\colon |v|\geq|\alpha|} \widehat{W}_{(v\alpha^{-1})\beta^\top,w} z^v \zeta^w z'^{\alpha^\top} \zeta'^{\beta^\top} + \sum_{v,w,\alpha,\beta\colon |v|<|\alpha|} \widehat{W}_{\beta^\top,w(\alpha v^{-1})^\top} z^v \zeta^w z'^{\alpha^\top} \zeta'^{\beta^\top} \\
&= \sum_{\gamma,\alpha,\beta,w\in\mathcal{F}_d} \widehat{W}_{\gamma\beta^\top,w} z^\gamma z^\alpha \zeta^w z'^{\alpha^\top} \zeta'^{\beta^\top} + \sum_{\gamma,v,\beta,w\colon \gamma\neq\emptyset} \widehat{W}_{\beta^\top,w\gamma} z^v \zeta^w z'^{v^\top} \zeta'^{\beta^\top} \\
&= \left(\widehat{W}(z,\zeta)\cdot\sum_\beta (z^{-1})^\beta \zeta'^{\beta^\top}\right)\cdot k_{\mathcal{F}_d}(z,z') + \widehat{W}(\zeta',\zeta)\cdot \sum_{\gamma,v\colon \gamma\neq\emptyset} (\zeta^{-1})^{\gamma^\top} z^v z'^{v^\top} z'^{\gamma^\top} \\
&= \left(\widehat{W}(z,\zeta)\cdot\sum_\beta (z^{-1})^\beta \zeta'^{\beta^\top}\right)\cdot k_{\mathcal{F}_d}(z,z') + \widehat{W}(\zeta',\zeta)\cdot k_{\mathcal{F}_d}(z,z')\cdot \sum_{\gamma\colon \gamma\neq\emptyset} (\zeta^{-1})^{\gamma^\top} z'^{\gamma^\top}
\end{aligned}$$

and formula (3.30) follows. □

Just as the spaces $\mathcal{L}_W$ with W a positive semidefinite noncommutative Toeplitz weight are closely connected with the representation of row isometries (see Section 3.2), the spaces $\mathcal{L}_W$ with W a positive semidefinite Cuntz weight are closely connected with the representation of row-unitary operators. Given a d-tuple of operators $\mathcal{U} = (\mathcal{U}_1, \dots, \mathcal{U}_d)$ on a Hilbert space $\mathcal{H}$, we say that $\mathcal{U}$ is *row-unitary* if the block row matrix $[\mathcal{U}_1 \ \ \dots \ \ \mathcal{U}_d] : \mathcal{H}\otimes\mathbb{C}^d \mapsto \mathcal{H}$ is unitary. In terms of the individual operators $\mathcal{U}_1, \dots, \mathcal{U}_d$, this simply means that the operators $\mathcal{U}_j$ are individually isometries with pairwise-orthogonal image spaces $\operatorname{im}\mathcal{U}_j$ having span equal to the whole space $\mathcal{H}$.

Suppose now that $\mathcal{U} = (\mathcal{U}_1, \dots, \mathcal{U}_d)$ is any row-unitary d-tuple on a Hilbert space $\mathcal{H}$ and that we are also given an operator (sometimes called a *scale*) $V\colon \mathcal{E} \mapsto \mathcal{H}$ from some coefficient Hilbert space $\mathcal{E}$ to $\mathcal{H}$. We proceed now to parallel the construction done in Section 3.2 for a row isometry but now for a row-unitary d-tuple. Let $z = (z_1, \dots, z_d)$ and $\zeta = (\zeta_1, \dots, \zeta_d)$ be two sets of noncommuting indeterminates which, however, do commute with each other, and define a Fourier representation operator $\Phi\colon \mathcal{H} \mapsto \mathcal{E}[[z,\zeta]]$ by

$$\Phi\colon h \mapsto \sum_{v,w \in \mathcal{F}_d} (\Phi_{v,w} h) z^v \zeta^w$$

where we have set $\Phi_{v,w}\colon \mathcal{H} \mapsto \mathcal{E}$ equal to

$$\Phi_{v,w}\colon h \mapsto V^* \mathcal{U}^w \mathcal{U}^{*v} h.$$

If we set $\Phi(z,\zeta) = \sum_{v,w \in \mathcal{F}_d} \Phi_{v,w} z^v \zeta^w$, then we see that the operator $\Phi\colon h \mapsto \Phi h$ amounts to multiplication of the vector h by the formal power series $\Phi(z,\zeta)$. Define a norm $\|\cdot\|_0$ on $\operatorname{im}\Phi$ as the lifted norm

$$\|\Phi(z) \cdot h\|_0 = \|Qh\|_{\mathcal{H}}$$

where Q is the orthogonal projection of $\mathcal{H}$ onto the orthogonal complement of the kernel of the map $h \mapsto \Phi(z,\zeta) \cdot h$. Then from the general considerations in Theorem 3.1 we see that $\operatorname{im}\Phi$ is a NFRKHS $\mathcal{H}(K)$ with kernel $K(z,\zeta;z',\zeta')$ given by $K(z,\zeta;z',\zeta') = \Phi(z,\zeta)\Phi(z',\zeta')^*$, or equivalently, $(\operatorname{im}\Phi, \|\cdot\|_0)$ is a space $\mathcal{L}_W$ with the weight $W = [W_{v,w;v',w'}]_{v,w,v',w' \in \mathcal{F}_d}$ given by

$$W_{v,w;v',w'} = \Phi_{v,w} \Phi^*_{v',w'} = V^* \mathcal{U}^w \mathcal{U}^{*v} \mathcal{U}^{v'^\top} \mathcal{U}^{*w'^\top} V.$$

The row-unitary property of $\mathcal{U}$ is equivalent to W being a Cuntz weight. Moreover, one can verify directly that

$$\Phi \mathcal{U}_j = \mathcal{U}_{W,j} \Phi \text{ for } j = 1, \dots, d$$

where we have set

$$\mathcal{U}_{W,j}\colon Wp \mapsto W S_j^R p = U_j^R W p \text{ for } p \in \mathcal{E}[z,\zeta] \text{ for } j = 1, \dots, d \tag{3.31}$$

with

$$\Phi \mathcal{U}_j^* = \mathcal{U}_{W,j}^* \Phi \text{ for } j = 1, \dots, d$$

where

$$\mathcal{U}_{W,j}^*\colon Wp \mapsto W U_j^{R[*]} p = S_j^{R[*]} W p \text{ for } p \in \mathcal{E}[z,\zeta]. \tag{3.32}$$

and where S_j^R and U_j^R are the generalized shift operators given by (3.25) and (3.26) with L^2-adjoints given by (3.27) and (3.28).

Conversely, if one starts with a positive semidefinite Cuntz weight W, one can define a d-tuple of operators $\mathcal{U}_W = (\mathcal{U}_{W,1}, \dots, \mathcal{U}_{W,d})$ on $\mathcal{L}_W$ by (3.31). By using the properties (3.18)–(3.20) of the Cuntz weight W, one can check directly that $\mathcal{U}_W$ is row-unitary on $\mathcal{L}_W$ with adjoints $\mathcal{U}_j^*$ given by (3.32) for $j = 1, \dots, d$. Furthermore, there is a canonical scale operator $V_W\colon \mathcal{E} \mapsto \mathcal{L}_W$:

$$V_W\colon e \mapsto We. \tag{3.33}$$

In general let us say that the collection

$$\mathfrak{S} = (\mathcal{U} = (\mathcal{U}_1, \dots, \mathcal{U}_d), V; \mathcal{H}, \mathcal{E})$$

is a *row-unitary scattering system* if $\mathcal{U}$ is a row-unitary d-tuple of operators on the Hilbert space $\mathcal{H}$ and V (the *scale* operator) is a bounded linear operator from the coefficient Hilbert space $\mathcal{E}$ into $\mathcal{H}$. The notion of when two such row-unitary scattering systems are *unitarily equivalent* is exactly parallel to that for the case of commuting d-variable unitary scattering systems (see Remark 2.12). We have the following analogue of Theorem 3.7.

Theorem 3.11.

1. *Let* $W = [W_{v,w;v',w'}]_{v,v',w,w' \in \mathcal{F}_d}$ *be a positive semidefinite Cuntz weight (so* W *is positive semidefinite and is determined by its symbol* $\widehat{W}(z,\zeta)$ *which satisfies* (3.22) *and* (3.23)*). Define operators* $\mathcal{U}_W = (\mathcal{U}_{W,1}, \dots, \mathcal{U}_{W,d})$ *as in* (3.31) *and define an operator* $V_W : \mathcal{E} \mapsto \mathcal{L}_W$ *by* (3.33). *Then the collection* $\mathfrak{S}_W$ *given by*
$$\mathfrak{S}_W = (\mathcal{U}_W = (\mathcal{U}_{W,1}, \dots, \mathcal{U}_{W,d}), V_W; \mathcal{L}_W, \mathcal{E})$$
is a row-unitary scattering system. The adjoint $\mathcal{U}_{W,j}^*$ *of* $\mathcal{U}_{W,j}$ *on* $\mathcal{L}_W$ *is given by* (3.32) *for* $j = 1, \dots, d$. *Moreover, we recover* W *via*
$$W_{v,w;v',w'} = V_W^* \mathcal{U}_W^w \mathcal{U}_W^{v*} \mathcal{U}_W^{v'^{\top}} \mathcal{U}_W^{*w'^{\top}} V.$$
2. *Conversely, if*
$$\mathfrak{S} = (\mathcal{U} = (\mathcal{U}_1, \dots, \mathcal{U}_d), V; \mathcal{H}, \mathcal{E})$$
is any row-unitary scattering system, define $W = [W_{v,w;v',w'}]_{v,w;v',w' \in \mathcal{F}_d}$ *by*
$$W_{v,w;v',w'} = V^* \mathcal{U}^w \mathcal{U}^{*v} \mathcal{U}^{v'^{\top}} \mathcal{U}^{*w'^{\top}} V. \tag{3.34}$$
Then W *is a positive semidefinite Cuntz weight and the map* $\Phi \colon \mathcal{H} \mapsto \mathcal{L}_W$ *given by*
$$\Phi \colon h \mapsto \Phi(z,\zeta)h \text{ where } \Phi(z,\zeta) = \sum_{v,w \in \mathcal{F}_d} (V^* \mathcal{U}^w \mathcal{U}^{*v} h) z^v \zeta^w$$
is a coisometry from $\mathcal{H}$ *onto* $\mathcal{L}_W$ *(with initial space equal to the smallest subspace reducing for* $\mathcal{U}$ *containing* $\operatorname{im} V$*) which intertwines* $\mathcal{U}$ *with* $\mathcal{U}_W$ *and* V *with* V_W*:*
$$\mathcal{U}_{W,j}\Phi = \Phi\mathcal{U}_j \text{ for } j = 1, \dots, d; \qquad V_W = \Phi V.$$
Hence the abstract row-unitary scattering system $\mathfrak{S}$ *is unitarily equivalent to the model row-unitary scattering system* $\mathfrak{S}_W$ *with* $W = [W_{v,w;v',w'}]_{v,w,v',w' \in \mathcal{F}_d}$ *as in* (3.34) *in case* $\operatorname{im} V$ *is a* $*$*-cyclic subspace for* $\mathcal{U}$.

Proof. This result is covered in detail in [5]. □

Minimal decompositions and generalized orthogonal complements for the space $\mathcal{L}_W$ work out as follows; this is the Cuntz analogue of Theorem 3.8.

Theorem 3.12. *Theorem* 2.13 *holds verbatim with "positive semidefinite Toeplitz weight" replaced by "positive semidefinite Cuntz weight". Explicitly, the Cuntz version of the converse statement in part* (2) *of Theorem* 2.13 *is as follows: Suppose that* W *is a positive semidefinite Cuntz weight and that* $\mathcal{L}_1$ *is a linear submanifold of* $\mathcal{L}_W$ *such that*

1. *$\mathcal{L}_1$ is a Hilbert space in the norm $\|\cdot\|_1$ and is contractively included in $\mathcal{L}_W$:*
$$\|f\|^2_{\mathcal{L}_W} \leq \|f\|^2_1 \text{ for all } f \in \mathcal{L}_1.$$

2. *$\mathcal{L}_1$ is invariant under each of the shift operators U_j^R and $S_j^{R[*]}$ with*
$$(U_j^R|_{\mathcal{L}_1})^* = S_j^{R[*]}|_{\mathcal{L}_1}$$
for $j = 1, \dots, d$ and the d-tuple
$$\mathcal{U}_1 = (U_1^R|_{\mathcal{L}_1}, \dots, U_d^R|_{\mathcal{L}_1})$$
is row-unitary on $\mathcal{L}_1$.

Then there is a positive semidefinite Cuntz weight $[W^1_{v,w;v',w'}]_{v,w,v',w' \in \mathcal{F}_d}$ such that $W^2 := [W_{v,w;v',w'} - W^1_{v,w;v';w'}]_{v,w,v',w' \in \mathcal{F}_d}$ is also positive semidefinite for which $(\mathcal{L}_1, \|\cdot\|_1)$ is equal to $\mathcal{L}_{W^1}$ isometrically.

Proof. The proof relies on the general theory of NFRKHSs together with one additional fact *if $\mathcal{L}_{W^1}$ is a NFRKHS (in the variable (z,ζ)) which is invariant under U_j^R and $S_j^{R[*]}$ with $S_j^{[*]} = (U_j^R|_{\mathcal{L}_1})^*$ for each $j = 1, \dots, d$, then the weight W is necessarily Haplitz, i.e., has properties* (3.18) *and* (3.19). To see this, note that the assumed invariance and adjoint properties give, on the one hand,
$$\langle U_j^R W p, W q\rangle_{\mathcal{L}_{W^1}} = \langle U_j^R W p, q\rangle_{L^2} = \langle p, W U_j^{R[*]} q\rangle_{L^2} \tag{3.35}$$
while on the other hand we have
$$\langle U_j^R W p, W q\rangle_{\mathcal{L}_{W^1}} = \langle W p, S_j^{R[*]} W q\rangle_{\mathcal{L}_{W^1}} = \langle p, S_j^{R[*]} W q\rangle_{L^2} \tag{3.36}$$
for all $p, q \in \mathcal{E}[z, \zeta]$. Equating (3.35) with (3.36) leads to the general intertwining relation $W U_j^{R[*]} = S_j^{R[*]} W$ (and hence also to $U_j^R W = W S_j^R$ by the L^2-inner-product selfadjointness of W) for all $j = 1, \dots, d$. By the equivalence between the intertwining conditions (3.24) (or (3.29)) with the Haplitz properties (3.18)–(3.19), we conclude that W^1 is Haplitz. This is enough to force the d-tuple $\mathcal{U}_W$ to be row-isometric (see [5]). Then the row-unitary property for $\mathcal{U}_{W^1}$ is equivalent to W^1 having the additional property (3.20), i.e., to W^1 being a Cuntz weight. This concludes the proof of Theorem 3.12. □

Remark 3.13. In [5, Section 4] decompositions of various classes of free-atomic representations of the Cuntz algebra $\mathcal{O}_d$ from [15] (and of the closely related permutative representations from [13]) were derived in terms of the associated model spaces $\mathcal{L}_W$. Although the NFRKHS point of view was not given there explicitly, this analysis can be seen as an application of the minimal decomposition theory in Theorem 3.12.

We have noted that the symbol $\widehat{W}(z,\zeta)$ of a positive semidefinite Cuntz weight W has a factorization $\widehat{W}(z,\zeta) = Y(\zeta)Y(z)^*$ and hence can be used as a reproducing kernel $K'(\zeta,\zeta') = \widehat{W}(\zeta',\zeta)$ to define a NFRKHS with elements equal to power series $f(\zeta) \in \mathcal{E}[[\zeta]]$ in the variables $\zeta = (\zeta_1, \dots, \zeta_d)$. Let us abuse notation somewhat and write $\mathcal{H}(\widehat{W})$ for this space $\mathcal{H}(K')$ (of power series in the variables ζ). The following proposition identifies the associated space $\mathcal{H}(K')$ as the "pure past" of $\mathcal{L}_W$.

Proposition 3.14. *Let W be a positive semidefinite Cuntz weight with symbol $\widehat{W}(z,\zeta)$. Then the NFRKHS $\mathcal{H}(\widehat{W})$ with reproducing kernel $\widehat{W}(\zeta',\zeta)$ (so elements of $\mathcal{H}(\widehat{W})$ are formal power series in ζ) can be identified with the "pure past" of $\mathcal{L}_W$. More precisely, the map $f(z,\zeta) \to f(0,\zeta) := \sum_{w \in \mathcal{F}_d} f_{\emptyset,w}\zeta^w$ is a coisometry from $\mathcal{L}_W$ onto $\mathcal{H}(\widehat{W})$ with cokernel equal to the closed span of the set $\{\mathcal{U}_W^{*w}We \colon w \in \mathcal{F}_d,\ e \in \mathcal{E}\}$.*

Proof. Using Theorem 3.11 we can write $W_{v,w;v',w'} = \Phi_{v,w}\Phi^*_{v',w'}$ where $\Phi_{v,w} = V^*\mathcal{U}^w\mathcal{U}^{*v}$ for some row-unitary scattering system $(\mathcal{U} = (\mathcal{U}_1,\dots,\mathcal{U}_d), V, \mathcal{H}, \mathcal{E})$. Since $\widehat{W}_{v,w} = W_{v,w,\emptyset,\emptyset}$, we get a factorization $\widehat{W}(z,\zeta) = Y(\zeta)Y(z)^*$ where $Y_w = V^*\mathcal{U}^w = \Phi_{\emptyset,w}$.

Next, by general principles we have coisometries $\varphi_1 \colon \mathcal{H} \to \mathcal{L}_W$ and $\varphi_2 \colon \mathcal{H} \to \mathcal{H}(\widehat{W})$ given by

$$\varphi_1 \colon h \to \sum_{v,w \in \mathcal{F}_d} (\Phi_{v,w}h) z^v \zeta^w,$$

$$\varphi_2 \colon h \to \sum_{w \in \mathcal{F}_d} (Y_w h)\zeta^w$$

with

$$\operatorname{coker}\varphi_1 = \underset{v,w \in \mathcal{F}_d}{\text{closed span}} \operatorname{im} \Phi^*_{v,w}|_{\mathcal{E}},$$

$$\operatorname{coker}\varphi_2 = \underset{w \in \mathcal{F}_d}{\text{closed span}} \operatorname{im} Y^*_w|_{\mathcal{E}}.$$

Then a generic element $f(z,\zeta)$ of $\mathcal{L}_W$ can be written as

$$f(z,\zeta) = \sum_{v,w \in \mathcal{F}_d} (\Phi_{v,w}\ell) z^v \zeta^w$$

where ℓ is a generic element of $\operatorname{coker}\varphi_1$. In this notation, $\varphi_1^* \colon f(z,\zeta) \to \ell$ and $\varphi_2 \colon \ell \to \sum_{w \in \mathcal{F}_d}(Y_w\ell)\zeta^w = \sum_{w \in \mathcal{F}_d}(\Phi_{\emptyset,w}\ell)\zeta^w = \sum_{w \in \mathcal{F}_d} f_{\emptyset,w}\zeta^w := f(0,\zeta)$. Thus we see that

$$\varphi_2\varphi_1^* \colon f(z,\zeta) \to f(0,\zeta).$$

From this formula we see that the kernel of $\varphi_2\varphi_1^*$ consists of $f \in \mathcal{L}_W$ for which $f_{\emptyset,w} = 0$ for all $w \in \mathcal{F}_d$. From the reproducing kernel property

$$\langle f, W(z^\emptyset \zeta^w e)\rangle_{\mathcal{L}_W} = \langle f_{\emptyset,w}, e\rangle_{\mathcal{E}}$$

we see that the cokernel of $\varphi_2\varphi_1^*$ is equal to the closed span of $\{W(z^\emptyset\zeta^w e) \colon w \in \mathcal{F}_d \text{ and } e \in \mathcal{E}\}$. From the observation

$$\mathcal{U}_W^{*w}(We) = W U^{R[*]w} e = W(z^\emptyset \zeta^{w^\top} e)$$

we see that this latter set can also be identified as the closure of $\{\mathcal{U}_W^{*w}We \colon w \in \mathcal{F}_d,\ e \in \mathcal{E}\}$, and the proposition follows. ∎

3.4. NFRKHSs associated with the noncommutative Schur class

Let $\mathcal{E}$ and $\mathcal{E}_*$ be two Hilbert spaces. We define the *Schur-class* $\mathcal{S}(\mathcal{E},\mathcal{E}_*)$ of functions to consist of analytic operator-valued functions $T\colon \mathbb{D} \mapsto \mathcal{L}(\mathcal{E},\mathcal{E}_*)$ with contractive values on $\mathbb{D}$: $\|T(z)\|_{\mathcal{L}(\mathcal{E},\mathcal{E}_*)} \le 1$ for all $z \in \mathbb{D}$. For any Schur-class function T, the associated kernel

$$K^S(\xi,\xi') = \frac{I - T(\xi)T(\xi')^*}{1-\xi\overline{\xi}'}$$

is a positive kernel over $\mathbb{D}$ with associated reproducing kernel Hilbert space often denoted as $\mathcal{H}(T)$, and has proven to be a major source of inspiration for study and for applications in a number of areas – see, e.g., [1]. A noncommutative generalization of the Schur class has been introduced more recently and a whole noncommutative analogue of the function theory on the unit disk is emerging – see, e.g., [19, 20, 21, 22, 15, 16, 5, 6]. Our purpose here is to make connections of these emerging ideas with the theory of NFRKHSs which we have introduced here.

In this section, we again assume that $z = (z_1,\dots,z_d)$ is a d-tuple of noncommuting indeterminates. We let $L^2(\mathcal{F}_d,\mathcal{E})$ denote the NFRKHS of formal power series

$$f(z) = \sum_{v\in\mathcal{F}_d} f_v z^v \in \mathcal{E}[[z]]$$

for which

$$\|f\|^2_{L^2(\mathcal{F}_d,\mathcal{E})} := \sum_{v\in\mathcal{F}_d} \|f_v\|^2_{\mathcal{E}} < \infty.$$

Now let $T(z) = \sum_{v\in\mathcal{F}_d} T_v z^v$ be an element of $\mathcal{L}(\mathcal{E},\mathcal{E}_*)[[z]]$ where $\mathcal{E}_*$ is another Hilbert space. Then for any $f \in \mathcal{E}[[z]]$, the multiplication of formal power series

$$T(z)\cdot f(z) = \sum_{v\in\mathcal{F}_d}\left(\sum_{v',v''\colon v'v''=v} T_{v'}f_{v''}\right) z^v$$

is well defined and yields an element $T(z)\cdot f(z)$ of $\mathcal{E}_*[[z]]$. We say that T is in the d-variable noncommutative Schur-class $\mathcal{S}_{nc,d}(\mathcal{E},\mathcal{E}_*)$ if $S(z)\cdot f(z) \in L^2(\mathcal{F}_d,\mathcal{E}_*)$ for all $f \in L^2(\mathcal{F}_d,\mathcal{E})$ with contraction of norms:

$$\|T(z)\cdot f(z)\|_{L^2} \le \|f\|_{L^2} \text{ for all } f \in L^2(\mathcal{F}_d,\mathcal{E}).$$

In this case there is a noncommutative analogue of the reproducing kernel Hilbert space $\mathcal{H}(T)$ described above.

Theorem 3.15. *Suppose that the formal power series $T(z) = \sum_{v\in\mathcal{F}_d} T_v z^v$ is in the noncommutative Schur class $\mathcal{S}_{nc,d}(\mathcal{E},\mathcal{E}_*)$. Then*

$$K^T(z,z') = (k_{\mathcal{F}_d}(z,z')I_{\mathcal{E}_*}) - T(z)\cdot k_{\mathcal{F}_d}(z,z')\cdot T(z')^* \in \mathcal{L}(\mathcal{E}_*)[[z,z']] \tag{3.37}$$

(where $k_{\mathcal{F}_d}(z,z') = \sum_{v\in\mathcal{F}_d} z^v z'^{v^\top}$ is the noncommutative Szegö kernel) is a positive kernel, and hence is the reproducing kernel for a NFRKHS $\mathcal{H}(K^T)$.

Proof. Define a submanifold $\mathcal{M}(T)$ of $L^2(\mathcal{F}_d, \mathcal{E}_*)$ by

$$\mathcal{M}(T) = T \cdot L^2(\mathcal{F}_d, \mathcal{E})$$

with norm

$$\|T \cdot f\|_{\mathcal{M}(T)} = \|Qf\|_{L^2}$$

where Q is the orthogonal projection of $L^2(\mathcal{F}_d, \mathcal{E})$ onto the orthogonal complement of the kernel of the map $f(z) \mapsto T(z) \cdot f(z)$. By Theorem 3.5 $\mathcal{M}(T)$ is a NFRKHS with reproducing kernel

$$K^{\mathcal{M}(T)}(z, z') = T(z)(k_{\mathcal{F}_d}(z, z')I_{\mathcal{E}})T(z')^*.$$

By the assumption that T is in the Schur class $\mathcal{S}_{nc,d}(\mathcal{E}, \mathcal{E}_*)$, we see that $\mathcal{M}(T)$ is contractively included in $L^2(\mathcal{F}_d, \mathcal{E}_*)$. By the theory of minimal decompositions and contractive inclusions adapted to this noncommutative situation as summarized in Theorem 3.4 (see [2, Theorem II p. 355] for the classical case), it follows that $K^T(z, z') = k_{\mathcal{F}_d}(z, z')I_{\mathcal{E}_*} - T(z)(k(z, z')I_{\mathcal{E}})T(z')^*$ is a positive kernel as asserted. □

One way that Schur-class formal power series $T \in \mathcal{S}_{nc,d}$ can arise is as the transfer function of a d-variable Cuntz unitary system or colligation. By a *d-variable Cuntz unitary colligation* we mean a unitary operator U of the form

$$U = \begin{bmatrix} A & B \\ C & D \end{bmatrix} = \begin{bmatrix} A & B_1 \\ \vdots & \vdots \\ A_d & B_d \\ C & D \end{bmatrix} : \begin{bmatrix} \mathcal{H} \\ \mathcal{E} \end{bmatrix} \mapsto \begin{bmatrix} \mathcal{H} \\ \vdots \\ \mathcal{H} \\ \mathcal{E}_* \end{bmatrix} \tag{3.38}$$

for some Hilbert spaces $\mathcal{H}$ (the *state space*), $\mathcal{E}$ (the *input space*) and $\mathcal{E}_*$ (the *output space*). The associated system has trajectories of the form (u, x, y) where u, x, y are vector-valued functions defined on the free semigroup $\mathcal{F}_d$:

$$\begin{aligned} u &: \mathcal{F}_d \mapsto \mathcal{E}, \\ x &: \mathcal{F}_d \mapsto \mathcal{H}, \\ y &: \mathcal{F}_d \mapsto \mathcal{E}_*. \end{aligned}$$

System trajectories are defined to be all solutions of the system equations associated with the colligation U:

$$\Sigma(U) \qquad \begin{cases} x(g_j v) &= A_j x(v) + B_j u(v) \text{ for } j = 1, \dots, d \\ y(v) &= Cx(v) + Du(v). \end{cases} \tag{3.39}$$

If we apply noncommutative Z-transform to the system equations

$$\{h(v)\}_{v \in \mathcal{F}_d} \mapsto \widehat{h}(z) = \sum_{v \in \mathcal{F}_d} h(v) z^v,$$

then the first of the system equations (3.39) transforms to

$$\sum_{v \in \mathcal{F}_d} x(g_j v) z^v = A_j \widehat{x}(z) + B_j \widehat{u}(z).$$

Multiplication on the left by z_j and then summing over $j = 1, \dots, d$ then gives

$$\widehat{x}(z) - x(\emptyset) = Z(z)A\widehat{x}(z) + Z(z)B\widehat{u}(z)$$

where we have set

$$Z(z) = \begin{bmatrix} z_1 I_{\mathcal{H}} & \cdots & z_d I_{\mathcal{H}} \end{bmatrix} : \mathcal{H} \otimes \mathbb{C}^d \mapsto \mathcal{H}.$$

We may then solve for $\widehat{x}(z)$ to get

$$\widehat{x}(z) = (I - Z(z)A)^{-1}x(\emptyset) + (I - Z(z)A)^{-1}Z(z)B\widehat{u}(z).$$

Substituting this expression into the Z-transform of the last of the system equations (3.39) then yields

$$\widehat{y}(z) = C(I - Z(z)A)^{-1}x(\emptyset) + T_{\Sigma(U)}(z) \cdot \widehat{u}(z) \tag{3.40}$$

where we have set $T_{\Sigma(U)}(z)$ equal to the *transfer function* of the system $\Sigma(U)$:

$$\begin{aligned} T_{\Sigma(U)}(z) &= D + C(I - Z(z)A)^{-1}Z(z)B \\ &= D + \sum_{j=1}^{d} \sum_{v \in \mathcal{F}_d} CA^v B_j \end{aligned} \tag{3.41}$$

and where we have set

$$A^v = A_{i_n} \cdots A_{i_1} \text{ if } v = g_{i_n} \cdots g_{i_1}, \qquad A^{\emptyset} = I_{\mathcal{H}}.$$

In particular, if we assume that we have zero initial condition ($x(\emptyset) = 0$), then we get simply that the Z-transform $\widehat{y}(z)$ of the output signal $\{y_v\}_{v \in \mathcal{F}_d}$ is equal to the transfer function times the Z-transform $\widehat{u}(z)$ of the input signal $\{u_v\}_{v \in \mathcal{F}_d}$:

$$\widehat{y}(z) = T_{\Sigma(U)}(z)\widehat{u}(z) \text{ if } x(\emptyset) = 0. \tag{3.42}$$

We also know from the assumed unitary property of U that

$$\left(\sum_{j=1}^{d} \|x(g_j v)\|^2_{\mathcal{H}} \right) - \|x(v)\|^2_{\mathcal{H}} = \|u(v)\|^2_{\mathcal{E}} - \|y(v)\|^2_{\mathcal{E}_*} \text{ for all } v \in \mathcal{F}_d. \tag{3.43}$$

If we assume that $x(\emptyset) = 0$ and sum up over all v with $|v| \le N$, the left-hand side of (3.43) telescopes and we are left with

$$0 \le \sum_{v \in \mathcal{F}_d :\ |v| = N+1} \|x(v)\|^2_{\mathcal{H}} = \sum_{v \in \mathcal{F}_d :\ |v| \le N} \left[\|u(v)\|^2_{\mathcal{E}} - \|y(v)\|^2_{\mathcal{E}_*} \right]$$

for all $N = 1, 2, \dots$. Under the assumption that $x(\emptyset) = 0$, it follows that $\widehat{y} \in L^2(\mathcal{F}_d, \mathcal{E}_*)$ whenever $\widehat{u} \in L^2(\mathcal{F}_d, \mathcal{E})$ with

$$\|T_{\Sigma(U)} \cdot \widehat{u}\|^2_{L^2(\mathcal{F}_d, \mathcal{E}_*)} = \|\widehat{y}\|^2_{L^2(\mathcal{F}_d, \mathcal{E}_*)} \le \|\widehat{u}\|^2_{L^2(\mathcal{F}_d, \mathcal{E})}$$

and we conclude that $T_{\Sigma(U)}(z)$ is in the noncommutative Schur class $\mathcal{S}_{nc,d}(\mathcal{E}, \mathcal{E}_*)$. Furthermore, when $T(z) \in \mathcal{S}_{nc,d}(\mathcal{E}, \mathcal{E}_*)$ is realized in the explicit form $T(z) =$

$T_{\Sigma(U)}(z)$, then the positivity of the kernel $K^T(z,z')$ (3.37) appearing in Theorem 3.15 is exhibited explicitly from the formula

$$\begin{aligned} k_{\mathcal{F}_d}(z,z')I_{\mathcal{E}_*} - T(z)(k_{\mathcal{F}_d}(z,z')I_{\mathcal{E}})T(z')^* &= C(I-Z(z)A)^{-1}(I-Z(z')^*A^*)^{-1}C^* \\ \text{if } T(z) = D + C(I-Z(z)A)^{-1}Z(z)B & \qquad (3.44)\end{aligned}$$

(see [6]).

The *realization question* is: *given $T \in \mathcal{S}_{nc,d}(\mathcal{E},\mathcal{E}_*)$, can one find a Cuntz unitary colligation U as in* (3.38) *so that T has the form $T(z) = T_{\Sigma(U)}(z)$ as in* (3.41)*?* The answer is now known to be affirmative – see [20] for a particular case and [6] for the general case. Here we present a new proof based on noncommutative reproducing kernel ideas which extends the "lurking isometry" approach for interpolation and realization problems much used of late for the multivariable, commutative setting (see [4]) to the noncommutative setting.

Theorem 3.16. *Suppose that $T = \sum_{v\in\mathcal{F}_d} T_v z^v$ is in the noncommutative Schur class $\mathcal{S}_{nc,d}(\mathcal{E},\mathcal{E}_*)$. Then there is a Cuntz unitary colligation U (3.38) so that $T(z) = T_{\Sigma(U)}(z)$, i.e., so that*

$$T_\emptyset = D, \qquad T_{vg_j} = CA^vB_j \text{ for all } v \in \mathcal{F}_d.$$

Proof. As already observed in Theorem 3.15, from the fact that $T \in \mathcal{S}_{nc,d}(\mathcal{E},\mathcal{E}_*)$ is in the Schur class $\mathcal{S}_{nc,d}(\mathcal{E},\mathcal{E}_*)$ we deduce that the kernel $K^T(z,z')$ (3.37) is a positive kernel. Hence, by Theorem 3.1 we have a factorization

$$K^T(z,z') = (k_{\mathcal{F}_d}(z,z')I_{\mathcal{E}_*}) - T(z)(k_{\mathcal{F}_d}(z,z')I_{\mathcal{E}})T(z')^* = H(z)H(z')^* \qquad (3.45)$$

for some $H \in \mathcal{L}(\mathcal{H}_0,\mathcal{E}_*)$ and some auxiliary Hilbert space $\mathcal{H}_0$. Multiply (3.45) on the left by z'_j and on the right by z_j to deduce that

$$\sum_{j=1}^d \left[(z'_j k_{\mathcal{F}_d}(z,z')z_j)I_{\mathcal{E}_*} - T(z)(z'_j k_{\mathcal{F}_d}(z,z')z_j I_{\mathcal{E}})T(z')^*\right] = \sum_{j=1}^d H(z)(z'_j z_j I_{\mathcal{H}})H(z')^*. \qquad (3.46)$$

Subtract (3.46) from (3.45) and use the easily verified identity

$$k_{\mathcal{F}_d}(z,z') - \sum_{j=1}^d z'_j k_{\mathcal{F}_d}(z,z')z_j = 1$$

to deduce that

$$I_{\mathcal{E}_*} - T(z)T(z')^* = H(z)H(z')^* - \sum_{j=1}^d H(z)z_j \cdot z'_j H(z')^*. \qquad (3.47)$$

Rearrange (3.47) in the form

$$\sum_{j=1}^d H(z)z_j \cdot z'_j H(z')^* + I_{\mathcal{E}_*} = H(z)H(z')^* + T(z)T(z')^*. \qquad (3.48)$$

This is an identity between elements of $\mathcal{E}_*[[z,z']]$. In general, an element $f \in \mathcal{E}_*[[z]]$ has the form $f(z) = \sum_{v\in\mathcal{F}_d} f_v z^v$ for coefficients $f_v \in \mathcal{E}_*$; for complicated expressions we shall use the notation $[f(z)]_v$ to denote the power-series coefficient f_v. Equating coefficients of $z^v z'^{v'^\top}$ in (3.47) yields

$$\sum_{j=1}^{d} H_{vg_j^{-1}} H^*_{v'g_j^{-1}} + \delta_{v,\emptyset}\cdot\delta_{v'\emptyset} I_{\mathcal{E}_*} = H_v H^*_{v'} + T_v T^*_{v'}. \tag{3.49}$$

It follows that the operator V given by

$$V\colon \begin{bmatrix} H^*_{v'g_1^{-1}} \\ \vdots \\ H^*_{v'g_d^{-1}} \\ \delta_{v',\emptyset} \end{bmatrix} e_* \mapsto \begin{bmatrix} H^*_{v'} \\ T^*_{v'} \end{bmatrix} e_* \tag{3.50}$$

extends by linearity and continuity to define an isometry from

$$\mathcal{D}_V := \text{closed span}\left\{ \begin{bmatrix} H^*_{v'g_1^{-1}} \\ \vdots \\ H^*_{v'g_d^{-1}} \\ \delta_{v',\emptyset} \end{bmatrix} e_* \colon v' \in \mathcal{F}_d, e_* \in \mathcal{E}_* \right\} \subset \begin{bmatrix} \mathcal{H}_0 \\ \vdots \\ \mathcal{H}_0 \\ \mathcal{E}_* \end{bmatrix}$$

onto

$$\mathcal{R}_V := \text{closed span}\left\{ \begin{bmatrix} H^*_{v'} \\ T^*_{v'} \end{bmatrix} e_* \colon v' \in \mathcal{F}_d, e_* \in \mathcal{E}_* \right\} \subset \begin{bmatrix} \mathcal{H}_0 \\ \mathcal{E} \end{bmatrix}.$$

We now extend the isometry V to a unitary operator

$$V_e\colon \begin{bmatrix} \mathcal{H} \\ \vdots \\ \mathcal{H} \\ \mathcal{E}_* \end{bmatrix} \to \begin{bmatrix} \mathcal{H} \\ \mathcal{E} \end{bmatrix},$$

where $\mathcal{H}$ is a Hilbert space containing $\mathcal{H}_0$ as a subspace. We may then write V_e in block form

$$V_e = \begin{bmatrix} A^* & C^* \\ B^* & D^* \end{bmatrix} = \begin{bmatrix} A_1^* & \dots & A_d^* & C^* \\ B_1^* & \dots & B_d^* & D^* \end{bmatrix} \colon \begin{bmatrix} \oplus_{k=1}^d \mathcal{H} \\ \mathcal{E}_* \end{bmatrix} \to \begin{bmatrix} \mathcal{H} \\ \mathcal{E} \end{bmatrix}.$$

Since V_e extends V as defined in (3.50), we see that

$$\begin{bmatrix} A^* & C^* \\ B^* & D^* \end{bmatrix} \colon \begin{bmatrix} H^*_{v'g_1^{-1}} \\ \vdots \\ H^*_{v'g_d^{-1}} \\ \delta_{v',\emptyset} I_{\mathcal{E}_*} \end{bmatrix} e_* \to \begin{bmatrix} H^*_{v'} \\ T^*_{v'} \end{bmatrix} e_*. \tag{3.51}$$

Multiplication of (3.51) by $z^{v'^{\top}}$ and taking the formal sum over all $v' \in \mathcal{F}_d$ then gives us

$$\begin{bmatrix} A^* & C^* \\ B^* & D^* \end{bmatrix} : \begin{bmatrix} z_1 H(z)^* \\ \vdots \\ z_d H(z)^* \\ I_{\mathcal{E}_*} \end{bmatrix} \mapsto \begin{bmatrix} H(z)^* \\ T(z)^* \end{bmatrix}. \tag{3.52}$$

(Here we suppress a tensoring of $\begin{bmatrix} A^* & C^* \\ B^* & D^* \end{bmatrix}$ with the identity operator on $\mathbb{C}[[z]]$ from the notation.) From the first block row of (3.52) we read off

$$A^* Z(z)^* H(z)^* + C^* = H(z)^*. \tag{3.53}$$

Solving (3.53) for $H(z)^*$ then gives

$$H(z)^* = (I - A^* Z(z)^*)^{-1} C^*.$$

Plugging this identity into the second block row of (3.53) then gives us

$$T(z)^* = D^* + B^* Z(z)^* (I - A^* Z(z)^*)^{-1} C^*.$$

Taking adjoints of this expression then gives us

$$T(z) = D + C(I - Z(z)A)^{-1} Z(z) B$$

from which we see that T has the form $T_{\Sigma(U)}$ with $U = V_e^*$ as wanted. This concludes the proof of Theorem 3.16 □

We give a similar reproducing-kernel proof of another realization theorem from [6].

Theorem 3.17. *Given*

$$T(z) = \sum_{v \in \mathcal{F}_d} T_v z^v \in \mathcal{L}(\mathcal{E}, \mathcal{E}_*)[[z]] \quad \textit{and} \quad X_c(\zeta, \zeta') = \sum_{i,j=1}^{d} \zeta_i X_c^{i,j}(\zeta, \zeta') \zeta'_j \in \mathcal{L}(\mathcal{E})[[\zeta, \zeta']],$$

then there is a Cuntz unitary colligation (3.38) such that U realizes $T(z)$ and $X_c(\zeta, \zeta')$ in the sense that

$$T(z) = D + C(I - Z(z)A)^{-1} Z(z) B \tag{3.54}$$

$$X_c(\zeta, \zeta') = B^* Z(\zeta)^* (I - Z(\zeta)^* A^*)^{-1} (I - Z(\zeta')A)^{-1} Z(\zeta') B \tag{3.55}$$

if and only if

1. *$X_c(\zeta, \zeta')$ is a positive kernel (in the variables $\zeta = (\zeta_1, \dots, \zeta_d)$), and*
2. *$X_c(\zeta, \zeta')$ satisfies the identity*

$$I - T(\zeta)^* T(\zeta') = \sum_{j=1}^{d} \zeta_j'^{-1} X_c(\zeta, \zeta') \zeta_j^{-1} - X_c(\zeta, \zeta'). \tag{3.56}$$

Proof. Suppose that $U = \begin{bmatrix} A & B \\ C & D \end{bmatrix}$ is a unitary colligation which realizes T and X as in (3.54) and (3.55). From (3.55) it is clear that $X_c(\zeta, \zeta')$ has the factorization

$$X_c(\zeta,\zeta') = H(\zeta)H(\zeta')^*$$

where $H(\zeta) = B^* Z(\zeta)^*(I - Z(\zeta)^* A^*)^{-1}$, and hence $X_c(\zeta, \zeta')$ is a positive kernel. Moreover, using that U is unitary one can verify that

$$I - T(\zeta)^* T(\zeta') = B^*(I - Z(\zeta)^* A^*)^{-1}(I - Z(\zeta)^* Z(\zeta'))(I - AZ(\zeta))^{-1} B$$

while, on the other hand,

$$\begin{aligned}
&\sum_{k=1}^{d} \zeta_k'^{-1} X_c(\zeta,\zeta')\zeta_k^{-1} - X_c(\zeta,\zeta') \\
&= \sum_{k=1}^{d} \zeta_k'^{-1} B^*(Z(\zeta)^*(I - Z^* Z(\zeta)^*)^{-1}(I - Z(\zeta)A)^{-1} Z(\zeta) B \zeta_k^{-1} \\
&\quad - B^* Z(\zeta)^*(I - A^* Z(\zeta)^*)^{-1}(I - Z(\zeta')A)^{-1} Z(\zeta') B \\
&= B^*(Z(\zeta)^*(I - A^* Z(\zeta)^*)^{-1} \left(\sum_{k=1}^{d} \zeta_k^{-1}\zeta_k'^{-1} - I \right) (I - Z(\zeta')A)^{-1} Z(\zeta') B \\
&= B^*(I - Z(\zeta)^* A^*)^{-1} Z(\zeta)^* \left(\sum_{k=1}^{d} \zeta_k^{-1}\zeta_k'^{-1} - 1 \right) Z(\zeta')(I - AZ(\zeta'))^{-1} B \\
&= B^*(I - Z(\zeta)^* A^*)^{-1}(I - Z(\zeta)^* Z(\zeta'))(I - AZ(\zeta'))^{-1} B
\end{aligned}$$

and the necessity in Theorem 3.17 is clear.

Conversely, suppose that conditions (1), (2) and (3) in the statement of the theorem are satisfied. By condition (2), we know that $X_c(\zeta, \zeta')$ has a factorization $X_c(\zeta, \zeta') = H^0(\zeta)H^0(\zeta)^*$ for an $H^0(\zeta) \in \mathcal{L}(\mathcal{H}_0, \mathcal{E})[[\zeta]]$ for some auxiliary Hilbert space $\mathcal{H}_0$. Since $X_c(\zeta, \zeta')$ has the special form $X_c(\zeta, \zeta') = \sum_{i,j} \zeta_i X^{i,j}(\zeta, \zeta')\zeta_j'$, necessarily $H^0(\zeta)$ has the form

$$H^0(\zeta) = H(\zeta)Z(\zeta)^* = \sum_{i=1}^{d} H^i(\zeta)\zeta_i$$

for some

$$H(\zeta) = \begin{bmatrix} H^1(\zeta) & \dots & H^d(\zeta) \end{bmatrix} \in \mathcal{L}(\oplus_{i=1}^d \mathcal{H}^0, \mathcal{E})[[\zeta]]$$

and $X_c^{i,j}(\zeta, \zeta')$ has the form

$$X_c^{i,j}(\zeta,\zeta') = H_i(\zeta)H_j(\zeta')^*$$

and then

$$X(\zeta,\zeta') = H(\zeta)Z(\zeta)^* Z(\zeta)H(\zeta)^*. \tag{3.57}$$

Using this form (3.57) of $X_c(\zeta,\zeta')$ we compute

$$\sum_{j=1}^{d} \zeta_j'^{-1} X_c(\zeta,\zeta')\zeta_j^{-1} = H(\zeta)Z(\zeta)^* \left(\sum_{j=1}^{d} \zeta_j^{-1}\zeta_j'^{-1}\right) Z(\zeta')H(\zeta')^* = H(\zeta)H(\zeta')^*.$$

The identity (3.56) therefore takes the form

$$I - T(\zeta)^*T(\zeta') = H(\zeta)H(\zeta')^* - H(\zeta)Z(\zeta)Z(\zeta')^*H(\zeta')^*.$$

Rewriting in terms of positive squares gives

$$H(\zeta)Z(\zeta)Z(\zeta')^*H(\zeta')^* + I = H(\zeta)H(\zeta')^* + T(\zeta)^*T(\zeta').$$

Equating coefficients of $\zeta^{v'}\zeta'^{v}$ then gives

$$\sum_{i,j=1}^{d} H^i_{v'g_i^{-1}}(H^j_{g_j^{-1}v^\top})^* + \delta_{v',\emptyset}\delta_{v,\emptyset}I = H_{v'}H^*_{v^\top} + T^*_{v'}T_v$$

(where we write $H(\zeta) = \sum_{v\in\mathcal{F}_d} H_v\zeta^v$ and $H^i(\zeta) = \sum_{v\in\mathcal{F}_d} H^i_v\zeta^v$). It follows that the operator U_0 given by

$$U_0\colon \begin{bmatrix} \sum_{j=1}^d (H_{g_j^{-1}v^\top})^* \\ \delta_{v,\emptyset} \end{bmatrix} e \mapsto \begin{bmatrix} (H_{v^\top})^* \\ T_v \end{bmatrix} e \text{ for } e\in\mathcal{E}$$

extends by linearity and continuity to define an isometry from

$$\mathcal{D}_{U_0} = \overline{\operatorname{span}}\left\{ \operatorname{im} \begin{bmatrix} \sum_{j=1}^d (H^j_{g_j^{-1}v^\top})^* \\ \delta_{v,\emptyset}I \end{bmatrix} : v\in\mathcal{F}_d \right\} \subset \begin{bmatrix} \mathcal{H}_0 \\ \mathcal{E} \end{bmatrix}$$

onto

$$\mathcal{R}_{U_0} = \overline{\operatorname{span}}\left\{ \operatorname{im} \begin{bmatrix} H^*_{v^\top} \\ T_v \end{bmatrix} : v\in\mathcal{F}_d, \right\} \subset \begin{bmatrix} \oplus_{j=1}^d \mathcal{H}_0 \\ \mathcal{E}_* \end{bmatrix}$$

Let

$$U \begin{bmatrix} A & B \\ C & D \end{bmatrix} : \begin{bmatrix} \mathcal{H} \\ \mathcal{E} \end{bmatrix} \mapsto \begin{bmatrix} \oplus_{j=1}^d \mathcal{H} \\ \mathcal{E}_* \end{bmatrix}$$

be any unitary extension of U_0, where $\mathcal{H} \supset \mathcal{H}_0$. Then in particular

$$\begin{bmatrix} A & B \\ C & D \end{bmatrix} : \begin{bmatrix} \sum_{j=1}^d H^*_{g_j{-1}v^\top} \\ \delta_{v,\emptyset} \end{bmatrix} e \mapsto \begin{bmatrix} H^*_{v^\top} \\ T_v \end{bmatrix} e \text{ for all } e\in\mathcal{E}.$$

Multiply by ζ'^v and take the sum over all $v\in\mathcal{F}_d$ to get

$$\begin{bmatrix} A & B \\ C & D \end{bmatrix} \begin{bmatrix} Z(\zeta')H(\zeta')^* \\ I \end{bmatrix} e = \begin{bmatrix} H(\zeta')^* \\ T(\zeta') \end{bmatrix} e \text{ for all } e\in\mathcal{E}. \tag{3.58}$$

(where we suppress a tensoring of the action of $\begin{bmatrix} A & B \\ C & D \end{bmatrix}$ with $I_{\mathbb{C}[[\zeta]]}$ from the notation). We can solve the first block equation in (3.58) for $H(\zeta')^*$ to get

$$H(\zeta')^* = (I - AZ(\zeta'))^{-1}B. \tag{3.59}$$

Plugging this into the second block equation then gives

$$T(\zeta') = CZ(\zeta')(I - AZ(\zeta'))^{-1}B + D$$

and we have realized T as the transfer function $T(z) = T_{\Sigma_U}(z)$ of the row-unitary system $\Sigma(U)$. Moreover, from (3.57) combined with (3.59) we have

$$\begin{aligned} X_c(\zeta, \zeta') &= H(\zeta)Z(\zeta)^* Z(\zeta')H(\zeta')^* \\ &= B^*(I - Z(\zeta)^* A^*)^{-1} Z(\zeta)^* Z(\zeta')(I - AZ(\zeta'))^{-1}B \\ &= B^* Z(\zeta)^*(I - A^* Z(\zeta)^*)^{-1}(I - Z(\zeta')A)^{-1} Z(\zeta')B \end{aligned}$$

and (3.55) follows as well. □

References

[1] D. Alpay, *Algorithme de Schur, espaces à Noyau Reproduisant et Théorie des Systèmes*, Panoramas et Synthèses (Numéro 6), Société Mathématique de France, 1998.

[2] N. Aronszajn, Theory of reproducing kernels, *Trans. Amer. Math. Soc.*, **68** (1950), 337–404.

[3] J.A. Ball, Factorization and model theory for contraction opertors with unitary part, *Memoirs of the American Mathematical Society* Vol. 13, Issue 1, No. 198, 1978.

[4] J.A. Ball, Linear systems, operator model theory and scattering: multivariable generalizations, in *Operator Theory and Its Applications (Winnipeg, MB, 1998)* (Ed. A.G. Ramm, P.N. Shivakumar and A.V. Strauss), 151–178, Fields Institute Communications Vol. 25, Amer. Math. Soc., Providence, 2000.

[5] J.A. Ball and V. Vinnikov, Functional models for representations of the Cuntz algebra, preprint.

[6] J.A. Ball and V. Vinnikov, Lax-Phillips scattering and conservative linear systems: a Cuntz-algebra multidimensional setting, preprint.

[7] C. Berg, The multidimensional moment problem and semigroups, in *Moments in Mathematics* (Ed. H.J. Landau), Vol. 37 Proceedings of the Symposia in Applied Mathematics, 110–124, Amer. Math. Soc., Providence, 1987.

[8] L. de Branges, Factorizaiton and invariant subspaces, *J. Math. Anal. Appl.* 29 (1970), 163–2000.

[9] L. de Branges, The expansion theorem for Hilbert spaces of analytic funcitons, in *Topics in Operator Theory Systems and Networks* (Ed. H. Dym and I. Gohberg), 75–107, OT 12 Birkhäuser-Verlag, Basel-Boston, 1984.

[10] L. de Branges, Factorization in Kreĭn spaces, *J. Functional Analysis* 124 (1994), 228–262.

[11] L. de Brangs and J. Rovnyak, *Square Summable Power Series*, Hold, Rinehart, Winston, New York, 1966.

[12] L. de Branges and J. Rovnyak, Canonical models in quantum scattering theory, in *Perturbation Theory and its Applications in Quantum Mechanics*, 296–391, Wiley, New York, 1966.

[13] O. Bratteli and P.E.T. Jorgensen, Iterated function systems and permutation representations of the Cuntz algebra, *Memoirs of the American Mathematical Society* Volume 139, Number 663 second of 5 numbers), 1999

[14] M. Cotlar and C. Sadosky, Integral representations of bounded Hankel forms defined in scattering systems with a multiparametric evolution group, in *Contributions to Operator Theory and Its Applications (Mesa, AZ, 1987)*, 357–375, OT 35, Birkhäuser-Verlag, Basel-Boston, 1988.

[15] K.R. Davidson and D.R. Pitts, Invariant subspaces and hyper-reflexivity for free semigroup algebras, *Proc. London Math. Soc.* 78 (1999), 401–430.

[16] K.R. Davidson and D.R. Pitts, The algebraic structure of non-commutative analytic Toeplitz algebras, *Math. Ann.* 311 (1998), 275–303.

[17] S.S. Boiko, V.K. Dubovoy and A.Ya. Kheifets, Measure Schur complements and spectral functions of unitary operators with respect to different scales, in *Operator Theory, System Theory and Related Topics (Beer-Sheva/Rehovot 1997)*, 89–138, OT 123, Birkhäuser-Verlag, Basel-Boston, 2001.

[18] S.G. Krantz, *Function Theory of Several Complex Variables*, Second Edition, AMS Chelsea Publishing, Amer. Math. Soc., Providence, 2001.

[19] G. Popescu, Models for infinite sequences of noncommuting operators, *Acta Sci. Math.* 53 (1989), 355–368.

[20] G. Popescu, Characteristic functions for infinite sequences of noncommuting operators, *J. Operator Theory* 22 (1989), 51–71.

[21] G. Popescu, Isometric dilations for infinite sequences of noncommuting operators, *Trans. Amer. Math. Soc.* 316 (1989), 523–536.

[22] G. Popescu, Multi-analytic operators on Fock spaces, *Math. Ann.* 303 (1995), 31–46.

[23] S. Saitoh, *Theory of Reproducing Kernels and its Applications*, vol. 189, Longman Scientific and Technical, 1988.

[24] L.A. Shulman, Perturbations of unitary transformations, *J. Math. Anal. Appl.* 28 (1969), 231–254.

[25] H.J. Woerdeman, Positive Carathéodory interpolation on the polydisc, *Integral Equations and Operator Theory* 42 (2002), 229–242.

Joseph A. Ball
Department of Mathematics
Virginia Tech
Blacksburg, Virginia 24061
e-mail: `ball@math.vt.edu`

Victor Vinnikov
Department of Mathematics
Ben Gurion University of the Negev
Beer-Sheva 84105, Israel
e-mail: `vinnikov@math.bgu.ac.il`

Operator Theory:
Advances and Applications, Vol. 143, 135–146

On Realizations of Rational Matrix Functions of Several Complex Variables II

M.F. Bessmertnyĭ

0°. We use notations and definitions from the first part [Bes] of this work. If M is a matrix (whose entries are complex numbers), then M' is the transpose matrix, and M^* is the Hermitian conjugate matrix. $\mathbb{C}^n = \mathbb{C} \times \cdots \times \mathbb{C}$ denotes the direct product of n copies of the complex plane $\mathbb{C}$; $z = (z_1, \ldots, z_n)$ denotes a point of $\mathbb{C}^n$. In the sequel we use the following subsets of $\mathbb{C}^n$:

$$\begin{aligned}
\mathcal{D}_R^+ &= \{z : z \in \mathbb{C}^n, \operatorname{Re} z_1 > 0, \ldots, \operatorname{Re} z_n > 0\} \\
\mathcal{D}_R^- &= \{z : z \in \mathbb{C}^n, \operatorname{Re} z_1 < 0, \ldots, \operatorname{Re} z_n < 0\} \\
\mathcal{D}_J^+ &= \{z : z \in \mathbb{C}^n, \operatorname{Im} z_1 > 0, \ldots, \operatorname{Im} z_n > 0\} \\
\mathcal{D}_J^- &= \{z : z \in \mathbb{C}^n, \operatorname{Im} z_1 < 0, \ldots, \operatorname{Im} z_n < 0\}.
\end{aligned}$$

A rational matrix function $f(z) = f(z_1, \ldots, z_n)$ of the complex variables $z_1, \ldots, z_n$ whose values are square matrices is said to be **positive** if the following *positivity conditions* hold:

$$\begin{aligned}
f(z) + f(z)^* &\geq 0 \quad \text{for} \quad z \in \mathcal{D}_R^+, \\
f(z) + f(z)^* &\leq 0 \quad \text{for} \quad z \in \mathcal{D}_R^-, \\
i\,(f(z)^* - f(z)) &\geq 0 \quad \text{for} \quad z \in \mathcal{D}_J^+, \\
i\,(f(z)^* - f(z)) &\leq 0 \quad \text{for} \quad z \in \mathcal{D}_J^-.
\end{aligned}$$

A matrix function of the variables $z = (z_1, \ldots, z_n)$ is said to be **real** if its entries take real values for z: $\operatorname{Im} z_1 = 0, \ldots, \operatorname{Im} z_n = 0$.

The class of all positive real rational matrix functions is denoted by $\mathcal{P}$.

A function $f(z)$ whose values are square matrices is said to be **non-degenerate** if $\det f(z) \not\equiv 0$.

2000 *Mathematics Subject Classification.* 47A48.
Key words and phrases. Realization theory, rational matrix functions.
This paper is a translation, prepared by V. Katsnelson, of a part of the second chapter from the author's Ph. D. thesis entitled "Functions of Several Variables in the Theory of Finite Linear Structures", Kharkov, 1982. The manuscript, entitled "Realization of Rational Functions of Several Variables by Means of a Long Resolvent" was deposited at VINITI (86 pages, submitted by the Kharkov University, July 8, 1981, No. 3352-81).

Let k and N be natural numbers, $k \leq N$. We use the $k \times N$ matrix π and its conjugate matrix π^*:

$$\pi = \begin{bmatrix} 1_k & 0 \end{bmatrix}, \quad \pi^* = \begin{bmatrix} 1_k \\ 0 \end{bmatrix}, \tag{1}$$

where 1_k is the identity matrix of dimension k. The matrix π acts naturally from the vector space $\mathbb{C}^N$ into the vector space $\mathbb{C}^k$ (whose vectors are columns of dimensions N and k respectively). The space $\mathbb{C}^k$ can be naturally considered as a subspace of the space $\mathbb{C}^N$, identifying a vector $[\xi_1, \dots, \xi_k]' \in \mathbb{C}^k$ with the vector $[\xi_1, \dots, \xi_k, 0, \dots, 0]' \in \mathbb{C}^N$. Under this agreement, the matrix π becomes an orthoprojector.

Let A be a $N \times N$ matrix, with the block partition

$$A = \begin{bmatrix} a_{11} & a_{12} \\ a_{21} & a_{22} \end{bmatrix}, \quad a_{11} \text{ is a } k \times k \text{ matrix}$$

Assuming that the matrix A is invertible, we relate to it the $k \times k$ matrix $\pi A^{-1} \pi^*$, π is 'the orthoprojector' introduced in (1). If the latter matrix is invertible we can consider the matrix $[\pi A^{-1} \pi^*]^{-1}$, who is a $k \times k$ matrix.

In the sequel, the following elementary result will be used:

If the matrix A is invertible, and its block a_{22} is also invertible, then the matrix $\pi A^{-1} \pi^*$ is invertible as well, and its inverse is the Schur complement of the block entry a_{22} in the matrix A:

$$[\pi A^{-1} \pi^*]^{-1} = a_{11} - a_{12}\, a_{22}^{-1}\, a_{21} \,.$$

In what follows, k is a fixed natural number: the original given matrix function is of dimension $k \times k$. To represent this matrix function, we have to extend the original space $\mathbb{C}^k$ to a bigger space $\mathbb{C}^N$. The number N is not given in advance, it appears in the construction, and depends on the given rational function.

1°. In view of Theorem 2.10 of [Bes], by consideration of positive real matrix functions, one can confine oneself to the case of non-degenerate matrix functions.

Let a $k \times k$ matrix function $f(z)$ be represented in the form

$$f(z) = [\pi (z_1 A_1 + \cdots + z_n A_n)^{-1} \pi^*]^{-1}, \tag{2}$$

where A_j ($j = 1, \dots, n$) are non-negative constant matrices of dimension $N \times N$.

It is clear that the matrix f is positive. If, moreover, the matrices A_j are real, then the matrix function f is real as well.

Function of the form (2), where A_j are non-negative real matrices of *rank one*, are the simplest representatives of the class $\mathcal{P}$ of positive real matrix functions. Identifying some variables in (2) (i.e., increasing the of matrices A_j), we obtain more complicated objects of the class $\mathcal{P}$.

Definition 1. *A rational matrix function $f(z)$ of dimension $k \times k$ is said to be* primary positive *if f admits the representation*

$$f(z) = [\pi(z_1 A_1 + \cdots + z_n A_n)^{-1} \pi^*]^{-1},$$

where π is an orthoprojector of rank k, introduced in (1), *and A_j are non-negative real matrices of rank one, of some dimension N, $N \geq k$.*

2°. The goal of this paper is to describe the class of primary positive matrix functions in convenient terms, and to construct the matrices A_j in the representation (2) for a given primary matrix function f. In the following three lemmas we obtain the necessary conditions for a real matrix function to be primary positive.

Lemma 1. *Let $\dfrac{P(z)}{q(z)}$ be a primary positive real matrix function (the polynomial matrix $P(z)$ and the polynomial $q(z)$ are coprime). Then the degrees of P and q with respect to each variable z_j (for fixed other variables) are not greater than one.*

Proof. By definition,

$$\frac{P(z)}{q(z)} = \left[\pi(z_1 A_1 + \ldots + A_n)^{-1} \pi^*\right]^{-1},$$

where π is an orthoprojector of rank k, and A_j are matrices of rank one. Since $A_j \geq 0$, there exists a non-degenerate matrix T,

$$T = \begin{bmatrix} t_{11} & t_{12} \\ 0 & t_{22} \end{bmatrix}$$

such that the matrix $T' A_j T$ is diagonal. Let us denote $B_\nu = T' A_\nu T, (\nu = 1, \ldots, n)$. The matrix

$$t_{11}' f(z) t_{11} = \left[\pi(z_1 B_1 + \ldots + z_n B_n)^{-1} \pi^*\right]^{-1}$$

is of degree not greater than one with respect to the variable z_j. The same property possesses the matrix $f(z)$ as well. □

Lemma 2. *Let $f(z) = \dfrac{P(z)}{q(z)}$ be a primary positive real matrix function of dimension $k \times k$. Then for every $j = 1, \ldots n$ there exist real polynomials $\psi_1^{(j)}(z), \ldots, \psi_k^{(j)}(z)$ such that*

$$q^2(z) \frac{\partial f}{\partial z_j}(z) = \begin{bmatrix} \psi_1^{(j)}(z) \\ \vdots \\ \psi_k^{(j)}(z) \end{bmatrix} \begin{bmatrix} \psi_1^{(j)}(z) & \ldots & \psi_k^{(j)}(z) \end{bmatrix}.$$

Proof. By definition,

$$f(z)=\frac{P(z)}{q(z)}=\left[\pi(z_1A_1+\ldots+z_nA_n)^{-1}\pi^*\right]^{-1}.$$

Then

$$\frac{\partial f}{\partial z_j}(z)=f(z)\pi A^{-1}(z)A_jA^{-1}(z)\pi^*f(z),$$

where

$$A(z)=z_1A_1+\ldots+z_nA_n.$$

Since A_j is a matrix of rank one,

$$A_j=\begin{bmatrix}\alpha_1\\ \vdots\\ \alpha_N\end{bmatrix}\begin{bmatrix}\alpha_1 & \ldots & \alpha_N\end{bmatrix}.$$

Denoting [1]

$$q(z)f(z)\pi A^{-1}(z)\begin{bmatrix}\alpha_1\\ \vdots\\ \alpha_N\end{bmatrix}=\begin{bmatrix}\psi_1^{(j)}(z)\\ \vdots\\ \psi_k^j(z)\end{bmatrix},$$

we obtain

$$q^2(z)\frac{\partial f(z)}{\partial z_j}=\begin{bmatrix}\psi_1^{(j)}(z)\\ \vdots\\ \psi_k^{(j)}(z)\end{bmatrix}\begin{bmatrix}\psi_1^{(j)}(z) & \ldots & \psi_k^{(j)}(z)\end{bmatrix}.$$

Since $q^2(z)\frac{\partial f}{\partial z_j}(z)$ is a polynomial matrix function, $\psi_\nu^{(j)}(z)$ $(\nu=1,\ldots k)$ are polynomials. □

Lemma 3. *Let* $f(z)=\dfrac{P(z)}{q(z)}$ *be a primary positive real matrix function (the polynomial matrix* $P(z)$ *and the polynomial* $q(z)$ *are coprime). Then for every* j, $\nu=1,\ldots,n$ *there exist polynomials* $\varphi_{j\nu}(z)$ *such that*

$$\left(\frac{\partial q}{\partial z_j}\right)^2\frac{\partial}{\partial z_\nu}\left(\frac{q}{\dfrac{\partial q}{\partial z_j}}\right)=\varphi_{j\nu}^2(z).$$

Proof. In view of the previous lemma, it is enough to prove that the scalar function $q/\frac{\partial q}{\partial z_j}$ is a primary real positive one. We remark that for every $j=1,\ldots,n$ the degree of the polynomial matrix $P(z)$ with respect to z_j is equal to precisely one. Indeed, if it is not so, then

$$f(z_j,z')=\frac{P(z')}{z_jq_1(z')+q_2(z')}.$$

---

[1] On the right-hand side, we omit the entries which are surely equal to zero.

From the definition of the class $\mathcal{P}$ it follows that $q_1(z') \equiv 0$, and $f(z)$ does not depend on z_j. According to the assumptions,

$$\frac{P(z)}{q(z)} = A_{11}(z) - A_{12}(z)A_{22}^{-1}(z)A_{12}(z)',$$

where

$$A(z) = \begin{bmatrix} A_{11}(z) & A_{12}(z) \\ A_{12}'(z) & A_{22}(z) \end{bmatrix} = z_1 A_1 + \cdots + z_n A_n\,.$$

Since

$$A_{22}^{-1}(z) = \frac{1}{\det A_{22}(z)} B_{22}(z),$$

where $B_{22}(z)$ is a polynomial matrix,

$$\frac{P(z)}{q(z)} = \frac{A_{11}(z)\det A_{22}(z) - A_{12}(z)B_{22}(z)A_{12}'(z)}{\det A_{22}(z)} = \frac{F(z)}{q_1(z)}\,.$$

Therefore, $q(z)$ is a divisor of a polynomial $\det A_{22}(z)$, i.e., there exists a polynomial $\varphi(z)$ such that

$$q(z)\varphi(z) = \det A_{22}(z)\,.$$

Then

$$F(z) = P(z)\varphi(z)\,.$$

This is possible only if $\varphi(z) =$ const, because the degree of $P(z)$ with respect to each variable is equal to one precisely, and the degree of $F(z)$ does not exceed one.

Thus, up to a constant factor, the equality

$$q(z) = \det A_{22}(z) = \det\left[z_1 A_1^{(22)} + \ldots + z_n A_n^{(22)}\right]$$

holds, where $A_j^{(22)} \geq 0$ are matrices of rank one. For every orthogonal matrix U, $\det A_{22}(z) = \det U A_{22}(z)U'$. Let us choose the matrix U in such a way, that

$$UA_j^{(22)}U' = \begin{bmatrix} \lambda_j & 0 \\ 0 & 0 \end{bmatrix} \qquad (\lambda_j > 0)\,. \tag{3}$$

Denoting $UA_k^{(22)}U'$ by B_k, we obtain

$$q(z) = \det[z_1 B_1 + \cdots + z_n B_n] = \det \begin{bmatrix} B_{11}(z) & B_{12}(z) \\ B_{12}'(z) & B_{22}(z) \end{bmatrix},$$

where the block entry $B_{11}(z)$ is of dimension 1×1. Differentiating the last identity with respect to z_j and taking into account the concrete form (3) of the matrix B_j, we get

$$\frac{\partial q}{\partial z_j}(z) = \lambda_j \det B_{22}(z)\,.$$

Therefore,

$$\lambda_j\, q / \frac{\partial q}{\partial z_j} = \left[\pi(z_1 B_1 + \cdots + z_n B_n)^{-1}\pi^*\right]^{-1},$$

where $\pi = \begin{bmatrix} 1 & 0 \\ 0 & 0 \end{bmatrix}$, and $B_k \geq 0$ are matrices of rank one. Thus, $q/\frac{\partial q}{\partial z_j}$ is a scalar primary positive real matrix function. Applying Lemma 2 to this function, we obtain the desired assertion. □

3°. Let us summarize. If $f(z) = \frac{P(z)}{q(z)}$ is a primary positive real matrix function of dimension $k \times k$, then the following conditions are satisfied:

(i) $f'(z) = f(z) = \overline{f(\overline{z})}$;

(ii) $f(\lambda z) = \lambda f(z) \quad \forall \lambda \in \mathbb{C}^1$;

(iii) for each $j = 1, \ldots, n$ the degrees of the polynomials $P(z)$ and $q(z)$ with respect to the variable z_j (for fixed other variables) is not greater than one;

(iv) for each j, $\nu = 1, \ldots, n$, there exist polynomials $\varphi_{j\nu}(z)$, $\psi_1^{(j)}(z), \ldots, \psi_k^{(j)}(z)$ such that

(a)

$$\left(\frac{\partial q}{\partial z_j}\right)^2 \frac{\partial}{\partial z_\nu}\left(\frac{q}{\frac{\partial q}{\partial z_j}}\right) = \varphi_{j\nu}^2(z);$$

(b)

$$q^2(z)\frac{\partial f}{\partial z_j}(z) \begin{bmatrix} \psi_1^{(j)}(z) \\ \vdots \\ \psi_k^{(j)}(z) \end{bmatrix} \begin{bmatrix} \psi_1^{(j)}(z) & \ldots & \psi_k^{(j)}(z) \end{bmatrix}.$$

Let us show that conditions (i)–(iv) are also sufficient for the function $f(z) = P(z)/q(z)$ to be primary positive real. First, we prove the following simple lemma.

Lemma 4. *Let $f(z)$ be a matrix function represented in the form*

$$f(z) = \left[\pi((A(z) + V(z))^{-1}\pi^*\right]^{[-1]},$$

where π is introduced in (1). *If $V(z)$ admits the representation*

$$V(z) = W_{11}(z) - W_{12}(z)W_{22}^{-1}(z)W_{12}'(z),$$

then the matrix function

$$B(z) = \begin{bmatrix} A(z) + W_{11}(z) & W_{12}(z) \\ W_{12}'(z) & W_{22}(z) \end{bmatrix}$$

is non-degenerate, and

$$f(z) = \left[\pi B^{-1}(z)\pi^*\right]^{[-1]}. \tag{4}$$

Proof. By the assumption, the matrix function

$$A(z)+V(z)=A(z)+W_{11}(z)-W_{12}(z)W_{22}^{-1}(z)W_{12}'(z)$$

is non-degenerate. From here it follows that the matrix function $B(z)$ is non-degenerate, and (4) holds. □

Theorem 1. *Let $f(z)$ be a non-degenerate matrix function of the form*

$$f(z)=\frac{P(z)}{q(z)},$$

where $P(z)$ is a polynomial matrix function (of dimension $k\times k$), and $q(z)$ is a polynomial, $P(z)$ and $q(z)$ are coprime.

For the function $f(z)$ be a primary positive real matrix function, it is necessary and sufficient, that the formulated above[2] *conditions* (i)–(iv) *are satisfied.*

Proof. The necessity of the conditions (i)–(iv) follows from Lemmas 1–3. Let us prove the sufficiency.

1. First of all, from the assumptions of the theorem it follows that $f(z)$ is a positive real matrix function. Indeed, according to condition (iii), for each $j=1,\dots,n$, $f(z)$ can be represented in the form

$$f(z)=\frac{P(z)}{q(z)}=\frac{z_jP_1+P_2}{z_jq_1+q_2},$$

where the polynomial matrices P_1, P_2 and polynomials q_1, q_2 do not depend on the variables z_j. Then

$$f(z)-f^*(z)=(z_j-\overline{z_j})\,\frac{P_1q_2-P_2q_1}{|q(z)|^2}\,.$$

If all the variables except z_j are real then, in view the condition (iv),

$$P_1q_2-P_2q_1=q^2\,\frac{\partial f}{\partial z_j}=\psi\cdot\psi'\geq 0\,.$$

Therefore, for each j, the inequality $i(f^*(z)-f(z))\geq 0$ holds for $\mathrm{Im}z_j\geq 0$ and real other variables. According to Theorem 2.4 from [Bes], the inequality

$$i(f^*(z)-f(z))\geq 0\quad\forall z\in\mathbb{D}_J^+$$

holds. The latter inequality (together with the conditions (i) and (ii) means that the function $f(z)$ is positive real.

2. Now let $f(z)=P(z)/q(z)$ be a positive real matrix function which satisfy the assumptions of the theorem. Then the degree of the polynomial matrix $P(z)$ with respect to each variable z_j (for fixed other variables) is either equal to the degree of the polynomial q with respect to z_j, or greater by one. (To prove this, we fix other variables on the appropriate positive half-axes, and consider the resulting function of the variable z_j.) *We can confine ourselves to the case in which the*

[2] At the beginning of **3°**

degrees of the numerator and of the denominator with respect to each variable are equal to one. Indeed, let the denominator does not contain the variable z_1:

$$f(z) = \frac{z_1 P_1(z') + P_2(z')}{q(z')} = z_1 A_1(z') + f_1(z') .$$

Since $f_1(z') = f(0, z')$, $f_1(z')$ satisfies the conditions (i)–(iv). We show that $A_1(z')$ is a non-negative *constant* matrix of rank one. On the one hand, since $A_1(z') = \frac{\partial f}{\partial z_1}$, $A_1(z')$ is holomorphic in $\mathbb{D}_R^+$. On the other hand, $A_1(z') \geq 0 \ \forall z' \in \mathbb{D}_R^+$, since for any fixed $z' \in \mathbb{D}_R^+$, $f(z)$ considered as a function of the variable z_1 satisfies the condition

$$f(z_1, z') + f^*(z_1, z') \geq 0, \quad \mathrm{Re} z_1 \geq 0 ,$$

and $A_1(z')$ is the residue of $f(z_1, z')$ at $z_1 = \infty$. However, a non-negative holomorphic function is constant (this follows, for example, from the Cauchy-Riemann equations). Thus, $A_1 \equiv \text{const}$.

Since $A_1(z') = \dfrac{\partial f}{\partial z_1}$, and, according to (iv), $q^2(z)\dfrac{\partial f}{\partial z_1}(z)$ is a matrix function of rank one, the matrix A_1 is of rank one. Therefore, if $q(z)$ does not depend on the variables $z_1, \dots, z_\nu$, then

$$f(z) = z_1 A_1 + \cdots + z_\nu A_\nu + f_0(z') ,$$

where A_j $(j = 1, \dots, \nu)$ are real non-negative constant matrices of rank one, and $f_0(z')$ satisfies the assumptions of the theorem, and moreover, the degrees of the numerator and of the denominator with respect to each variable (for fixed other variables) are equal to one.

If we represent $f_0(z')$ in the form

$$f_0(z') = a_{11}(z') - a_{12}(z') a_{22}^{-1}(z') a_{12}(z') ,$$

where

$$A(z') = z_{\nu+1} A_{\nu+1} + \cdots + z_n A_n ,$$

then, in view of Lemma 4,

$$f(z) = \left[\pi(z_1 A_1 + \cdots + z_n A_n)^{-1} \pi^*\right]^{-1} ,$$

where A_j $(j = 1, \dots, \nu)$ are extended by zeros to matrices of the appropriate dimension.

3. Thus, it is enough to prove the theorem only for those matrix functions, whose numerator and denominator are of degree one with respect to each variable (for fixed other variables). Let $f(z)$ be such a function. Then

$$f(z) = \frac{P(z)}{q(z)} = \frac{z_1 P_1(z')) + P_2(z')}{z_1 q_1(z') + q_2(z')} , \tag{5}$$

where $P_j(z')$ $(j = 1, 2)$ are polynomial matrix functions of dimension 2×2; $q_j(z')$ $(j = 1, 2)$ are polynomials. Let us rewrite $f(z)$ in the form

$$f(z) = \frac{P_1(z')}{q_1(z')} - \frac{q_2P_1 - q_1P_2}{q_1^2\left(z_1 + \frac{q_2}{q_1}\right)}. \tag{6}$$

It is clear that

$$q_2P_1 - q_1P_2 = q^2\frac{\partial f}{\partial z_j}.$$

Therefore

$$q_2P_1 - q_1P_2 = \begin{bmatrix} \psi_1^{(1)}(z) \\ \vdots \\ \psi_k^{(1)}(z) \end{bmatrix} \begin{bmatrix} \psi_1^{(1)}(z) & \cdots & \psi_k^{(1)}(z) \end{bmatrix}.$$

Let us construct the matrix function $V(z')$ (of dimension $(k+1) \times (k+1)$:

$$V(z') = \begin{bmatrix} v_{11} & v_{12} \\ v_{12}' & v_{22} \end{bmatrix} = \left[\begin{array}{ccc|c} & & & \frac{\psi_1^{(1)}}{q_1} \\ & \frac{P_1}{q_1} & & \vdots \\ & & & \frac{\psi_k^{(1)}}{q_1} \\ \hline \frac{\psi_1^{(1)}}{q_1} & \cdots & \frac{\psi_k^{(1)}}{q_1} & \frac{q_2}{q_1} \end{array}\right]$$

In view of (6),

$$f(z) = v_{11}(z') - v_{12}(z')[z_1 + v_{22}(z')]^{-1}v_{12}'(z').$$

Applying Lemma 4, we obtain

$$f(z_1, z') = \left\{\pi_k\left(\begin{bmatrix} 0_k & 0 \\ 0 & 1 \end{bmatrix} z_1 + \begin{bmatrix} v_{11} & v_{12} \\ v_{12}' & v_{22} \end{bmatrix}\right)^{-1}\pi_k^*\right\}^{-1}.$$

If we show that $V(z')$ satisfies conditions (i)–(iv), then we can apply the same procedure to $V(z')$. Continuing this process, after finitely many steps, we come to a matrix function of the form $\sum z_jA_j$, where $A_j \geq 0$ are matrices of rank one. Then, applying Lemma 4 successively as many times as necessary, we obtain the representation

$$f(z) = \left[\pi_k(z_1A_1 + \cdots + z_nA_n)^{-1}\pi_k^*\right]^{-1},$$

where $A_j \geq 0\,(j = 1, \ldots, n)$ are matrices of rank one.

Let us show that $V(z')$ satisfies conditions (i)–(iv). Conditions (i)–(iii) for $V(z')$ are evident. To prove (iv), we represent $V(z')$ in the form

$$V(z') = \frac{F(z')}{q(z')},$$

where

$$F(z') = \left[\begin{array}{c|c} P_1(z') & \begin{matrix} \psi_1^{(1)}(z') \\ \vdots \\ \psi_k^{(1)}(z') \end{matrix} \\ \hdashline \psi_1^{(1)}(z') \dots \psi_k^{(1)}(z') & q_2(z') \end{array}\right] = \begin{bmatrix} P_1 & \vec{\psi}\,' \\ \vec{\psi} & q_2 \end{bmatrix}.$$

The direct calculation gives:

$$q_1^2 \frac{\partial V}{\partial z_j} = \left[\begin{array}{c|c} q_1 \dfrac{\partial P_1}{\partial z_j} - \dfrac{\partial q_1}{\partial z_j} P_1 & q_1 \dfrac{\partial \vec{\psi}\,'}{\partial z_j} - \dfrac{\partial q_1}{\partial z_j} \vec{\psi}\,' \\ \hdashline q_1 \dfrac{\partial \vec{\psi}}{\partial z_j} - \dfrac{\partial q_1}{\partial z_j} \vec{\psi} & q_1 \dfrac{\partial q_2}{\partial z_j} - \dfrac{\partial q_1}{\partial z_j} q_2 \end{array}\right]. \tag{7}$$

First of all, we remark that

$$q_1 \frac{\partial q_2}{\partial z_j} - q_2 \frac{\partial q_1}{\partial z_j} = \varphi_2^2(z'),$$

where $\varphi_2(z')$ is a polynomial. Indeed,

$$\frac{q_2}{q_1} = \frac{q(z)}{\dfrac{\partial q}{\partial z_j}(z)} \Bigg|_{z_1 = 0}.$$

Therefore,

$$q_1 \frac{\partial q_2}{\partial z_j} - q_2 \frac{\partial q_1}{\partial z_j} = \left(\frac{\partial q_1}{\partial z_1}\right)^2 \frac{\partial}{\partial z_j}\left(\frac{q}{\dfrac{\partial q}{\partial z_1}}\right)\Bigg|_{z_1 = 0} = \varphi_{1j}^2(z)\Bigg|_{z_1 = 0} = \varphi_2^2(z').$$

On the other hand,

$$\frac{P_1(z')}{q_1(z')} = \lim_{z_1 \to \infty} \frac{P(z)}{q(z)}.$$

Therefore

$$\frac{q_1\dfrac{\partial P_1}{\partial z_j}-\dfrac{\partial q_1}{\partial z_j}P_1}{q_1^2(z')}=\frac{\partial}{\partial z_j}\left[\lim_{z_1\to\infty}\frac{P(z)}{q(z)}\right]=\lim_{z_1\to\infty}\frac{\partial}{\partial z_j}\left[\frac{P(z)}{q(z}\right]$$

$$=\lim_{z_1\to\infty}\frac{\begin{bmatrix}\psi_1^{(j)}\\ \vdots\\ \psi_k^{(j)}\end{bmatrix}\begin{bmatrix}\psi_1^{(j)} & \cdots & \psi_k^{(j)}\end{bmatrix}}{q^2(z)}=\frac{\begin{bmatrix}\varphi_1^{(j)}\\ \vdots\\ \varphi_k^{(j)}\end{bmatrix}\begin{bmatrix}\varphi_1^{(j)} & \cdots & \varphi_k^{(j)}\end{bmatrix}}{q_1^2(z')}.$$

Thus,

$$q_1\frac{\partial P_1}{\partial z_j}-\frac{\partial q_1}{\partial z_j}P_1=\overrightarrow{\varphi}_1'\cdot\overrightarrow{\varphi}_1\,,$$

where $\overrightarrow{\varphi}=\left[\varphi_1^{(j)}\cdots\varphi_k^{(j)}\right]$ is a polynomial vector. Now, (7) can be rewritten in the form

$$q_1^2\frac{\partial V}{\partial z_j}=\left[\begin{array}{c|c}\overrightarrow{\varphi}_1'\cdot\overrightarrow{\varphi}_1 & q_1\dfrac{\partial\overrightarrow{\psi}'}{\partial z_j}-\dfrac{\partial q_1}{\partial z_j}\overrightarrow{\psi}' \\ \hline q_1\dfrac{\partial\overrightarrow{\psi}}{\partial z_j}-\dfrac{\partial q_1}{\partial z_j}\overrightarrow{\psi} & \varphi_2\cdot\varphi_2\end{array}\right].$$

Hence, to prove the property (iv), it is enough to show that

$$q_1\frac{\partial\overrightarrow{\psi}'}{\partial z_j}-\frac{\partial q_1}{\partial z_j}\overrightarrow{\psi}'=\overrightarrow{\varphi}_1'\cdot\varphi_2\,. \tag{8}$$

Since from the equality

$$\begin{bmatrix}\alpha_1\\ \vdots\\ \alpha_N\end{bmatrix}\cdot\begin{bmatrix}\alpha_1 & \cdots & \alpha_N\end{bmatrix}=\begin{bmatrix}\beta_1\\ \vdots\\ \beta_N\end{bmatrix}\cdot\begin{bmatrix}\beta_1 & \cdots & \beta_N\end{bmatrix}$$

it follows that either $\alpha_j=\beta_j$ $(j=1,\dots N)$, or $\alpha_j=-\beta_j$ $(j=1,\dots N)$, to prove (8) it is enough to prove that

$$\left(q_1\frac{\partial\overrightarrow{\psi}'}{\partial z_j}-\frac{\partial q_1}{\partial z_j}\overrightarrow{\psi}'\right)\cdot\left(q_1\frac{\partial\overrightarrow{\psi}}{\partial z_j}-\frac{\partial q_1}{\partial z_j}\overrightarrow{\psi}\right)=\overrightarrow{\varphi}_1'\varphi_2^2\overrightarrow{\varphi}_1\,. \tag{9}$$

Let us establish the latter equality. On the one hand,

$$\overrightarrow{\varphi}_1'\varphi_2^2\overrightarrow{\varphi}_1=\left(q_1\frac{\partial q_2}{\partial z_j}-q_2\frac{\partial q_1}{\partial z_j}\right)\left(q_1\frac{\partial P_1}{\partial z_j}-\frac{\partial q_1}{\partial z_j}P_1\right)$$

$$=q_1^2\frac{\partial q_2}{\partial z_j}\frac{\partial P_1}{\partial z_j}-q_1q_2\frac{\partial q_1}{\partial z_j}\frac{\partial P_1}{\partial z_j}-q_1\frac{\partial q_1}{\partial z_j}\frac{\partial q_2}{\partial z_j}P_1+q_2\left(\frac{\partial q_1}{\partial z_j}\right)^2P_1\,.$$

Before the consideration of the product on the left-hand side of (9), we obtain some auxiliary identities. As

$$q_2 P_1 - q_1 P_2 = \overrightarrow{\psi}'\overrightarrow{\psi}\,, \tag{10}$$

differentiating (10) with respect to z_j, we get

$$q_2 \frac{\partial P_1}{\partial z_j} + \frac{\partial q_2}{\partial z_j} P_1 - q_1 \frac{\partial P_2}{\partial z_j} - \frac{\partial q_1}{\partial z_j} P_2 = \overrightarrow{\psi}' \frac{\partial \overrightarrow{\psi}}{\partial z_j} + \frac{\partial \overrightarrow{\psi}'}{\partial z_j} \overrightarrow{\psi}\,. \tag{11}$$

Differentiating (11) with respect to z_j and taking into account that $\frac{\partial^2 \overrightarrow{\psi}}{\partial z_j} \equiv 0$, we obtain

$$\frac{\partial q_2}{\partial z_j}\frac{\partial P_1}{\partial z_j} - \frac{\partial q_1}{\partial z_j}\frac{\partial P_2}{\partial z_j} = \frac{\partial \overrightarrow{\psi}'}{\partial z_j} \cdot \frac{\partial \overrightarrow{\psi}}{\partial z_j}\,. \tag{12}$$

Using (10), (11), (12), it is not difficult to see that

$$\begin{aligned} &\left(q_1 \frac{\partial \overrightarrow{\psi}'}{\partial z_j} - \frac{\partial q_1}{\partial z_j} \overrightarrow{\psi}' \right) \left(q_1 \frac{\partial \overrightarrow{\psi}}{\partial z_j} - \frac{\partial q_1}{\partial z_j} \overrightarrow{\psi} \right) \\ &= q_1^2 \frac{\partial q_2}{\partial z_j}\frac{\partial P_1}{\partial z_j} - q_1 q_2 \frac{\partial q_1}{\partial z_j}\frac{\partial P_1}{\partial z_j} - q_1 \frac{\partial q_1}{\partial z_j}\frac{\partial q_2}{\partial z_j} P_1 + q_2 \left(\frac{\partial q_1}{\partial z_j} \right)^2 P_1\,. \end{aligned}$$

This coincides with (9). Therefore,

$$q_1 \frac{\partial \overrightarrow{\psi}'}{\partial z_j} - \frac{\partial q_1}{\partial z_j} \overrightarrow{\psi}' = \overrightarrow{\varphi}_1' \cdot \varphi_2\,,$$

and the representation

$$q_1^2 \frac{\partial V}{\partial z_j} = \begin{bmatrix} \overrightarrow{\varphi}_1' \\ \varphi_2 \end{bmatrix} \begin{bmatrix} \overrightarrow{\varphi}_1 & \varphi_2 \end{bmatrix}.$$

holds. The second equality (iv.b) of (iv) for the function $V(z')$ is established. The first identity (iv.a) can be established without difficulties. □

References

[Bes] Bessmertnyĭ, M.F. *On realization of rational matrix functions of several complex variables*, 157–185 in: *Interpolation Theory, System Theory and Related Topics* (D. Alpay, I. Gohberg and V. Vinnikov – editors.) (Operator Theory. Advances and Applications: **OT 134**), Birkhäuser Verlag, Basel · Boston · Berlin, 2002.

M.F. Bessmertnyĭ
Svobody Square, 4
Department of Mathematics, Faculty of Physics
Kharkov National University
61077, Kharkov, Ukraine

Operator Theory:
Advances and Applications, Vol. 143, 147–169

Bergman Projection and Weighted Holomorphic Functions

Der-Chen Chang, Robert Gilbert, and Jingzhi Tie

Abstract. In this paper, we first survey some regularities and irregularities resulting from the effects of weighted Bergman projections on decoupled and worm domains in $\mathbb{C}^{n+1}$. In the second part of the paper, we characterize weighted Bergman spaces with the help of the weighted Bergman kernel. Then we show that the generalized Cesáro operator is bounded on Bloch, Bergman and Hardy spaces.

Let Ω be a bounded, pseudo-convex domain of finite type in $\mathbb{C}^{n+1}$. We assume that $\partial\Omega$ is smooth.

The behavior of the Bergman kernel (even with respect to a smooth weight function) defined on either a decoupled or a convex domain of finite type in $\mathbb{C}^{n+1}$ has been studied by many mathematicians (see, *e.g.*, [**6**, **7**, **10**, **18**, **25**, **26**, **29**]). A significant example of a decoupled domain is the non-isotropic Siegel upper half-space

$$\mathcal{U}_{n+1} = \left\{ \mathbf{z} \in \mathbb{C}^{n+1} : \operatorname{Im}(z_{n+1}) > \sum_{k=1}^{n} a_k |z_k|^2 \right\}$$

with $a_k > 0$ for $k = 1, \ldots, n$. It is known that $\mathcal{U}_{n+1}$ is biholomorphic to the unit ball

$$B_n = \left\{ \mathbf{w} \in \mathbb{C}^{n+1} : |w_{n+1}|^2 + \sum_{k=1}^{n} |w_k|^2 < 1 \right\}$$

via the generalized Cayley transform:

$$w_{n+1} = \frac{i - z_{n+1}}{i + z_{n+1}}, \qquad w_k = \frac{2i\sqrt{a_k}\, z_k}{i + z_{n+1}}, \qquad k = 1, \ldots, n$$

$$z_{n+1} = i\left(\frac{1 - w_{n+1}}{1 + w_{n+1}}\right), \quad z_k = \frac{1}{\sqrt{a_k}} \frac{w_k}{1 + w_{n+1}}, \quad k = 1, \ldots, n.$$

This research is partially supported by a grant from National Science Foundation and a William Fulbright Research Grant.
1999 Mathematics Subject Classification: Primary: 32W20; Secondary: 35H10, 35N15.
keywords and phrases: Bergman projection, Bergman spaces, decoupled domains, worm domains, weighted functions, Cesáro operator, Bloch spaces, Hardy spaces.

The domain $\mathcal{U}_{n+1}$ plays an important role in analysis on strongly pseudo-convex domains (see [**4**, **32**]). In fact, techniques that were developed on the domain $\mathcal{U}_{n+1}$ inspired analysis on decoupled domains, and even convex domains. In Sections 1 and 2, we will survey the local geometry of decoupled domains as well as worm domains and apply those results to estimate weighted Bergman kernels. Then we will return to the model domain $\mathcal{U}_{n+1}$, and discuss its applications on both weighted holomorphic functions and generalized Cesáro operators in Section 3.

1. Local geometry of a decoupled domain

Let $\Omega \subset\subset \mathbb{C}^{n+1}$ be a regular pseudo-convex domain. We say Ω is decoupled of finite type near a point $p \in \partial\Omega$ if there exists a holomorphic coordinate system $(z_1, \ldots, z_{n+1})$ mapping ζ to $\mathbf{0}$ and a neighborhood U_p of ζ onto a neighborhood U of $\mathbf{0}$ and smooth, subharmonic, but not harmonic, functions $\{f_k\}_{k=1,\ldots,n}$, $f : \mathbb{C} \to \mathbb{R}$ with $f_k(0) = 0$, and each f_k vanishing to finite order at 0, such that

$$\left\{\mathbf{z} \in U : \rho(\mathbf{z}) = \operatorname{Im}(z_{n+1}) + \sum_{k=1}^{n} f_k(z_k) > 0\right\} \simeq \Omega \cap U_p. \tag{1.1}$$

Let us denote by $m_k(p)$ the order of vanishing of f_k at 0.

The finiteness condition here is equivalent to finite type in the case of a real analytic, pseudo-convex, hypersurface $\mathcal{M} \subset \mathbb{C}^{n+1}$, since the Levi form of Ω is diagonalizable (see Kohn [**21**]). When the Levi form is diagonalizable, the finite type is also equivalent to the condition introduced in [**22**].

We will assume that Ω is a bounded domain which is decoupled of finite type near each point of $\partial\Omega$ of non-strict pseudo-convexity, denote by

$$(m_1, \ldots, m_n) = \max_{p \in \partial\Omega}(m_1(p), \ldots, m_n(p)).$$

Let $p \in \partial\Omega$. If p is a point of non-strict pseudo-convexity, let f_k be the functions given by (1.1) and let

$$Z_k = \frac{\partial}{\partial z_k} + i\frac{\partial f_k}{\partial z_k}\frac{\partial}{\partial z_{n+1}}$$

for $k = 1, \ldots, n$ and

$$Z_{n+1} = N + iT.$$

Then $\{Z_1, \ldots, Z_n\}$ form a basis of the sub-bundle $T^{(1,0)}(\partial\Omega)$ of the complex tangent bundle

$$T(\partial\Omega) \otimes \mathbb{C} = T^{(1,0)}(\partial\Omega) \oplus T^{(1,0)}(\partial\Omega) \oplus (\mathbb{C} \times \partial\Omega)$$

near the point p and Z_{n+1} is a complex normal vector field, *i.e.*, Z_{n+1} is a $(1,0)$-vector filed which is $\mathcal{C}^\infty(\bar{\Omega})$ satisfying $Z_{n+1}(\rho) = 1$ on the boundary $\partial\Omega$. It means that Z_{n+1} goes outward from Ω, and $T = \dfrac{\partial}{\partial t}$, $t = \operatorname{Ree}(z_{n+1})$, is transverse to the complex tangent space at the boundary. As usual, we call T the "missing direction". As in the paper [**28**], for q in a neighborhood of p and for $0 < \delta \leq 1$, we may construct n functions $\mu_k(q,\delta)$ for $k=1,\dots,n$, given the size in the Z_k direction of the biggest polydisc included in Ω.

We will recall some basic properties of the μ_k's (see, *e.g.*, [**9, 12, 24, 29**]). For every ℓ-tuple of integers $(i_1,\dots,i_\ell)$ define smooth functions $\lambda^{(j)}_{i_1,\dots,i_\ell}$, $\alpha^{(j)}_{i_1,\dots,i_\ell}$ and $\beta^{(j)}_{i_1,\dots,i_\ell}$, in a neighborhood of p by the equation

$$[X_{i_\ell},[\cdots,[X_{i_1},X_{i_2}]\cdots]]=\lambda^{(j)}_{i_1,\dots,i_\ell}T+\alpha^{(j)}_{i_1,\dots,i_\ell}Z_j+\beta^{(j)}_{i_1,\dots,i_\ell}\bar{Z}_j$$

with $X_{i_\nu}\in\{Z_j,\bar{Z}_j\}$ for $\nu=1,\dots,\ell$ and $j=1,\dots,n$.

For each integer $2\le\ell\le m_j$, we define a smooth function $\Lambda^{(\ell)}_j$ in a neighborhood of p by the equation

$$\Lambda^{(\ell)}_j(p)=\left[\sum_{\nu\le\ell}|\lambda^{(j)}_{i_1,\dots,i_\nu}(p)|^2\right]^{\frac{1}{2}}.$$

Then we set

$$\Lambda_j(p,\delta)=\sum_{\ell=2}^{m_j}\Lambda^{(\ell)}_j(p)\delta^\ell.$$

Let $h\mapsto\mu_j(p,h)$ be the inverse function of $\delta\mapsto\Lambda_j(p,\delta)$. Then one has

$$\mu_j(p,h)\approx\min_{2\le\ell\le m_j}\left(\frac{h}{\Lambda^{(\ell)}_j(p)}\right)^{\frac{1}{\ell}}\approx\left(\sum_{\ell=2}^{m_j}\left(\Lambda^{(\ell)}_j(p)\right)^{\frac{1}{\ell}}\cdot h^{-\frac{1}{\ell}}\right)^{-1}.$$

For $\mathbf{x},\mathbf{y}\in\partial\Omega$ near p, we may consider the distance $d(\mathbf{x},\mathbf{y})$ on $\partial\Omega$ (see [**28**]) defined by

$$\begin{aligned}d(\mathbf{x},\mathbf{y})=\min\Big\{\delta>0\,|\,\gamma:[0,1]\to\partial\Omega,\ \gamma(0)=\mathbf{x},\gamma(1)=\mathbf{y}\text{ such that}\\ \gamma'(s)=\sum_{j=1}^{n}(a_j(s)Z_j+b_j(s)\bar{Z}_j)+c(s)T\text{ with}\\ |a_j(s)|\approx|b_j(s)|\le\mu_j(\mathbf{x},\delta),\ |c(s)|\le\delta\Big\}.\end{aligned}$$

We may use coordinate patches over $\partial\Omega$ to obtain a globally defined pseudo-metric. This pseudo-metric is well defined and by the Campbell-Hausdorff formula, one can show that the corresponding balls $\mathcal{B}(\mathbf{x},\delta)$ are equivalent, up to a holomorphic change of coordinates, to the polydisc we mentioned above (see [**24**]). Furthermore, we have the additional property of the functions μ_k's that $\mu_k(\mathbf{x},\delta)\approx\mu_k(\mathbf{y},\delta)$ for $k=1,\dots,n$ if $\mathbf{x}\in\mathcal{B}(\mathbf{y},\delta)$. It follows that for each $\mathbf{x}\in\partial\Omega$, the family of balls $\mathcal{B}(\mathbf{x},\delta)=\{\mathbf{y}\in\partial\Omega:d(\mathbf{x},\mathbf{y})<\delta\}$ on the boundary make the boundary $\partial\Omega$ a "space of homogeneous type". This is because the family of balls $\{\mathcal{B}_\delta\}$ satisfies the Vitali doubling property and the engulfing property with respect to dilation of the radii δ.

Without loss of generality, we assume that $p = \mathbf{0}$ and $U_p = U$. For $\mathbf{z}, \mathbf{w} \in U$, we define

$$D(\mathbf{z}, \mathbf{w}) = d(\pi(\mathbf{z}), \pi(\mathbf{w})) + \rho(\mathbf{z}) + \rho(\mathbf{w}).$$

For any $\mathbf{w} \in U$, we define the "tent" over the ball

$$\mathcal{B}(\pi(w), \delta) = \{\mathbf{z} \in \partial\Omega : \ d(\mathbf{z}, \pi(\mathbf{w}) \leq \delta\}$$

as the set

$$\mathcal{B}^{\#}(\mathbf{w}, \delta) = \bar{\Omega} \cap \{\mathbf{z} \in \mathbb{C}^{n+1} : \ D(\mathbf{z}, \mathbf{w}) \leq \delta\}.$$

Then we have

$$\mathrm{Vol}_\alpha[\mathcal{B}^{\#}(\mathbf{w}, \delta)] \approx \left[\prod_{j=1}^{n} \mu_j^2(\mathbf{w}, \delta)\right] \times \delta^{2+\alpha}.$$

Remark 1.1. *Let $\Omega \subset\subset \mathbb{C}^{n+1}$ be a bounded, smoothly convex domain of finite type, we still can realize the local geometry in Ω near a point $p \in \partial\Omega$ (see* **[24, 25, 26]***). Here we follow the notations and ideas in* **[25, 26]***. Let $p \in \partial\Omega$ be a point of finite type. Let ρ be a defining function for Ω. For $q \in \Omega \cap U$, U a small neighborhood of p, and $\vec{v} \in S^{2n+1}$, let*

$$\sigma(q, \vec{v}, \delta) = dist_{\vec{v}}(\mathcal{M}_\delta, q)$$

be the distance from q to the level set $\mathcal{M}_\delta = \{\mathbf{z} \in \mathbb{C}^{n+1} : \rho(\mathbf{z}) = \delta\}$ along the complex line generated by the vector $\vec{v}$. Denote $\tau_1(q, \delta) = \delta$ and let $\vec{n}_1 \in S^{2n+1}$ be a vector so that $\sigma(q, \vec{n}_1, \delta) = \tau_1(q, \delta) = \delta$. Let us parameterize this direction and denote it as z_1. Denote

$$\tau_2(q, \delta) = \max\{\sigma(q, \vec{v}, \delta) : \vec{v} \perp \vec{n}_1\},$$

and let $\vec{n}_2$ be a vector such that $\sigma(q, \vec{v}, \delta) = \tau_2(q, \delta)$. Similarly, we parameterize this direction as z_2. Inductively, denote

$$\tau_{k+1}(q, \delta) = \max\{\sigma(q, \vec{v}, \delta), \ \vec{v} \perp \mathcal{V}(\vec{n}_1, \ldots, \vec{n}_k)\}$$

for $k = 2, \ldots, n$. Here $\mathcal{V}(\vec{n}_1, \ldots, \vec{n}_k)$ is the linear span of the vectors $\{\vec{n}_1, \ldots, \vec{n}_k\}$. Note that

$$\tau_1(q, \delta) < \tau_{n+1}(q, \delta) \leq \cdots \leq \tau_2(q, \delta).$$

Define the polydisc

$$P_\delta(q) = \{\mathbf{z} \in U : \ |z_j| < \tau_j(q, \delta), \quad j = 1, \ldots, n+1\}.$$

Using this construction, we know that there is a constant $C > 0$, independent of $q \in \Omega \cap U$, such that $CP_\delta(q) \subset \{\mathbf{z} \in U : \ \rho(\mathbf{z}) < \delta + \rho(q)\}$. More precisely, for $q \in \Omega$, $P_\delta(q)$ is the polydisc centered at the point q just reaching out to the hypersurface $\mathcal{M}_{\delta+\rho(q)}$. It follows from **[26]** *that the boundary $\partial\Omega$ a "space of homogeneous type".*

2. Regular and irregular properties for Bergman kernel

Let dm denote the Lebesgue measure in $\mathbb{C}^{n+1}$, and let dm_α, $\alpha \in \mathbb{Z}_+$ denote the weighted measure

$$dm_\alpha(\mathbf{z}) = \rho^\alpha(\mathbf{z})dm(\mathbf{z}).$$

The space $L^p(dm_\alpha)$ consists of all Lebesgue measurable functions f defined on Ω satisfying

$$\|f\|^p_{L^p(dm_\alpha)} = \int_\Omega |f(\mathbf{z})|^p \rho^\alpha(\mathbf{z})dm(\mathbf{z}) < \infty.$$

It is easy to see that for $1 \le p < \infty, L^p(dm_\alpha)$ is a Banach space with norm $\|\cdot\|_{L^p(dm_\alpha)}$ and for $0 < p < 1$, $L^p(dm_\alpha)$ is an F-space under the metric

$$d(f, g) = \|f - g\|^p_{L^p(dm_\alpha)}.$$

Definition 2.1. *For $0 < p < \infty$, and $\alpha \in \mathbb{Z}_+$*

$$\mathcal{A}^p(dm_\alpha) = L^p(dm_\alpha) \cap \mathcal{H}(\Omega)$$

is the weighted Bergman space.

For $1 \le p < \infty$, $\mathcal{A}^p(dm_\alpha)$ is a closed subspace of $L^p(dm_\alpha)$ (see, *e.g.*, p. 14 in [**16**]).

For $p = 2$ we define the weighted Bergman projection, *i.e.*, the orthogonal projection

$$\mathbf{B}_\alpha : L^2(dm_\alpha) \to \mathcal{A}^2(dm_\alpha), \qquad \alpha \in \mathbb{Z}_+$$

Denote $B_\alpha(\mathbf{z}, \mathbf{w})$ the weighted Bergman kernel, *i.e.*,

$$\mathbf{B}_\alpha(f)(\mathbf{z}) = \int_\Omega B_\alpha(\mathbf{z}, \mathbf{w})f(\mathbf{w})dm_\alpha(\mathbf{w}) = \int_\Omega B_\alpha(\mathbf{z}, \mathbf{w})f(\mathbf{w})\rho^\alpha(\mathbf{w})dm(\mathbf{w}).$$

The weighted Bergman projection operator is smooth away from the diagonal $\Sigma = \{(\mathbf{z}, \mathbf{w}) \in \partial\Omega \times \partial\Omega : \mathbf{z} = \mathbf{w}\}$ on the boundary. Hence, the corresponding weighted Bergman kernel function and its derivatives are uniformly bounded on compact subsets of Ω, we may concentrate the behavior of the kernel function near a tubular neighborhood U of $\partial\Omega$ in $\bar{\Omega}$ where the projection $\pi : U \to \partial\Omega$ is well defined.

The first theorem we state here is the pointwise estimates of the weighted Bergman kernel $B_\alpha(\mathbf{z}, \mathbf{w})$ for $(\mathbf{z}, \mathbf{w}) \in \bar{\Omega} \times \bar{\Omega} \setminus \Sigma$. The proof of this theorem can be found in several papers, *e.g.*, [**6, 10, 11, 29**].

Theorem 2.2. *Let $\alpha \in \mathbb{Z}_+$. For each ℓ, there exists a constant C_ℓ so that if $Y_1, \dots, Y_\ell$ are vector fields (acts either on $\mathbf{z}$ or $\mathbf{w}$ variable), each of which is one of $\{Z_j, \bar{Z}_j\}_{j=1}^{n+1}$, with β_k either Z_k or $\bar{Z}_k$, $k = 1, \dots, n$, and $\beta_{n+1} = \ell - (\beta_1 + \cdots + \beta_n)$ either Z_{n+1} or $\bar{Z}_{n+1}$, then*

$$|Y_1 \cdots Y_\ell B_\alpha(\mathbf{z}, \mathbf{w})| \le C_\ell \frac{\left[\prod_{j=1}^n \mu_j^{-\beta_j}(\mathbf{z}, D(\mathbf{z}, \mathbf{w}))\right] \times D^{-\beta_{n+1}}(\mathbf{z}, \mathbf{w})}{Vol_\alpha[\mathcal{B}\#(\mathbf{z}, D(\mathbf{z}, \mathbf{w}))]}$$

for any $(\mathbf{z}, \mathbf{w}) \in \bar{\Omega} \times \bar{\Omega} \setminus \Sigma$.

When Ω is a smooth, bounded, convex domain of finite type, we may imitate the techniques in [**25**, **26**] to obtain the following result.

Theorem 2.3. *Let $\alpha \in \mathbb{Z}_+$ and let $\Omega = \{\mathbf{z} \in \mathbb{C}^{n+1} : \rho(\mathbf{z}) < 0\}$ be a smoothly bounded, convex domain of finite type in $\mathbb{C}^{n+1}$. Then for any multi-index β, there exists a constant C_β such that for all $\mathbf{z} \in \Omega$ we have*

$$\left| \frac{\partial^{|\beta|}}{\partial z_1^{\beta_1} \cdots \partial z_{n+1}^{\beta_{n+1}}} B_\alpha(\mathbf{z}, \mathbf{w}) \right| \leq \frac{C_\beta}{\tau_1^{2+\alpha+\beta_1}(\mathbf{z}, \tau_1(\mathbf{z}))} \prod_{j=2}^{n+1} \frac{1}{\tau_j^{\beta_j+2}(\mathbf{z}, \tau_j(\mathbf{z}))},$$

for all $\mathbf{w} \in \mathcal{B}^{\#}(\mathbf{z}, \delta)$. Here the coordinates $(z_1, \ldots, z_{n+1})$ are constructed related to the vanishing order of ρ near the boundary $\partial\Omega$.

For $k \in \mathbb{Z}_+$, we define $L_k^p(dm_\alpha)$ as the usual Sobolev space related to the measure dm_α with a norm

$$\|f\|_{L_k^p(dm_\alpha)} = \sum_{|\mathbf{m}_1|+|\mathbf{m}_2|=0}^{k} \left\| \frac{\partial^{\mathbf{m}} f}{\partial \mathbf{z}^{\mathbf{m}_1} \partial \bar{\mathbf{z}}^{\mathbf{m}_2}} \right\|_{L^p(dm_\alpha)}$$

Similar to [**12**], let $\mathcal{S}_k^p(dm_\alpha)$ be the collection of all $f \in L^p(dm_\alpha)$ such that

$$\mathcal{P}(Z, \bar{Z}) f \in L^p(dm_\alpha)$$

for all (noncommutative) polynomial $\mathcal{P}$ in Z_j and $\bar{Z}_j$, $j = 1, \ldots, n$, of degree at most k. To define a norm, let $\mathcal{V}_k$ be any (fixed) basis for the vector space of these polynomials and set

$$\|f\|_{\mathcal{S}_k^p(dm_\alpha)} = \sum_{\mathcal{P} \in \mathcal{V}_k} \|\mathcal{P}(Z, \bar{Z}) f\|_{L^p(dm_\alpha)}.$$

For more details concerning the spaces $\mathcal{S}_k^p(dm_\alpha)$, readers may consult Stein's book [605–610, **31**].

Now we define

$$L_{k,T}^p(dm_\alpha) = \{f \in L^p(dm_\alpha) : T^\ell f \in L^p(dm_\alpha),\ 1 \leq \ell \leq k\},$$

and for each $j = 1, \ldots, n$,

$$L_{k,Z_j}^p(dm_\alpha) = \{f \in L^p(dm_\alpha) : Z_j^\ell f \in L^p(dm_\alpha),\ 1 \leq \ell \leq k\}.$$

We have the following theorem:

Theorem 2.4. *The weighted Bergman projection maps*

$$L_{k,T}^p(dm_\alpha) \to L_k^p(dm_\alpha) \quad \text{and} \quad L_{k,Z_j}^p(dm_\alpha) \to L_k^p(dm_\alpha)$$

for $1 < p < \infty$ and $j = 1, \ldots, n$. In particular, the isotropic and anisotropic spaces, $L_k^p(dm_\alpha)$ and $\mathcal{S}_k^p(dm_\alpha)$ are left invariant by the weighted Bergman projection $\mathbf{B}_\alpha$.

Now let us turn to Lipschitz estimates for $\mathbf{B}_\alpha$. Let us recall the spaces $\Lambda_{T,\mathrm{as}}^\beta(\Omega)$ as well as $\Lambda_{Z_j,\mathrm{as}}^\beta(\Omega)$, $j = 1, \ldots, n$ as follows. More precisely, for a function

g of one real variable, we define $\Delta_h^k g(t)$ as the kth symmetric difference, which is obtained by induction from

$$\Delta_h g(t) = g(t+h) - g(t).$$

Let $\Psi_T = \Psi_T(\cdot, \xi)$ be the integral curve of T such that $\Psi_T(0) = \xi$, and let

$$\Delta_{T,h}^k f(\xi) = \Delta_h^k (f \circ \Psi_T)(0).$$

The space $\Lambda_{T,\mathrm{as}}^\beta(\Omega)$ is the space of functions f which are asymptotically Lipschitz in the T direction, *i.e.*, let

$$\Omega_\varepsilon = \{\mathbf{z} \in \Omega : \rho(\mathbf{z}) < \varepsilon\},$$

where ρ denotes the distance to the boundary, and assume that, for $\xi \in \Omega_{\varepsilon_0}$, $\Psi_T(\cdot, \xi)$ is defined for $|t| < \varepsilon_0 < \varepsilon$.

Definition 2.5. *A function $f \in L^\infty(\Omega)$ belongs to $\Lambda_{T,as}^\beta(\Omega)$ if there exist an integer k, $k > \beta$ and a constant $C > 0$ such that*

$$|\Delta_h^k f(\xi)| \le C|h|^{\ell-k}(\rho(\xi) + |h|)^\beta, \quad \ell = 0, \dots, k$$

for every $\xi \in \Omega_{\varepsilon_0}$ and $|h| < \frac{\varepsilon_0}{k}$.

We may define the spaces $\Lambda_{Z_j,\mathrm{as}}^\beta(\Omega)$, $j = 1, \dots, n$ in a similar way. We have the following definition.

Definition 2.6. *A function $f \in L^\infty(\Omega)$ belongs to $\Lambda_{Z_j,as}^\beta(\Omega)$ if there exist an integer k, $k > \beta$ and a constant $C > 0$ such that*

$$|\Delta_{Z_j,h}^k f(\xi)| = |\Delta_h^k (f \circ \Psi_{Z_j})(0)| \le C|h|^{\ell-k}(\rho(\xi) + |h|)^\beta, \quad \ell = 0, \dots, k$$

for every $\xi \in \Omega_{\varepsilon_0}$ and $|h| < \frac{\varepsilon_0}{k}$.

It can be shown (see [**6**]) that a function $f \in L^\infty(\Omega)$ belongs to $\Lambda_{T,\mathrm{as}}^\beta(\Omega)$ if and only if there exist an integer k, $k > \beta$ and a constant C such that, for every $\ell \ge 0$, f can be written as $b + g_1 + g_2$, where g_2 is supported outside $\Omega_{\varepsilon_0/2}$, b and g_1 supported in Ω_ε, and

$$|b(\xi)| \le C(\rho(\xi) + 2^{-\ell})^\beta, \quad \sum_{j=0}^k |T^j g_1(\xi)| \le C 2^{k\ell}(\rho(\xi) + 2^{-\ell})^\beta. \tag{2.1}$$

Similar to (2.1), it can be shown that a function $f \in L^\infty(\Omega)$ belongs to $\Lambda_{Z_j,\mathrm{as}}^\beta(\Omega)$ if and only if there exist an integer k, $km > \beta$ and a constant C such that, for every $\ell \ge 0$, f can be written as $b + g_1 + g_2$, where g_2 is supported outside $\Omega_{\varepsilon_0/2}$, b and g_1 supported in Ω_ε, and

$$|b(\xi)| \le C(\rho(\xi) + 2^{-\ell})^\beta, \quad (\mu_j(\xi, 2^{-\ell}))^k \sum_{i=0}^k |Z_j^i g_1(\xi)| \le C 2^{k\ell}(\rho(\xi) + 2^{-\ell})^\beta. \tag{2.2}$$

Now we state our theorem on Lipschitz estimates for the operator $\mathbf{B}_\alpha$ as follows.

Theorem 2.7. *Let $\alpha \in \mathbb{Z}_+$. Then the weighted Bergman projection $\mathbf{B}_\alpha$ maps the spaces $\Lambda^\beta_{T,as}(\Omega)$ and $\Lambda^\beta_{T,as}(\Omega)$, $j = 1, \ldots, n$ into $\Lambda^\beta(\Omega)$.*

Remark 2.8. *Let $\alpha \in \mathbb{Z}_+$. We may follow the idea in [**26**] to prove that the weighted Bergman projection on a smoothly bounded, convex domain of finite type maps the L^p-Sobolev space $L^p_k(\Omega)$ to itself for $1 < p < \infty$ and $k \in \mathbb{Z}_+$. Similarly, $\mathbf{B}_\alpha : \Lambda^\beta(\Omega) \to \Lambda^\beta(\Omega)$ is bounded for $0 < \beta < 1$.*

From what we have discussed, we know that the weighted Bergman projection preserves smooth functions defined on Ω. In fact, local regularity (*i.e.*, $\mathbf{B}_\alpha(L^p_{k,\mathrm{loc}}(\Omega)) \subset L^p_{k,\mathrm{loc}}(\Omega)$) is always true if Ω is pseudo-convex of finite type (see [**9**]). Now let us turn to a quite different situation.

Let us consider the following domain in $\mathbb{C}^2$:

$$\mathcal{W}_\eta = \left\{ \mathbf{z} \in \mathbb{C}^2 : \rho(z_1, z_2) = \left| z_1 - e^{i \log |z_2|} \right| - \sqrt{1 - \varphi(\log |z_2|)} < 0 \right\} \tag{2.3}$$

where φ is a smooth function such that

$$\varphi(\xi) = \begin{cases} 1 & \xi = \eta_1, \\ 0 & \eta_2 \le \xi \le \eta_3, \quad 0 < \varphi(\xi) < 1 \text{ if } \eta_1 < \xi < \eta_2 \text{ and } \eta_3 < \xi < \eta_4 \\ 1 & \xi = \eta_4. \end{cases} \tag{2.4}$$

Here $\eta_1 = -\frac{\pi}{2} - \eta$, $\eta_2 = -\frac{\pi}{2}$, $\eta_3 = \frac{\pi}{2}$, and $\eta_4 = \frac{\pi}{2} + \eta$ for some positive number η. We assume further that $\dfrac{d^2\varphi(\xi)}{d\xi^2} + \varphi(\xi) \ge 0$. From a result of Kiselman [**20**], we first rewrite (2.3) as follows

$$|z_1|^2 - 2\mathrm{Re}(z_1 e^{-i \log |z_2|}) + \varphi(\log |z_2|) < 0$$

or in the local complex variable $\zeta = e^{-i \log(z_2)}$ as

$$|z_1|^2 - 2\mathrm{Re}(z_1 e^{i \arg \zeta}) + \varphi(-\arg \zeta) < 0.$$

After multiplying by the modulus $|\zeta|$ of ζ, the above inequality becomes

$$|\zeta z_1^2| - 2\mathrm{Re}(z_1 \zeta) + \varphi(-\arg \zeta)|\zeta| < 0.$$

Rewrite $\log \zeta = u + iv$. The plurisubharmonicity of the left-hand side of the above inequality follows from

$$\begin{aligned} \left(\frac{d^2}{du^2} + \frac{d^2}{dv^2} \right) [\varphi(-\arg \zeta)|\zeta|] &= \left(\frac{d^2}{du^2} + \frac{d^2}{dv^2} \right) [\varphi(-v) e^u] \\ &= [\varphi''(-v) + \varphi(-v)] e^u \ge 0. \end{aligned}$$

This implies that the domain $\mathcal{W}_\eta$ is pseudo-convex everywhere. For a fixed z_2, $\mathcal{W}_\eta$ is a disc in $\mathbb{C}$ centered at $(z_1, e^{i \log |z_2|})$ and radius $\sqrt{1 - \varphi(\log |z_2|)}$. This is the so-called worm domain which was first invented by Diederich and Fornaess [**15**] in 1977. It can be shown that $\overline{\mathcal{W}_\eta}$ has no Stein neighborhood bases, *i.e.*, there exists $\delta > 0$ such that

$$\overline{\mathcal{W}_\eta} \neq \cap_{\varepsilon > 0} \Omega_\varepsilon$$

where Ω_ε are strongly pseudo-convex domains with

$$\Omega_\varepsilon \subset \{\mathbf{z} \in \mathbb{C}^2 : \operatorname{dist}(\mathbf{z}, \mathcal{W}_\eta) < \varepsilon\}$$

for all $\varepsilon < \delta$.

Furthermore, it can be checked that $\mathcal{W}_\eta$ is weakly pseudo-convex everywhere and strongly pseudo-convex at points with $z_1 \neq 0$. The first important breakthrough for the irregularity property of the Bergman projection $\mathbf{B}_0$ was obtained by Barrett [**2**] in 1992.

Theorem 2.9. *For each $\eta > 0$ there exists $t_\eta \in \mathbb{R}^+$ such that for any worm domain $\mathcal{W}_\eta$, the Bergman projection $\mathbf{B}_0$ does not satisfy the following estimate:*

$$\|\mathbf{B}_0(f)\|_{L^2_s(\mathcal{W}_\eta)} \le C\|f\|_{L^2_s(\mathcal{W}_\eta)}, \quad \text{for all} \quad s \ge t_\eta.$$

Moreover, $t_\eta \to 0$ as $\eta \to \infty$.

The crucial step is to analyze the behavior of the Bergman kernel on the nonsmooth domains

$$\mathcal{W}'_\eta = \left\{\mathbf{z} \in \mathbb{C}^2 : \left|z_1 + e^{i\log|z_2|}\right| < 1 \quad \text{and} \quad -\eta < \log|z_2| < \eta\right\}.$$

In this case, $\mathbf{B}_0$ not only fails to preserve the Sobolev space $L^2_{t_\eta}$, but even fails to map $C^\infty(\overline{\mathcal{W}'_\eta})$ to $L^2_{t_\eta}$. However, even though we know that the Bergman projection does not preserve the Sobolev space L^2_η; nevertheless the problem for C^∞ is still open. This open problem was solved by Christ [**13, 14**] in 1996. He solved the problem by looking at the $\bar\partial$-Neumann operator. The Bergman projection is related to the $\bar\partial$-Neumann problem via the following formula (see [**21, 22**]):

$$\mathbf{B}_0 = \mathbf{I} - \bar\partial^* N \bar\partial,$$

where $\mathbf{I}$ denotes the identity operator and N denotes the $\bar\partial$-Neumann operator, *i.e.*, the solving operator of the following problem:

$$\begin{aligned} \Box u = (\bar\partial\bar\partial^* + \bar\partial^*\bar\partial)u = f \quad &\text{in} \quad \Omega \\ u \lrcorner \bar\partial\rho = 0 \quad &\text{on} \quad \partial\Omega \\ \bar\partial u \lrcorner \bar\partial\rho = 0 \quad &\text{on} \quad \partial\Omega \end{aligned}$$

It is known that in $\mathbb{C}^2$, for any exponent s, $\mathbf{B}_0$ preserves (scalar-valued) L^2_s if and only if N preserves ((0,1)-form-valued) L^2_s; $\mathbf{B}_0$ preserves $C^\infty(\bar\Omega)$ if and only if N does so (see Boas and Straube [**8**]). Now we may state Christ's remarkable result as follows.

Theorem 2.10. *For each smoothly bounded worm domain $\mathcal{W}$ in $\mathbb{C}^2$ and a datum $f \in C^\infty_{(0,1)}(\overline{\mathcal{W}})$, there exists an s such that the unique solution $u \in L^2_{(0,1)}(\mathcal{W})$ of the $\bar\partial$-Neumann problem does not belong to $(0,1)$-form-valued $L^2_s(\overline{\mathcal{W}})$.*

Combining Theorem 2.9 and Boas-Straube's result, we know that for each smoothly bounded worm domain, there exists an s such that the Bergman projection operator $\mathbf{B}_0(C^\infty(\bar\Omega)) \not\subset L^2_s(\Omega)$. Recently, Siu [**31**] proved that on certain worm domains the operator $\mathbf{B}_0$ even does not preserve Lipschitz spaces. Here is Siu's theorem.

Theorem 2.11. *Let k, m_1 and m_2 be positive integers such that $\frac{m_1}{k}$ is an odd integer and $\frac{m_2}{k}$ is an even integer. Let $\eta_2 = 2m_1\pi$ and $\eta_3 = 2m_2\pi$. Let $\eta_1 = \eta_2 - \frac{1}{4}$ and $\eta_4 = \eta_3 + \frac{1}{4}$. Define the function $\varphi(\xi)$ for $\xi \in [\eta_1, \eta_4]$ as follows:*

$$\varphi(\xi) = \begin{cases} 4^{2k}(\xi - \eta_2)^{2k} & \eta_1 \le \xi \le \eta_2, \\ 0, & \eta_2 \le \xi \le \eta_3, \\ 4^{2k}(\xi - \eta_3)^{2k} & \eta_3 \le \xi \le \eta_4. \end{cases}$$

Let $\mathcal{W}$ be the worm domain

$$\left\{\mathbf{z} \in \mathbb{C}^2 : \rho(z_1, z_2) = \left|z_1 - e^{i\log|z_2|}\right| - \sqrt{1 - \varphi(\log|z_2|)} < 0\right\}.$$

Then the Bergman projection operator $\mathbf{B}_0(f) \notin \Lambda^\beta(\overline{\mathcal{W}})$ for any $\beta > 0$ whenever $f \in \Lambda^\beta(\overline{\mathcal{W}})$.

By assumption, we know that $\varphi''(\xi) + \varphi(\xi) \ge 0$. It follows that $\mathcal{W}$ is weakly pseudo-convex everywhere and is strongly pseudo-convex at boundary points with $z_1 \ne 0$. Since

$$\frac{d\varphi}{d\xi}(\eta_1) = \frac{d\varphi}{d\xi}(\eta_2) = 8k,$$

it follows that the boundary of $\mathcal{W}$ is C^{2k-1}. Moreover, in the same paper **[31]**, Siu constructed worm domains with smooth boundary which the Bergman projection operator does not map smooth L^2 functions into $C^k(\bar{\Omega})$.

Theorem 2.12. *Let $\ell \in \mathbb{Z}_+$ and let η be a positive number. Let*

$$\eta_1 = -\frac{\pi}{\ell+2} - \eta, \quad \eta_2 = -\frac{\pi}{\ell+2}, \quad \eta_3 = \frac{\pi}{\ell+2} \quad \text{and} \quad \eta_4 = \frac{\pi}{\ell+2} + \eta.$$

Then there exists a smooth function $0 \le \varphi(\xi) \le 1$ on the interval $[\eta_1, \eta_4]$ satisfying (2.4) such that the boundary of the domain $\mathcal{W}_\eta$ is smooth and is weakly pseudo-convex everywhere and strongly pseudo-convex at points with $z_1 \ne 0$ and the Bergman projection operator $\mathbf{B}_0$ maps some function which is smooth up to the boundary of Ω to a function which is not $C^{\ell+\varepsilon}$ up to the boundary of Ω for any $\varepsilon > 0$.

3. Weighted holomorphic function spaces and Cesáro operators

When $\Omega = B_n = \{\mathbf{z} = (z_1, \dots, z_n) \in \mathbb{C}^n : |\mathbf{z}| < 1\}$, the unit ball in $\mathbb{C}^n$, the weighted Bergman projection has the following form:

$$\begin{aligned} \mathbf{B}_\alpha(f)(\mathbf{z}) &= \frac{(\alpha+1)\cdots(\alpha+n)}{n!} \int_{B_n} \frac{f(\mathbf{w})}{(1 - \mathbf{z}\cdot\bar{\mathbf{w}})^{(n+1+\alpha)}}(1 - |\mathbf{w}|^2)^\alpha dm(\mathbf{w}) \\ &= \int_{B_n} f(\mathbf{w}) B_\alpha(\mathbf{z}, \mathbf{w})(1 - |\mathbf{w}|^2)^\alpha dm(\mathbf{w}), \end{aligned}$$

where

$$B_\alpha(\mathbf{z},\mathbf{w}) = \frac{(\alpha+1)\cdots(\alpha+n)}{n!}\frac{1}{(1-\mathbf{z}\cdot\bar{\mathbf{w}})^{n+1+\alpha}}$$

is the weighted Bergman kernel. With help of this kernel, we obtain the following theorem. The proof of this theorem can be found in [**3**].

Theorem 3.1. *Let $1 \le p < \infty$, $\alpha > -1$, N is a fixed positive integer and $\mathbf{k} = (k_1,\ldots,k_n) \in (\mathbb{Z}_+)^n$. Let f be a holomorphic function defined on the unit ball B_n in $\mathbb{C}^n$. Then $f \in \mathcal{A}^p(dm_\alpha)$ if and only if*

$$(1-|\mathbf{z}|^2)^{|\mathbf{k}|}\frac{\partial^{|\mathbf{k}|}f}{\partial z_1^{k_1}\cdots\partial z_n^{k_n}}(\mathbf{z}) \in L^p(dm_\alpha), \quad \textit{for all} \quad |\mathbf{k}| = N.$$

Moreover,

$$\|f\|_{L^p(dm_\alpha)} \approx \left(\sum_{|\mathbf{k}|=0}^{N-1}\left|\frac{\partial^{|\mathbf{k}|}f(\mathbf{0})}{\partial z_i^{k_1}\cdots\partial z_n^{k_n}}\right| + \sum_{|\mathbf{k}|=N}\left\|(1-|\mathbf{z}|^2)^N\frac{\partial^N f}{\partial z_i^{k_1}\cdots\partial z_n^{k_n}}\right\|_{L^p(dm_\alpha)}\right).$$

In case of the polydisc $D_n = \{\mathbf{z} = (z_1,\ldots,z_n) \in \mathbb{C}^n : |z_j| < 1,\ j = 1,\ldots,n\}$ in $\mathbb{C}^n$, it is easy to see that the kernel $B_{\vec{\alpha}}(\mathbf{z},\mathbf{w})$ for the weighted Bergman projection $\mathbf{B}_{\vec{\alpha}} : L^2(dm_{\vec{\alpha}}) \to \mathcal{A}^2(dm_{\vec{\alpha}})$ is

$$B_{\vec{\alpha}}(\mathbf{z},\mathbf{w}) = \prod_{j=1}^{n}\frac{(\alpha_j+1)}{(1-z_j\bar{w}_j)^{\alpha_j+2}}.$$

The results of Section 2 implies that the operator $\mathbf{B}_{\vec{\alpha}}$ can be extended as a bounded operator from $L^p(dm_{\vec{\alpha}})$ onto $\mathcal{A}^p(dm_{\vec{\alpha}})$. Moreover, we have the following theorem.

Theorem 3.2. *Let $1 \le p < \infty$, $\alpha_j > -1$, $j = 1,\ldots,n$ and N be a fixed positive integer and let $\mathbf{k} = (k_1,\ldots,k_n) \in (\mathbb{Z}_+)^n$. Let f be a holomorphic function defined on the polydisc D_n in $\mathbb{C}^n$. Then for $\vec{\alpha} = (\alpha_1,\ldots,\alpha_n)$, $f \in \mathcal{A}^p(dm_{\vec{\alpha}})$ if and only if*

$$\left[\prod_{j=1}^{n}(1-|z_j|^2)^{k_j}\right]\frac{\partial^{|\mathbf{k}|}f}{\partial z_1^{k_1}\cdots\partial z_n^{k_n}}(z) \in L^p(dm_{\vec{\alpha}}), \qquad \textit{for all} \quad |\mathbf{k}| = N.$$

Moreover,

$$\|f\|_{\mathcal{A}^p(dm_{\vec{\alpha}})} \approx \sum_{|\mathbf{k}|=0}^{N-1}\left|\frac{\partial^{|\mathbf{k}|}f}{\partial z_1^{k_1}\cdots\partial z_n^{k_n}}(\mathbf{0})\right| + \sum_{|\mathbf{k}|=N}\left\|\left[\prod_{j=1}^{n}(1-|z_j|^2)^{k_j}\right]\frac{\partial^N f}{\partial z_1^{k_1}\cdots\partial z_n^{k_n}}\right\|_{L^p(dm_{\vec{\alpha}})}$$

Theorems 3.1 and 3.2 generalize a classical result of Hardy and Littlewood (see, *e.g.*, [**19**] and [p. 81, **17**]). Furthermore, we may consider these two theorems in a more general setting, *i.e.*, replace the weight functions $(1-|\mathbf{z}|^2)^\alpha$ or

$\prod_{j=1}^{n}(1-|z_j|^2)^{\alpha_j}$ by radial weight functions $\omega(\mathbf{z})=\omega(|\mathbf{z}|)$ or $\vec{\omega}(\mathbf{z})=\prod_{j=1}^{n}\omega_j(|z_j|)$ on B_n or D_n respectively. For a weight function ω, we define its distortion function τ as follows:

$$\tau(\rho)=\tau_\omega(\rho)=\frac{1}{\omega(\rho)}\int_\rho^1 \omega(s)ds, \qquad 0\le\rho<1.$$

Then a weight function ω is admissible (see [**30**]) if it satisfies the following conditions:

(1) There exists a positive constant C_ω such that

$$\omega(\rho)\ge\frac{C_\omega}{1-\rho}\int_\rho^1\omega(s)ds, \qquad 0\le\rho<1;$$

(2) ω is differentiable and there exists a constant C'_ω such that

$$\frac{d\omega}{d\rho}\le\frac{C'_\omega}{1-\rho}\omega(\rho), \qquad 0\le\rho<1;$$

(3) For each sufficiently small $\varepsilon>0$, there exists a positive number δ_ε such that

$$\sup_{0\le\rho<1}\frac{\omega(\rho)}{\omega[\rho+\varepsilon\tau(\rho)]}\le\delta_\varepsilon.$$

For admissible weight functions $\omega(\mathbf{z})$ and $\vec{\omega}(\mathbf{z})$, denote the weighted Bergman spaces as follows:

$$\mathcal{A}^p_\omega(B_n)=\left\{f\in\mathcal{H}(B_n):\int_{B_n}|f(\mathbf{z})|^p\omega(\mathbf{z})dm(\mathbf{z})<\infty\right\},$$

and

$$\mathcal{A}^p_{\vec{\omega}}(D_n)=\left\{f\in\mathcal{H}(D_n):\int_{D_n}|f(z_1,\dots,z_n)|^p\prod_{j=1}^{n}\omega_j(z_j)dm(z_j)<\infty\right\}.$$

Here we state two theorems concerning the space $\mathcal{A}^p_\omega(B_n)$ and $\mathcal{A}^p_{\vec{\omega}}(D)n)$. The proofs of these two theorems can be found in [**3**, **34**].

Theorem 3.3. *Suppose $0<p<\infty$ and ω is an admissible weight function with distortion function τ. Then there exists a universal constant C depending on n, ω and p such that*

$$|f(\mathbf{0})|^p+\int_0^1\left(\int_{\partial B_n}\sum_{j=1}^{n}\left|\frac{\partial f}{\partial z_j}(\rho\xi)\right|^p d\sigma(\xi)\right)\tau^p(\rho)\omega(\rho)d\rho$$
$$\approx\int_0^1\left(\int_{\partial B_n}|f(\rho\xi)|^p d\sigma(\xi)\right)\omega(\rho)d\rho=\|f\|^p_{\mathcal{A}^p_\omega(B_n)},$$

for all holomorphic function f defined on B_n.

Theorem 3.4. *Assume that $\omega_j(z_j)$, $j = 1, \dots, n$, are admissible weight functions on the unit disc D_1 with distortion function $\tau_j(z_j)$. Let $N \in \mathbb{N}$ be fixed, we have*

(1) *For $0 < p < \infty$*

$$\int_{D_n} \left| \frac{\partial^{|\mathbf{k}|} f}{\partial z_1^{k_1} \cdots \partial z_n^{k_n}}(\mathbf{z}) \right|^p \prod_{j=1}^n \tau_j^{k_j}(z_j)\omega(z_j)dm(z_j)$$
$$\leq \int_{D_n} |f(z_1, \dots, z_n)|^p \prod_{j=1}^n \omega_j(z_j)dm(z_j).$$

Moreover, there exists a constant $C_{\vec{\omega},p,N,n}$ such that

$$\left(\sum_{|\mathbf{k}|=0}^{N-1} \left| \frac{\partial^{|\mathbf{k}|} f}{\partial z_1^{k_1} \cdots \partial z_n^{k_n}}(\mathbf{0}) \right| + \sum_{|\mathbf{k}|=N} \left\| \left[\prod_{j=1}^n \tau_j^{k_j}(z_j) \right] \frac{\partial^{|\mathbf{k}|} f}{\partial z_1^{k_1} \cdots \partial z_n^{k_n}}(\mathbf{z}) \right\|_{L^p_{\vec{\omega}}(D_n)} \right)$$
$$\leq C_{\vec{\omega},p,N,n} \|f\|^p_{\mathcal{A}_{\vec{\omega}}(D_n)}.$$

(2) *For $1 < p < \infty$, if $\left[\prod_{j=1}^n \tau_j^{k_j}(z_j) \right] \frac{\partial^{|\mathbf{k}|} f}{\partial z_1^{k_1} \cdots \partial z_n^{k_n}}(\mathbf{z}) \in L^p_{\vec{\omega}}(D_n)$ for all $|\mathbf{k}| = N$, then $f \in \mathcal{A}^p_{\vec{\omega}}(D_n)$. Moreover, there exists a constant $C'_{\vec{\omega},p,N,n}$ such that*

$$\|f\|^p_{\mathcal{A}_{\vec{\omega}}(D_n)}$$
$$\leq C'_{\vec{\omega},p,N,n} \left(\sum_{|\mathbf{k}|=0}^{N-1} \left| \frac{\partial^{|\mathbf{k}|} f(\mathbf{0})}{\partial z_1^{k_1} \dots \partial z_n^{k_n}} \right| + \sum_{|\mathbf{k}|=N} \left\| \left[\prod_{j=1}^n \tau_j^{k_j}(z_j) \right] \frac{\partial^{|\mathbf{k}|} f}{\partial z_1^{k_1} \dots \partial z_n^{k_n}}(\mathbf{z}) \right\|_{L^p_{\vec{\omega}}(D_n)} \right).$$

Open problem: Is the conclusion (2) true for $0 < p \leq 1$?

For $\gamma \in \mathbb{C}$ with $\mathrm{Ree}(\gamma) > -1$ and $k \in \mathbb{Z}_+$, we define A_k^γ as the kth coefficient in the following expression

$$\frac{1}{(1-x)^{\gamma+1}} = \sum_{k=0}^\infty A_k^\gamma x^k.$$

It is easy to see that

$$A_k^\gamma = \frac{(\gamma+1)(\gamma+2)\cdots(\gamma+k)}{k!}.$$

We define the generalized sliced Cesáro operator $\mathcal{C}_\gamma$ on the unit ball B_n as follows:

$$\mathcal{C}_\gamma(f)(\mathbf{z}) = \sum_{n=0}^\infty \left(\frac{1}{A_n^{\gamma+1}} \sum_{k=0}^n A_{n-k}^\gamma a_k(\xi) \right) w^n$$
$$= (\gamma+1) \int_0^1 \frac{f_\xi(tw)}{(1-tw)^{\gamma+1}} (1-t)^\gamma dt,$$

where

$$f(\mathbf{z}) = f(w\xi) = f_\xi(w) = \sum_{k=0}^{\infty} a_k(\xi) w^k \in \mathcal{H}(B_n),$$

with $\xi \in \partial B_n$ and $w \in D_1 = \{z \in \mathbb{C} : |z| \le 1\}$, the closed unit disc in $\mathbb{C}^1$. When $\gamma = 0$, the operator $\mathcal{C}_0$ is the classical Cesáro operator. This operator was first introduced in [**33**] and has been studied by many mathematicians, *e.g.*, [**1**] and [**36**].

For $\alpha > 0$, let $\mathcal{B}_\alpha(B_n)$ be the α-Bloch space of all holomorphic functions on the unit ball B_n in $\mathbb{C}^n$ such that

$$N(f)_{\mathcal{B}_\alpha} = \sup_{\mathbf{z} \in B_n} |f(\mathbf{z})|(1 - |\mathbf{z}|)^\alpha < \infty.$$

Our first theorem related to the generalized Cesáro operator $\mathcal{C}_\gamma$ is the following:

Theorem 3.5. *Let $\alpha > 1$. Then the generalized sliced Cesáro operator $\mathcal{C}_\gamma$ is bounded on the space $\mathcal{B}_\alpha(B_n)$. More precisely, there exists a universal constant C depending on n, γ but independent of f such that*

$$N(\mathcal{C}_\gamma(f))_{\mathcal{B}_\alpha} \le C N(f)_{\mathcal{B}_\alpha}.$$

Proof. Without loss of generality, we may assume that $\gamma > -1$. Let $f \in \mathcal{B}_\alpha(B_n)$. Then

$$\begin{aligned} |\mathcal{C}_\gamma(f)(\mathbf{z})| &\le |\gamma + 1| \int_0^1 \frac{|f_\xi(tw)|}{(1 - t|w|)^{\gamma+1}} (1-t)^\gamma dt \\ &\le C N(f)_{\mathcal{B}_\alpha} \int_0^1 \frac{(1-t)^\gamma}{(1 - t|w|)^{\gamma+\alpha+1}} dt \\ &\le C N(f)_{\mathcal{B}_\alpha} \int_0^1 \frac{1}{(1 - t|w|)^{\alpha+1}} dt \\ &\le C \frac{N(f)_{\mathcal{B}_\alpha}}{(1 - |w|)^\alpha}. \end{aligned}$$

Because $|w| = |\mathbf{z}|$, the conclusion of the theorem follows immediately. □

When we consider the operator $\mathcal{C}_0$, we may apply the above method to recover the following stronger results in [**3**].

Theorem 3.6. *The sliced Cesáro operator*

$$\mathcal{C}_0(f)(\mathbf{z}) = \mathcal{C}_0(f_\xi)(w) = \sum_{k=0}^{\infty} \left[\frac{1}{k+1} \sum_{\ell=0}^{k} a_\ell(\xi) \right] w^k$$

is bounded on the weighted Bergman space $\mathcal{A}^p(dm_\alpha)$ for $0 < p < \infty$.

Now let us turn to the polydisc D_n. For $\alpha > 0$, let α-Bloch space $\mathcal{B}_\alpha(D_n)$ is the collection of all holomorphic functions f on D_n such that

$$\mathcal{N}(f)_\alpha = \sup_{\mathbf{z} \in D_n} \sum_{j=1}^{n} (1 - |z_j|^2)^\alpha \left| \frac{\partial f}{\partial z_j}(\mathbf{z}) \right| < \infty.$$

We also denote $\mathcal{B}_{\vec{\alpha}}(D_n)$ the space of all holomorphic functions f on D_n such that

$$N(f)_{\mathcal{B}_{\vec{\alpha}}} = \sup_{\mathbf{z}\in D_n} |f(\mathbf{z})| \prod_{j=1}^{n} (1-|z_j|)^{\alpha_j} < \infty,$$

where $\vec{\alpha} = (\alpha_1, \ldots, \alpha_n)$, $\alpha_j > 0$ for $j = 1, \ldots, n$.

It can be shown that when $n = 1$, $f \in \mathcal{B}_{\alpha}(D_1)$ if and only if $N(f)_{\mathcal{B}_{\alpha-1}} < \infty$.

Now we may define a family of Cesáro operators $\mathcal{C}_{\vec{\gamma}}$ for functions defined on D_n. Let $\vec{\gamma} = (\gamma_1, \ldots, \gamma_n) \in \mathbb{C}^n$ with $\mathrm{Re}(\gamma_j) > -1$ for $j = 1, \ldots, n$. Then the generalized Cesáro operator $\mathcal{C}_{\vec{\gamma}}$ is defined as follows:

$$\mathcal{C}_{\vec{\gamma}}(f)(\mathbf{z}) = \sum_{|\vec{\alpha}|=0}^{\infty} \left(\frac{\sum_{\vec{\beta}\le\vec{\alpha}} a_{\vec{\alpha}-\vec{\beta}} \prod_{j=1}^{n} A_{\alpha_j-\beta_j}^{\gamma_j}}{\prod_{j=1}^{n} A_{\alpha_j}^{\gamma_j+1}} \right) \mathbf{z}^{\alpha},$$

where $f(\mathbf{z}) = \sum_{|\vec{\alpha}|=0}^{\infty} a_{\vec{\alpha}} z_1^{\alpha_1} \cdots z_n^{\alpha_n} = \sum_{|\vec{\alpha}|=0}^{\infty} a_{\alpha_1\cdots\alpha_n} z_1^{\alpha_1} \cdots z_n^{\alpha_n}$ is holomorphic on D_n. A simple calculation shows that

$$\mathcal{C}_{\vec{\gamma}}(f)(\mathbf{z}) = C_{\vec{\gamma}} \int_0^1 \cdots \int_0^1 f(t_1 z_1, \ldots, t_n z_n) \prod_{j=1}^{n} \frac{(1-t_j)^{\gamma_j}}{(1-t_j z_j)^{\gamma_j+1}} dt_j, \tag{3.1}$$

where $C_{\vec{\gamma}} = \prod_{j=1}^{n} (\gamma_j + 1)$.

Theorem 3.7. *Let $\vec{\alpha} = (\alpha_1, \ldots, \alpha_n)$ with $\alpha_j > 1$ for $j = 1, \ldots, n$. Then the generalized Cesáro operator $\mathcal{C}_{\vec{\gamma}}$ is bounded on the space $\mathcal{B}_{\vec{\alpha}}(D_n)$. More precisely, there exists a universal constant C depending on n, $\vec{\gamma}$, $\vec{\alpha}$ but independent of f such that*

$$N(\mathcal{C}_{\vec{\gamma}}(f))_{\mathcal{B}_{\vec{\alpha}}(D_n)} \le C N(f)_{\mathcal{B}_{\vec{\alpha}}(D_n)}.$$

Proof. Once again, we may assume that $\gamma_j > -1$ for $j = 1, \ldots, n$. Let f be a function in the Bloch space $\mathcal{B}_{\vec{\alpha}}(D_n)$. Then we have

$$\begin{aligned}
|\mathcal{C}_{\vec{\gamma}}(f)(\mathbf{z})| &\le C_{\vec{\gamma}} \int_0^1 \cdots \int_0^1 |f(t_1 z_1, \ldots, t_n z_n)| \prod_{j=1}^{n} \frac{(1-t_j)^{\gamma_j}}{|1-t_j z_j|^{\gamma_j+1}} dt_j \\
&\le C_{\vec{\gamma}} \int_0^1 \cdots \int_0^1 |f(t_1 z_1, \ldots, t_n z_n)| \prod_{j=1}^{n} \frac{(1-t_j|z_j|)^{\alpha_j}(1-t_j)^{\gamma_j}}{(1-t_j|z_j|)^{\gamma_j+\alpha_j+1}} dt_j \\
&\le C'_{\vec{\gamma}} N(f)_{\mathcal{B}_{\vec{\alpha}}} \int_0^1 \cdots \int_0^1 \prod_{j=1}^{n} \frac{(1-t_j)^{\gamma_j}}{(1-t_j|z_j|)^{\gamma_j+\alpha_j+1}} dt_j \\
&= C'_{\vec{\gamma}} N(f)_{\mathcal{B}_{\vec{\alpha}}} \prod_{j=1}^{n} \int_0^1 \frac{(1-t_j)^{\gamma_j}}{(1-t_j|z_j|)^{\gamma_j+\alpha_j+1}} dt_j
\end{aligned}$$

$$\le C'_{\vec{\gamma}} N(f)_{\mathcal{B}_{\vec{\alpha}}} \prod_{j=1}^{n} \int_0^1 \frac{1}{|1-t_j z_j|^{\alpha_j+1}} dt_j$$

$$= C'_{\vec{\gamma}} N(f)_{\mathcal{B}_{\vec{\alpha}}} \prod_{j=1}^{n} \frac{1}{\alpha_j |z_j|} \frac{1-(1-|z_j|)^{\alpha_j/n}}{(1-|z_j|)^{\alpha_j}}$$

$$\le C'_{\vec{\gamma}} N(f)_{\mathcal{B}_{\vec{\alpha}}} \prod_{j=1}^{n} \frac{1}{(1-|z_j|)^{\alpha_j}},$$

the result follows immediately. □

When $\vec{\gamma} = \vec{\mathbf{0}}$, we use the above technique to recover the following stronger result in [**3**].

Theorem 3.8. *Let $\vec{\alpha} = (\alpha_1, \dots, \alpha_n)$ with $\alpha_j > -1$ for $j = 1, \dots, n$. Then the Cesáro operator $\mathcal{C}_{\vec{0}}$ is bounded on $\mathcal{A}^p(dm_{\vec{\alpha}})$ for $0 < p < \infty$.*

Finally, let us discuss results on Hardy spaces. The Hardy space $H^p(D_n)$, $0 < p < \infty$, is defined by

$$H^p(D_n) = \{f \in \mathcal{H}(D_n), \quad \|f\|_{H^p(D_n)} < \infty\},$$

where

$$\|f\|^p_{H^p(D_n)} = \sup_{0 \le r_j < 1} \int_{-\pi}^{\pi} \cdots \int_{-\pi}^{\pi} |f(r_1 e^{i\theta_1}, \dots, r_n e^{i\theta_n})|^p \prod_{j=1}^{n} d\theta_j.$$

We next state the second theorem related to the Cesáro operator as follows.

Theorem 3.9. *Let $0 < p < \infty$, $\vec{\gamma} = (\gamma_1, \dots, \gamma_n)$ with $\mathrm{Re}(\gamma_j) > -1$ for $j = 1, \dots, n$. Then the generalized Cesáro operator $\mathcal{C}_{\vec{\gamma}}$ is bounded on the Hardy spaces $H^p(D_n)$. More precisely, there exists a universal constant C depending on n, $\vec{\gamma}$ but independent of f such that*

$$\|\mathcal{C}_{\vec{\gamma}}(f)\|_{H^p(D_n)} \le C \cdot \|f\|_{H^p(D_n)}.$$

Proof. Case (1): $0 < p \le 1$. We may use the techniques in [**27, 35**] to prove the result. Without loss of generality, we assume that $\gamma_j > -1$ for $j = 1, \dots, n$. Let $f \in H^p(D_n)$. Denote $\lambda_k = 1 - 2^{-k}$ for $k \in \mathbb{Z}_+$. Then we have

$$\int_{[-\pi,\pi]^n} \left|\mathcal{C}_{\vec{\gamma}}(f)(r_1 e^{i\theta_1}, \dots, r_n e^{i\theta_n})\right|^p \prod_{j=1}^{n} d\theta_j$$

$$\le C \int_{[-\pi,\pi]^n} \left| \int_{[0,1]^n} |f(\rho_1 r_1 e^{i\theta_1} \cdots, \rho_n r_n e^{i\theta_n}) \prod_{j=1}^{n} \frac{(1-\rho_j)^{\gamma_j} d\rho_j}{(1-\rho_j r_j e^{i\theta_j})^{\gamma_j+1}|} \right|^p \prod_{j=1}^{n} d\theta_j$$

$$\leq C \sum_{|\mathbf{k}|=1}^{\infty} \int_{[-\pi,\pi]^n} \int_{\lambda_{k_1-1}}^{\lambda_{k_1}} \cdots \int_{\lambda_{k_n-1}}^{\lambda_{k_n}}$$

$$\times \left(|f(\rho_1 r_1 e^{i\theta_1}, \cdots, \rho_n r_n e^{i\theta_n})| \prod_{j=1}^{n} \frac{(1-\rho_j)^{\gamma_j} d\rho_j}{|(1-\rho_j r_j e^{i\theta_j})^{\gamma_j+1}|} \right)^p \prod_{j=1}^{n} d\theta_j$$

$$\leq C \sum_{|\mathbf{k}|=1}^{\infty} \left[\prod_{j=1}^{n} \frac{1}{2^{pk_j(\gamma_j+1)}} \right] \int_{[-\pi,\pi]^n} \sup_{\Lambda_1} \left| \frac{f(\rho_1 r_1 e^{i\theta_1} \cdots, \rho_n r_n e^{i\theta_n})}{\prod_{j=1}^{n} (1-\rho_j r_j e^{i\theta_j})^{\gamma_j+1}} \right|^p \prod_{j=1}^{n} d\theta_j$$

$$\leq C \sum_{|\mathbf{k}|=1}^{\infty} \left[\prod_{j=1}^{n} \frac{1}{2^{pk_j(\gamma_j+1)}} \right] \int_{[-\pi,\pi]^n} \sup_{\Lambda_2} \left| \frac{f(\rho_1 r_1 e^{i\theta_1} \cdots, \rho_n r_n e^{i\theta_n})}{\prod_{j=1}^{n} (1-\rho_j r_j e^{i\theta_j})^{\gamma_j+1}} \right|^p \prod_{j=1}^{n} d\theta_j \quad (3.2)$$

where

$$\Lambda_1 = \{(\rho_1, \cdots, \rho_n) : \rho_j \in [\lambda_{k_j-1}, \lambda_{k_j}], \quad \text{for } j = 1, \cdots, n\}$$
$$\Lambda_2 = \{(\rho_1, \cdots, \rho_n) : \rho_j \in [0, \lambda_{k_j}], \quad \text{for } j = 1, \cdots, n\}$$

By a result of Bochner [**5**], the last line of above is bounded by

$$C \sum_{|\mathbf{k}|=1}^{\infty} \frac{1}{2^{p\sum_{j=1}^{n} k_j(\gamma_j+1)}} \int_{[-\pi,\pi]^n} \left| \frac{f(\lambda_{k_1} r_1 e^{i\theta_1} \cdots, \lambda_{k_n} r_n e^{i\theta_n})}{\prod_{j=1}^{n} (1-\lambda_{k_j} r_j e^{i\theta_j})^{\gamma_j+1}} \right|^p \prod_{j=1}^{n} d\theta_j$$

$$\leq C \sum_{|\mathbf{k}|=1}^{\infty} \int_{\lambda_{k_1}}^{\lambda_{k_1+1}} \cdots \int_{\lambda_{k_n}}^{\lambda_{k_n+1}} \int_{[-\pi,\pi]^n} \left| \frac{f(\rho_1 r_1 e^{i\theta_1} \cdots, \rho_n r_n e^{i\theta_n})}{\prod_{j=1}^{n} (1-\rho_j r_j e^{i\theta_j})^{\gamma_j+1}} \right|^p \prod_{j=1}^{n} d\theta_j$$

$$\times \prod_{j=1}^{n} (1-\rho_j)^{p(\gamma_j+1)-1} d\rho_j$$

$$\leq C \int_{[0,1]^n} \int_{[-\pi,\pi]^n} \left| \frac{f(\rho_1 r_1 e^{i\theta_1} \cdots, \rho_n r_n e^{i\theta_n})}{\prod_{j=1}^{n} (1-\rho_j r_j e^{i\theta_j})^{\gamma_j+1}} \right|^p \prod_{j=1}^{n} d\theta_j \prod_{j=1}^{n} (1-\rho_j)^{p(\gamma_j+1)-1} d\rho_j$$

Now we may choose $q > 1$ so that $\max_{j=1,\dots,n}\{1 - p(\gamma_j+1)\} < q^{-1}$. Applying Hölder's inequality with exponents q and $q' = \frac{q}{q-1}$, one has

$$\int_{[-\pi,\pi]^n} \left| \frac{f(\rho_1 r_1 e^{i\theta_1} \cdots, \rho_n r_n e^{i\theta_n})}{\prod_{j=1}^{n} (1-\rho_j r_j e^{i\theta_j})^{\gamma_j+1}} \right|^p \prod_{j=1}^{n} d\theta_j$$

$$\leq C \left(\int_{[-\pi,\pi]^n} |f(\rho_1 r_1 e^{i\theta_1}, \dots, \rho_n r_n e^{i\theta_n})|^{pq} \prod_{j=1}^{n} d\theta_j \right)^{\frac{1}{q}}$$

$$\times \left(\int_{[-\pi,\pi]^n} \prod_{j=1}^{n} \frac{d\theta_j}{|1-\rho_j r_j e^{i\theta_j}|^{(\gamma_j+1)pq'}} \right)^{\frac{1}{q'}}$$

$$\leq C \left| \int_{[-\pi,\pi]^n} |f(\rho_1 r_1 e^{i\theta_1}, \ldots, \rho_n r_n e^{i\theta_n})|^{pq} \prod_{j=1}^n d\theta_j \right|^{\frac{1}{q}} \prod_{j=1}^n (1-\rho_j r_j)^{-p(\gamma_j+1)+1-\frac{1}{q}}$$

$$\leq C \left(\int_{[-\pi,\pi]^n} |f(\rho_1 r_1 e^{i\theta_1}, \ldots, \rho_n r_n e^{i\theta_n})|^{pq} \prod_{j=1}^n d\theta_j \right)^{\frac{1}{q}} \prod_{j=1}^n (1-\rho_j)^{-p(\gamma_j+1)+1-\frac{1}{q}}. \tag{3.3}$$

Here we use an estimate from p. 65 of [**17**]:

$$\int_{-\pi}^{\pi} |1 - te^{i\theta}|^{\varepsilon} d\theta \leq C(1-t)^{1-\varepsilon} \quad \text{for } 0 \leq t < 1,\ 1 < \varepsilon < \infty.$$

Combining (3.2) and (3.3), one has

$$\int_{[-\pi,\pi]^n} |\mathcal{C}_{\vec{\gamma}}(f)(r_1 e^{i\theta_1}, \ldots, r_n e^{i\theta_n})|^p \prod_{j=1}^n d\theta_j$$

$$\leq C \int_{[0,1]^n} \left(\int_{[-\pi,\pi]^n} |f(\rho_1 r_1 e^{i\theta_1}, \ldots, \rho_n r_n e^{i\theta_n})|^{pq} \prod_{j=1}^n d\theta_j \right)^{\frac{1}{q}} \prod_{j=1}^n (1-\rho_j)^{-\frac{1}{q}} d\rho_j$$

$$\leq C \int_{[-\pi,\pi]^n} |f(r_1 e^{i\theta_1}, \ldots, r_n e^{i\theta_n})|^p \prod_{j=1}^n d\theta_j.$$

Here we apply a well-known inequality of Hardy and Littlewood [**18**]: For $0 < p < \infty$, $1 < q < \infty$ and $0 \leq r < 1$,

$$\int_0^1 \left(\int_{-\pi}^{\pi} |f(\rho r e^{i\theta})|^{pq} d\theta \right)^{\frac{1}{q}} (1-\rho)^{-\frac{1}{q}} d\rho \leq C \int_{-\pi}^{\pi} |f(re^{i\theta})|^p d\theta$$

whenever $f \in \mathcal{H}(D_1)$. This completes the proof of Case (1).

Case (2): $1 < p < \infty$. Without loss of generality, we may assume that $\gamma_j > -1$ for $j = 1, \ldots, n$. The problem reduces to proving that

$$\int_{-\pi}^{\pi} \cdots \int_{-\pi}^{\pi} |\mathcal{C}_{\vec{\gamma}}(f)(r_1 e^{i\theta_1}, \ldots, r_n e^{i\theta_n})|^p d\theta_1 \cdots d\theta_n$$

$$\leq C \int_{-\pi}^{\pi} \cdots \int_{-\pi}^{\pi} |f(r_1 e^{i\theta_1}, \ldots, r_n e^{i\theta_n})|^p \prod_{j=1}^n d\theta_j,$$

for all $f \in H(D_n)$. For $0 < r < 1$, denote

$$f_{\vec{r}}(e^{i\phi_1}, \ldots, e^{i\phi_n}) = f(r_1 e^{i\phi_1}, \ldots, r_n e^{i\phi_n}).$$

Then for $0 < \rho_j < 1$ with $j = 1, \ldots, n$, $f(\rho_1 r_1 e^{i\phi_1}, \ldots, \rho_n r_n e^{i\phi_n})$ can be written as follows;

$$f(\rho_1 r_1 e^{i\phi_1}, \ldots, \rho_n r_n e^{i\phi_n}) = \frac{1}{(2\pi)^n} \int_{-\pi}^{\pi} \cdots \int_{-\pi}^{\pi} f_{\vec{r}}(e^{i\phi_1}, \ldots, e^{i\phi_n}) \prod_{j=1}^{n} P(\rho_j, \phi_j - \theta_j) d\phi_j \tag{3.4}$$

where

$$P(\rho_j, \phi_j - \theta_j) = \frac{1 - \rho_j^2}{1 - 2\rho_j \cos(\phi_j - \theta_j) + \rho_j^2}$$

is the Poisson kernel on the unit disc. Combining (3.1) and (3.4), one has

$$C_{\vec{\gamma}}(f)(r_1 e^{i\theta_1}, \ldots, r_n e^{i\theta_n}) = \left[\prod_{j=1}^{n} \frac{(\gamma_j + 1)}{2\pi}\right] \int_{[-\pi,\pi]^n} K_{\vec{\gamma}}(\vec{r}, \vec{\theta}, \vec{\phi}) f(r_1 e^{i(\theta_1+\phi_1)}, \ldots, r_n e^{i(\theta_n+\phi_n)}) \prod_{j=1}^{n} d\phi_j,$$

where

$$K_{\vec{\gamma}}(\vec{r}, \vec{\theta}, \vec{\phi}) = \prod_{j=1}^{n} \int_0^1 \frac{(1+\rho_j)(1-\rho_j)^{\gamma_j+1}}{(1 - 2\rho_j \cos\phi_j + \rho_j^2)(1 - r_j \rho_j e^{i\theta_j})^{\gamma_j+1}} d\rho_j.$$

Using an estimate in [p. 621, **1**], we know that there exists a constant $c_{\vec{\gamma}}$ such that

$$|K_{\vec{\gamma}}(\vec{r}, \vec{\theta}, \vec{\phi})| \le c_{\vec{\gamma}} \prod_{j=1}^{n} \int_0^1 \frac{\nu^{\gamma_j+1}}{(\nu^2 + \phi_j^2)(\nu^2 + \theta_j^2)^{\frac{\gamma_j+1}{2}}} d\nu$$

for $|\theta_j| \le \pi$, $|\phi_j| \le \pi$, $j = 1, \ldots, n$. By a result in [**1**], one has

$$\int_0^1 \frac{\nu^{\gamma_j+1}}{(\nu + \phi_j^2)(\nu^2 + \theta_j^2)^{\frac{\gamma_j+1}{2}}} d\nu \le C'_{\gamma_j} \frac{\Theta^{\gamma_j}(\phi_j/\theta_j)}{|\theta_j|}$$

for $\phi_j \in \mathbb{R}$ and $\theta_j \ne 0$. We know that there exists a constant $C'_{\vec{\gamma}}$ such that

$$\int_0^1 \cdots \int_0^1 \prod_{j=1}^{n} \frac{\nu^{\gamma_j+1}}{(\nu + \phi_j^2)(\nu^2 + \theta_j^2)^{\frac{\gamma_j+1}{2}}} d\nu_j \le C'_{\vec{\gamma}} \prod_{j=1}^{n} \frac{\Theta^{\gamma_j}(\phi_j/\theta_j)}{|\theta_j|}.$$

It follows that

$$|K_{\vec{\gamma}}(\vec{r}, \vec{\theta}, \vec{\phi})| \le C'_{\vec{\gamma}} \prod_{j=1}^{n} \frac{\Theta^{\gamma_j}(\phi_j/\theta_j)}{|\theta_j|}$$

for $0 < |\theta_j| \le \pi$, $|\phi_j| \le \pi$, $0 < r_j < 1$ and $j = 1, \ldots, n$. Here

$$\Theta^{\gamma}(\nu) = \begin{cases} \frac{1+|\nu|^{\gamma}}{1+|\nu|} & \text{if } \; -1 < \gamma < 0 \\ \frac{\log(2+|\nu|^{-1})}{1+|\nu|} & \text{if } \; \gamma = 0 \\ \frac{1}{1+|\nu|} & \text{if } \; \gamma > 0. \end{cases}$$

It is easy to see that

$$C_{\vec{\gamma},p} = 2^{\frac{n}{p}} \int_{\mathbb{R}^n} \prod_{j=1}^{n} \frac{\Theta^{\gamma_j}(\nu_j)}{|1+\nu_j|^{\frac{1}{p}}} d\nu_j < \infty. \tag{3.5}$$

Let us define the function $F_{\vec{\nu}}$ as follows:

$$F_{\vec{\nu}}(r_1e^{i\theta_1},\dots,r_ne^{i\theta_n}) = \begin{cases} f(r_1e^{i(\nu_1+1)\theta_1},\dots,r_ne^{i(\nu_n+1)\theta_n}) & \text{for some } j\ |\nu_j\theta_j| \le \pi \\ 0 & \text{otherwise} \end{cases}$$

Therefore,

$$\begin{aligned} &|\mathcal{C}_{\vec{\gamma}}(f)(r_1e^{i\theta_1},\dots,r_ne^{i\theta_n})| \\ \le & A_{\vec{\gamma}} \int_{[-\pi,\pi]^n} |K_{\vec{\gamma}}(\vec{r},\vec{\theta},\vec{\phi})||f(r_1e^{i(\theta_1+\phi_1)},\dots,r_ne^{i(\theta_n+\phi_n)})| \prod_{j=1}^{n} d\phi_j \\ \le & B_{\vec{\gamma}} \int_{[-\pi,\pi]^n} \left|f(r_1e^{i(\theta_1+\phi_1)},\dots,r_ne^{i(\theta_n+\phi_n)})\right| \prod_{j=1}^{n} \frac{\Theta^{\gamma_j}(\phi_j/\theta_j)}{|\theta_j|} d\phi_j \\ = & B_{\vec{\gamma}} \int_{\mathbb{R}^n} \left|F_{\vec{\nu}}(r_1e^{i(\nu_1+1)\theta_1},\dots,r_ne^{i(\nu_n+1)\theta_n})\right| \prod_{j=1}^{n} \Theta^{\gamma_j}(\nu_j) d\nu_j. \end{aligned}$$

Then by Minkowski's inequality, one has

$$\begin{aligned} &\left(\int_{[-\pi,\pi]^n} |\mathcal{C}_{\vec{\gamma}}(f)(r_1e^{i\theta_1},\dots,r_ne^{i\theta_n})|^p \prod_{j=1}^{n} d\theta_j\right)^{\frac{1}{p}} \\ = & \left(\int_{[-\pi,\pi]^n} \left|\int_{\mathbb{R}^n} |F_{\vec{\nu}}(r_1e^{i\theta_1},\dots,r_ne^{i\theta_n})| \prod_{j=1}^{n} \Theta^{\gamma_j}(\nu_j) d\nu_j\right|^p \prod_{j=1}^{n} d\theta_j\right)^{\frac{1}{p}} \\ \le & \int_{\mathbb{R}^n} \left(\int_{[-\pi,\pi]^n} |F_{\vec{\nu}}(r_1e^{i\theta_1},\dots,r_ne^{i\theta_n})|^p \prod_{j=1}^{n} d\theta_j\right)^{\frac{1}{p}} \prod_{j=1}^{n} \Theta^{\gamma_j}(\nu_j) d\nu_j. \end{aligned} \tag{3.6}$$

Now let us turn to the inner integral of (3.6).

$$\begin{aligned} &\int_{[-\pi,\pi]^n} \left|F_{\vec{\nu}}(r_1e^{i\theta_1},\dots,r_ne^{i\theta_n})\right|^p \prod_{j=1}^{n} d\theta_j \\ = & \int_{\Gamma} \left|f(r_1e^{i(\nu_1+1)\theta_1},\dots,r_ne^{i(\nu_n+1)\theta_n})\right|^p \prod_{j=1}^{n} d\theta_j \end{aligned}$$

$$\leq \prod_{j=1}^{n} \frac{1}{|\nu_j + 1|} \int_{\Gamma} |f(r_1 e^{i\theta_1}, \ldots, r_n e^{i\theta_n})|^p \prod_{j=1}^{n} d\theta_j$$

$$\leq \prod_{j=1}^{n} \frac{1}{|\nu_j + 1|} \int_{\otimes_{j=1}^{n}\{\theta_j : |\theta_j| \leq 2\pi\}} |f(r_1 e^{i\theta_1}, \ldots, r_n e^{i\theta_n})|^p \prod_{j=1}^{n} d\theta_j$$

$$\leq \prod_{j=1}^{n} \frac{2^n}{|\nu_j + 1|} \int_{[-\pi,\pi]^n} |f(r_1 e^{i\theta_1}, \ldots, r_n e^{i\theta_n})|^p \prod_{j=1}^{n} d\theta_j. \tag{3.7}$$

Here the set $\Gamma = \otimes_{j=1}^{n}\{\theta_j : |\nu_j \theta_j| \leq \pi\} \cap \{\theta_j : |\theta_j| \leq \pi\}$. Therefore, by (3.5), we have

$$\left(\int_{[-\pi,\pi]^n} |\mathcal{C}_{\vec{\gamma}}(f)(r_1 e^{i\theta_1}, \ldots, r_n e^{i\theta_n})|^p \prod_{j=1}^{n} d\theta_j \right)^{\frac{1}{p}}$$

$$\leq 2^{\frac{n}{p}} \int_{\mathbb{R}^n} \left(\int_{[-\pi,\pi]^n} |f(r_1 e^{i\theta_1}, \ldots, r_n e^{i\theta_n})|^p \prod_{j=1}^{n} d\theta_j \right)^{\frac{1}{p}} \prod_{j=1}^{n} \frac{\Theta^{\gamma_j}}{|\nu_j + 1|^{\frac{1}{p}}} d\nu_j$$

$$\leq C_{\vec{\gamma},p} \left(\int_{[-\pi,\pi]^n} |f(r_1 e^{i\theta_1}, \ldots, r_n e^{i\theta_n})|^p \prod_{j=1}^{n} d\theta_j \right)^{\frac{1}{p}}.$$

This completes the proof of the theorem. □

For $0 < p, q < \infty$, we define the weighted Bergman space $\mathcal{A}^{p,q}_{\vec{\alpha},\vec{\beta}}(D_n)$ is the set of all holomorphic functions defined on D_n such that

$$\|f\|_{\mathcal{A}^{p,q}_{\vec{\alpha},\vec{\beta}}(D_n)}$$

$$= \left[\int_{[0,1]^n} \left(\int_{[-\pi,\pi]^n} |f(r_1 e^{i\theta_1}, \ldots, r_n e^{i\theta_n})|^p \prod_{j=1}^{n} d\theta_j \right)^q \prod_{j=1}^{n} (1 - r_j)^{\alpha_j} r_j^{\beta_j} dr_j \right]^{\frac{1}{q}} < \infty.$$

Here $\vec{\alpha}, \vec{\beta} \in (\mathbb{Z}_+)^n$.

Using Theorem 3.9, we have the following corollary.

Corollary 3.10. *The generalized Cesáro operator is bounded on the space $\mathcal{A}^{p,q}_{\vec{\alpha},\vec{\beta}}(D_n)$ for $1 < p < \infty$ and $0 < q < \infty$. More precisely, there exists a universal constant C depending on n, $\vec{\gamma}$, $\vec{\alpha}$ and $\vec{\beta}$ but independent of f such that*

$$\|\mathcal{C}_{\vec{\gamma}}(f)\|_{\mathcal{A}^{p,q}_{\vec{\alpha},\vec{\beta}}(D_n)} \leq C \cdot \|f\|_{\mathcal{A}^{p,q}_{\vec{\alpha},\vec{\beta}}(D_n)}.$$

References

[1] K.F. Andersen, *Cesáro averaging operators on Hardy spaces,* Proc. Royal Soc. Edinburgh, **126A** (1996), 617–624.

[2] D.E. Barrett, *Behaviour of the Bergman projection on the Diederich-Fornaess worm,* Acta. Math. **168** (1992), 1–10.

[3] G. Benke & D.C. Chang, *A note on weighted Bergman spaces and the Cesáro operator,* Nagoya Math. J., **159** (2000), 25–43.

[4] C. Berenstein, D.C. Chang *and* J. Tie, *Laguerre Calculus and Its Applications in the Heisenberg Group,* AMS/IP Series in Advanced Mathematics #**22**, International Press, Cambridge, Massachusetts, (2001).

[5] S. Bochner, *Classes of holomorphic functions of several variables in circular domains,* Proc. Natl. Acad. Sci. UAS, **46** (1960), 721–723.

[6] A. Bonami, D.C. Chang *and* S. Grellier, *Commutation Properties and Lipschitz estimates for the Bergman and Szegö projections,* Math. Zeit. **223** (1996), 275–302.

[7] A. Bonami & S. Grellier, *Weighted Bergman projections in domains of finite type in* $\mathbb{C}^2$ Contemporary Math. **189** (1995), 65–80.

[8] H.P. Boas & E.J. Straube, *Equivalence of regularity for the Bergman projection and the $\bar\partial$-Neumann operator,* Manuscripta Math., **67** (1990), 25–33.

[9] D. Catlin, *Subelliptic estimates for the $\bar\partial$-Neumann problem on pseudoconvex domains* Ann. of Math., **126** (1987), 131–191.

[10] D.C. Chang & S. Grellier, *Régularité de la projection de Szegö dans domaines découplés de type fini de* $\mathbb{C}^n$ C. R. Acad. Sci. Paris **315** (1992), 1365–1370.

[11] D.C. Chang & B.Q. Li, *Sobolev and Lipschitz Estimates for weighted Bergman projections* Nagoya Mathematical Journal, **147** (1997), 147–178.

[12] D.C. Chang, A. Nagel & E.M. Stein, *Estimates for the $\bar\partial$-Neumann problem in pseudoconvex domains of finite type in* $\mathbf{C}^2$ Acta Mathemtica, **169** (1992), 153–228.

[13] M. Christ, *Global C^∞-irregularity of the $\bar\partial$-Neumann problem for worm domains,* J. Amer. Math. Soc., **9** (1996), 1171–1185.

[14] M. Christ, *Remarks on global irregularity in the $\bar\partial$-Neumann problem,* Several Complex Variables, MSRI Publications, **37** (1999), 161–198.

[15] K. Diederich & J.E. Fornaess, *Pseudoconvex domains: an example with nontrivial Nebenhülle,* Math. Ann., **225** (1977), 275–292.

[16] A.E. Djrbashian *and* F.A. Shamoian, *Topics in the Theory of A^p_α Spaces,* Teubner Verlagsgesellschaft, Leipzig (1988).

[17] P. Duren, *Theory of H^p Spaces,* Academic Press, New York,(1970).

[18] C.L. Fefferman, *The Bergman kernel and biholomorphic mappings of pseudo-convex domains,* Invent. Math., **26** (1974), 1–65.

[19] G.H. Hardy *and* J.E. Littlewood, *Some properties of fractional integrals II,* Math. Zeit., **34** (1932), 403–439.

[20] C. Kiselman, *A study of the Bergman projection in certain Hartogs domains,* Proc. Symp. Pure Math., AMS, **52** (1991), 219–231.

[21] J.J. Kohn, *Boundaries of complex manifolds,* Proceedings of Conference on Complex Analysis, Minneapolis (1964), 81–94.

[22] J.J. Kohn, *Boundary behaviour of $\bar\partial$ on weakly pseudo-convex manifolds of dimension two,* J. Differential Geom., **6** (1972), 523–542.

[23] J.D. McNeal, *Local geometry of decoupled pseudoconvex domains,* Complex Analysis (Wuppertal, 1990) (K. Diederich, ed.), (1991), 223–230.

[24] J.D. McNeal, *Convex domains of finite type,* J. Funct. Anal., **108** (1992), 361–373.

[25] J.D. McNeal, *Estimates on the Bergman kernels of convex domains,* Adv. Math., **109** (1996), 108–139.

[26] J.D. McNeal & E.M. Stein, *Mapping properties of the Bergman projection on convex domains of finite type,* Duke Math. J., **73** (1994), 177–199.

[27] J. Miao, *The Cesáro operator is bounded* H^p *for* $0 < p < 1$, Proc. Amer. Math. Soc., **116** (1992), 1077–1079.

[28] A. Nagel, E.M. Stein & S. Wainger, *Balls and metrics defined by vector fields I: basic properties,* Acta Math., **155** (1985), 103–147.

[29] A. Nagel, J.-P. Rosay, E.M. Stein & S. Wainger, *Estimates for the Bergman and Szegö kernels in* $\mathbb{C}^2$, Ann. Math., **129** (1989), 113–149.

[30] A.G. Siskakis, *The Cesáro operator is bounded on* H^1, Proceedings of AMS, **110** (1990), 461–462.

[31] Y.T. Siu, *Non-Hölder property of Bergman projection of smooth worm domain,* in **Aspects of Mathematics** (editor: N. Mok), University of Hong Kong Press, Hong Kong, (2001), 264–304.

[32] E.M. Stein, *Harmonic Analysis: Real-Variable Methods, Orthogonality, and Oscillatory Integrals,* Princeton University Press, Princeton, New Jersey, (1993).

[33] K. Stempak, *Cesáro averaging operators,* Proc. Royal Soc. Edinburgh, **124A**(1994), 121–126.

[34] S. Stević, *Weighted integrals of holomorphic functions in* $\mathbb{C}^n$, Preprint, (2001).

[35] S. Stević, *Cesáro averaging operators,* Preprint, (2001).

[36] J. Xiao, *Cesáro-type operators on Hardy, BMOA and Bloch spaces,* Arch. Math.,**68** (1997), 398–406.

Der-Chen Chang
Department of Mathematics,
Georgetown University,
Washington, DC 20057-1233
e-mail: chang@math.georgetown.edu

Robert Gilbert
Department of Mathematics,
University of Delaware,
Newark, DE 19716-2553
e-mail: gilbert@math.udel.edu

Jingzhi Tie
Department of Mathematics,
The University of Georgia,
Athens, GA 30602-7403
e-mail: jtie@math.uga.edu

Operator Theory:
Advances and Applications, Vol. 143, 171–212

Linear Fractional Transformations, Riccati Equations and Bitangential Interpolation, Revisited

Harry Dym

Abstract. A characterization of the set of linear fractional transformations $(w_{11}\varepsilon + w_{12})(w_{21}\varepsilon + w_{22})^{-1}$, where the wij are the block entries of a j_{pq} inner matrix-valued function and ε runs over the set of $p \times q$ matrix-valued functions that are holomorphic and contractive in either the open unit disk or the open upper (or right) half-plane is derived. This formula leads to transparent proofs of a number of bitangential interpolation problems. The analysis uses methods based on the special class of reproducing kernel Hilbert spaces that were introduced and extensively studied by L. de Branges. Connections with Riccati equations are exploited.

1. Introduction

Let Ω_+ stand for either the open unit disk $\mathbb{D}$, or the open upper half plane $\mathbb{C}_+$, or the open right half-plane Π_+. Let $\mathcal{S}^{p\times q}(\Omega_+)$ denote the Schur class of $p \times q$ mvf's (matrix-valued functions) that are both holomorphic and contractive in Ω_+ and let

$$j_{pq} = \begin{bmatrix} I_p & 0 \\ 0 & -I_q \end{bmatrix}, \text{ where } \quad p \geq 1, q \geq 1 \text{ and } p + q = m.$$

It is well known that if

$$W(\lambda) = \begin{bmatrix} w_{11}(\lambda) & w_{12}(\lambda) \\ w_{21}(\lambda) & w_{22}(\lambda) \end{bmatrix}$$

is a j_{pq}-inner mvf with respect to Ω_+, with diagonal blocks $w_{11}(\lambda)$ of size $p \times p$ and $w_{22}(\lambda)$ of size $q \times q$, then the linear fractional transformation

$$T_W[\varepsilon] = (w_{11}\varepsilon + w_{12})(w_{21}\varepsilon + w_{22})^{-1} \tag{1.1}$$

2000 *Mathematics Subject Classification.* 46E22, 47A57, 30E05.

Key words and phrases. J-inner, linear fractional transformations, Schur class, reproducing kernel Hilbert spaces, Lyapunov-Stein equations, Riccati equations, bitangential interpolation.
The author thanks Renee and Jay Weiss for endowing the Chair that supports his research.

maps every mvf $\varepsilon \in \mathcal{S}^{p\times q}(\Omega_+)$ into $\mathcal{S}^{p\times q}(\Omega_+)$; see, e.g., [D1]. A more difficult problem is to furnish a useful description of the set

$$T_W[\mathcal{S}^{p\times q}] = \{T_W[\varepsilon] : \varepsilon \in \mathcal{S}^{p\times q}\}. \tag{1.2}$$

In this paper we shall use RKHS (reproducing kernel Hilbert space) methods to obtain such a description. The main theorem is Theorem 3.1, which is established in Section 3. In order to make the paper reasonably self contained, a number of preliminary facts about RKHS's and a number of examples of such spaces are presented in Section 2. Applications of the main theorem to bitangential interpolation in the closed upper half-plane and the closed unit disk are furnished in Sections 4 and 5, respectively.

Most of the material that is covered in this paper is well known. Bitangential interpolation problems have been extensively covered in the literature by a variety of methods, see, e.g., the monographs [BGR], [DFK], [D1], [FF], [KrN], [Sak] and the papers [ABDS1]–[ABDS3], [Ar], [BaH], [D4], [BD], [Du], [KKY], [KhY] and [Kh1], and the references cited therein. The emphasis and methods employed in the present paper are somewhat different. To begin with, the focus, as we have already noted, is to obtain a description of the set $T_W[\mathcal{S}^{p\times q}(\Omega_+)]$. Moreover, in keeping with the central theme of this volume, we shall obtain this description by using RKHS methods. (Some of the estimates that come into play can also be obtained by the method of fundamental matrix inequalities that originates with Potapov; see, e.g., [Ka] for an introduction and [BD] for applications of this method to bitangential interpolation.) Part of the proof of the main theorem is adapted from an argument due to A.L. Sakhnovich [Sak-A] that was used to study a class of linear fractional transformations based on a different signature matrix that is connected with the Carathéodory class, rather than the Schur class. Sakhnovich worked in $\mathbb{C}_+$. We also extend this method to the setting of the disk, which requires different estimates. In our analysis we exploit the RKHS's $\mathcal{H}(s)$ and $\mathcal{H}_*(s)$. The first of these spaces was characterized in an elegant way by de Branges and Rovnyak [BrR]; the latter is in some sense a mirror image of the former, which seems to be less popular (even though each of these spaces can be identified with range of an operator of the form $(I - TT^*)^{1/2}$ endowed with the appropriate norm, for a suitably chosen contraction T). Details on both of these spaces and others will be provided in Section 2. Another novelty of the exhibited approach is the use of Riccati equations to characterize some finite-dimensional reproducing kernel Hilbert spaces that are connected with bitangential interpolation problems with singular Pick matrices.

1.1. Notation

With the aid of some flexible notation that is summarized in the next table, the analysis in each of the three classical regions (and their closures) can be carried out in much the same way.

We shall also set

$$\Omega_- = \mathbb{C} \setminus (\Omega_+ \cup \Omega_0) = \{\omega \in \mathbb{C} : \rho_\omega(\omega) < 0\}. \tag{1.3}$$

Ω_+	$\mathbb{D}$	$\mathbb{C}_+$	Π_+
$\rho_\omega(\lambda)$	$1-\lambda\overline{\omega}$	$-2\pi i(\lambda-\overline{\omega})$	$2\pi(\lambda+\overline{\omega})$
ω^o	$1/\overline{\omega}$	$\overline{\omega}$	$-\overline{\omega}$
Ω_0	$\mathbb{T}$	$\mathbb{R}$	$i\mathbb{R}$
$\langle f,g\rangle$	$\frac{1}{2\pi}\int_0^{2\pi} g(e^{i\theta})^* f(e^{i\theta})d\theta$	$\int_{-\infty}^{\infty} g(x)^* f(x)dx$	$\int_{-\infty}^{\infty} g(iy)^* f(iy)dy$

TABLE 1. The classical regions

Most of the other notation that we use will be fairly standard:
mvf for matrix-valued function, vvf for vector-valued function, $\sigma(A)$ for the spectrum of the matrix A, $\mathcal{N}_A$ for its nullspace and $\mathcal{R}_A$ for its range. If $f(\lambda)$ is a mvf, then

$f^\#(\lambda) = f(\lambda^o)^*$, $\mathfrak{H}_f$ denotes the set of points at which f is holomorphic,

$$\Re f(\lambda) = \frac{f(\lambda)+f(\lambda)^*}{2} \quad \text{and} \quad \Im f(\lambda) = \frac{f(\lambda)-f(\lambda)^*}{2i}.$$

An $m\times m$ constant matrix J is said to be a signature matrix if it is both Hermitian and unitary with respect to the standard inner product in $\mathbb{C}^m$. Thus, if $J \neq \pm I_m$, then J is unitarily equivalent to j_{pq} with

$$p = \text{ rank } (I_m + J) \text{ and } q = \text{ rank } (I_m - J).$$

An $m \times m$ mvf $U(\lambda)$ is said to belong to the class $\mathcal{U}(J,\Omega_+)$ if it is J-inner with respect to Ω_+, i.e., if it is meromorphic in Ω_+ and

1. $U(\lambda)^*JU(\lambda) \leq J$ for every point $\lambda \in \Omega_+ \cap \mathfrak{H}_U$
2. $U(\mu)^*JU(\mu) = J$ for a.e. point $\mu \in \Omega_0$.

The values $U(\mu)$ on the boundary Ω_0 of Ω_+ exist as nontangential limits, since the inequality in the first item forces $U(\lambda)$ to be of bounded type in Ω_+; see, e.g., [D1].

Every J-inner mvf $U(\lambda)$ admits a pseudo-meromorphic extension to Ω_- by the rule

$$U(\omega) = JU^\#(\omega)^{-1}J = JU(\omega^o)^{-*}J$$

where ω^o, the reflection of ω about Ω_0, is defined in Table 1. An important role will be played by the mvf

$$\Delta_s(\mu) = \begin{bmatrix} I_p & -s(\mu) \\ -s(\mu)^* & I_q \end{bmatrix} \text{ for a.e. } \mu \in \Omega_0, \tag{1.4}$$

which is positive semidefinite for every $s \in \mathcal{S}^{p\times q}(\Omega_+)$.

1.2. R_α invariance

A major role in this subject is played by the generalized backwards shift operator R_α that acts on matrix-valued meromorphic functions $F(\lambda)$ by the rule

$$R_\alpha F(\lambda) = \frac{F(\lambda) - F(\alpha)}{\lambda - \alpha} \tag{1.5}$$

for every point $\alpha \in \mathfrak{H}_F$. In the next chapter we shall consider Hilbert spaces of vector-valued functions that are invariant under the action of R_α for one or more appropriately chosen points $\alpha \in \mathbb{C}$.

2. Reproducing kernel Hilbert spaces

A Hilbert space $\mathcal{H}$ of complex-valued functions that are defined on a set Ω is said to be a RKHS if for each point $\omega \in \Omega$ the evaluation operator that sends

$$f \in \mathcal{H} \to f(\omega) \in \mathbb{C}$$

is a bounded linear functional. Then, by the Riesz representation theorem, there exists a unique vector $k_\omega \in \mathcal{H}$ such that

$$f(\omega) = \langle f, k_\omega \rangle_{\mathcal{H}} \ . \tag{2.1}$$

Moreover,

$$\max\{|f(\omega)| : f \in \mathcal{H} \text{ and } \|f\|_{\mathcal{H}} \leq 1\} = \|k_\omega\|_{\mathcal{H}} \ . \tag{2.2}$$

Conversely, it follows readily from the Cauchy-Schwarz inequality that if there exists a vector $k_\omega \in \mathcal{H}$ such that (2.1) holds for every $f \in \mathcal{H}$, then point evaluation is a bounded linear functional. Thus, *$\mathcal{H}$ is a RKHS if and only if there exists a a vector $k_\omega \in \mathcal{H}$ such that* (2.1) *holds for each point $\omega \in \Omega$ and every $f \in \mathcal{H}$.* Continuing in the same spirit, we shall say that a Hilbert space $\mathcal{H}$ of $m \times 1$ vvf's (vector-valued functions) on a set Ω is a RKHS if there exists an $m \times m$ mvf $K_\omega(\lambda)$ on $\Omega \times \Omega$ such that the following two conditions hold for every point $\omega \in \Omega$ and every vector $u \in \mathbb{C}^m$:

1. $K_\omega u \in \mathcal{H}$.
2. $\langle f, K_\omega u\rangle_{\mathcal{H}} = u^* f(\omega)$ for every $f \in \mathcal{H}$. (2.3)

The main facts are:

- The RK is unique.
- $K_\alpha(\beta)^* = K_\beta(\alpha)$.

- $K_\omega(\lambda) \succeq 0$, i.e., the kernel $K_\omega(\lambda)$ is positive in the sense that for every choice of points $\omega_1, \ldots, \omega_n \in \Omega$, vectors $u_1, \ldots, u_n \in \mathbb{C}^m$ and every integer $n \geq 1$,

$$\sum_{i,j=1}^{n} u_i^* K_{\omega_j}(\omega_i) u_j \geq 0 \,. \tag{2.4}$$

All of these statements are fairly easy consequences of formula (2.3).

- An $m \times 1$ vvf f that is defined on Ω belongs to the RKHS $\mathcal{H}$ with RK $K_\omega(\lambda)$ if and only if there exists a constant $\gamma > 0$ such that the kernel

$$L_\omega(\lambda) = K_\omega(\lambda) - \frac{f(\lambda)f(\omega)^*}{\gamma}$$

is a positive kernel in the sense described in the previous item. In this instance

$$\inf\{\gamma > 0 : K_\omega(\lambda) - \frac{f(\lambda)f(\omega)^*}{\gamma} \succeq 0\} = \|f\|_{\mathcal{H}}^2 \,. \tag{2.5}$$

There is also a useful fact that is due to Aronszajn [Arz] (for the scalar case).

- Let $K_\omega(\lambda)$ be an $m \times m$ matrix-valued function on $\Omega \times \Omega$ that is positive in the sense described above. Then there is a unique RKHS $\mathcal{H}$ of $m \times 1$ vvf's on Ω with $K_\omega(\lambda)$ as its RK.

2.1. Examples of RKHS's

Example 1. Orthogonal polynomials with respect to a weight on the unit circle.

Discussion: Let $\varphi_j(\zeta)$, $j = 0, \ldots, n$, be monic polynomials of degree j that are orthogonal with respect to a weight $\delta(e^{i\theta}) \geq 0$ on the unit circle:

$$\begin{aligned} \langle \varphi_j, \varphi_j \rangle &= \frac{1}{2\pi} \int_0^{2\pi} \varphi_j(e^{i\theta}) \varphi_k(e^{i\theta})^* \delta(e^{i\theta}) d\theta \\ & \qquad \begin{cases} = 0 & \text{if } j \neq k \\ > 0 & \text{if } j = k \,. \end{cases} \end{aligned}$$

Then,

$$\mathcal{M}_n = \text{span}\{\varphi_j(\lambda) : 0 \leq j \leq n\}$$

is a RKHS with RK

$$K_\omega(\lambda) = \sum_{j=0}^{n} \frac{\varphi_j(\lambda)\varphi_j(\omega)^*}{\|\varphi_j\|^2} \,.$$

The well-known formula

$$\sum_{j=0}^{n} \frac{\varphi_j(\lambda)\varphi_j(\omega)^*}{\|\varphi_j\|^2} = \frac{\psi_{n+1}(\lambda)\psi_{n+1}(\omega)^* - \varphi_{n+1}(\lambda)\varphi_{n+1}(\omega)^*}{(1 - \lambda\overline{\omega})\|\varphi_{n+1}\|^2},$$

in which $\psi_k(\lambda) = \lambda^k \psi_k^{\#}(\lambda)$ is the reversed polynomial, is another manifestation of the uniqueness of the reproducing kernel.

Example 2. The Bergman space $A^2(\mathbb{D})$ is the set of functions $f(\lambda)$ that are holomorphic in $\mathbb{D}$ and meet the bound

$$\int_0^1 rdr \int_0^{2\pi} |f(re^{i\theta})|^2 d\theta < \infty.$$

Discussion: It is readily checked that $A^2(\mathbb{D})$ is a RKHS with respect to the inner product

$$\langle f, g\rangle_{A^2} = \frac{1}{\pi}\int_0^1 rdr \int_0^{2\pi} g(re^{i\theta})^* f(re^{i\theta})d\theta = \sum_{k=0}^{\infty} \frac{\widehat{g}(k)^*\widehat{f}(k)}{k+1},$$

where $\widehat{f}(k)$ denotes the k'th coefficient in the Taylor series expansion of $f(\lambda)$. The RK is given by the formula

$$K_\omega(\lambda) = \frac{1}{\rho_\omega(\lambda)^2} = \left(\sum_{j=0}^{\infty} \lambda^j \overline{\omega}^j\right)^2, \quad \lambda, \omega \in \mathbb{D}.$$

For additional information and generalizations to weighted Bergman spaces, see [HKZ].

Example 3. The Dirichlet space $\mathcal{D}$ is the set of functions $f(\lambda)$ that are holomorphic in $\mathbb{D}$ such that $f'(\lambda)$ belongs to the Bergman space.

Discussion: $\mathcal{D}$ is a RKHS with respect to the inner product

$$\begin{aligned}\langle f, g\rangle_{\mathcal{D}} &= \frac{1}{\pi}\int_0^1 rdr \int_0^{2\pi} g'(re^{i\theta})^* f'(re^{i\theta})d\theta + \frac{1}{2\pi}\int_0^{2\pi} g(e^{i\theta})^* f(e^{i\theta})d\theta \\ &= \sum_{k=0}^{\infty}(k+1)\widehat{g}(k)^*\widehat{f}(k)\end{aligned}$$

with RK given by

$$K_\omega(\lambda) = \frac{-1}{\lambda\overline{\omega}}\log\{\rho_\omega(\lambda)\} = \sum_{j=0}^{\infty} \frac{\lambda^j\overline{\omega}^j}{j+1}, \quad \lambda, \omega \in \mathbb{D}.$$

The indicated boundary values in the last integral exist as nontangential limits by Fatou's lemma, which is applicable because every $f \in \mathcal{D}$ belongs to the Hardy space $H_2(\mathbb{D})$.

Example 4. The vector Hardy space $H_2^r(\Omega_+)$ of $r \times 1$ vector-valued functions with components in the scalar Hardy space $H_2(\Omega_+)$ ($= H_2^r(\Omega_+)$ with $r = 1$).

Discussion: It is readily checked that $u/\rho_\omega(\lambda)$ belongs to $H_2^r(\Omega_+)$ as a function of λ for every choice of $\omega \in \Omega_+$ and $u \in \mathbb{C}^r$. Moreover, the requisite identity

$$\left\langle f, \frac{u}{\rho_\omega}\right\rangle = u^* f(\omega)$$

for $f \in H_2^r(\Omega_+)$ (with respect to the standard inner product given in the table) is just Cauchy's formula for the Hardy spaces $H_2(\Omega_+)$. Thus, we see that $H_2^r(\Omega_+)$ is a RKHS with RK

$$K_\omega(\lambda) = \frac{I_r}{\rho_\omega(\lambda)}\ , \quad \lambda, \omega \in \Omega_+ \ .$$

Example 5. The vector Hardy space $H_2^r(\Omega_+)^\perp$ of $r \times 1$ vector-valued functions with components in the scalar Hardy space $H_2(\Omega_+)^\perp$ ($= H_2^r(\Omega_+)^\perp$ with $r = 1$).

Discussion: The superscript $\perp$ signifies the orthogonal complement with respect to the inner product that is specified in terms of integrals over Ω_0 in the table. The requisite identity (2.3) is again verified with the aid of Cauchy's formula when the RK

$$K_\omega(\lambda) = -\frac{I_r}{\rho_\omega(\lambda)}\ , \quad \lambda, \omega \in \Omega_- .$$

We remark that if Ω_- is equal to either the open lower half-plane $\mathbb{C}_-$ or the open left half-plane Π_-, then $H_2^r(\Omega_+)^\perp = H_2^r(\Omega_-)$. However, if $\Omega_- = \{\lambda \in \mathbb{C} : |\lambda| > 1\}$, then $H_2^r(\Omega_+)^\perp = H_2^r(\Omega_-) \ominus \mathbb{C}^r$ and, in this case, the RK of $H_2^r(\Omega_-)$ is

$$K_\omega(\lambda) = -\lambda\overline{\omega}\frac{I_r}{\rho_\omega(\lambda)} = \sum_{j=0}^{\infty}(\lambda\overline{\omega})^{-j} I_r, \ \lambda, \omega \in \mathbb{C} \setminus (\mathbb{D} \cup \mathbb{T}) \ .$$

2.2. Some positive kernels

It is well known that:

1. $s \in \mathcal{S}^{p\times q}(\Omega_+) \Longrightarrow \{I_p - s(\lambda)s(\omega)^*\}/\rho_\omega(\lambda) \succeq 0$ in $\Omega_+ \times \Omega_+$.
2. $s \in \mathcal{S}^{p\times q}(\Omega_+) \Longrightarrow \{s^\#(\lambda)s^\#(\omega)^* - I_q\}/\rho_\omega(\lambda) \succeq 0$ in $\Omega_- \times \Omega_-$.
3. $U \in \mathcal{U}(J, \Omega_+) \Longrightarrow \{J - U(\lambda)JU(\omega)^*\}/\rho_\omega(\lambda) \succeq 0$ in $\mathfrak{H}_U \times \mathfrak{H}_U$. (2.6)

Therefore, by the last item in the property list furnished above, each of the indicated kernels is the RK (reproducing kernel) of exactly one RKHS. These spaces will be designated $\mathcal{H}(s), \mathcal{H}_*(s)$ and $\mathcal{H}(U)$, respectively. The next set of examples supply more detailed descriptions of these spaces.

Example 6. The RKHS $\mathcal{H}(s)$, $s \in \mathcal{S}^{p\times q}(\Omega_+)$.

Discussion: The very elegant description of the space $\mathcal{H}(s)$ that is supplied below is due to de Branges and Rovnyak [BrR]. Let $s \in \mathcal{S}^{p\times q}(\Omega_+)$ and for $f \in H_2^p(\Omega_+)$ let

$$\alpha_f = \sup\{\|f + sg\|^2 - \|g\|^2 : g \in H_2^q(\Omega_+)\} \ .$$

Then the space

$$\mathcal{H}(s) = \{f \in H_2^p(\Omega_+) : \alpha_f < \infty\}$$

is a RKHS with respect to the norm

$$\|f\|_{\mathcal{H}(s)} = \sqrt{\alpha_f} \ .$$

The RK of $\mathcal{H}(s)$ is given by the formula

$$K_\omega(\lambda) = \frac{I_p - s(\lambda)s(\omega)^*}{\rho_\omega(\lambda)}\ , \quad \lambda, \omega \in \Omega_+ \ .$$

The space $\mathcal{H}(s)$ can also be identified as the range of $(I-TT^*)^{1/2}$ acting on $H_2^p(\Omega_+)$ equipped with the norm

$$\|(I-TT^*)^{1/2}f\|_{range} = \inf\{\|f-g\| : g \in H_2^p(\Omega_+) \text{ and } (I-TT^*)^{1/2}g = 0\},$$

where T denotes the linear operator that is defined by the rule

$$T : f \in H_2^q(\Omega_+) \longrightarrow sf \in H_2^p(\Omega_+).$$

It is instructive to verify that if $s \in \mathcal{S}^{p\times q}(\Omega_+)$ is inner, i.e., if $s(\mu)$ is isometric a.e. on Ω_0, then

$$\mathcal{H}(s) = H_2^p(\Omega_+) \ominus sH_2^q(\Omega_+) .$$

Example 7. The RKHS $\mathcal{H}_*(s)$, $s \in \mathcal{S}^{p\times q}(\Omega_+)$.

Discussion: The space $\mathcal{H}_*(s)$ is basically the mirror image of the space $\mathcal{H}(s)$ discussed in the last example.

Let $s \in \mathcal{S}^{p\times q}(\Omega_+)$ and for $f \in H_2^q(\Omega_+)^\perp$ let

$$\beta_f = \sup\{\|f + s^\# g\|^2 - \|g\|^2 : g \in H_2^p(\Omega_+)^\perp\} .$$

Then the space

$$\mathcal{H}_*(s) = \{f \in H_2^q(\Omega_+)^\perp : \beta_f < \infty\}$$

is a RKHS with respect to the norm

$$\|f\|_{\mathcal{H}_*(s)} = \sqrt{\beta_f} .$$

The RK of $\mathcal{H}_*(s)$ is given by the formula

$$K_\omega(\lambda) = \frac{s^\#(\lambda)s^\#(\omega)^* - I_q}{\rho_\omega(\lambda)} \quad , \quad \lambda, \omega \in \Omega_- .$$

The space $\mathcal{H}(s)$ can also be identified as the range of $(I-TT^*)^{1/2}$ acting on $H_2^q(\Omega_+)^\perp$ equipped with the norm

$$\|(I-TT^*)^{1/2}f\|_{range} = \inf\{\|f-g\| : g \in H_2^q(\Omega_+)^\perp \text{ and } (I-TT^*)^{1/2}g = 0\},$$

where T denotes the linear operator that is defined by the rule

$$T : f \in H_2^p(\Omega_+)^\perp \longrightarrow s^\# f \in H_2^q(\Omega_+)^\perp.$$

It is instructive to verify that if $s \in \mathcal{S}^{p\times q}$ is co-inner, i.e., if $s(\mu)$ is co-isometric a.e. on Ω_0, then

$$\mathcal{H}_*(s) = H_2^q(\Omega_+)^\perp \ominus s^\# H_2^p(\Omega_+)^\perp .$$

Example 8. The RKHS $\mathcal{H}(W)$, $W \in \mathcal{U}(j_{pq})$.

Discussion: If $W \in \mathcal{U}(j_{pq})$, $p \geq 1$, $q \geq 1$, $p+q = m$, then the space $\mathcal{H}(W)$ can be characterized in terms of the block entries of the Potapov-Ginzburg transform $s \in \mathcal{S}^{m\times m}$ of $W(\lambda)$:

$$s(\lambda) = \begin{bmatrix} s_{11}(\lambda) & s_{12}(\lambda) \\ s_{21}(\lambda) & s_{22}(\lambda) \end{bmatrix} = \begin{bmatrix} w_{11}(\lambda) & w_{12}(\lambda) \\ 0 & I_q \end{bmatrix} \begin{bmatrix} I_p & 0 \\ w_{21}(\lambda) & w_{22}(\lambda) \end{bmatrix}^{-1} \tag{2.7}$$

as follows: Let

$$f = \begin{bmatrix} g \\ h \end{bmatrix}$$

have components g and h of height p and q, respectively. Then $f \in \mathcal{H}(W)$ if and only if

1. $[I_p \quad -s_{12}]f \in H_2^p(\Omega_+)$
2. $[-(s_{12})^\# \quad I_q]f \in H_2^q(\Omega_+)^\perp$
3. $s_{11}^\# g \in H_2^p(\Omega_+)^\perp$
4. $s_{22}h \in H_2^q(\Omega_+)$.

Moreover, if these conditions are met, then

$$\|f\|^2_{\mathcal{H}(W)} = \|g - s_{12}h\|^2 + \|s_{22}h\|^2 = \langle \Delta_{s_{12}} f, f\rangle\ . \tag{2.8}$$

A proof of this characterization can be found in [D1]. More precisely, the forward direction is stated explicitly as Theorem 2.7. The other direction is easily filled in thanks to Theorem 2.4 of [D1], but is not spelled out explicitly. This characterization of $\mathcal{H}(W)$ was also obtained by other methods in [Ara].

We shall have particular interest in the RKHS's $\mathcal{H}(U)$ for mvf's $U(\lambda)$ that are J-inner with respect to Ω_+. A fundamental theorem of de Branges [Br] characterizes RKHS's with RK's of the form (2.6).

Theorem 2.1. *Let $\mathcal{H}$ be a RKHS of $\mathbb{C}^m$-valued functions that are holomorphic in an open subset Ω of $\mathbb{C}$ that is symmetric with respect to $\mathbb{R}$. Then the RK $K_\omega(\lambda)$ of $\mathcal{H}$ can be expressed in the form*

$$K_\omega(\lambda) = \frac{J - U(\lambda)JU(\omega)^*}{\rho_\omega(\lambda)} \tag{2.9}$$

for some choice of $m \times m$ mvf $U(\lambda)$ that is holomorphic in Ω and some signature matrix J if and only if the following two conditions are met:

1. *$R_\alpha f \in \mathcal{H}$ for every choice of $\alpha \in \Omega$ and $f \in \mathcal{H}$.*
2. *The identity*

$$\langle R_\alpha f, g\rangle_{\mathcal{H}} - \langle f, R_\beta g\rangle_{\mathcal{H}} - (\alpha - \overline{\beta})\langle R_\alpha f, R_\beta g\rangle_{\mathcal{H}} = 2\pi i g(\beta)^* J f(\alpha) \tag{2.10}$$

holds for every choice of α, β in Ω and f, g in $\mathcal{H}$.

Moreover, if $\Omega \cap \mathbb{R} \neq \emptyset$, then the function $U(\lambda)$ that appears in (2.10) is uniquely specified by the formula

$$U(\lambda) = I_m - \rho_\mu(\lambda)K_\mu(\lambda)J \tag{2.11}$$

for some point $\mu \in \Omega \cap \mathbb{R}$, up to a constant J-unitary factor on the right.

In his original formulation, de Branges assumed that $\Omega \cap \mathbb{R} \neq \emptyset$. This condition was later shown to be superfluous by Rovnyak [Ro]. An analogue of this theorem for $\Omega_0 = \mathbb{T}$ was worked out by Ball [Ba]. A general formulation that includes both these theorems and others in a unified framework that covers Krein spaces too is presented in [AlD2]. In particular, for the other two classical choices of Ω_+, the preceding theorem can be modified as follows:

Theorem 2.2. *Theorem* 2.1 *is valid if* $\mathbb{R}$ *is replaced by* Ω_0, *providing that* $\rho_\omega(\lambda)$ *is defined as in the table and the structural identity* (2.10) *is replaced by:*

$$\langle (I+\alpha R_\alpha)f, (I+\beta R_\beta)g\rangle_{\mathcal{H}} - \langle R_\alpha f, R_\beta g\rangle_{\mathcal{H}} = g(\beta)^* J f(\alpha) \qquad (2.12)$$

if $\Omega_+ = \mathbb{D}$ *and*

$$\langle R_\alpha f, g\rangle_{\mathcal{H}} + \langle f, R_\beta g\rangle_{\mathcal{H}} + (\alpha+\beta^*)\langle R_\alpha f, R_\beta g\rangle_{\mathcal{H}} = -2\pi\, g(\beta)^* J f(\alpha) \qquad (2.13)$$

if $\Omega_+ = \Pi_+$,

2.3. Some instructive finite-dimensional examples

Example 9. Let $F(\lambda)$ be an $m \times n$ matrix-valued meromorphic function in $\mathbb{C}$ and assume that the columns of $F(\lambda)$ are linearly independent in the sense that if $u \in \mathbb{C}^n$ and $F(\lambda)u = 0$ for all points λ in some nonempty open subset of $\mathbb{C}$, then $u = 0$. Let $P > 0$ be any positive definite $n \times n$ matrix. Then the space

$$\mathcal{M} = \{F(\lambda)u : u \in \mathbb{C}^n\}$$

endowed with the inner product

$$\langle Fu, Fv\rangle_M = v^* P u \qquad (2.14)$$

is a RKHS.

Discussion: It is readily checked that

$$K_\omega(\lambda) = F(\lambda)P^{-1}F(\omega)^*$$

is the RK for $\mathcal{M}$.

The role of the two conditions in Theorems 2.1 and 2.2 becomes particularly transparent when $\mathcal{H}$ is finite-dimensional. Indeed, if the n-dimensional space $\mathcal{M}$ considered in Example 9 is R_α invariant for some point α in the domain of analyticity of $F(\lambda)$, then,

$$\frac{F(\lambda)-F(\alpha)}{\lambda-\alpha} = F(\lambda)E$$

for some $n \times n$ matrix $E = E_\alpha$ which depends upon α but not upon λ. Therefore,

$$F(\lambda)\{I_n + (\alpha-\lambda)E\} = F(\alpha)\ ,$$

which implies that $F(\lambda)$ must be a rational function of the form

$$F(\lambda) = V(M-\lambda N)^{-1}\ . \qquad (2.15)$$

Thus, R_α invariance forces the elements in $\mathcal{M}$ to be rational functions of the form (2.15). Moreover, since

$$(R_\beta F)(\lambda) = F(\lambda)N(M-\beta N)^{-1} \qquad (2.16)$$

for every point β at which the matrix $M - \beta N$ is invertible (i.e., for every $\beta \in \mathfrak{H}_F$, the domain of analyticity of F), it is readily checked that, for each of the three special choices of Ω_+ under consideration, the associated structural identity reduces to a matrix equation for P. The equation is obtained by working out (2.10), (2.12) and (2.13) with the aid of (2.14) and (2.16). In other words:

In a finite-dimensional R_α invariant RKHS $\mathcal{M}$ *with an invertible Gram matrix P, each of the structural identities* (2.10), (2.12) *and* (2.13) *is equivalent to a Lyapunov-Stein equation for P.*

This last conclusion seems to have been first established explicitly in [D2] by a considerably lengthier calculation. The present more appealing argument is taken from [D3] and [D4].

The formulas simplify if $F(\lambda)$ is either holomorphic at zero or holomorphic at infinity:

1. M is invertible if and only if $F(\lambda)$ is holomorphic at zero. In this case we can divide through by M and reexpress $F(\lambda)$ as
$$F(\lambda) = C(I_n - \lambda A)^{-1}. \tag{2.17}$$
2. N is invertible if and only if $F(\lambda)$ is holomorphic at infinity and tends to zero as $|\lambda|$ tends to infinity. In this case we can divide through by N and reexpress $F(\lambda)$ as
$$F(\lambda) = C(\lambda I_n - A)^{-1}. \tag{2.18}$$

Theorem 2.3. *Let $\mathcal{M}$ denote the finite-dimensional RKHS that was introduced in Example 9 and let $F(\lambda)$ be given by* (2.15). *Then the RK of $\mathcal{M}$ can be expressed in the form*
$$K_\omega(\lambda) = \frac{J - U(\lambda)JU(\omega)^*}{\rho_\omega(\lambda)} \tag{2.19}$$
with $\rho_\omega(\lambda)$ as in the table if and only if the positive definite matrix P is a solution of the equation
$$M^*PM - N^*PN = V^*JV \text{ when } \Omega_+ = \mathbb{D}\ , \tag{2.20}$$
$$M^*PN - N^*PM = 2\pi i V^*JV \text{ when } \Omega_+ = \mathbb{C}_+\ , \tag{2.21}$$
$$M^*PN + N^*PM = -2\pi V^*JV \text{ when } \Omega_+ = \Pi_+\ . \tag{2.22}$$
Moreover, in each of these cases $U(\lambda)$ is uniquely specified up to a J unitary constant multiplier on the right by the formula
$$U(\lambda) = I_m - \rho_\mu(\lambda)F(\lambda)P^{-1}F(\mu)^*J \tag{2.23}$$
for any choice of the point $\mu \in \Omega_0 \cap \mathfrak{H}_F$.

It is well to note that formula (2.23) is a realization formula for $U(\lambda)$, and that in the usual notation of (2.17) and (2.18) it depends only upon A, C and P. It can be reexpressed in one of the standard A, B, C, D forms by elementary manipulations. A very general class of realization formulas of the form (2.23) and extensions thereof may be found in [AlD3].

Remark 2.4. Examples of RKHS's when P is only positive semidefinite but an associated Riccati equation is satisfied are furnished in Theorems 4.6 and 5.6.

3. A basic theorem

This chapter is devoted to obtaining a useful description of the set $T_W[\mathcal{S}^{p\times q}]$. In addition to formula (1.1), we shall make use of two other formulas:

$$T_W[\varepsilon] = (w_{11}^{\#} + \varepsilon w_{12}^{\#})^{-1}(w_{21}^{\#} + \varepsilon w_{22}^{\#}), \tag{3.1}$$

and a formula in terms of the entries in the Potapov-Ginzburg transform (2.7) of $W(\lambda)$ that is often referred to as the Redheffer transform:

$$T_W[\varepsilon] = s_{12} + s_{11}\varepsilon(I_q - s_{21}\varepsilon)^{-1}s_{22}. \tag{3.2}$$

The main theorem is:

Theorem 3.1. *Let $s \in \mathcal{S}^{p\times q}(\Omega_+)$ and let $W \in \mathcal{U}(j_{pq}, \Omega_+)$. Then $s \in T_W[\mathcal{S}^{p\times q}]$ if and only if the following three conditions are met:*

(B1) $[I_p \quad -s]f$ *belongs to the Hardy space $H_2^p(\Omega_+)$ for every choice of $f \in \mathcal{H}(W)$.*

(B2) $[-s^{\#} \quad I_q]f$ *belongs to the Hardy space $H_2^q(\Omega_+)^{\perp}$ for every choice of $f \in \mathcal{H}(W)$.*

(B3) $\langle \Delta_s f, f\rangle \leq \langle f, f\rangle_{\mathcal{H}(W)}$ *for every choice of $f \in \mathcal{H}(W)$.*

The proof of the theorem is lengthy and is established in the next two subsections, one for each direction. We shall take advantage of the following definition and two preliminary theorems:

Definition 3.2. A $p \times m$ mvf $L(\lambda)$ that is meromorphic in Ω_+ is said to be (Ω_+, J) admissible if

$$\sum_{i,k=1}^{n} \frac{\xi_k^* L(\omega_k) J L(\omega_i)^* \xi_i}{\rho_{\omega_i}(\omega_k)} \geq 0 \tag{3.3}$$

for every choice of vectors $\xi_1, \ldots, \xi_n$ in $\mathbb{C}^p$, points $\omega_1, \ldots, \omega_n$ in $\Omega_+ \cap \mathfrak{H}_L$ and every integer $n \geq 1$.

Theorem 3.3. *Let $L(\lambda) = [M(\lambda) \quad N(\lambda)]$ be a $k \times m$ mvf of bounded type in Ω_+ with components $M(\lambda)$ of size $k \times p$ and $N(\lambda)$ of size $k \times q$. Then $L(\lambda)$ is (Ω_+, j_{pq}) admissible if and only if*

$$N(\lambda) = M(\lambda)\varepsilon(\lambda)$$

for some mvf $\varepsilon \in \mathcal{S}^{p\times q}(\Omega_+)$.

Proof. See, e.g., Theorem 3.7 of [D1]. (The assumption that $L(\lambda)$ is of bounded type means that it can be expressed in the form $L(\lambda) = F(\lambda)/f(\lambda)$, where $F \in \mathcal{S}^{k\times m}(\Omega_+)$ and $f \in \mathcal{S}^{1\times 1}(\Omega_+)$.) □

Theorem 3.4. *Let $s \in \mathcal{S}^{p\times q}(\Omega_+)$ and let $W \in \mathcal{U}(j_{pq}, \Omega_+)$. Then $s \in T_W[\mathcal{S}^{p\times q}]$ if and only if $[I_p \quad -s]W$ is (Ω_+, j_{pq}) admissible.*

Proof. Suppose first that $s = T_W[\varepsilon]$ for some mvf $\varepsilon \in \mathcal{S}^{p\times q}(\Omega_+)$. Then, with the help of formula (3.1), it is readily checked that

$$\begin{aligned}[I_p \quad -s]W &= (w_{11}^{\#}+\varepsilon w_{12}^{\#})^{-1}[(w_{11}^{\#}+\varepsilon w_{12}^{\#}) \quad w_{21}^{\#}+\varepsilon w_{22}^{\#}]j_{pq}W \\ &= (w_{11}^{\#}+\varepsilon w_{12}^{\#})^{-1}[I_p \quad \varepsilon]W^{\#}j_{pq}W \\ &= (w_{11}^{\#}+\varepsilon w_{12}^{\#})^{-1}[I_p \quad -\varepsilon].\end{aligned}$$

This leads easily to the conclusion that $[I_p \quad -s]W$ is (Ω_+, j_{pq}) admissible. The converse follows easily from the preceding theorem upon writing $[I_p \quad -s]W = [M \quad N]$ and carrying through the requisite calculations (with $N = -M\varepsilon$). □

Remark 3.5. Since $[I_p \quad s(\lambda)]$ has rank p in Ω_+, $[I_p \quad -s]W$ is (Ω_+, j_{pq}) admissible if and only if the mvf $X = [I_p \quad -s]W$ meets the condition $X(\omega)j_{pq}X(\omega)^* \geq 0$ for every point $\omega \in \Omega_+ \cap \mathfrak{H}_W$. For other notions of admissibility and a theorem of de Branges-Rovnyak that is related to this circle of ideas, see Section 6 of [AlD1].

3.1. If (B1)–(B3) are in force, then ...

The first two lemmas show that if assumptions (B1)–(B3) are in force, then the operator of multiplication by $[I_p \quad -s]$ is a contractive map of $\mathcal{H}(W)$ into $\mathcal{H}(s)$ and that the operator of multiplication by $[-s^{\#} \quad I_q]$ is a contractive map of $\mathcal{H}(W)$ into $\mathcal{H}_*(s)$. Thus, we could in fact invoke the theorem of de Branges-Rovnyak alluded to above to make the proof of Theorem 3.10 a little shorter. However, for the convenience of the reader, we shall take the long route and fill in all the details.

Lemma 3.6. *If assumptions (B1)–(B3) are in force and $s \in \mathcal{S}^{p\times q}(\Omega_+)$, then*

$$[I_p \quad -s]f \in \mathcal{H}(s)$$

and

$$\|[I_p \quad -s]f\|_{\mathcal{H}(s)} \leq \|f\|_{\mathcal{H}(W)}$$

for every $f \in \mathcal{H}(W)$.

Proof. Let $f \in \mathcal{H}(W)$ be written as

$$f(\lambda) = \begin{bmatrix} g(\lambda) \\ h(\lambda) \end{bmatrix}, \tag{3.4}$$

with components $g(\lambda)$ of height p and $h(\lambda)$ of height q. Then, the decomposition

$$\begin{aligned}\Delta_s(\mu) &= \begin{bmatrix} I_p & -s(\mu) \\ -s(\mu)^* & I_q \end{bmatrix} \\ &= \begin{bmatrix} I_p \\ -s(\mu)^* \end{bmatrix}[I_p \quad -s(\mu)] + \begin{bmatrix} 0 & 0 \\ 0 & I_q - s(\mu)^*s(\mu) \end{bmatrix},\end{aligned}$$

which is valid for a.e. $\mu \in \Omega_0$, implies that

$$\langle \Delta_s f, f\rangle = \|g - sh\|^2 + \langle (I_q - s^*s)h, h\rangle.$$

Therefore, $(I_q - s^*s)^{1/2}h \in L_2^q(\Omega_0)$ by (B3). Next, since

$$g - sh \in H_2^p(\Omega_+),\ -s^{\#}g + h \in H_2^q(\Omega_+)^{\perp} \text{ and } (I_q - s^*s)h \in L^2(\Omega_0),$$

we see that for every $\varphi \in H_2^q(\Omega_+)$,

$$\|g - sh + s\varphi\|^2 = \|g - sh\|^2 + 2\Re\langle g - sh, s\varphi\rangle + \|s\varphi\|^2$$

and

$$\begin{aligned}\langle g - sh, s\varphi\rangle &= \langle s^*g - h, \varphi\rangle + \langle (I_q - s^*s)h, \varphi\rangle \\ &= \langle (I_q - s^*s)h, \varphi\rangle.\end{aligned}$$

Thus,

$$\|g - sh + s\varphi\|^2 - \|\varphi\|^2 = \langle \Delta_s f, f\rangle - \langle (I_q - s^*s)(h - \varphi), (h - \varphi)\rangle$$

for every $\varphi \in H_2^q(\Omega_+)$. Therefore, by the de Branges-Rovnyak characterization of $\mathcal{H}(s)$ that is cited in Example 6, $g - sh \in \mathcal{H}(s)$ and

$$\|g - sh\|^2_{\mathcal{H}(s)} \leq \langle \Delta_s f, f\rangle.$$

This serves to complete the proof, since $\langle \Delta_s f, f\rangle \leq \langle f, f\rangle_{\mathcal{H}(W)}$, by (B3). □

Lemma 3.7. *If assumptions (B1)–(B3) are in force and $s \in \mathcal{S}^{p\times q}(\Omega_+)$, then*

$$[-s^\# \quad I_q]f \in \mathcal{H}_*(s)$$

and

$$\|[-s^\# \quad I_q]f\|_{\mathcal{H}_*(s)} \leq \|f\|_{\mathcal{H}(W)}$$

for every $f \in \mathcal{H}(W)$.

Proof. Let $f \in \mathcal{H}(W)$ be written as in formula (3.4) with components $g(\lambda)$ of height p and $h(\lambda)$ of height q. Then the decomposition

$$\begin{aligned}\Delta_s(\mu) &= \begin{bmatrix} I_p & -s(\mu) \\ -s(\mu)^* & I_q \end{bmatrix} \\ &= \begin{bmatrix} -s(\mu) \\ I_q \end{bmatrix} [-s(\mu)^* \quad I_q] + \begin{bmatrix} I_p - s(\mu)s(\mu)^* & 0 \\ 0 & 0 \end{bmatrix},\end{aligned} \tag{3.5}$$

which is valid for a.e. $\mu \in \Omega_0$, implies that

$$\langle \Delta_s f, f\rangle = \|-s^*g + h\|^2 + \langle (I_p - ss^*)g, g\rangle.$$

Therefore, $(I_p - ss^*)^{1/2}g \in L_2^p(\Omega_0)$ by (B3). Next, since

$$g - sh \in H_2^p(\Omega_+), \quad -s^\# g + h \in H_2^q(\Omega_+)^\perp \text{ and } (I_p - ss^*)g \in L_2^p(\Omega_0),$$

we see that for every $\psi \in H_2^q(\Omega_+)^\perp$,

$$\|-s^*g + h + s^*\psi\|^2 = \|-s^*g + h\|^2 + 2\Re\langle -s^*g + h, s^*\psi\rangle + \|s^*\psi\|^2$$

and

$$\begin{aligned}\langle -s^*g + h, s^*\psi\rangle &= \langle -g + sh, \psi\rangle + \langle -ss^*g + g, \psi\rangle \\ &= \langle (I_p - ss^*)g, \psi\rangle.\end{aligned}$$

Thus,

$$\| - s^*g + h + s^*\psi\|^2 - \|\psi\|^2 = \langle \Delta_s f, f\rangle - \langle (I_p - ss^*)(g-\psi), (g-\psi)\rangle$$

for every $\psi \in H_2^q(\Omega_+)^\perp$. Therefore, by the discussion in Example 7, $-s^\# g + h \in \mathcal{H}_*(s)$ and

$$\| - s^*g + h\|^2_{\mathcal{H}_*(s)} \leq \langle \Delta_s f, f\rangle .$$

This serves to complete the proof, since $\langle \Delta_s f, f\rangle \leq \langle f, f\rangle_{\mathcal{H}(W)}$, by (B3). □

Lemma 3.8. *Let $s \in \mathcal{S}^{p\times q}(\Omega_+)$ and $W \in \mathcal{U}(j_{pq}, \Omega_+)$ be a pair of mvf's that satisfy the conditions (B1)-(B3). Then $[I_p \quad - s(\lambda)]W(\lambda)$ is (Ω_+, j_{pq}) admissible. In particular,*

$$[I_p \quad - s(\omega)]W(\omega)j_{pq}W(\omega)^* \begin{bmatrix} I_p \\ -s(\omega)^* \end{bmatrix} \geq 0$$

for every point $\omega \in \mathfrak{H}_W \cap \mathbb{C}_+$.

Proof. Let $K_\omega(\lambda)$ denote the RK for the RKHS $\mathcal{H}(W)$, let $\omega_1, \ldots, \omega_n \in \mathbb{C}_+$, let $\Lambda^s_\omega(\lambda)$ denote the RK for the RKHS $\mathcal{H}(s)$ and set

$$Z(\lambda) = [I_p \quad - s(\lambda)],\ f(\lambda) = \sum_{i=1}^n K_{\omega_i}(\lambda)u_i \text{ and } g(\lambda) = \sum_{k=1}^n \Lambda^s_{\omega_k}\xi_k .$$

Then

$$\begin{aligned} \|f\|^2_{\mathcal{H}(W)} &= \sum_{i,k=1}^n u_k^* K_{\omega_i}(\omega_k)u_i \\ &= \sum_{k=1}^n u_k^* f(\omega_k)\ , \end{aligned}$$

and hence, upon setting

$$u_k = Z(\omega_k)^*\xi_k$$

and invoking Lemma 3.6, we obtain

$$\begin{aligned} \|f\|^2_{\mathcal{H}(W)} &= \sum_{k=1}^n \xi_k^* Z(\omega_k) f(\omega_k) \\ &= \langle Zf, g\rangle_{\mathcal{H}(s)} \\ &\leq \|Zf\|_{\mathcal{H}(s)}\|g\|_{\mathcal{H}(s)} \\ &\leq \|f\|_{\mathcal{H}(W)}\|g\|_{\mathcal{H}(s)} \end{aligned}$$

Therefore,

$$\|f\|^2_{\mathcal{H}(W)} \leq \|g\|^2_{\mathcal{H}(s)}\ ,$$

which is equivalent to the asserted admissibility, since

$$\begin{aligned}\|g\|^2_{\mathcal{H}(s)} &= \sum_{i,k=1}^{n} \xi_k^* \left\{\frac{I_p - s(\omega_k)s(\omega_i)^*}{\rho_{\omega_i}(\omega_k)}\right\}\xi_i \\ &= \sum_{i,k=1}^{n} \xi_k^* \frac{Z(\omega_k)j_{pq}Z(\omega_i)^*}{\rho_{\omega_i}(\omega_k)}\xi_i \\ &= \sum_{i,k=1}^{n} \frac{u_k^* j_{pq} u_i}{\rho_{\omega_i}(\omega_k)}\end{aligned}$$

and

$$\|f\|^2_{\mathcal{H}(W)} = \sum_{i,k=1}^{n} u_k^* \left\{\frac{j_{pq} - W(\omega_k)j_{pq}W(\omega_i)^*}{\rho_{\omega_i}(\omega_k)}\right\} u_i \,. \qquad \square$$

The next lemma, which in some sense is a reflected version of the preceding lemma is added for the sake of completeness. It is not needed in the sequel.

Lemma 3.9. *Let $s \in \mathcal{S}^{p\times q}(\Omega_+)$ and $W \in \mathcal{U}(j_{pq}, \Omega_+)$ be a pair of mvf's that satisfy the conditions (B1)–(B3). Then*

$$\sum_{i,k=1}^{n} \frac{\eta_k^*[-s(\omega_k)^* \quad I_q]W^\#(\omega_k)^* j_{pq} W^\#(\omega_i)\begin{bmatrix} -s(\omega_i) \\ I_q \end{bmatrix}\eta_i}{\rho_{\omega_k}(\omega_i)} \le 0$$

for every set of vectors $\eta_1, \dots, \eta_n$ in $\mathbb{C}^q$ and points $\omega_1, \dots, \omega_n$ in $\mathbb{C}_+ \cap \mathfrak{H}_W$ and every positive integer n. In particular,

$$[-s(\omega)^* \quad I_q]W^\#(\omega)^* j_{pq} W^\#(\omega)\begin{bmatrix} -s(\omega) \\ I_q \end{bmatrix} \le 0$$

for every point $\omega \in \mathfrak{H}_{W^\#} \cap \mathbb{C}_+$.

Proof. The proof is modeled on the proof of the last lemma, except that now, although we still let $K_\omega(\lambda)$ denote the RK for the RKHS $\mathcal{H}(W)$, we let $\widetilde{\Lambda}^s_\omega(\lambda)$ denote the RK for the RKHS $\mathcal{H}_*(s)$ and set

$$Z(\lambda) = [-s^\#(\lambda) \quad I_q],\ f(\lambda) = \sum_{i=1}^{n} K_{\alpha_i}(\lambda)u_i \text{ and } g(\lambda) = \sum_{k=1}^{n} \widetilde{\Lambda}^s_{\alpha_k}(\lambda)\eta_k,$$

where $\alpha_k = \omega_k^o$ and $\omega_k \in \Omega_+$ for $k = 1, \dots, n$. Then

$$\begin{aligned}\|f\|^2_{\mathcal{H}(W)} &= \sum_{i,k=1}^{n} u_k^* K_{\alpha_i}(\alpha_k)u_i \\ &= \sum_{k=1}^{n} u_k^* f(\alpha_k)\ ,\end{aligned}$$

and hence, upon setting

$$u_k = Z(\alpha_k)^* \eta_k$$

and invoking Lemma 3.7, we obtain

$$\begin{aligned} \|f\|^2_{\mathcal{H}(W)} &= \sum_{k=1}^{n} \eta_k^* Z(\alpha_k) f(\alpha_k) \\ &= \langle Zf, g \rangle_{\mathcal{H}_*(s)} \\ &\leq \|Zf\|_{\mathcal{H}_*(s)} \|g\|_{\mathcal{H}_*(s)} \\ &\leq \|f\|_{\mathcal{H}(W)} \|g\|_{\mathcal{H}_*(s)} \,. \end{aligned}$$

Therefore,

$$\|f\|^2_{\mathcal{H}(W)} \leq \|g\|^2_{\mathcal{H}_*(s)} \,,$$

which is equivalent to the asserted inequality, since

$$\begin{aligned} \|g\|^2_{\mathcal{H}_*(s)} &= \sum_{i,k=1}^{n} \eta_k^* \left\{ \frac{s^\#(\alpha_k) s^\#(\alpha_i)^* - I_q}{\rho_{\alpha_i}(\alpha_k)} \right\} \eta_i \\ &= \sum_{i,k=1}^{n} \eta_k^* \frac{Z(\alpha_k) j_{pq} Z(\alpha_i)^*}{\rho_{\alpha_i}(\alpha_k)} \eta_i \\ &= \sum_{i,k=1}^{n} \frac{u_k^* j_{pq} u_i}{\rho_{\alpha_i}(\alpha_k)} \end{aligned}$$

and

$$\|f\|^2_{\mathcal{H}(W)} = \sum_{i,k=1}^{n} u_k^* \left\{ \frac{j_{pq} - W(\alpha_k) j_{pq} W(\alpha_i)^*}{\rho_{\alpha_i}(\alpha_k)} \right\} u_i \,.$$

The final step of the calculations for the case $\Omega_+ = \mathbb{D}$ is a shade more complicated than for the other two, because $\rho_{\alpha_i}(\alpha_k) = -\rho_{\omega_k}(\omega_i)$ if $\Omega_+ = \mathbb{C}_+$ or Π_+, whereas $\rho_{\alpha_i}(\alpha_k) = -(\overline{\omega_k}\omega_i)^{-1} \rho_{\omega_k}(\omega_i)$ if $\Omega_+ = \mathbb{D}$. The extra factors of $\omega_1, \ldots, \omega_n$ and their conjugates, can be absorbed in the η's. □

Theorem 3.10. *Let $s \in \mathcal{S}^{p\times q}(\Omega_+)$ and $W \in \mathcal{U}(j_{pq}, \Omega_+)$ be a pair of mvf's that satisfy the conditions (B1)–(B3). Then*

$$s \in T_W[\mathcal{S}^{p\times q}] \,.$$

Proof. Lemma 3.8 implies that the mvf

$$[M(\lambda) \quad N(\lambda)] = [I_p \quad -s(\lambda)] W(\lambda)$$

with blocks $M(\lambda)$ of size $p \times p$ and $N(\lambda)$ of size $p \times q$ is (Ω_+, j_{pq}) admissible and therefore, by Theorem 3.3, there exists a mvf $\varepsilon \in \mathcal{S}^{p\times q}(\Omega_+)$ such that $N(\lambda) = -M(\lambda)\varepsilon(\lambda)$. Thus, upon writing $W(\lambda)$ in the standard block form and multiplying out, we obtain two formulas:

$$\begin{aligned} w_{11}(\lambda) - s(\lambda) w_{21}(\lambda) &= M(\lambda) \\ w_{12}(\lambda) - s(\lambda) w_{22}(\lambda) &= -M(\lambda)\varepsilon(\lambda) \,. \end{aligned}$$

But this in turn implies that

$$w_{11}(\lambda)\varepsilon(\lambda) + w_{12}(\lambda) = s(\lambda)\{w_{21}(\lambda)\varepsilon(\lambda) + w_{22}(\lambda)\}$$

and hence as

$$\det\{(w_{21}\varepsilon + w_{22})(\lambda)\} \not\equiv 0 \ ,$$

that $s = T_W[\varepsilon]$ for some $\varepsilon \in \mathcal{S}^{p\times q}(\Omega_+)$, as claimed. □

3.2. If $s \in T_W[\mathcal{S}^{p\times q}]$, then ...

Lemma 3.11. *Let $W \in \mathcal{U}(j_{pq}, \Omega_+)$ and let $s \in T_W[\mathcal{S}^{p\times q}]$. Then:*

1. *The inequality*

$$[-s(\omega)^* \quad I_q]W^\#(\omega)^* j_{pq} W^\#(\omega) \begin{bmatrix} -s(\omega) \\ I_q \end{bmatrix} \leq 0 \tag{3.6}$$

is in force for every point $\omega \in \Omega_+ \cap \mathfrak{H}_{W^\#}$.

2. *The inequality*

$$[I_p \quad -s(\omega)]W(\omega) j_{pq} W(\omega)^* \begin{bmatrix} I_q \\ -s(\omega)^* \end{bmatrix} \geq 0 \tag{3.7}$$

is in force for every point $\omega \in \Omega_+ \cap \mathfrak{H}_W$.

Proof. Let $s = T_W[\varepsilon]$ for some mvf $\varepsilon \in \mathcal{S}^{p\times q}(\Omega_+)$. Then

$$\det\{w_{21}\varepsilon + w_{22}\} \not\equiv 0 \ \text{ in } \ \Omega_+$$

and, by formula (1.1),

$$\begin{bmatrix} s(\lambda) \\ I_q \end{bmatrix} (w_{21}(\lambda)\varepsilon(\lambda) + w_{22}(\lambda)) = W(\lambda) \begin{bmatrix} \varepsilon(\lambda) \\ I_q \end{bmatrix}$$

for $\lambda \in \Omega_+ \cap \mathfrak{H}_W$. Therefore, as

$$j_{pq} W^\#(\lambda) j_{pq} \begin{bmatrix} s(\lambda) \\ I_q \end{bmatrix} = \begin{bmatrix} \varepsilon(\lambda) \\ I_q \end{bmatrix} (w_{21}(\lambda)\varepsilon(\lambda) + w_{22}(\lambda))^{-1}$$

and

$$\begin{bmatrix} \varepsilon(\lambda) \\ I_q \end{bmatrix}^* j_{pq} \begin{bmatrix} \varepsilon(\lambda) \\ I_q \end{bmatrix} \leq 0 \ ,$$

it is readily seen that

$$\begin{bmatrix} s(\lambda) \\ I_q \end{bmatrix}^* j_{pq} W^\#(\lambda)^* j_{pq} W^\#(\lambda) j_{pq} \begin{bmatrix} s(\lambda) \\ I_q \end{bmatrix} \leq 0 \ .$$

But this is equivalent to the first asserted inequality.

The second inequality follows from Theorem 3.4. It may also be obtained directly from the second formula (3.1) for linear fractional transformations and the observation that $s = T_W[\varepsilon]$ means that

$$[I_p \quad \varepsilon(\lambda)]W^\#(\lambda) = (w_{11}^\# + \varepsilon w_{12}^\#)[I_p \quad s(\lambda)] \ ,$$

which in turn implies that

$$[I_p \quad \varepsilon(\lambda)] = (w_{11}^\# + \varepsilon w_{12}^\#)[I_p \quad s(\lambda)] j_{pq} W(\lambda) j_{pq}$$

and hence, as

$$[I_p \quad \varepsilon(\lambda)]j_{pq}\begin{bmatrix} I_p \\ \varepsilon(\lambda)^* \end{bmatrix} \geq 0$$

for $\lambda \in \mathbb{C}_+$, that inequality (3.7) holds. □

Theorem 3.12. *Let $W \in \mathcal{U}(j_{pq}, \Omega_+)$, let $s \in T_W[\mathcal{S}^{p\times q}]$ and let $\Delta_s(\mu)$ be defined by formula (1.4). Then*

$$[I_p \quad -s]f \in H_2^p(\Omega_+) \tag{3.8}$$

and

$$\langle \Delta_s f, f\rangle \leq \|f\|^2_{\mathcal{H}(W)} \tag{3.9}$$

for every $f \in \mathcal{H}(W)$.

Proof. Suppose first that $\Omega_+ = \mathbb{C}_+$, fix $f \in \mathcal{H}(W)$ and define

$$\psi(\omega) = -i\langle R_\omega f - K_{\overline{\omega}}u, f\rangle_{\mathcal{H}(W)} \tag{3.10}$$

for every point $\omega \in \mathbb{C}_+ \cap \mathfrak{H}_W \cap \mathfrak{H}_{W^\#}$, where $u \in \mathbb{C}^m$ is chosen to be of the form

$$u = -2\pi i \begin{bmatrix} -s(\omega) \\ I_q \end{bmatrix} \xi \tag{3.11}$$

for some $\xi \in \mathbb{C}^q$. Then, the inequality (3.6) implies that

$$\begin{aligned} u^* K_{\overline{\omega}}(\overline{\omega})u &= \frac{u^*\{j_{pq} - W(\overline{\omega})j_{pq}W(\overline{\omega})^*\}u}{\rho_{\overline{\omega}}(\overline{\omega})} \\ &\leq \frac{u^* j_{pq} u}{\rho_{\overline{\omega}}(\overline{\omega})}. \end{aligned}$$

Therefore,

$$\begin{aligned} \|R_\omega f - K_{\overline{\omega}}u\|^2_{\mathcal{H}(W)} &= \|R_\omega f\|^2_{\mathcal{H}(\omega)} - 2\Re u^*(R_\omega f)(\overline{\omega}) + u^* K_{\overline{\omega}}(\overline{\omega})u \\ &\leq \|R_\omega f\|^2_{\mathcal{H}(\omega)} - \frac{u^*\{f(\overline{\omega}) - f(\omega)\}}{\overline{\omega} - \omega} + \frac{\{f(\overline{\omega}) - f(\omega)\}^* u}{\overline{\omega} - \omega} \\ &\quad + \frac{u^* j_{pq} u}{\rho_{\overline{\omega}}(\overline{\omega})}. \end{aligned}$$

Next, the following special case of the de Branges identity (2.10)

$$\langle R_\omega f, f\rangle_{\mathcal{H}(W)} - \langle f, R_\omega f\rangle_{\mathcal{H}(W)} - (\omega - \overline{\omega}))\|R_\omega f\|^2_{\mathcal{H}(W)} = 2\pi i f(\omega)^* j_{pq} f(\omega) \tag{3.12}$$

for $\Omega_+ = \mathbb{C}_+$ allows us to write

$$\frac{\psi(\omega) + \psi(\omega)^*}{-i(\omega - \overline{\omega})} = \|R_\omega f\|^2_{\mathcal{H}(W)} + \frac{2\pi i f(\omega)^* j_{pq} f(\omega) + u^* f(\overline{\omega}) - f(\overline{\omega})^* u}{\omega - \overline{\omega}} \tag{3.13}$$

and hence, upon using this formula to eliminate the term $\|R_\omega f\|^2$ in the last inequality and setting $u = -2\pi i v$, that

$$\begin{aligned} \rho_\omega(\omega)\|R_\omega f - K_{\overline{\omega}}u\|^2_{\mathcal{H}(W)} &\leq 2\pi\{\psi(\omega) + \psi(\omega)^*\} \\ &\quad -4\pi^2\{v^* f(\omega) + f(\omega)^* v + v^* j_{pq} v + f(\omega)^* j_{pq} f(\omega)\}. \end{aligned}$$

Thus, upon setting

$$v=\begin{bmatrix} v_1 \\ v_2 \end{bmatrix}, f=\begin{bmatrix} f_1 \\ f_2 \end{bmatrix} \quad \text{and} \quad \xi = f_2(\omega) ,$$

the last term in curly brackets becomes

$$v_1^*f_1+v_2^*f_2+f_1^*v_1+f_2^*v_2+v_1^*v_1-v_2^*v_2+f_1^*f_1-f_2^*f_2=(f_1-sf_2)^*(f_1-sf_2)$$

and we obtain the inequality

$$\begin{aligned} \rho_\omega(\omega)\|R_\omega f-K_{\overline{\omega}}u\|^2_{\mathcal{H}(W)} &+ 4\pi^2\{(f_1(\omega)-s(\omega)f_2(\omega))^*(f_1(\omega)-s(\omega)f_2(\omega)\} \\ &\leq 2\pi\{\psi(\omega)+\psi(\omega)^*\} , \end{aligned} \tag{3.14}$$

which implies the following two inequalities as special cases:

$$\|R_\omega f-K_{\overline{\omega}}u\|^2_{\mathcal{H}(W)} \leq \frac{\psi(\omega)+\psi(\omega)^*}{-i(\omega-\overline{\omega})} \tag{3.15}$$

$$2\pi(f_1(\omega)-s(\omega)f_2(\omega))^*(f_1(\omega)-s(\omega)f_2(\omega)) \leq \psi(\omega)+\psi(\omega)^* . \tag{3.16}$$

The first inequality implies that the meromorphic function $\psi(\omega)$ must in fact be a holomorphic function of Carathéodory class in $\mathbb{C}_+$. Moreover, for any $\gamma \in \mathbb{R}$,

$$\begin{aligned} 0 &\leq \|\gamma f-i(R_\omega f-K_{\overline{\omega}}u)\|^2_{\mathcal{H}(W)} \\ &= \gamma^2\|f\|^2_{\mathcal{H}(W)}+\gamma\{\psi(\omega)+\psi(\omega)^*\}+\|R_\omega f-K_{\overline{\omega}}u\|^2_{\mathcal{H}(W)} \\ &\leq \gamma^2\|f\|^2_{\mathcal{H}(W)}+\gamma\{\psi(\omega)+\psi(\omega)^*\}-\frac{\psi(\omega)+\psi(\omega)^*}{i(\omega-\overline{\omega})} . \end{aligned}$$

Consequently, upon setting $\omega=\mu+i\nu$ and $\gamma=-1/\nu$, we obtain

$$\nu\,\Re\psi(\omega) \leq \|f\|^2_{\mathcal{H}(W)} \cdot \tag{3.17}$$

Similar considerations based on setting $\gamma=\pm 1/\nu$ in the inequality

$$0 \leq \|\gamma f-(R_\omega f-K_{\overline{\omega}}u)\|^2_{\mathcal{H}(W)}$$

and invoking (3.17) lead to the auxiliary conclusion that

$$|\nu\,\Im\psi(\omega)| \leq \|f\|^2_{\mathcal{H}(W)} \cdot \tag{3.18}$$

Therefore, in the Nevanlinna representation formula

$$\psi(\lambda)=i\alpha-i\beta\lambda+\frac{1}{\pi i}\int_{-\infty}^{\infty}\left\{\frac{1}{\mu-\lambda}-\frac{\mu}{1+\mu^2}\right\}d\sigma(\mu)$$

for $\lambda\in\mathbb{C}_+$, $\beta=0$ and σ is subject to the bound

$$\frac{b^2}{\pi}\int_{-\infty}^{\infty}\frac{d\sigma(\mu)}{\mu^2+b^2}=b\,\Re\psi(ib) \leq \|f\|^2_{\mathcal{H}(W)} \cdot \tag{3.19}$$

Thus, upon letting $b\uparrow\infty$, we obtain the bound

$$\frac{1}{\pi}\int_{-\infty}^{\infty}d\sigma(\mu) \leq \|f\|^2_{\mathcal{H}(W)} \cdot \tag{3.20}$$

Moreover, the identity

$$\psi(a+ib)+\psi(a+ib)^* = \frac{2b}{\pi}\int_{-\infty}^{\infty}\frac{1}{(\mu-a)^2+b^2}d\sigma(\mu)$$

(which prevails since $\beta = 0$ in the Nevanlinna representation) implies that

$$\frac{1}{2}\int_{-\infty}^{\infty}\{\psi(a+ib)+\psi(a+ib)^*\}da = \int_{-\infty}^{\infty}d\sigma(\mu) \leq \pi\|f\|^2_{\mathcal{H}(W)} \tag{3.21}$$

for every $b > 0$ and hence, in view of the bound (3.16), that (3.8) holds.

Next, the identity (3.13) implies that

$$\begin{aligned}\psi(\omega)+\psi(\omega)^* &\geq 2\pi f(\omega)^* j_{pq} f(\omega) - iu^* f(\overline{\omega}) + if(\overline{\omega})^* u \\ &= 2\pi\{f_1(\omega)^* f_1(\omega) - f_2(\omega)^* f_2(\omega) - f_2(\omega)^* s(\omega)^* f_1(\overline{\omega}) \\ &\quad + f_2(\omega)^* f_2(\overline{\omega}) - f_1(\overline{\omega})^* s(\omega) f_2(\omega) + f_2(\overline{\omega})^* f_2(\omega)\}\,.\end{aligned}$$

Thus, as the last term

$$2\pi\{\cdots\} \to 2\pi f(a)^*\Delta_s(a) f(a)$$

as $\omega = a+ib$ tends to a, the desired conclusion (3.9) now drops out easily via the inequality (3.21). This completes the proof of the theorem for $\Omega_+ = \mathbb{C}_+$, since

$$\begin{aligned}\int_{-\infty}^{\infty}d\sigma &= \frac{1}{2}\int_{-\infty}^{\infty}\{\psi(a+ib)+\psi(a+ib)^*\}da \\ &= \liminf_{b\downarrow 0}\frac{1}{2}\int_{-\infty}^{\infty}\{\psi(a+ib)+\psi(a+ib)^*\}da \\ &\geq \frac{1}{2}\int_{-\infty}^{\infty}\liminf_{b\downarrow 0}\{\psi(a+ib)+\psi(a+ib)^*\}da \\ &\geq \pi\langle\Delta_s f, f\rangle.\end{aligned}$$

The estimates for this case were, as we have already noted, adapted from [Sak-A]. The proof for $\Omega_+ = \Pi_+$ is almost exactly the same, up to minor changes of notation.

It remains to consider the case $\Omega_+ = \mathbb{D}$. The strategy is much the same as for the preceding cases. However, the details are a little different. In this setting, we define

$$\psi(\omega) = \langle(\frac{1}{2}I+\omega R_\omega)f - K_{\omega^o}u, f\rangle_{\mathcal{H}(W)}$$

for $f \in \mathcal{H}(W), \omega \in \mathbb{D}\cap\mathfrak{H}_{W^\#}\cap\mathfrak{H}_W$, $\omega \neq 0$ and $\omega^o = 1/\overline{\omega}$. Then

$$\begin{aligned}\psi(\omega)+\psi(\omega)^* &= \langle f,f\rangle_{\mathcal{H}(W)} + \langle\omega R_\omega f, f\rangle_{\mathcal{H}(W)} \\ &\quad + \langle f, \omega R_\omega f\rangle_{\mathcal{H}(\omega)} - u^* f(\omega^o) - f(\omega^o)^* u\,.\end{aligned}$$

Therefore, by the de Branges structural identity (2.12) for $\Omega_+ = \mathbb{D}$,

$$\psi(\omega)+\psi(\omega)^* = (1-|\omega|^2)\|R_\omega f\|^2_{\mathcal{H}(W)} + f(\omega)^* j_{pq} f(\omega) - u^* f(\omega^o) - f(\omega^o)^* u\,. \tag{3.22}$$

Moreover, if the vector u is chosen of the form (3.11), then, in view of the inequality (3.6),

$$u^* K_{\omega^o}(\omega^o) u \le \frac{u^* j_{pq} u}{\rho_{\omega^o}(\omega^o)}$$

and hence

$$\begin{aligned}\|R_\omega f - \overline{\omega^o} K_{\omega^o} u\|^2_{\mathcal{H}(W)} &= \|R_\omega f\|^2_{\mathcal{H}(W)} - 2\Re \omega^o u^* (R_\omega f)(\omega^o) + |\omega^o|^2 u^* K_{\omega^o}(\omega^o) u \\ &\le \|R_\omega f\|^2_{\mathcal{H}(W)} - \frac{u^*\{f(\omega^o) - f(\omega)\} + \{f(\omega^o) - f(\omega)\}^* u}{1 - |\omega|^2} - \frac{u^* j_{pq} u}{1 - |\omega|^2} .\end{aligned}$$

Thus, upon invoking formula (3.22) in order to eliminate the term $\|R_\omega f\|^2_{\mathcal{H}(W)}$ in the last inequality, we obtain

$$\begin{aligned}&\rho_\omega(\omega) \|R_\omega f - \overline{\omega^o} K_{\omega^o} u\|^2_{\mathcal{H}(W)} \\ &\qquad \le \psi(\omega) + \psi(\omega)^* - \{f(\omega)^* j_{pq} f(\omega) - u^* f(\omega) - f(\omega)^* u + u^* j_{pq} u\}.\end{aligned}$$

Now, setting

$$f(\omega) = \begin{bmatrix} f_1(\omega) \\ f_2(\omega) \end{bmatrix} \quad \text{and} \quad u = -\begin{bmatrix} -s(\omega) \\ I_q \end{bmatrix} f_2(\omega)$$

much as before, the term in curly brackets simplifies to

$$\{\ \} = \{f_1(\omega) - s(\omega) f_2(\omega)\}^* \{f_1(\omega) - s(\omega) f_2(\omega)\} .$$

Consequently, in place of (3.14), we now obtain

$$\begin{aligned}&\rho_\omega(\omega) \|R_\omega f - \overline{\omega^o} K_{\omega^o} u\|^2_{\mathcal{H}(W)} + \{f_1(\omega) - s(\omega) f_2(\omega)\}^* \{f_1(\omega) - s(\omega) f_2(\omega)\} \\ &\le \psi(\omega) + \psi(\omega)^* \end{aligned} \tag{3.23}$$

for all points $\omega \in \mathbb{D} \cap \mathfrak{H}_{W^\#} \cap \mathfrak{H}_W$ apart from $\omega = 0$. This once again displays $\psi(\lambda)$ as a holomorphic function of Carathéodory class, only this time in $\mathbb{D}$. Thus, by the Herglotz representation,

$$\psi(\omega) = i\alpha + \frac{1}{2\pi} \int_0^{2\pi} \frac{e^{i\theta} + \omega}{e^{i\theta} - \omega} d\sigma(\theta) \tag{3.24}$$

for some real number α and some finite measure $d\sigma$ on $\mathbb{T}$. The inequality

$$\begin{aligned}\{f_1(\omega) - s(\omega) f_2(\omega)\}^* \{f_1(\omega) - s(\omega) f_2(\omega)\} &\le \psi(\omega) + \psi(\omega)^* \\ &= \frac{1}{\pi} \int_0^{2\pi} \frac{1 - |\omega|^2}{|e^{i\theta} - \omega|^2} d\sigma(\theta)\end{aligned}$$

leads easily to the conclusion that the integral of the left-hand side over a circle of radius r, $0 < r < 1$, is uniformly bounded independently of the choice of r and hence that $f_1 - s f_2 \in H_2^p(\mathbb{D})$.

Next, since

$$(1 - |\omega|^2) \|R_\omega f\|^2_{\mathcal{H}(W)} \ge 0,$$

it follows from (3.22) that

$$\liminf_{r \uparrow 1} \{\psi(re^{i\theta}) + \psi(re^{i\theta})^*\} \ge f(e^{i\theta})^* \Delta_s(e^{i\theta}) f(e^{i\theta}) \quad \text{a.e. on } \mathbb{T}.$$

Thus,

$$\begin{aligned}
\int_0^{2\pi} d\sigma &= \lim_{r\uparrow 1}\frac{1}{2}\int_0^{2\pi}\{\psi(re^{i\theta})+\psi(re^{i\theta})^*\}d\theta \\
&\geq \frac{1}{2}\int_0^{2\pi}\liminf_{r\uparrow 1}\{\psi(re^{i\theta})+\psi(re^{i\theta})^*\}d\theta \\
&\geq \frac{1}{2}\int_0^{2\pi} f(e^{i\theta})^*\Delta_s(e^{i\theta})f(e^{i\theta})d\theta \quad = \quad \pi\langle\Delta_s f, f\rangle.
\end{aligned}$$

The next step is to obtain a bound on$\langle\Delta_s F, f\rangle$ in terms of $\|f\|_{\mathcal{H}(W)}$ with the aid of formula (3.22), the inequality (3.23) and the observation that if $\gamma\in\mathbb{R}$, then

$$\begin{aligned}
\|\gamma f-(\omega R_\omega f-K_{\omega^o}u)\|_{\mathcal{H}(W)}^2 &= \gamma^2\|f\|_{\mathcal{H}(W)}^2+\|\omega R_\omega f-K_{\omega^o}u\|_{\mathcal{H}(W)}^2 \\
&\quad -2\gamma\Re\langle(\frac{1}{2}I+\omega R_\omega)f-K_{\omega^o}u-\frac{1}{2}f, f\rangle_{\mathcal{H}(W)} \\
&= \gamma^2\|f\|_{\mathcal{H}(W)}^2-\gamma\{\psi(\omega)+\psi(\omega)^*-\|f\|_{\mathcal{H}(W)}^2\} \\
&\quad +|\omega|^2\|R_\omega f-\overline{\omega^o}K_{\omega^o}u\|_{\mathcal{H}(W)}^2 \\
&\leq (\gamma^2+\gamma)\|f\|_{\mathcal{H}(W)}^2-\gamma\{\psi(\omega)+\psi(\omega)^*\} \\
&\quad +\frac{|\omega|^2}{1-|\omega|^2}\{\psi(\omega)+\psi(\omega)^*\}\,.
\end{aligned}$$

The last inequality implies that

$$\left(\gamma-\frac{|\omega|^2}{1-|\omega|^2}\right)\{\psi(\omega)+\psi(\omega)^*\}\leq(\gamma^2+\gamma)\|f\|_{\mathcal{H}(W)}^2$$

and hence, upon setting $\gamma=2|\omega|^2/(1-|\omega|^2)$, and simplifying, that

$$\frac{1-|\omega|^2}{1+|\omega|^2}\,\frac{\psi(\omega)+\psi(\omega)^*}{2}\leq\|f\|_{\mathcal{H}(W)}^2.$$

This bound is not good enough for our purposes. The better bound

$$\frac{\sqrt{1-|\omega|^2}-|\omega|}{\sqrt{1-|\omega|^2}+|\omega|}\{\psi(\omega)+\psi(\omega)^*\}\leq\|f\|_{\mathcal{H}(W)}^2 \text{ for } |\omega|^2<1/2,$$

is obtained by choosing

$$\gamma=\frac{|\omega|}{\sqrt{1-|\omega|^2}}.$$

But this in turn implies that

$$\frac{1}{2\pi}\int_0^{2\pi}d\sigma=\frac{\psi(0)+\psi(0)^*}{2}\leq\frac{1}{2}\|f\|_{\mathcal{H}(W)}^2$$

and hence that

$$\langle\Delta_s f, f\rangle\leq\|f\|_{\mathcal{H}(W)}^2\,. \qquad \square$$

Theorem 3.13. *Let $W \in \mathcal{U}(j_{pq}, \Omega_+)$ and let $s \in T_W[\mathcal{S}^{p\times q}]$. Then*

$$[-s^\# \quad I_q]f \in H_2^q(\Omega_+)^\perp$$

for every $f \in \mathcal{H}(W)$.

Proof. Suppose first that $\Omega_+ = \mathbb{C}_+$ and then fix $f \in \mathcal{H}(W)$ and let

$$\varphi(\omega) = i\langle R_{\overline{\omega}} f - K_\omega u, f\rangle_{\mathcal{H}(W)}$$

and

$$u = 2\pi i \begin{bmatrix} I_p \\ -s(\omega)^* \end{bmatrix} f_1(\overline{\omega})$$

for some point $\omega \in \mathbb{C}_+ \cap \mathfrak{H}_W \cap \mathfrak{H}_{W^\#}$. Then, by another application of the special case of the de Branges identity (2.10) for $\Omega_+ = \mathbb{C}_+$, but with ω replaced by $\overline{\omega}$,

$$\varphi(\omega) + \varphi(\omega)^* = i\{(\overline{\omega} - \omega)\|R_{\overline{\omega}} f\|^2_{\mathcal{H}(W)} + 2\pi i f(\overline{\omega})^* j_{pq} f(\overline{\omega}) + u^* f(\omega) - f(\omega)^* u\}\,. \tag{3.25}$$

Moreover, in view of the inequality (3.7),

$$\begin{aligned} \|R_{\overline{\omega}} f - K_\omega u\|^2_{\mathcal{H}(W)} &= \|R_{\overline{\omega}} f\|^2_{\mathcal{H}(W)} - 2\,\Re u^*(R_{\overline{\omega}} f)(\omega) + u^* K_\omega(\omega) u \\ &\le \|R_{\overline{\omega}} f\|^2_{\mathcal{H}(W)} - 2\,\Re u^*(R_{\overline{\omega}} f)(\omega) + \frac{u^* j_{pq} u}{\rho_\omega(\omega)}\,. \end{aligned} \tag{3.26}$$

Thus, upon using the identity (3.25) to eliminate the term $\|R_{\overline{\omega}} f\|^2_{\mathcal{H}(W)}$ in the last inequality, we see that

$$\begin{aligned} &\rho_\omega(\omega)\|R_{\overline{\omega}} f - K_\omega u\|^2_{\mathcal{H}(W)} + 4\pi^2(-s(\omega)^* f_1(\overline{\omega}) + f_2(\overline{\omega}))^*(-s(\omega)^* f_1(\overline{\omega}) + f_2(\overline{\omega})) \\ &\quad \le 2\pi\{\varphi(\omega) + \varphi(\omega)^*\}\,. \end{aligned}$$

In particular, this implies that the meromorphic function $\varphi(\lambda)$ must be holomorphic in $\mathbb{C}_+$ and that

$$2\pi(-s^\#(\overline{\omega}) f_1(\overline{\omega}) + f_2(\overline{\omega}))^*(-s^\#(\overline{\omega}) f_1(\overline{\omega}) + f_2(\overline{\omega})) \le \varphi(\omega) + \varphi(\omega)^* \tag{3.27}$$

for every point $\omega \in \mathbb{C}_+ \cap \mathfrak{H}_{W^\#}$. Moreover, much the same sort of analysis that led to the inequalities (3.17) and (3.18) yields the bounds

$$\nu\,\Re\phi(\omega) \le \|f\|^2_{\mathcal{H}(W)} \quad \text{and} \quad |\nu\,\Im\phi(\omega)| \le \|f\|^2_{\mathcal{H}(W)}\,,$$

for $\omega = \mu + i\nu \in \mathbb{C}_+$. Therefore, just as in the proof of the formula (3.21), it follows that

$$\frac{1}{2}\int_{-\infty}^{\infty} \{\phi(a+ib) + \phi(a+ib)^*\} da \le \pi\|f\|^2_{\mathcal{H}(W)}\,.$$

But this bound coupled with the inequality (3.27) serves to complete the proof for $\Omega_+ = \mathbb{C}_+$. The analysis for $\Omega_+ = \mathbb{\Pi}_+$ is much the same.

It remains to consider the case $\Omega_+ = \mathbb{D}$. To obtain the needed bounds, fix $f \in \mathcal{H}(W)$, a point $\omega \in \mathbb{D} \cap \mathfrak{H}_W \cap \mathfrak{H}_{\#W}$ and set

$$\varphi(\omega) = -\langle(\frac{1}{2}I + \omega^\circ R_{\omega^\circ})f - K_\omega u, f\rangle_{\mathcal{H}(W)}$$

with

$$u = \begin{bmatrix} I_p \\ -s(\omega)^* \end{bmatrix} f_1(\omega^o) .$$

Then, with the aid of the de Branges identity for $\Omega_+ = \mathbb{D}$, we can reexpress

$$\begin{aligned} \varphi(\omega) + \varphi(\omega)^* &= -\{\|f\|^2_{\mathcal{H}(W)} + \langle \omega^o R_{\omega^o} f, f\rangle_{\mathcal{H}(W)} + \langle f, \omega^o R_{\omega^o} f\rangle_{\mathcal{H}(W)}\} \\ &\quad + u^* f(\omega) + f(\omega)^* u \end{aligned}$$

as

$$\begin{aligned} \varphi(\omega) + \varphi(\omega)^* &= (|\omega^o|^2 - 1)\|R_{\omega^o} f\|^2_{\mathcal{H}(W)} - f(\omega^o)^* j_{pq} f(\omega^o) \\ &\quad + u^* f(\omega) + f(\omega)^* u . \end{aligned} \tag{3.28}$$

Furthermore, since

$$u^* K_\omega(\omega) u \leq \frac{u^* j_{pq} u}{\rho_\omega(\omega)} ,$$

we have the bound

$$\begin{aligned} \|\omega^o R_{\omega^o} f - K_\omega u\|^2_{\mathcal{H}(W)} &\leq |\omega^o|^2 \|R_{\omega^o} f\|^2_{\mathcal{H}(W)} - 2\Re u^* \omega^o (R_{\omega^o} f)(\omega) + \frac{u^* j_{pq} u}{\rho_\omega(\omega)} \\ &= \frac{1}{|\omega|^2} \|R_{\omega^o} f\|^2_{\mathcal{H}(W)} + \frac{2\Re u^* \{f(\omega) - f(\omega^o)\} + u^* j_{pq} u}{1 - |\omega|^2} . \end{aligned}$$

Therefore, upon using formula (3.28) to eliminate the term $\|R_{\omega^o} f\|^2_{\mathcal{H}(W)}$ in the last inequality, we obtain the inequality

$$\begin{aligned} (1 - |\omega|^2)\|\omega^o R_{\omega^o} f - K_\omega u\|^2_{\mathcal{H}(W)} &\leq \varphi(\omega) + \varphi(\omega)^* + f(\omega^o)^* j_{pq} f(\omega^o) \\ &\quad - u^* f(\omega^o) - f(\omega^o)^* u + u^* j_{pq} u \end{aligned}$$

and hence, upon substituting the given formula for the vector u and rearranging terms, we obtain the inequality

$$\begin{aligned} &(-s(\omega)^* f_1(\omega^o) + f_2(\omega^o))^*(-s(\omega)^* f_1(\omega^o) + f_2(\omega^o)) \\ &\leq \varphi(\omega) + \varphi(\omega)^* . \end{aligned}$$

Therefore, since

$$\int_0^{2\pi} \{\varphi(re^{i\theta}) + \varphi(re^{i\theta})^*\} d\theta$$

is bounded uniformly in r for $0 \leq r < 1$, it follows that $[-s^\# \quad I_q] f \in H_2^q(\mathbb{C} \setminus (\mathbb{D} \cup \mathbb{T}))$. However, we wish to show a little more, i.e., that $[-s^\# \quad I_q] f \in H_2^q(\mathbb{D})^\perp$.

By the characterization of $\mathcal{H}(W)$ that is presented in Example 8, we know that $[-s_{12}^\# \quad I_q] f \in H_2^q(\mathbb{D})^\perp$. Therefore, the desired conclusion will prevail if and only if $(s - s_{12})^\# g \in H_2^q(\mathbb{D})^\perp$, where g designates the top p entries in f, just as in the example. By formula (3.2),

$$s - s_{12} = s_{11}\varepsilon(I_q - s_{21}\varepsilon)^{-1} s_{22}.$$

From the characterization of $\mathcal{H}(W)$ in Example 8 we also know that $s_{11}^\# g \in H_2^q(\mathbb{D})^\perp$. Moreover, as $s_{21}(\lambda)$ is strictly contractive for every point $\lambda \in \mathbb{D}$, it follows that $I_q - s_{21}\varepsilon$ is an outer mvf in $H_\infty^{q\times q}(\mathbb{D})$ and hence that $\det(I_q - s_{21}\varepsilon)$ is an outer scalar function in $H_\infty(\mathbb{D})$. Thus the desired result follows from (a suitably transposed version of) the Smirnov maximum principle; see, e.g., Lemma 2.1, Theorem 2.1 and the corollary on P 397 of [ArD] for the justification of the facts cited in this last argument. □

3.3. The proof of the main theorem

The proof of Theorem 3.1 is now an immediate consequence of Theorems 3.10, 3.12 and 3.13. □

It is worth noting that the conclusion of Theorem 3.1 is valid regardless of the dimension of the space $\mathcal{H}(W)$, finite or infinite.

4. Bitangential interpolation in the closed upper half-plane

In this section we shall apply Theorem 3.1 to study bitangential interpolation problems with data points in the closed upper half-plane $\overline{\mathbb{C}_+}$. The problems of interest can be formulated most simply and elegantly in terms of three complex matrices $C \in \mathbb{C}^{m\times n}$, $A \in \mathbb{C}^{n\times n}$, $P \in \mathbb{C}^{n\times n}$ that satisfy the following assumptions:[1]

(A1) $\operatorname{rank}\begin{bmatrix} A - \lambda I_n \\ C \end{bmatrix} = n \quad \text{for every} \quad \lambda \in \mathbb{R}\,.$

(A2) P is a positive semidefinite solution of the Lyapunov equation

$$A^*P - PA = 2\pi i C^* j_{pq} C, \tag{4.1}$$

where $p \geq 1,\ q \geq 1,$ and $p + q = m$.

The supplementary notation

$$F(\lambda) = C(\lambda I_n - A)^{-1}\ ,\ \lambda \in \mathbb{C}\setminus\sigma(A) \tag{4.2}$$

$$P_s = \int_{-\infty}^{\infty} F(\mu)^*\Delta_s(\mu)F(\mu)d\mu \tag{4.3}$$

will prove useful in this section. (The meaning of these symbols will be modified in the next section.)

Let $\widehat{\mathcal{S}}(C, A, P)$ denote the set of mvf's $s \in \mathcal{S}^{p\times q}(\mathbb{C}_+)$ that meet the following three conditions.

(C1) $[I_p \quad -s]Fu$ belongs to the Hardy space $H_2^p(\mathbb{C}_+)$ for every $u \in \mathbb{C}^n$.

(C2) $[-s^\# \quad I_q]Fu$ belongs to the Hardy space $H_2^q(\mathbb{C}_+)^\perp$ $(= H_2^q(\mathbb{C}_-))$ for every $u \in \mathbb{C}^n$.

(C3) $P_s \leq P$.

---

[1] We shall also assume that A is an upper triangular matrix in Jordan form. This involves no loss of generality, as will be explained shortly.

The mvf's that belong to the set $\widehat{\mathcal{S}}(C, A, P)$ are the solutions of the bitangential interpolation problem alluded to earlier. To clarify the connection with interpolation, consider for example the choice

$$C = \begin{bmatrix} C_1 \\ C_2 \end{bmatrix} = \begin{bmatrix} \xi_1 & \cdots & \xi_n \\ \eta_1 & \cdots & \eta_n \end{bmatrix} \text{ and } A = \operatorname{diag}\{\overline{\alpha_1}, \ldots, \overline{\alpha_n}\} ,$$

where

$$\xi_j \in \mathbb{C}^p \text{ and } \eta_j \in \mathbb{C}^q \text{ for } j = 1, \ldots, n ,$$

$$\alpha_j \in \mathbb{C}_+ \text{ for } j = 1, \ldots, \ell \text{ and } \alpha_j \in \mathbb{C}_- \text{ for } j = \ell + 1, \ldots, n .$$

Then a mvf $s \in \mathcal{S}^{p \times q}$ satisfies condition (C1) if and only if

$$\frac{\xi_j - s(\lambda)\eta_j}{\lambda - \overline{\alpha_j}} \text{ is holomorphic in } \mathbb{C}_+ \text{ for } j = 1, \ldots, n .$$

For $j = 1, \ldots, \ell$, this is automatically so. For $j = \ell + 1, \ldots, n$, condition (C1) is met if and only if

$$\xi_j = s(\overline{\alpha_j})\eta_j \text{ for } j = \ell + 1, \ldots, n .$$

Similarly, $s(\lambda)$ will meet condition (C2) if and only if

$$\frac{-s^{\#}(\lambda)\xi_j + \eta_j}{\lambda - \overline{\alpha_j}} \text{ is holomorphic in } \mathbb{C}_- \text{ for } j = 1, \ldots, n ,$$

or, equivalently, if and only if

$$\frac{\xi_j^* s(\lambda) - \eta_j^*}{\lambda - \alpha_j} \text{ is holomorphic in } \mathbb{C}_+ \text{ for } j = 1, \ldots, n .$$

But this is so if and only if

$$\xi_j^* s(\alpha_j) = \eta_j^* \text{ for } j = 1, \ldots, \ell .$$

Derivatives may intervene in the interpolation conditions if A is not restricted to be a diagonal matrix. For additional examples and discussion of the origins of this formulation, see, e.g., [D4] and [D8] and, for even more general formulations, [KKY], [KhY] [Kh1] and [Kh2].

Our objective is to describe the set $\widehat{\mathcal{S}}(C, A, P)$. There is in fact a hierarchy of such problems that depend on the distribution of the spectrum of A and upon whether or not P is invertible. Thus, in order of increasing complexity, the natural choices are:

1. Either $\sigma(A) \subset \mathbb{C}_+$ or $\sigma(A) \subset \mathbb{C}_-$.
2. $\sigma(A) \cap \sigma(A^*) = \emptyset$.
3. $\sigma(A) \cap \mathbb{R} = \phi$.
4. No constraints on $\sigma(A)$.

Moreover, each of this cases can be considered both with P invertible and P singular.

Let

$$\mathfrak{U} = \bigcap_{k=0}^{n-1} \mathcal{N}_{\{CA^k\}} . \tag{4.4}$$

Lemma 4.1. *If assumptions (A1) and (A2) are in force, then*

$$\mathfrak{U} \subset \mathcal{N}_P \ . \tag{4.5}$$

Thus, if (A1) and (A2) are in force, and P is invertible, then the pair (C, A) is observable.

Proof. The subspace $\mathfrak{U}$ is invariant under A. Thus, if $\mathfrak{U} \neq \{0\}$, then there exists a basis for $\mathfrak{U}$ that is made up of Jordan chains for A. Therefore, there exists a set of vectors $u_1, \ldots, u_\ell$ in $\mathfrak{U}$ and a point $\alpha \in \mathbb{C}$ such that

$$Au_j = \alpha u_j + u_{j-1} \ \text{ for } \ j = 1, \ldots, \ell \ ,$$

where u_0 is the zero vector in $\mathbb{C}^n$. Thus,

$$\begin{aligned} (\overline{\alpha} - \alpha) u_1^* P u_1 &= u_1^*(A^*P - PA)u_1 \\ &= 2\pi i u_1^*(C^* j_{pq} C) u_1 = 0 \ . \end{aligned}$$

Moreover, since $Cu_1 = 0$, $Au_1 = \alpha u_1$ and $u_1 \neq 0$, Assumption (A1) guarantees that $\alpha \notin \mathbb{R}$. Consequently, as $P \geq 0$, it follows that $u_1 \in \mathcal{N}_P$.

The next step is to observe that

$$\begin{aligned} (\overline{\alpha} - \alpha) u_2^* P u_2 &= (\alpha u_2 + u_1)^* P u_2 - u_2^* P(\alpha u_2 + u_1) \\ &= u_2^*(A^*P - PA)u_2 = 0 \ . \end{aligned}$$

Therefore, $u_2 \in \mathcal{N}_P$. A simple inductive argument in this spirit serves to prove that all the vectors $u_1, \ldots, u_\ell$ in this chain belong to $\mathcal{N}_P$. The same argument holds for all such Jordan chains. □

4.1. P is invertible

If (A1) and (A2) are in force and P is invertible, then the preceding lemma combined with the discussion of Example 9 serve to guarantee that the space

$$\mathcal{M} = \{F(\lambda)u : u \in \mathbb{C}^n\}$$

endowed with the inner product

$$\langle Fu, Fv \rangle_{\mathcal{M}} = v^* P u$$

is a de Branges space $\mathcal{H}(W)$, i.e., it is a RKHS with RK

$$K_\omega(\lambda) = F(\lambda) P^{-1} F(\omega)^* = \frac{j_{pq} - W(\lambda) j_{pq} W(\omega)^*}{\rho_\omega(\lambda)}, \tag{4.6}$$

where

$$W(\lambda) = I_m - 2\pi i C(\lambda I_n - A)^{-1} P^{-1} C^* j_{pq}. \tag{4.7}$$

This last formula is obtained by letting $\mu \to \infty$ in formula (2.23), with $F(\lambda)$ given by (4.2) and then checking that formula (4.6) holds with $F(\lambda)$ as in (4.2) and $W(\lambda)$ as in (4.7). Theorem 3.1 states that a mvf $s \in T_W[\mathcal{S}^{p\times q}]$ if and only if $f(\lambda) = F(\lambda)u$ meets the conditions (C1), (C2) and

$$\langle \Delta_s f, f \rangle \leq \langle f, f \rangle_{\mathcal{H}(W)}.$$

Thus, as

$$\langle \Delta_s f, f \rangle = u^* P_s u \quad \text{and} \quad \langle f, f \rangle_{\mathcal{H}(W)} = u^* P u,$$

these three conditions are seen to be exactly the same as those defining $\widehat{\mathcal{S}}(C, A, P)$. We have thus established the following theorem:

Theorem 4.2. *If (A1) and (A2) are in force, and P is invertible, then $\widehat{\mathcal{S}}(C, A, P) = T_W[\mathcal{S}^{p\times q}]$, where $W(\lambda)$ is the j_{pq}-inner mvf with respect to $\mathbb{C}_+$ that is defined by formula (4.7).*

4.2. P is singular and a related Riccati equation is solvable

Our next objective is to obtain a description of the set $\widehat{\mathcal{S}}(C, A, P)$ when P is singular. We shall first do so under the following extra assumption:
(A3) There exists an $n \times n$ complex matrix X such that:

1. $X \geq 0$.
2. X solves the Riccati equation

$$XA^* - AX = 2\pi i XC^* JCX. \tag{4.8}$$

3. $XPX = X$.
4. $PXP = P$.

If P is invertible, then the choice $X = P^{-1}$ fulfills all four requirements. If P is not invertible, then there is no guarantee that such an X exists. If it does, then in view of (1), (3) and (4), it must be a positive semidefinite pseudo-inverse of P that has the same rank as P. But that is not enough, $\mathcal{R}_X$, the range of X, must be invariant under A:

Lemma 4.3. *If (A1) and (A2) are in force and X is any $n \times n$ Hermitian matrix such that $XPX = X$, then the following conditions are equivalent:*

1. $AX = XPAX$.
2. $A\mathcal{R}_X \subseteq \mathcal{R}_X$, *i.e.*, $AX = XA_0$ *for some matrix* A_0.
3. X *is a solution of the Riccati equation* (4.8).
4. $AX = XZAX$ *for any matrix* Z *such that* $XZX = X$.

Proof. The implications (1) $\Rightarrow$ (2), (4) $\Rightarrow$ (2) and (3) $\Rightarrow$ (2) are self-evident. Suppose next that (2) is in force and $XZX = X$. Then

$$XZAX = XZXA_0 = XA_0 = AX .$$

Therefore, (2) $\Rightarrow$ (4) and (2) $\Rightarrow$ (1).

Finally, upon multiplying both sides of the Lyapunov equation (4.1) by X, we obtain

$$XA^*PX - XPAX = 2\pi i XC^* JCX .$$

But this leads easily to the conclusion that (1) $\Rightarrow$ (3). □

Lemma 4.4. *If assumptions (A1)–(A3) are in force*[2], *then*

$$\mathbb{C}^n = \mathcal{N}_P + \mathcal{R}_X \ . \tag{4.9}$$

Proof. Let $u \in \mathcal{N}_P \cap \mathcal{R}_X$. Then $Pu = 0$ and $u = Xy$ for some $y \in \mathbb{C}^n$. Therefore,

$$u = Xy = XPXy = XPu = 0\,.$$

This proves that $\mathcal{N}_P \cap \mathcal{R}_X = \{0\}$, i.e., the sum is direct. To complete the proof, it suffices to observe that

$$n = \dim \mathcal{N}_P + \dim \mathcal{R}_P = \dim \mathcal{N}_P + \dim \mathcal{R}_X \ ,$$

since $\operatorname{rank} P = \operatorname{rank} X$. □

Lemma 4.5. *If assumptions (A1)–(A3) are in force, then*

$$\mathfrak{U} \cap \mathcal{R}_X = \{0\}\ .$$

Proof. Let $u \in \mathfrak{U} \cap \mathcal{R}_X$. Then $u = Xy$ for some $y \in \mathbb{C}^n$. Moreover,

$$Pu = PXy = 0$$

by Lemma 4.1. Therefore,

$$y^*XPXy = y^*Xy = 0$$

and hence, as $X \geq 0$, we conclude that $u = \{0\}$, as claimed. □

The preceding analysis, combined with the discussion following Example 9, leads to the following conclusion:

Theorem 4.6. *If assumptions (A1), (A2) and (A3) are in force and $F(\lambda)$ is given by formula* (4.2), *then the space*

$$\mathcal{M}_X = \{F(\lambda)Xu : u \in \mathbb{C}^n\}$$

endowed with the inner product

$$\langle FXu, FXv\rangle_{\mathcal{M}_X} = v^*Xu$$

is a de Branges space $\mathcal{H}(W_X)$, i.e., it is a RKHS with RK

$$K_\omega(\lambda) = \frac{j_{pq} - W_X(\lambda) j_{pq} W_X(\omega)^*}{\rho_\omega(\lambda)},$$

where

$$W_X(\lambda) = I_m - 2\pi i C(\lambda I_n - A)^{-1} X C^* j_{pq}. \tag{4.10}$$

[2] The lemma is formulated this way in order to ease the exposition. However, the conclusion is valid for any pair of $n \times n$ matrices P and X of the same rank for which $XPX = X$.

Proof. By Lemmas 4.1 and 4.5, $F(\lambda)Xy = 0$ for every point $\lambda \in \mathfrak{H}_F$ if and only if $Xy = 0$. Therefore, $\mathcal{M}_X$ is a Hilbert space with respect to the indicated inner product. Moreover, it is in fact a RKHS with RK $K_\omega(\lambda) = F(\lambda)XF(\omega)^*$. To complete the proof, it is enough to verify that

$$F(\lambda)XF(\omega)^* = \frac{j_{pq} - W_X(\lambda)j_{pq}W_X(\omega)^*}{\rho_\omega(\lambda)}$$

for all points $\lambda, \omega \in \mathfrak{H}_F$. But this is a straightforward computation based on the Riccati equation (4.8). □

Thus, Theorem 3.1 implies that $s \in T_{W_X}[\mathcal{S}^{p\times q}]$ if and only if it meets the following three conditions:

(D1) $[I_p \quad -s]FXy$ belongs to the Hardy space $H_2^p(\mathbb{C}_+)$ for every $y \in \mathbb{C}^n$.

(D2) $[-s^\# \quad I_q]FXy$ belongs to the Hardy space $H_2^q(\mathbb{C}_+)^\perp$ $(= H_2^q(\mathbb{C}_-))$ for every $y \in \mathbb{C}^n$.

(D3) $XP_sX \le X$.

Thus, as the conditions (C1)–(C3) are more restrictive than the conditions (D1)–(D3), we see that if P is singular and the extra assumption (A3) is in force, then

$$\widehat{\mathcal{S}}(C, A, P) \subset T_W[\mathcal{S}^{p\times q}].$$

Moreover, the inclusion may be proper. The first step in understanding what restrictions must be imposed in order to obtain a match rests on the direct sum decomposition exhibited in Lemma 4.4

Lemma 4.7. *If assumptions (A1)–(A3) are in force, then*

$$P_s \le P \iff XP_sX \le X \text{ and } P_sv = 0 \text{ for every } v \in \mathcal{N}_P.$$

Proof. If the two conditions on the right are met, then clearly

$$(v + Xy)^*P_s(v + Xy) = y^*XP_sXy \le y^*XPXy = (v + Xy)^*P(v + Xy).$$

Therefore, in view of the direct sum decomposition (4.9), the inequality $P_s \le P$ must also be valid. This serves to complete the proof, since the other direction is self-evident. □

The next step is to explore the significance of the supplementary constraint

$$P_sv = 0 \quad \text{for every } v \in \mathcal{N}_P. \tag{4.11}$$

Lemma 4.8. *If assumptions (A1)–(A3) are in force, then*

(D1a) $[I_p \quad -s]Fv = 0$ *for every vector* $v \in \mathcal{N}_P$.

(D2a) $[-s^\# \quad I_q]Fv = 0$ *for every vector* $v \in \mathcal{N}_P$.

Proof. The inequality $P_s \le P$ implies that

$$\int_{-\infty}^{\infty} v^*F(\mu)^*\Delta_s(\mu)F(\mu)v d\mu = 0$$

for every vector $v \in \mathcal{N}_P$. Therefore, since $\Delta_s(\mu) \geq 0$, we see that

$$\Delta_s(\mu)F(\mu)v = 0 \quad \text{for a.e. point } \mu \in \mathbb{R}.$$

But this in turn implies that (D1a) and (D2a) are in force, as claimed. □

From this point on, the analysis is much the same as in [BD] and [D8]. Therefore, the next three lemmas are stated without proof.

Lemma 4.9. *If (A1)–(A3) are in force and $s = T_W[\varepsilon]$ for $W(\lambda) = W_X(\lambda)$ and some $\varepsilon \in \mathcal{S}^{p\times q}$, then*

$$[I_p \quad -s(\lambda)]F(\lambda)u = 0$$

for every point $\lambda \in \mathbb{C}_+$ and every vector $u \in \mathcal{N}_P$ if and only if

$$[I_p \quad \varepsilon(\lambda)]j_{pq}Cu = 0 \tag{4.12}$$

for every point $\lambda \in \mathbb{C}_+$ and every vector $u \in \mathcal{N}_P$.

Lemma 4.10. *If (A1)–(A3) are in force and $s = T_W[\varepsilon]$ for $W(\lambda) = W_X(\lambda)$ and some mvf $\varepsilon \in \mathcal{S}^{p\times q}$, then*

$$[-s^\#(\lambda) \quad I_q]F(\lambda)u = 0$$

for every point $\lambda \in \mathbb{C}_-$ and every vector $u \in \mathcal{N}_P$ if and only if

$$u^*C^*j_{pq}\begin{bmatrix} \varepsilon(\lambda) \\ I_q \end{bmatrix} = 0 \tag{4.13}$$

for every point $\lambda \in \mathbb{C}_+$ and every vector $u \in \mathcal{N}_P$.

Lemma 4.11. *If assumptions (A1)–(A3) are in force, then $\{Cu : u \in \mathcal{N}_P\}$ is a j_{pq}-neutral subspace of $\mathbb{C}^m$.*

The rest of the argument is self-contained and instructive, and therefore it is presented even though it can be found in the sources cited earlier. Let $w_1, \ldots, w_\nu$ be an orthogonal basis for the j_{pq}-neutral subspace $\{Cu : u \in \mathcal{N}_P\}$ such that

$$\|w_j\|^2 = 2 \ , \ \text{for } j = 1, \ldots, \nu \ .$$

Then

$$w_j^* j_{pq} w_i = 0 \ , \ \text{for } i, j = 1, \ldots, \nu$$

and

$$w_j^* w_i = 0 \ , \ \text{for } i, j = 1, \ldots, \nu \text{ and } i \neq j \ .$$

Hence, upon writing

$$w_j^* = [x_j^* \quad y_j^*]$$

with $x_j \in \mathbb{C}^p$ and $y_j \in \mathbb{C}^q$, we see that

$$\{x_j\}, \ j = 1, \ \ldots, \nu, \text{ is an orthonormal family in } \mathbb{C}^p$$

and

$$\{y_j\}, \ j = 1, \ \ldots, \nu, \text{ is an orthonormal family in } \mathbb{C}^q \ .$$

Therefore,

$$\nu \leq \min\{p, q\} \ . \tag{4.14}$$

Moreover, the condition (4.12) holds, if and only if

$$x_j = \varepsilon(\lambda)y_j \ , \ \text{for } \lambda \in \mathbb{C}_+ \text{ and } j = 1, \dots, \nu \ , \tag{4.15}$$

whereas condition (4.13) holds if and only if

$$x_j^* \varepsilon(\lambda) = y_j^* \ , \ \text{for } \lambda \in \mathbb{C}_+ \text{ and } j = 1, \dots, \nu \ . \tag{4.16}$$

However, since $\varepsilon \in \mathcal{S}^{p\times q}$, it follows from the maximum principle that these two conditions are equivalent, see, e.g., the discussion on pages 7–9 of [D1] for a more detailed explanation, if need be.

Let $U = [U_1 \quad U_2]$ be a $p \times p$ unitary matrix with $U_2 = [x_1 \cdots x_\nu]$ and let $V = [V_1 \quad V_2]$ be a $q \times q$ unitary matrix with $V_2 = [y_1 \cdots y_\nu]$. Then the condition (4.15) holds if and only if

$$U_2 = \varepsilon(\lambda)V_2 \ , \ \text{for } \lambda \in \mathbb{C}_+ \ ,$$

whereas the equivalent second condition (4.16) holds if and only if

$$U_2^* \varepsilon(\lambda) = V_2^* \ , \ \text{for } \lambda \in \mathbb{C}_+$$

Thus, under condition (4.15),

$$\begin{aligned} \varepsilon(\lambda) &= UU^*\varepsilon(\lambda)VV^* \\ &= U \begin{bmatrix} U_1^*\varepsilon(\lambda)V_1 & U_1^*\varepsilon(\lambda)V_2 \\ U_2^*\varepsilon(\lambda)V_1 & U_2^*\varepsilon(\lambda)V_2 \end{bmatrix} V^* \\ &= U \begin{bmatrix} U_1^*\varepsilon(\lambda)V_1 & U_1^*U_2 \\ V_2^*V_1 & V_2^*V_2 \end{bmatrix} V^* , \end{aligned}$$

which reduces to

$$\varepsilon(\lambda) = U \begin{bmatrix} \widetilde{\varepsilon}(\lambda) & 0 \\ 0 & I_\nu \end{bmatrix} V^* , \tag{4.17}$$

where

$$\widetilde{\varepsilon}(\lambda) = U_1^*\varepsilon(\lambda)V_1$$

is an arbitrary element in $\mathcal{S}^{(p-\nu)\times(q-\nu)}$.

Observe that if formula (4.17) is used to define $\varepsilon(\lambda)$ for any $\widetilde{\varepsilon} \in \mathcal{S}^{(p-\nu)\times(q-\nu)}$, then $\varepsilon \in \mathcal{S}^{p\times q}$ and meets the requirement (4.15).

Thus, the analysis of the preceding sections combined with the calculations in this section, serves to establish most of the following result:

Theorem 4.12. *If assumptions (A1)–(A3) are in force, then there exist a pair of unitary matrices $U \in \mathbb{C}^{p\times p}$ and $V \in \mathbb{C}^{q\times q}$ such that*

$$\widehat{\mathcal{S}}(C, A, P) = \left\{ T_{W_X} \left[U \begin{bmatrix} \widetilde{\varepsilon} & 0 \\ 0 & I_\nu \end{bmatrix} V^* \right] : \widetilde{\varepsilon} \in \mathcal{S}^{(p-\nu)\times(q-\nu)}(\mathbb{C}_+) \right\} . \tag{4.18}$$

Moreover,

$$\nu = \operatorname{rank}\{P + C_1^*C_1\} - \operatorname{rank}P = \operatorname{rank}\{P + C_2^*C_2\} - \operatorname{rank}P, \tag{4.19}$$

where C_1 and C_2 are the top and bottom block components of C of sizes $p \times n$ and $q \times n$, respectively (i.e., $C^* = [C_1^* \quad C_2^*]$).

Proof. To obtain formula (4.19), let L denote the linear operator that maps $u \in \mathcal{N}_P$ into $Cu \in \mathbb{C}^m$. Then the dimension of the nullspace $\mathcal{N}_L$ of L is related to ν by the standard formula

$$\dim\mathcal{N}_P = \nu + \dim\mathcal{N}_L.$$

Thus, as

$$\mathcal{N}_L = \mathcal{N}_C \cap \mathcal{N}_P = \mathcal{N}_{P+C^*C}$$

and, as follows from the Lyapunov equation (4.1),

$$\mathcal{N}_{P+C^*C} = \mathcal{N}_{P+C_1^*C_1} = \mathcal{N}_{P+C_2^*C_2},$$

it is readily seen that

$$\nu = \dim\mathcal{N}_P - \dim\mathcal{N}_{P+C_1^*C_1} = \dim\mathcal{N}_P - \dim\mathcal{N}_{P+C_2^*C_2}.$$

The last two formulas are equivalent to the formulas in (4.19). Everything else is covered by the preceding discussion. □

4.3. P is singular and (A3) is not assumed to be in force

The preceding theorem gives a complete description of the set $\widehat{\mathcal{S}}(C, A, P)$ when assumptions (A1)–(A3) are all in force. Our objective, however, is to describe this set when only (A1) and (A2) are assumed to hold. If $\sigma(A)\cap\mathbb{R} = \emptyset$, then it is possible to modify the off diagonal entries of the given upper triangular matrix A a little to a new upper triangular matrix A_0 in such a way that $\widehat{\mathcal{S}}(C, A, P) = \widehat{\mathcal{S}}(C, A_0, P)$ and (A1)–(A3) hold for the triple (C, A_0, P). The details, which rest on a theorem that is established in [BD], are provided in [D8]. The remaining case, when only (A1) and (A2) are in force and $\sigma(A) \cap \mathbb{R} \neq \emptyset$, will be considered elsewhere.

5. Bitangential interpolation in the closed unit disk

There is an extensive literature on bitangential interpolation problems in the unit disk; see, e.g., the references cited in the introduction to this paper. In this section we shall invoke Theorem 3.1 to obtain a number of results for this class of problems, mostly known, in a fairly transparent way. Our starting point is the formulation in [BD], but with fewer restrictions. In particular, we start with four given matrices $M, N, P \in \mathbb{C}^{n\times n}$ and $C \in \mathbb{C}^{m\times n}$ that are subject to the following assumptions:

(E1) $\det(M - \lambda N) \neq 0$ except for at most finitely many points.

(E2) $\operatorname{rank}\begin{bmatrix} M-\mu N \\ C \end{bmatrix} = n$ for every $\mu \in \mathbb{T}$.

(E3) P is a positive semidefinite solution of the Lyapunov-Stein equation

$$M^*PM - N^*PN = C^*j_{pq}C. \tag{5.1}$$

Let

$$G(\lambda) = M - \lambda N, \quad F(\lambda) = C(M-\lambda N)^{-1}, \tag{5.2}$$

$$P_s = \frac{1}{2\pi}\int_0^{2\pi} F(e^{i\theta})^*\Delta_s(e^{i\theta})F(e^{i\theta})d\theta \tag{5.3}$$

and let $\widehat{\mathcal{S}}(M,N,P,C)$ denote the set of mvf's $s \in \mathcal{S}^{p\times q}(\mathbb{D})$ that meet the following three conditions:

(F1) $[I_p \quad -s]Fu$ belongs to the Hardy space $H_2^p(\mathbb{D})$ for every $u \in \mathbb{C}^n$.

(F2) $[-s^\# \quad I_q]Fu$ belongs to the Hardy space $H_2^q(\mathbb{D})^\perp$ for every $u \in \mathbb{C}^n$.

(F3) $P_s \leq P$.

Let

$$\mathfrak{U}_o = \{u \in \mathbb{C}^n : F(\lambda)u = 0 \text{ for every point } \lambda \in \mathfrak{H}_F\}. \tag{5.4}$$

Lemma 5.1. *If the conditions (E1)–(E3) are in force and $F(\lambda)$ is given by formula (5.2), then*

$$\mathfrak{U}_o \subset \mathcal{N}_P. \tag{5.5}$$

Proof. Fix a point $\alpha \in \mathbb{C}$ at which the matrix $G(\alpha)$ is invertible. Then

$$\begin{aligned} G(\lambda) &= C(M-\alpha N-(\lambda-\alpha)N)^{-1} && (5.6)\\ &= CG(\alpha)^{-1}(I_n-(\lambda-\alpha)NG(\alpha)^{-1})^{-1}. && (5.7)\end{aligned}$$

Therefore,

$$\mathfrak{U}_o = \{u \in \mathbb{C}^n : CG(\alpha)^{-1}(NG(\alpha)^{-1})^j u = 0 \text{ for } j = 0,\ldots,n-1\}. \tag{5.8}$$

In particular, this last identity exhibits the fact that $\mathfrak{U}_o$ is invariant under the action of $NG(\alpha)^{-1}$. Thus it has a basis made up of Jordan chains of $NG(\alpha)^{-1}$.

Suppose, for example, that

$$NG(\alpha)^{-1}u_j = \beta u_j + u_{j-1} \text{ for } j = 1,\ldots,\ell\,,$$

where

$$u_0 = 0 \text{ and } u_j \in \mathfrak{U}_o \text{ for } j = 1,\ldots,\ell\,.$$

Then, upon setting $w_1 = G(\alpha)^{-1}u_1$, it is readily seen that

$$Cw_1 = 0 \text{ and } Nw_1 = \beta G(\alpha)w_1 = \beta(M-\alpha N)w_1\,.$$

Therefore,

$$(1+\beta\alpha)Nw_1 = \beta Mw_1$$

and hence, if $\beta \neq 0$, then we can set

$$\mu = \frac{1+\beta\alpha}{\beta}$$

to obtain

$$\mu Nw_1 = Mw_1 \tag{5.9}$$

and subsequently, upon invoking the Lyapunov-Stein equation (5.1),

$$(|\mu|^2-1)w_1^*N^*PNw_1 = 0\,. \tag{5.10}$$

The two identities

$$(M-\mu N)w_1 = 0 \text{ and } Cw_1 = 0$$

guarantee that $|\mu| \neq 1$, because otherwise, by assumption (E2), $w_1 = 0$, which contradicts the choice of $u_1 = G(\alpha)w_1$ as an eigenvector of the matrix $NG(\alpha)^{-1}$. Consequently, Formulas (5.9) and (5.10) imply that

$$Nw_1 \in \mathcal{N}_P \ , \ Mw_1 \in \mathcal{N}_P$$

and hence that

$$u_1 = (M - \alpha N)w_1 \ \text{ also belongs to } \mathcal{N}_P \ .$$

If $\beta = 0$, then the preceding formulas imply that

$$Nw_1 = 0 \ \text{ and } \ Mw_1 \in \mathcal{N}_P \ ,$$

which again leads to the conclusion that $u_1 \in \mathcal{N}_P$.

Suppose next that $\ell \geq 2$ and set $w_2 = G(\alpha)^{-1}u_2$ so that

$$Cw_2 = 0 \ \text{ and } \ Nw_2 = \beta G(\alpha)w_2 + u_1 \ .$$

Then, if $\beta \neq 0$, the latter formula can be rewritten as

$$\mu N w_2 = Mw_2 + \beta^{-1}u_1$$

and hence, since $u_1 \in \mathcal{N}_P$ and $Cw_2 = 0$, a short calculation based on the Lyapunov-Stein equation (5.1) leads to the conclusion that

$$(1 - |\mu|^2)w_2^* N^* P N w_2 = 0 \ .$$

Thus, since $|\mu| \neq 1$, we can conclude just as for w_1, that

$$Nw_2 \in \mathcal{N}_P \ , \ Mw_2 \in \mathcal{N}_P$$

and therefore that

$$u_2 = (M - \alpha N)w_2 \ \text{ also belongs to } \ \mathcal{N}_P \ .$$

Next, if $\beta = 0$, then the formulas

$$Nw_2 = u_1 \ \text{ and } \ w_2^*(M^*PM - N^*PN)w_2 = 0$$

clearly imply that $Nw_2 \in \mathcal{N}_P$ and $Mw_2 \in \mathcal{N}_P$ and hence that $u_2 \in \mathcal{N}_P$ in this case also. The remaining steps in the proof are completed in much the same way. □

5.1. P is invertible

Theorem 5.2. *If (E1)–(E3) are in force and P is invertible and $F(\lambda)$ is given by formula* (5.2), *then the space*

$$\mathcal{M} = \{F(\lambda)u : u \in \mathbb{C}^n\}$$

endowed with the inner product

$$\langle Fu, Fv\rangle_{\mathcal{M}} = v^* P u$$

is a de Branges space $\mathcal{H}(W)$, i.e., it is a RKHS with RK

$$K_\omega(\lambda) = \frac{j_{pq} - W(\lambda)j_{pq}W(\omega)^*}{\rho_\omega(\lambda)},$$

where

$$W(\lambda) = I_m - (1 - \lambda\overline{\mu})C(M - \lambda N)^{-1}P^{-1}(M^* - \overline{\mu}N^*)^{-1}C^* j_{pq}, \qquad (5.11)$$

and μ is any point in $\mathbb{T}$ at which $G(\mu)$ is invertible.

Proof. Lemma 5.1 guarantees that $\mathfrak{U}_o = \{0\}$ when P is invertible and hence that $\mathcal{M}$ is a RKHS. The rest then follows from Theorem 2.3. □

Theorem 3.1 states that a mvf $s \in T_W[\mathcal{S}^{p\times q}]$ if and only if $f(\lambda) = F(\lambda)u$ meets the conditions (F1), (F2) and

$$\langle \Delta_s f, f\rangle \leq \langle f, f\rangle_{\mathcal{H}(W)}.$$

Thus, as

$$\langle \Delta_s f, f\rangle = u^* P_s u \quad \text{and} \quad \langle f, f\rangle_{\mathcal{H}(W)} = u^* P u,$$

these three conditions are exactly the same as those defining $\widehat{\mathcal{S}}(M, N, C, P)$. We have thus established the following theorem:

Theorem 5.3. *If assumptions (E1)–(E3) are in force, and P is invertible, then $\widehat{\mathcal{S}}(M, N, C, P) = T_W[\mathcal{S}^{p\times q}]$, where $W(\lambda)$ is the j_{pq}-inner mvf with respect to $\mathbb{D}$ that is defined by formula (5.11).*

5.2. P is singular, but M and N satisfy an invariance condition

In this subsection we shall allow P to be singular, but shall impose an extra assumption:

(E4) There exists an $n \times n$ complex matrix X such that:

1. $X \geq 0$.
2. $XPX = X$.
3. $PXP = P$.
4. There exist a pair of matrices M_o and N_o such that

$$MX = XM_o, \quad NX = XN_o$$

   and $M_o - \lambda N_o$ is invertible for all but finitely many points $\lambda \in \mathbb{C}$.

If P is invertible, then the choice $X = P^{-1}$ fulfills all four requirements. If P is not invertible, then there is no guarantee that such an X exists. If it does, then in view of (1), (2) and (3), it must be a positive semidefinite pseudo-inverse of P that has the same rank as P.

Lemma 5.4. *Let assumptions (E1)–(E4) be in force and let μ be a point $\in \mathbb{T}$ at which $G(\mu)$ is invertible. Then X is a solution of the Riccati equation*

$$MG(\mu)^{-1}X + XG(\mu)^{-*}N^*\overline{\mu} = XG(\mu)^{-*}C^* j_{pq} G(\mu)^{-1}X \qquad (5.12)$$

Proof. This is a straightforward calculation that is based on the Lyapunov-Stein equation (5.1) and items (2) and (4) of (E4). □

Remark 5.5. A number of other Riccati equations can be obtained under assumptions (E1)–(E4). In particular, it is readily checked that X is a solution of the Riccati equation

$$XM^*X^\dagger MX - XN^*X^\dagger NX = C^*j_{pq}C, \tag{5.13}$$

where $X^\dagger$ denotes the Moore-Penrose inverse of X.

Yet another form of the Riccati equation can be obtained by first rewriting the Lyapunov-Stein equation (5.1) as

$$(M-\alpha N)^*P(M-\alpha N) - (\overline{\alpha}M - N)^*P(\overline{\alpha}M-N) = (1-|\alpha|^2)C^*j_{pq}C \tag{5.14}$$

and then, upon choosing a point $\alpha \in \mathbb{D}$ so that the matrices $(M-\alpha N)$ and $(\overline{\alpha}M - N)$ are both invertible, dividing through, to get the new equation

$$P - (A_\alpha)^*PA_\alpha = (C_\alpha)^*j_{pq}C_\alpha, \tag{5.15}$$

where

$$A_\alpha = (\overline{\alpha}M - N)(M-\alpha N)^{-1} \text{ and } C_\alpha = \sqrt{(1-|\alpha|^2}C(M-\alpha N)^{-1}.$$

This choice of α insures that the matrix A_α is invertible. In fact, we can and will assume that α is chosen so that the two matrices $M_o - \alpha N_o$ and $\overline{\alpha}M_o - N_o$ are also invertible and hence that $A_\alpha X = X\widetilde{A}_\alpha$, where $\widetilde{A}_\alpha = (\overline{\alpha}M_o - N_o)(M_o-\alpha N_o)^{-1}$ is also invertible. Then, upon multiplying the preceding equation by X on both sides and rearranging terms, we obtain

$$A_\alpha X = X(A_\alpha)^{-*}(I_m - (C_\alpha)^*j_{pq}C_\alpha X),$$

which exhibits the requisite invariance of $\mathcal{R}_X$. If $I_m - (C_\alpha)^*j_{pq}C_\alpha X$ is invertible, then X is a solution of

$$X = A_\alpha X(I_m - (C_\alpha)^*j_{pq}C_\alpha X)^{-1}A_\alpha^*. \tag{5.16}$$

For additional discussion of the connection between Riccati equations of this form and RKHS's, see [D5].

Next, returning to the main theme, it is readily checked via (4) of (E4) that

$$(R_\alpha F)(\lambda) = F(\lambda)N(M-\alpha N)^{-1}X = F(\lambda)XN_o(M_o - \alpha N_o)^{-1} \tag{5.17}$$

for every point $\alpha \in \mathbb{C}$ at which both of the indicated inverses exist.

Theorem 5.6. *If the assumptions (E1)–E(4) are in force and $F(\lambda)$ is given by formula* (5.2), *then the space*

$$\mathcal{M}_X = \{F(\lambda)Xu : u \in \mathbb{C}^n\}$$

endowed with the inner product

$$\langle FXu, FXv\rangle_{\mathcal{M}_X} = v^*Xu$$

is a de Branges space $\mathcal{H}(W_X)$, i.e., it is a RKHS with RK

$$K_\omega(\lambda) = \frac{j_{pq} - W_X(\lambda)j_{pq}W_X(\omega)^*}{\rho_\omega(\lambda)}, \tag{5.18}$$

where

$$W_X(\lambda) = I_m - (1-\lambda\overline{\mu})C(M-\lambda N)^{-1}X(M^*-\overline{\mu}N^*)^{-1}C^*j_{pq} \tag{5.19}$$

and μ is a point in $\mathbb{T}$ at which $G(\mu)$ is invertible.

Proof. The proof that

$$F(\lambda)Xy = 0 \implies Xy = 0$$

rests on Lemma 5.1 and the direct sum decomposition (4.9), which is valid when (E4) is in force. This guarantees that the space $\mathcal{M}_X$ is a RKHS. Moreover, by formula (5.17), it is R_α invariant for all but finitely many points $\alpha \in \mathbb{C}$. The rest then follows from Theorem 2.3. One can also check directly that the right-hand side of formula (5.18) is equal to $F(\lambda)XF(\omega)^*$ with the help of the Riccati equation (5.12). □

Thus, Theorem 3.1 implies that $s \in T_{W_X}[\mathcal{S}^{p\times q}]$ if and only if it meets the following three conditions:

(G1) $[I_p \quad -s]FXy$ belongs to the Hardy space $H_2^p(\mathbb{D})$ for every $y \in \mathbb{C}^n$.

(G2) $[-s^\# \quad I_q]FXy$ belongs to the Hardy space $H_2^q(\mathbb{D})^\perp$ for every $y \in \mathbb{C}^n$.

(G3) $XP_sX \leq X$.

Thus, as the conditions (F1)–(F3) are more restrictive than the conditions (G1)–(G3), we see that if P is singular and the extra assumption (E4) is in force, then

$$\widehat{\mathcal{S}}(M,N,C,P) \subset T_{W_X}[\mathcal{S}^{p\times q}].$$

Moreover, the inclusion may be proper. Equality is achieved by restricting the linear fractional transformation T_{W_X} to an appropriate subset, much as before. We skip the details because they can be easily filled in by either adapting the analysis of the preceding section to the present setting, or by simply looking at [BD] for an analogous development. The two treatments have much in common and both of them make essential use of the direct sum decomposition (4.9), which is, as we have already noted, valid when (E4) is imposed. Either way, the main conclusion, which is also presented as Theorem 1.2 of [BD], is:

Theorem 5.7. *If assumptions (E1)–(E4) are in force, then there exist a pair of unitary matrices $U \in \mathbb{C}^{p\times p}$ and $V \in \mathbb{C}^{q\times q}$ such that*

$$\widehat{\mathcal{S}}(M,N,C,P) = \left\{ T_{W_X}\left[U \begin{bmatrix} \widetilde{\varepsilon} & 0 \\ 0 & I_\nu \end{bmatrix} V^* \right] : \widetilde{\varepsilon} \in \mathcal{S}^{(p-\nu)\times(q-\nu)}(\mathbb{D}) \right\}. \tag{5.20}$$

Moreover,

$$\nu = \operatorname{rank}\{N^*PN + C_1^*C_1\} - \operatorname{rank}P = \operatorname{rank}\{M^*PM + C_2^*C_2\} - \operatorname{rank}P, \tag{5.21}$$

where C_1 and C_2 are the top and bottom block components of C of sizes $p \times n$ and $q \times n$, respectively (i.e., $C^* = [C_1^* \quad C_2^*]$).

5.3. Final remarks

In the last subsection we have used Theorem 3.1 to obtain a complete description of the set $\widehat{\mathcal{S}}(M,N,C,P)$ when assumptions (E1)–(E4) are in force. Our objective, however, was to obtain this description under the less restrictive assumption that only (E1)–(E3) are in force. If the mvf $G(\lambda)$ is invertible for every point $\lambda \in \mathbb{T}$, then, as demonstrated in [BD], one can always modify the offdiagonal entries of the upper triangular matrices M and N a little to obtain new matrices M_1 and N_1 such that $\widehat{\mathcal{S}}(M,N,C,P) = \widehat{\mathcal{S}}(M_1,N_1,C,P)$ and (E1)–(E4) hold for the new set of data M_1, N_1, C, P. Thus,

If assumptions (E1), (E3) *and*

(E2a) rank $(M - \mu N) = n$ for every $\mu \in \mathbb{T}$,

are in force, then the conclusions of Theorem 5.7 *are all valid.*

The remaining cases, when only (E1)–(E3) are in force and (E2a) fails, will be considered elsewhere.

References

[ABDS1] D. Alpay, P. Bruinsma, A. Dijksma and H. de Snoo, Interpolation problems, extensions of symmetric operators and reproducing kernel spaces. I. in: *Topics in matrix and operator theory (Rotterdam, 1989), Oper. Theory Adv. Appl.*, **50**, Birkhäuser, Basel, 1991, 35–82.

[ABDS2] D. Alpay, P. Bruinsma, A. Dijksma and H. de Snoo, Interpolation problems, extensions of symmetric operators and reproducing kernel spaces. II, *Integral Equations Operator Theory,* **14** (1991), no. 4, 465–500.

[ABDS3] D. Alpay, P. Bruinsma, A. Dijksma and H. de Snoo, Addendum: Interpolation problems, extensions of symmetric operators and reproducing kernel spaces. II, *Integral Equations Operator Theory* **15** (1992), no. 3, 378–388.

[AlD1] D. Alpay and H. Dym, Hilbert spaces of analytic functions, inverse scattering and operator models. I, *Integral Equations Operator Theory,* **7** (1984), 589–741.

[AlD2] D. Alpay and H. Dym, On a new class of structured reproducing kernel spaces, *J. Funct. Anal.*, **111** (1993), 1–28.

[AlD3] D. Alpay and H. Dym, On a new class of realization formulas and their application, *Linear Algebra Appl.*, **241–243** (1996), 3–84.

[Ar] D.Z. Arov, The generalized bitangent Carathéodory-Nevanlinna-Pick problem and (j, J_0)-inner matrix functions, *Russian Acad. Sci. Izv. Math.* **42** (1994), no. 1, 1–26.

[Ara] Z.D. Arova, The functional model of a j-unitary node with a given j-inner characteristic matrix function, *Integral Equations Operator Theory,* **28** (1997), 1–16.

[ArD] D.Z. Arov and H. Dym, J-inner matrix functions, interpolation and inverse problems for canonical systems, I: Foundations, *Integral Equations Operator Theory* **29** (1997), No. 4, 373–454.

[Arz] N. Aronszajn, Theory of reproducing kernels, *Trans. Amer. Math. Soc.*, **68**(1950), 337–404.

[BGR] J.A. Ball, I. Gohberg and L. Rodman, *Interpolation of Rational Matrix Functions*, Birkhäuser, Basel, 1990.

[Ba] J.A. Ball, Models for non contractions, *J. Math. Anal. Apl.*, **52** (1975), 235–254.

[BaH] J.A. Ball and J.W. Helton, Interpolation problems of Pick-Nevanlinna and Loewner types for meromorphic matrix functions, *Integral Equations Operator Theory*, **9** (1986), 155–203.

[Bh] R. Bhatia, *Matrix Analysis*, Springer, New York, 1997.

[BD] V. Bolotnikov and H. Dym, On degenerate interpolation, entropy and extremal problems for matrix Schur functions, *Integral Equations Operator Theory*, **32** (1998), 367–435.

[Br] L. de Branges, Some Hilbert spaces of analytic functions I, *Trans. Amer. Math. Soc.*, **106** (1963), 445–668.

[BrR] L. de Branges and J. Rovnyak, Canonical models in quantum scattering theory, in: *Perturbation Theory and its Applications in Quantum Mechanics* (C. Wilcox, ed.) Wiley, New York, 1966, 295–392.

[Du] V.K. Dubovoj, Indefinite metric in the Schur interpolation problem for analytic matrix functions IV, *Theor. Funktsii, Func. Anal. i Prilozen*, **42** (1984), 46–57.

[DFK] V.K. Dubovoj, B. Fritzsche and B. Kirstein, *Matricial Version of the Classical Schur Problem*, B.G. Teubner, Stuttgart, 1992.

[D1] H. Dym, *J-Contractive Matrices, Reproducing Kernel Hilbert Spaces and Interpolation*, CBMS Regional Conference Series, American Mathematical Society, **71** Providence, R.I., 1989.

[D2] H. Dym, On reproducing kernel spaces, J-unitary matrix functions, interpolation and displacement rank, in: *The Gohberg Anniversary Collection*, **OT41**, Birkhäuser (1989), 173–239.

[D3] H. Dym, Shifts, realizations and interpolation, redux, in: *Oper. Theory: Adv. Appl.*, **OT73** (1994), 182–243.

[D4] H. Dym, A Basic Interpolation Problem, in: *Holomorphic Spaces* (S. Axler, J. McCarthy and D. Sarason, eds.), Cambridge University Press, Cambridge, England, 1998, 381–423.

[D5] H. Dym, On Riccati equations and reproducing kernel spaces, in: *Recent Advances in Operator Theory, Oper. Theory: Adv. Appl.*, **OT124**, Birkhäuser, Basel, 2001, 189–215.

[D6] H. Dym, Reproducing kernels and Riccati equations, *Int. J. Appl. Math. Comp. Science*, **11** (2001), 35–53.

[D7] H. Dym, Structured matrices, reproducing kernels and interpolation, in: *Structured Matrices in Mathematics, Computer Science and Engineering I*, Contemporary Mathematics (V. Olshevsky, ed.), **Vol. 280** Amer. Math. Soc., Providence, R.I., 2001, 3–29.

[D8] H. Dym, Riccati equations and interpolation, in: *Fast Algorithms*, Contemporary Mathematics (V. Olshevsky, ed.), Amer. Math. Soc., Providence, R.I., in press.

[FF] C. Foias and A. Frazho, *The Commuting Lifting Approach to Interpolation Problems*, Birkhäuser, Basel, 1990.

[HKZ] H. Hedenmalm, B. Korenblum and K. Zhu, *Theory of Bergman Spaces*, Springer, New York, 2000.

[KKY] V.E. Katsnelson, A.Ya. Kheifets and P.M. Yuditskii, An abstract interpolation problem and the extension theory of isometric operators, in: *Topics in Interpolation Theory, Oper. Theory Adv. Appl.*, **95**, Birkhäuser, Basel, 1997, 283–298.

[KrN] M.G. Krein and A.A. Nudelman, *The Markov Moment Problem and Extremal Problems*, Transl. Math. Monographs, **50** Amer. Math. Soc., Providence, R.I., 1977.

[Ka] V.E. Katsnelson, On transformations of Potapov's fundamental matrix identity, in: *Topics in Interpolation Theory, Oper. Theory Adv. Appl.*, (H. Dym, B. Fritzsche, V. Katsnelson and B. Kirstein, eds.) **95**, Birkhäuser, Basel, 1997, 253–281.

[Kh1] A.Ya. Kheifets, The abstract interpolation problem and its applications, in: *Holomorphic Spaces* (S. Axler, J. McCarthy and D. Sarason, eds.), Cambridge University Press, Cambridge, England, 1998, 351–379.

[Kh2] A.Ya. Kheifets, Abstract interpolation scheme for harmonic functions, in: *Interpolation Theory, Systems Theory and Related Topics, Oper. Theory Adv. Appl.,* (D. Alpay, I. Gohberg and V. Vinnikov, eds.) **134**, Birkhäuser, Basel, 2002, 287–317.

[KhY] A.Ya. Kheifets and P.M. Yuditskii, An analysis and extension of V.P. Potapov's approach to problems with applications to the generalized bi-tangential Schur-Nevanlinna-Pick problem and J-inner-outer factorization, in: *Matrix and Operator Valued Functions, Oper. Theory Adv. Appl.*, **72**, Birkhäuser, Basel, 1994, 133–161.

[RR] M. Rosenblum and J. Rovnyak, *Topics in Hardy Classes and Univalent Functions*, Birkhäuser, Basel, 1994.

[Ro] J. Rovnyak, Characterizations of spaces $\mathbf{K}(M)$, unpublished manuscript, 1968.

[Sak] L.A. Sakhnovich, *Interpolation Theory and its Applications*, Kluwer, Dordrecht, 1997.

[Sak-A] A.L. Sakhnovich, Spectral functions of a canonical system of order $2n$, Math. USSR Sbornik, **71** (1992), no. 2, 119–128.

Harry Dym
Department of Mathematics
The Weizmann Institute of Science
Rehovot 76100, Israel
e-mail: dym@wisdom.weizmann.ac.il

Operator Theory:
Advances and Applications, Vol. 143, 213–253

A Class of Matrix-valued Schrödinger Operators with Prescribed Finite-band Spectra

Fritz Gesztesy and Lev A. Sakhnovich

Abstract. We construct a class of matrix-valued Schrödinger operators with prescribed finite-band spectra of maximum spectral multiplicity. The corresponding matrix potentials are shown to be stationary solutions of the KdV hierarchy. The methods employed in this paper rely on matrix-valued Herglotz functions, Weyl–Titchmarsh theory, pencils of matrices, and basic inverse spectral theory for matrix-valued Schrödinger operators.

1. Introduction

While basic aspects of inverse spectral theory for matrix-valued Schrödinger operators were established some time ago, finer properties such as isospectral sets (manifolds) of potentials, for instance, in the periodic or algebro-geometric finite-band cases, are still in their infancy. This paper intends to make a modest contribution to this circle of ideas. More precisely, given a closed set $\Sigma \subset \mathbb{R}$ of the type

$$\Sigma = \left\{ \bigcup_{j=0}^{n-1} [E_{2j}, E_{2j+1}] \right\} \cup [E_{2n}, \infty), \tag{1.1}$$

where

$$\{E_\ell\}_{0 \leq \ell \leq 2n} \subseteq \mathbb{R},\ n \in \mathbb{N},\ \text{with } E_\ell < E_{\ell+1},\ 0 \leq \ell \leq 2n-1, \tag{1.2}$$

we construct $m \times m$ matrix-valued Schrödinger operators $H_\Sigma = -d^2/dx^2 \mathcal{I}_m + \mathcal{Q}_\Sigma$ in $L^2(\mathbb{R})^{m\times m}$ ($\mathcal{I}_m$ the identity matrix in $\mathbb{C}^{m\times m}$, $m \in \mathbb{N}$) with prescribed spectrum Σ,

$$\operatorname{spec}(H_\Sigma) = \Sigma, \tag{1.3}$$

of uniform spectral multiplicity $2m$. The constructed matrix potentials $\mathcal{Q}_\Sigma$ will turn out to be reflectionless in the sense discussed in [22], [39], and [55], that

2000 *Mathematics Subject Classification.* Primary: 34A55, 34B27, 34L40, 35Q53, Secondary: 34B20, 34B24, 34L05.
Key words and phrases. Matrix-valued Schrödinger operators, matrix potentials, finite-band spectra, Weyl–Titchmarsh matrices.

is, the half-line Weyl–Titchmarsh matrices $\mathcal{M}_{\pm,\Sigma}(z,x)$ associated with H_Σ, the half-lines $[x,\pm\infty)$, and a Dirichlet boundary condition at $x\in\mathbb{R}$, satisfy

$$\lim_{\varepsilon\downarrow 0}\mathcal{M}_{+,\Sigma}(\lambda+i\varepsilon,x)=\lim_{\varepsilon\downarrow 0}\mathcal{M}_{-,\Sigma}(\lambda-i\varepsilon,x), \tag{1.4}$$
$$\lambda\in\bigcup_{j=0}^{n-1}(E_{2j},E_{2j+1})\cup(E_{2n},\infty),\ x\in\mathbb{R}.$$

Especially, $\mathcal{M}_{+,\Sigma}(\cdot,x)$ is the analytic continuation of $\mathcal{M}_{-,\Sigma}(\cdot,x)$ through the set Σ, and vice versa. In other words, $\mathcal{M}_{+,\Sigma}(\cdot,x)$ and $\mathcal{M}_{-,\Sigma}(\cdot,x)$ are the two branches of an analytic matrix-valued function $\mathcal{M}_\Sigma(\cdot,x)$ on the two-sheeted Riemann surface of $\big(\prod_{\ell=0}^{2n}(z-E_\ell)\big)^{1/2}$. These facts imply a purely absolutely continuous spectrum Σ of the associated Schrödinger operator H_Σ of uniform (maximal) multiplicity $2m$.

Before we turn to a brief description of the contents of each section, it seems appropriate to mention some of the pertinent results and especially, the most recent activities in connection with (inverse) spectral theory of matrix-valued Schrödinger operators. The basic Weyl–Titchmarsh theory of singular Hamiltonian systems and their basic spectral theory were developed by Hinton and Shaw, Kogan and Rofe-Beketov, Orlov, and others (see, e.g., [42], [44]–[48], [53], [54], [56], [57], [80], [90], [91], [93], [96], [100, Ch. 9], and the references therein). Various aspects of direct spectral theory, including investigations of the nature of the spectrum involved, (regularized) trace formulas, uniqueness theorems, etc., appeared in [10], [12]–[15], [19], [22], [23], [34], [36], [38], [59], [81], [88], [89]. General asymptotic expansions of Weyl–Titchmarsh matrices as the (complex) spectral parameter tends to infinity under optimal regularity assumptions on the coefficients are of relatively recent origin and can be found in [19], [20] (see also [94], [106]). The inverse scattering formalism for Schrödinger operators has been studied by a variety of authors and we refer, for instance, to [1], [2], [3], [7], [74], [76], [77], [107]. General inverse spectral theory, the existence of transformation operators, etc., are discussed in [67], [68], [84], [93], [96], [97], [99], [100], and the references therein. Inverse monodromy problems have recently been discussed in [4], [5], [6], [15], [67], [68], [95], [100], and the literature cited therein. More specific inverse spectral problems, such as compactness of the isospectral set of periodic Schrödinger operators [13], special isospectral matrix-valued Schrödinger operators, and Borg-type uniqueness theorems (for periodic coefficients as well as eigenvalue problems on compact intervals) were recently studied in [17], [18], [20], [22], [25], [50], [51], [67], [68], [100], [101], [102], [103]. Moreover, direct spectral theory in the particular case of periodic Schrödinger operators (i.e., Floquet theory and the like) has been studied in [12], [14], [22], [24], [25], [33], [52], [58], [59], [85], [101], [102], [109]–[111], with many more pertinent references to be found therein. Apart from Floquet theoretic applications in connection with Schrödinger operators already briefly touched upon, we also need to mention applications to random Schrödinger operators associated with strips as discussed, for instance, in [16], [53], [55], and especially to nonabelian completely integrable systems. Since the literature associated with the latter topic is of enormous proportions, we can only refer to a few pertinent publications, such as, [7], [8], [11], [26], [29], [69], [70], [74], [78], [79], [82], [92], [94]–[96], [98]. The interested reader will find a wealth of additional material in these references.

Section 2 summarizes basic results in Weyl–Titchmarsh theory and some elements of inverse spectral theory for matrix-valued Schrödinger operators. Polynomial pencils of matrices are briefly reviewed in Section 3. In Section 4 we present our principal new result, the construction of $m \times m$ matrix-valued Schrödinger operators H_Σ with spectrum Σ (cf. (1.1)) and uniform maximal spectral multiplicity $2m$. In our final Section 5 we prove that $\mathcal{Q}_\Sigma$ satisfies a stationary KdV equation (in fact, we explicitly identify the first equation in the stationary KdV hierarchy satisfied by $\mathcal{Q}_\Sigma$) and derive matrix-valued trace formulas for $\mathcal{Q}_\Sigma$ and higher-order KdV invariants.

2. Basic facts on Weyl–Titchmarsh theory

In this section we briefly recall basic elements of the Weyl–Titchmarsh theory for matrix-valued Schrödinger operators. Throughout this paper all matrices will be considered over the field of complex numbers $\mathbb{C}$, and the corresponding linear space of $k \times \ell$ matrices will be denoted by $\mathbb{C}^{k\times\ell}$, $k, \ell \in \mathbb{N}$. Moreover, $\mathcal{I}_k$ denotes the identity matrix in $\mathbb{C}^{k\times k}$, $\mathcal{M}^*$ the adjoint (i.e., complex conjugate transpose), $\mathcal{M}^t$ the transpose of a matrix $\mathcal{M}$, $\operatorname{diag}(m_1, \dots, m_k) \in \mathbb{C}^{k\times k}$ a diagonal $k \times k$ matrix, and $\mathrm{AC}_{\mathrm{loc}}(\mathbb{R})$ denotes the set of locally absolutely continuous functions on $\mathbb{R}$. The spectrum, point spectrum (the set of eigenvalues), essential spectrum, absolutely continuous spectrum, and singularly continuous spectrum of a self-adjoint linear operator T in a separable complex Hilbert space are denoted by $\operatorname{spec}(T)$, $\operatorname{spec}_{\mathrm{p}}(T)$, $\operatorname{spec}_{\mathrm{ess}}(T)$, $\operatorname{spec}_{\mathrm{ac}}(T)$, $\operatorname{spec}_{\mathrm{sc}}(T)$, respectively.

The basic assumption for this section will be the following.

Hypothesis 2.1.

(i) *Fix* $m \in \mathbb{N}$, *suppose* $\mathcal{Q} = \mathcal{Q}^* \in L^1_{\mathrm{loc}}(\mathbb{R})^{m\times m}$ *and introduce the differential expression*

$$\mathcal{L} = -\mathcal{I}_m \frac{d^2}{dx^2} + \mathcal{Q}, \quad x \in \mathbb{R}. \tag{2.1}$$

(ii) *Suppose* $\mathcal{L}$ *is in the limit point case at* $\pm\infty$.

Given Hypothesis 2.1 (i) we consider the matrix-valued Schrödinger equation

$$-\psi''(z, x) + \mathcal{Q}(x)\psi(z, x) = z\psi(z, x) \text{ for a.e. } x \in \mathbb{R}, \tag{2.2}$$

where $z \in \mathbb{C}$ plays the role of a spectral parameter and ψ is assumed to satisfy

$$\psi(z, \cdot), \psi'(z, \cdot) \in \mathrm{AC}_{\mathrm{loc}}(\mathbb{R})^{m\times m}. \tag{2.3}$$

Throughout this paper, x-derivatives are abbreviated by a prime $'$.

Let $\Psi(z,x,x_0)$ be a $2m\times 2m$ normalized fundamental system of solutions of (2.2) at some $x_0\in\mathbb{R}$ which we partition as

$$\Psi(z,x,x_0)=\begin{pmatrix}\theta(z,x,x_0) & \phi(z,x,x_0)\\ \theta'(z,x,x_0) & \phi'(z,x,x_0)\end{pmatrix}. \tag{2.4}$$

Here $'$ denotes d/dx, $\theta(z,x,x_0)$ and $\phi(z,x,x_0)$ are $m\times m$ matrices, entire with respect to $z\in\mathbb{C}$, and normalized according to

$$\Psi(z,x_0,x_0)=\mathcal{I}_{2m}, \tag{2.5}$$

that is,

$$\theta(z,x_0,x_0)=\phi'(z,x_0,x_0)=\mathcal{I}_m,\quad \phi(z,x_0,x_0)=\theta'(z,x_0,x_0)=0. \tag{2.6}$$

In this context, we briefly recall a set of formulas needed later in Section 4. Introducing

$$\mathcal{J}=\begin{pmatrix}0 & -\mathcal{I}_m\\ \mathcal{I}_m & 0\end{pmatrix}, \tag{2.7}$$

one infers

$$\Psi(\bar z,x,x_0)^*\mathcal{J}\Psi(z,x,x_0)=\mathcal{J}, \tag{2.8}$$

which implies $\mathcal{J}\Psi(z,x,x_0)(\Psi(\bar z,x,x_0)\mathcal{J})^*=\mathcal{I}_{2m}$ and hence

$$\Psi(z,x,x_0)\mathcal{J}\Psi(\bar z,x,x_0)^*=\mathcal{J}. \tag{2.9}$$

Writing out (2.8) and (2.9) explicitly yields

$$\theta'(\bar z,x,x_0)^*\theta(z,x,x_0)-\theta(\bar z,x,x_0)^*\theta'(z,x,x_0)=0, \tag{2.10}$$
$$\phi'(\bar z,x,x_0)^*\phi(z,x,x_0)-\phi(\bar z,x,x_0)^*\phi'(z,x,x_0)=0, \tag{2.11}$$
$$\phi'(\bar z,x,x_0)^*\theta(z,x,x_0)-\phi(\bar z,x,x_0)^*\theta'(z,x,x_0)=\mathcal{I}_m, \tag{2.12}$$
$$\theta(\bar z,x,x_0)^*\phi'(z,x,x_0)-\theta'(\bar z,x,x_0)^*\phi(z,x,x_0)=\mathcal{I}_m, \tag{2.13}$$

and

$$\phi(z,x,x_0)\theta(\bar z,x,x_0)^*-\theta(z,x,x_0)\phi(\bar z,x,x_0)^*=0, \tag{2.14}$$
$$\phi'(z,x,x_0)\theta'(\bar z,x,x_0)^*-\theta'(z,x,x_0)\phi'(\bar z,x,x_0)^*=0, \tag{2.15}$$
$$\phi'(z,x,x_0)\theta(\bar z,x,x_0)^*-\theta'(z,x,x_0)\phi(\bar z,x,x_0)^*=\mathcal{I}_m, \tag{2.16}$$
$$\theta(z,x,x_0,\alpha)\phi'(\bar z,x,x_0)^*-\phi(z,x,x_0,\alpha)\theta'(\bar z,x,x_0)^*=\mathcal{I}_m. \tag{2.17}$$

Next we consider the spaces

$$N(z,\pm\infty)=\{\phi\in L^2((c,\pm\infty))^m\mid -\phi''+\mathcal{Q}\phi=z\phi \text{ a.e. on } (c,\pm\infty)\}, \tag{2.18}$$

for some $c\in\mathbb{R}$ and $z\in\mathbb{C}$. (Here $(\phi,\psi)_{\mathbb{C}^n}=\sum_{j=1}^n\overline{\phi}_j\psi_j$ denotes the standard scalar product in $\mathbb{C}^n$, abbreviating $\chi\in\mathbb{C}^n$ by $\chi=(\chi_1,\cdots,\chi_n)^t$.) Both dimensions of the spaces in (2.18), $\dim_{\mathbb{C}}(N(z,\infty))$ and $\dim_{\mathbb{C}}(N(z,-\infty))$, are constant for $z\in\mathbb{C}_\pm=\{\zeta\in\mathbb{C}\mid \pm\mathrm{Im}(\zeta)>0\}$ (see, e.g., [54]). One then recalls that $\mathcal{L}$ in (2.1) is in the limit point case at $\pm\infty$ whenever

$$\dim_{\mathbb{C}}(N(z,\pm\infty))=m \text{ for all } z\in\mathbb{C}\backslash\mathbb{R}. \tag{2.19}$$

Since the potential $\mathcal{Q}_\Sigma$ to be constructed in Section 4 will automatically lead to the limit point case at $\pm\infty$, we decided to limit our considerations mainly to this situation. In this context we note the well-known fact that if $\mathcal{L}$ in (2.1) is in the limit point case at $\pm\infty$, then the $m\times m$ Weyl–Titchmarsh matrices associated with $\mathcal{L}$, the half-lines $[x,\pm\infty)$, and a Dirichlet boundary condition at x, are given by

$$\mathcal{M}_\pm(z,x_0)=\Psi'_\pm(z,x,x_0)\Psi_\pm(z,x,x_0)^{-1}\big|_{x=x_0},\quad z\in\mathbb{C}\backslash\mathbb{R}, \tag{2.20}$$

where $\Psi_\pm$ satisfy $(\mathcal{L}-z\mathcal{I}_m)\Psi_\pm(z,\cdot,x_0)=0$ and

$$\Psi_\pm(z,\cdot,x_0)\in L^2([x_0,\pm\infty))^{m\times m}. \tag{2.21}$$

The actual normalization of $\Psi_\pm(z,\cdot,x_0)$ is clearly irrelevant and hence $\Psi_\pm(z,\cdot,x_0)$ can be replaced by $\Psi_\pm(z,\cdot,x_0)C$, where C is any nonsingular $m\times m$ matrix.

For later reference we summarize the principal results on $\mathcal{M}_\pm(z,x_0)$ in the next theorem. First we recall the following definition.

Definition 2.2. *A map $\mathcal{M}\colon \mathbb{C}_+\to\mathbb{C}^{n\times n}$, $n\in\mathbb{N}$, extended to $\mathbb{C}_-$ by $\mathcal{M}(\bar z)=\mathcal{M}(z)^*$ for all $z\in\mathbb{C}_+$, is called an $n\times n$ Herglotz matrix*[1] *if it is analytic on $\mathbb{C}_+$ and* $\mathrm{Im}(\mathcal{M}(z))\geq 0$ *for all $z\in\mathbb{C}_+$.*

Here we denote $\mathrm{Im}(\mathcal{M})=(\mathcal{M}-\mathcal{M}^*)/2i$ and $\mathrm{Re}(\mathcal{M})=(\mathcal{M}+\mathcal{M}^*)/2$.

In the scalar context $n=1$, the condition $\mathrm{Im}(\mathcal{M}(z))\geq 0$ in Definition 2.2 can be replaced by $\mathrm{Im}(m(z))>0$ for the corresponding scalar counterpart $m(z)$.

Theorem 2.3 ([20], [39], [44], [45], [48], [55]). *Assume Hypothesis* 2.1 *and suppose that $z\in\mathbb{C}\backslash\mathbb{R}$, $x_0\in\mathbb{R}$. Then*

(i) *$\pm\mathcal{M}_\pm(z,x_0)$ is an $m\times m$ matrix-valued Herglotz function of maximal rank. In particular,*

$$\mathrm{Im}(\pm\mathcal{M}_\pm(z,x_0))>0,\quad z\in\mathbb{C}_+, \tag{2.22}$$

$$\mathcal{M}_\pm(\overline{z},x_0)=\mathcal{M}_\pm(z,x_0)^*, \tag{2.23}$$

$$\mathrm{rank}(\mathcal{M}_\pm(z,x_0))=m, \tag{2.24}$$

$$\lim_{\varepsilon\downarrow 0}\mathcal{M}_\pm(\lambda+i\varepsilon,x_0)\ \textit{exists for a.e.}\ \lambda\in\mathbb{R}. \tag{2.25}$$

Local singularities of $\pm\mathcal{M}_\pm(z,x_0)$ and $\mp\mathcal{M}_\pm(z,x_0)^{-1}$ are necessarily real and at most of first order in the sense that

$$\mp\lim_{\epsilon\downarrow 0}(i\epsilon\,\mathcal{M}_\pm(\lambda+i\epsilon,x_0))\geq 0,\quad \lambda\in\mathbb{R}, \tag{2.26}$$

$$\pm\lim_{\epsilon\downarrow 0}i\epsilon\mathcal{M}_\pm(\lambda+i\epsilon,x_0)^{-1}\geq 0,\quad \lambda\in\mathbb{R}. \tag{2.27}$$

---

[1] There appears to be considerable confusion in the literature since Nevanlinna, Pick, Nevanlinna–Pick matrix, in addition to Herglotz matrix, are also in use. In part these discrepancies can be traced back to the use of the upper half-plane $\mathbb{C}_+$ versus the open unit disk D, and in some cases the geographical location of the author in question determines the preferred notation. Following a tradition in mathematical physics, we adopt the notion of Herglotz functions in this paper.

(ii) $\pm\mathcal{M}_\pm(z,x_0)$ *admit the representations*

$$\pm\mathcal{M}_\pm(z,x_0) = \mathrm{Re}(\pm\mathcal{M}_\pm(\pm i,x_0)) + \int_{\mathbb{R}} d\Omega_\pm(\lambda,x_0)\left((\lambda-z)^{-1} - \lambda(1+\lambda^2)^{-1}\right), \tag{2.28}$$

where

$$\int_{\mathbb{R}} \|d\Omega_\pm(\lambda,x_0)\|_{\mathbb{C}^{m\times m}} (1+\lambda^2)^{-1} < \infty \tag{2.29}$$

and

$$\Omega_\pm((\lambda,\mu],x_0) = \lim_{\delta\downarrow 0}\lim_{\varepsilon\downarrow 0}\frac{1}{\pi}\int_{\lambda+\delta}^{\mu+\delta} d\nu\, \mathrm{Im}(\pm\mathcal{M}_\pm(\nu+i\varepsilon,x_0)). \tag{2.30}$$

(iii) *Define the* $2m\times m$ *matrices*

$$\begin{aligned}\Psi_\pm(z,x,x_0) &= \begin{pmatrix}\psi_\pm(z,x,x_0)\\ \psi'_\pm(z,x,x_0)\end{pmatrix}\\ &= \begin{pmatrix}\theta(z,x,x_0) & \phi(z,x,x_0)\\ \theta'(z,x,x_0) & \phi'(z,x,x_0)\end{pmatrix}\begin{pmatrix}\mathcal{I}_m\\ \mathcal{M}_\pm(z,x_0)\end{pmatrix},\end{aligned} \tag{2.31}$$

then

$$\mathrm{Im}(\mathcal{M}_\pm(z,x_0)) = \mathrm{Im}(z)\int_{x_0}^{\pm\infty} dx\, \psi_\pm(z,x,x_0)^*\psi_\pm(z,x,x_0). \tag{2.32}$$

(iv) *Denote by* $C_\varepsilon \subset \mathbb{C}_+$ *the open sector with vertex at zero, symmetry axis along the positive imaginary axis, and opening angle* ε, *with* $0<\varepsilon<\pi/2$. *Then*

$$\mathcal{M}_\pm(z,x_0) \underset{\substack{|z|\to\infty\\ z\in C_\varepsilon}}{=} \pm z^{1/2}\mathcal{I}_m + o(1). \tag{2.33}$$

Necessary and sufficient conditions for $\mathcal{M}_\pm(z,x_0)$ to be the half-line $m\times m$ Weyl–Titchmarsh matrix associated with a Schrödinger operator on $[x_0,\pm\infty)$ in terms of the corresponding measures $\Omega_\pm(\cdot,x_0)$ in the Herglotz representation (2.28) of $\mathcal{M}_\pm(z,x_0)$ can be derived using the matrix-valued extension of the classical inverse spectral theory approach due to Gelfand and Levitan [32], as worked out by Rofe-Beketov [84]. The following result describes sufficient conditions for a monotonically nondecreasing matrix function to be the matrix spectral function of a half-line Schrödinger operator. It extends well-known results in the scalar case $m=1$ (cf. [64, Sects. 2.5, 2.9], [65], [75, Sect. 26.5], [105]).

Theorem 2.4 ([84]). *Let* Ω_+ *be a monotonically nondecreasing* $m\times m$ *matrix-valued function on* $\mathbb{R}$ *satisfying the following two conditions.*

(i) *Whenever* $f\in C([x_0,\infty))^{m\times 1}$ *with compact support contained in* $[x_0,\infty)$ *and*

$$\int_{\mathbb{R}} F(\lambda)^* d\Omega_+(\lambda)\, F(\lambda) = 0, \text{ then } f=0 \text{ a.e.}, \tag{2.34}$$

where

$$F(\lambda) = \lim_{R\uparrow\infty}\int_{x_0}^{R} dx\, \frac{\sin(\lambda^{1/2}(x-x_0))}{\lambda^{1/2}} f(x), \quad \lambda\in\mathbb{R}. \tag{2.35}$$

(ii) *Define*

$$\widetilde{\Omega}_+(\lambda) = \begin{cases} \Omega_+(\lambda) - \frac{2}{3\pi}\lambda^{3/2}, & \lambda \geq 0, \\ \Omega_+(\lambda), & \lambda < 0 \end{cases} \tag{2.36}$$

and assume the limit

$$\lim_{R\uparrow\infty} \int_{-\infty}^{R} d\widetilde{\Omega}_+(\lambda)\, \frac{\sin(\lambda^{1/2}(x-x_0))}{\lambda^{1/2}} = \Phi(x) \tag{2.37}$$

exists with $\Phi \in L^\infty([x_0,R])^{m\times m}$ *for all* $R > x_0$. *Moreover, suppose that for some* $r \in \mathbb{N}_0$, $\Phi^{(r+1)} \in L^1([x_0,R])^{m\times m}$ *for all* $R > x_0$, *and* $\Phi(x_0) = 0$. *Then* Ω_+ *is the matrix spectral function of a self-adjoint Schrödinger operator* H_+ *in* $L^2([x_0,\infty))^m$ *associated with the* $m \times m$ *matrix-valued differential expression* $\mathcal{L}_+ = -d^2/dx^2\mathcal{I}_m + \mathcal{Q}$, $x > x_0$, *with a Dirichlet boundary condition at* x_0, *a self-adjoint boundary condition at* ∞ *(if necessary), and a self-adjoint potential matrix* $\mathcal{Q}$ *with* $\mathcal{Q}^{(r)} \in L^1([x_0,R])^{m\times m}$ *for all* $R > x_0$.

Next, assuming Hypothesis 2.1, we introduce the self-adjoint Schrödinger operator H in $L^2(\mathbb{R})^m$ by

$$H = -\mathcal{I}_m \frac{d^2}{dx^2} + \mathcal{Q}, \tag{2.38}$$
$$\mathrm{dom}(H) = \{g \in L^2(\mathbb{R})^m \mid g, g' \in \mathrm{AC}_{\mathrm{loc}}(\mathbb{R})^m;\ (-g'' + \mathcal{Q}g) \in L^2(\mathbb{R})^m\}.$$

The resolvent of H then reads

$$((H-z)^{-1}f)(x) = \int_{\mathbb{R}} dx'\, \mathcal{G}(z,x,x') f(x'), \quad z \in \mathbb{C}\backslash\mathbb{R},\ f \in L^2(\mathbb{R})^m, \tag{2.39}$$

with the Green's matrix $\mathcal{G}(z,x,x')$ of H given by

$$\mathcal{G}(z,x,x') = \psi_\mp(z,x,x_0)[\mathcal{M}_-(z,x_0) - \mathcal{M}_+(z,x_0)]^{-1}\psi_\pm(\overline{z},x',x_0)^*, \quad x \lesseqgtr x',\ z \in \mathbb{C}\backslash\mathbb{R}. \tag{2.40}$$

Introducing

$$\mathcal{N}_\pm(z,x_0) = \mathcal{M}_-(z,x_0) \pm \mathcal{M}_+(z,x_0), \quad z \in \mathbb{C}\backslash\mathbb{R}, \tag{2.41}$$

the $2m \times 2m$ Weyl–Titchmarsh function $\mathcal{M}(z,x_0)$ associated with H on $\mathbb{R}$ is then given by

$$\begin{aligned} \mathcal{M}(z,x_0) &= \big(\mathcal{M}_{j,j'}(z,x_0)\big)_{j,j'=1,2} \\ &= \begin{pmatrix} \mathcal{M}_\pm(z,x_0)\mathcal{N}_-(z,x_0)^{-1}\mathcal{M}_\mp(z,x_0) & \mathcal{N}_-(z,x_0)^{-1}\mathcal{N}_+(z,x_0)/2 \\ \mathcal{N}_+(z,x_0)\mathcal{N}_-(z,x_0)^{-1}/2 & \mathcal{N}_-(z,x_0)^{-1} \end{pmatrix}, \\ &\qquad z \in \mathbb{C}\backslash\mathbb{R}. \end{aligned} \tag{2.42}$$

The basic results on $\mathcal{M}(z,x_0)$ then read as follows.

Theorem 2.5 ([39], [44], [45], [48], [55]). *Assume Hypothesis* 2.1 *and suppose that* $z \in \mathbb{C}\backslash\mathbb{R}$, $x_0 \in \mathbb{R}$. *Then,*

(i) $\mathcal{M}(z, x_0)$ *is a matrix-valued Herglotz function of rank* $2m$ *with representation*

$$\mathcal{M}(z, x_0) = \mathrm{Re}(\mathcal{M}(i, x_0)) + \int_{\mathbb{R}} d\Omega(\lambda, x_0)\, \big((\lambda - z)^{-1} - \lambda(1 + \lambda^2)^{-1}\big), \tag{2.43}$$

where

$$\int_{\mathbb{R}} \|d\Omega(\lambda, x_0)\|_{\mathbb{C}^{2m\times 2m}}\, (1 + \lambda^2)^{-1} < \infty \tag{2.44}$$

and

$$\Omega((\lambda, \mu], x_0) = \lim_{\delta\downarrow 0} \lim_{\varepsilon\downarrow 0} \frac{1}{\pi} \int_{\lambda+\delta}^{\mu+\delta} d\nu\, \mathrm{Im}(\mathcal{M}(\nu + i\varepsilon, x_0)). \tag{2.45}$$

(ii) $z \in \mathbb{C}\backslash\mathrm{spec}(H)$ *if and only if* $\mathcal{M}(z, x_0)$ *is holomorphic near* z.

Here $\mathrm{spec}(T)$ denotes the spectrum of T. Later on we will denote by $\mathrm{spec}_{\mathrm{ac}}(T)$ the absolutely continuous spectrum of T.

Finally, we state the following characterization of $\mathcal{M}(z, x_0)$ to be used later on. In the scalar context $m = 1$ this has been used by Rofe-Beketov [86], [87] (see also [64, Sect. 7.3]).

Theorem 2.6 ([86], [87]). *Assume Hypothesis* 2.1, *suppose that* $z \in \mathbb{C}\backslash\mathbb{R}$, $x_0 \in \mathbb{R}$, *and let* $\ell, r \in \mathbb{N}_0$. *Then the following assertions are equivalent.*

(i) $\mathcal{M}(z, x_0)$ *is the* $2m \times 2m$ *Weyl–Titchmarsh matrix associated with a Schrödinger operator* H *in* $L^2(\mathbb{R})^m$ *of the type* (2.38) *with an* $m \times m$ *matrix-valued potential* $\mathcal{Q} \in L^1_{\mathrm{loc}}(\mathbb{R})$ *and* $\mathcal{Q} \in C^\ell((-\infty, x_0))$ *and* $\mathcal{Q} \in C^r((x_0, \infty))$.

(ii) $\mathcal{M}(z, x_0)$ *is of the type* (2.42) *with* $\mathcal{M}_\pm(z, x_0)$ *being half-line* $m \times m$ *Weyl–Titchmarsh matrices on* $[x_0, \pm\infty)$ *corresponding to a Dirichlet boundary condition at* x_0 *and a self-adjoint boundary condition at* $-\infty$ *and/or* ∞ *(if any) which are associated with an* $m \times m$ *matrix-valued potential* $\mathcal{Q}$ *satisfying* $\mathcal{Q} \in C^\ell((-\infty, x_0))$ *and* $\mathcal{Q} \in C^r((x_0, \infty))$, *respectively.*

If (i) or (ii) holds, then the $2m \times 2m$ *matrix-valued spectral measure* $\Omega(\cdot, x_0)$ *associated with* $\mathcal{M}(z, x_0)$ *is determined by* (2.42) *and* (2.45).

Next, we consider variations of the reference point $x_0 \in \mathbb{R}$. In analogy to (2.20), we note that in the case where the Schrödinger differential expression $\mathcal{L}$ is in the limit point case at $\pm\infty$,

$$\mathcal{M}_\pm(z, x) = \Psi'_{\pm,x}(z, x, x_0)\Psi_\pm(z, x, x_0)^{-1}, \quad z \in \mathbb{C}\backslash\mathbb{R}, \tag{2.46}$$

represents the corresponding half-line Weyl–Titchmarsh matrix on $[x, \pm\infty)$, $x \in \mathbb{R}$, with $\Psi_\pm(z, \cdot, x_0)$ defined in (2.31). Again the actual normalization of $\Psi_\pm$ is, of course, irrelevant. Since $\Psi_\pm$ satisfies the second-order linear $m \times m$ matrix-valued differential equation (2.2), $\mathcal{M}_\pm$ in (2.46) satisfies the matrix-valued Riccati-type equation (independently of any limit point assumptions at $\pm\infty$)

$$\mathcal{M}'_\pm(z, x) + \mathcal{M}_\pm(z, x)^2 = \mathcal{Q}(x) - z\mathcal{I}_m, \quad x \in \mathbb{R},\ z \in \mathbb{C}\backslash\mathbb{R}. \tag{2.47}$$

The asymptotic high-energy behavior of $\mathcal{M}_\pm(z,x)$ as $|z| \to \infty$ has recently been determined in [20] under minimal smoothness conditions on $\mathcal{Q}$ and without assuming that $\mathcal{L}$ is in the limit point case at $\pm\infty$. Here we recall just a special case of the asymptotic expansion proved in [20] which is most suited for our discussion in Section 4. We denote by $C_\varepsilon \subset \mathbb{C}_+$ the open sector with vertex at zero, symmetry axis along the positive imaginary axis, and opening angle ε, with $0 < \varepsilon < \pi/2$.

Theorem 2.7 ([20]). *Fix $x_0 \in \mathbb{R}$ and let $x \geq x_0$. In addition to Hypothesis 2.1 suppose that $\mathcal{Q} \in C^\infty([x_0, \pm\infty))^{m\times m}$ and that $\mathcal{L}$ is in the limit point case at $\pm\infty$. Let $\mathcal{M}_\pm(z,x)$, $x \geq x_0$, be defined as in (2.46). Then, as $|z| \to \infty$ in C_ε, $\mathcal{M}_\pm(z,x)$ has an asymptotic expansion of the form* ($\mathrm{Im}(z^{1/2}) > 0$, $z \in \mathbb{C}_+$)

$$\mathcal{M}_\pm(z,x) \underset{\substack{|z|\to\infty \\ z\in C_\varepsilon}}{=} \pm i\mathcal{I}_m z^{1/2} + \sum_{k=1}^{N} \mathcal{M}_{\pm,k}(x) z^{-k/2} + o(|z|^{-N/2}), \quad N \in \mathbb{N}. \tag{2.48}$$

The expansion (2.48) is uniform with respect to $\arg(z)$ for $|z| \to \infty$ in C_ε and uniform in x as long as x varies in compact subsets of $[x_0, \infty)$. The expansion coefficients $\mathcal{M}_{\pm,k}(x)$ can be recursively computed from

$$\begin{aligned}
&\mathcal{M}_{\pm,1}(x) = \mp\frac{i}{2}\mathcal{Q}(x), \quad \mathcal{M}_{\pm,2}(x) = \frac{1}{4}\mathcal{Q}'(x), \\
&\mathcal{M}_{\pm,k+1}(x) = \pm\frac{i}{2}\Big(\mathcal{M}'_{\pm,k}(x) + \sum_{\ell=1}^{k-1} \mathcal{M}_{\pm,\ell}(x)\mathcal{M}_{\pm,k-\ell}(x)\Big), \quad k \geq 2.
\end{aligned} \tag{2.49}$$

The asymptotic expansion (2.48) can be differentiated to any order with respect to x.

Remark 2.8.

(i) *Due to the recursion relation (2.49), the coefficients $\mathcal{M}_{\pm,k}$ are universal polynomials in $\mathcal{Q}$ and its x-derivatives (i.e., differential polynomials in $\mathcal{Q}$). That the asymptotic expansion (2.48) can be differentiated to arbitrary order in x follows from repeated use of the Riccati-type equation (2.47).*

(ii) *In the case where $\mathcal{Q}$ and its x-derivatives are in $L^1(\mathbb{R})^{m\times m}$, or in the case where $\mathcal{Q}$ is periodic and hence Floquet theory applies, the proof of the existence of an asymptotic expansion of the type (2.48) follows in a routine manner by iterating appropriate Volterra-type integral equations. The general case, however, is intricate as is evident from the treatment in [20].*

Finally, in addition to (2.40), one infers for the $2m \times 2m$ Weyl–Titchmarsh function $\mathcal{M}(z,x)$ associated with H on $\mathbb{R}$ in connection with arbitrary half-lines $[x, \pm\infty)$, $x \in \mathbb{R}$,

$$\mathcal{M}(z,x) = \big(\mathcal{M}_{j,j'}(z,x)\big)_{j,j'=1,2}, \quad z \in \mathbb{C}\backslash\mathbb{R}, \tag{2.50}$$

$$\begin{aligned}
\mathcal{M}_{1,1}(z,x) &= \mathcal{M}_\pm(z,x)[\mathcal{M}_-(z,x) - \mathcal{M}_+(z,x)]^{-1}\mathcal{M}_\mp(z,x) \\
&= \psi'_+(z,x,x_0)[\mathcal{M}_-(z,x_0) - \mathcal{M}_+(z,x_0)]^{-1}\psi'_-(\overline{z},x,x_0)^*,
\end{aligned} \tag{2.51}$$

$$\begin{aligned}\mathcal{M}_{1,2}(z,x) &= 2^{-1}[\mathcal{M}_-(z,x) - \mathcal{M}_+(z,x)]^{-1}[\mathcal{M}_-(z,x) + \mathcal{M}_+(z,x)] \\ &= \psi_+(z,x,x_0)[\mathcal{M}_-(z,x_0) - \mathcal{M}_+(z,x_0)]^{-1}\psi'_-(\overline{z},x,x_0)^*, \quad (2.52)\\ \mathcal{M}_{2,1}(z,x) &= 2^{-1}[\mathcal{M}_-(z,x) + \mathcal{M}_+(z,x)][\mathcal{M}_-(z,x) - \mathcal{M}_+(z,x)]^{-1} \\ &= \psi'_+(z,x,x_0)[\mathcal{M}_-(z,x_0) - \mathcal{M}_+(z,x_0)]^{-1}\psi_-(\overline{z},x,x_0)^*, \quad (2.53)\\ \mathcal{M}_{2,2}(z,x) &= [\mathcal{M}_-(z,x) - \mathcal{M}_+(z,x)]^{-1} \\ &= \psi_+(z,x,x_0)[\mathcal{M}_-(z,x_0) - \mathcal{M}_+(z,x_0)]^{-1}\psi_-(\overline{z},x,x_0)^*. \quad (2.54)\end{aligned}$$

Introducing the convenient abbreviation,

$$\mathcal{M}(z,x) = \begin{pmatrix} \mathfrak{h}(z,x) & -\mathfrak{g}_2(z,x) \\ -\mathfrak{g}_1(z,x) & \mathfrak{g}(z,x) \end{pmatrix}, \quad z \in \mathbb{C}\backslash\mathbb{R},\ x \in \mathbb{R}, \qquad (2.55)$$

one then verifies from (2.50)–(2.55) and from $\mathcal{M}(\overline{z},x)^* = \mathcal{M}(z,x)$, $\mathcal{M}_\pm(\overline{z},x)^* = \mathcal{M}_\pm(z,x)$ that

$$\begin{aligned}&\mathfrak{g}(\overline{z},x)^* = \mathfrak{g}(z,x), \quad \mathfrak{g}_2(\overline{z},x)^* = \mathfrak{g}_1(z,x), \quad \mathfrak{h}(\overline{z},x)^* = \mathfrak{h}(z,x), && (2.56)\\ &\mathfrak{g}(z,x)\mathfrak{g}_1(z,x) = \mathfrak{g}_2(z,x)\mathfrak{g}(z,x), && (2.57)\\ &\mathfrak{h}(z,x)\mathfrak{g}_2(z,x) = \mathfrak{g}_1(z,x)\mathfrak{h}(z,x), && (2.58)\\ &\mathfrak{g}(z,x) = [\mathcal{M}_-(z,x) - \mathcal{M}_+(z,x)]^{-1}, && (2.59)\\ &\mathfrak{g}(z,x)\mathfrak{h}(z,x) - \mathfrak{g}_2(z,x)^2 = -(1/4)\mathcal{I}_m, && (2.60)\\ &\mathfrak{h}(z,x)\mathfrak{g}(z,x) - \mathfrak{g}_1(z,x)^2 = -(1/4)\mathcal{I}_m, && (2.61)\\ &\mathcal{M}_\pm(z,x) = \mp(1/2)\mathfrak{g}(z,x)^{-1} - \mathfrak{g}(z,x)^{-1}\mathfrak{g}_2(z,x) && (2.62)\\ &\qquad\qquad\ \ \mp (1/2)\mathfrak{g}(z,x)^{-1} - \mathfrak{g}_1(z,x)\mathfrak{g}(z,x)^{-1}. && (2.63)\end{aligned}$$

Moreover, the Riccati-type equations (2.47) imply the following results needed in Section 4.

Lemma 2.9. *Let $z \in \mathbb{C}\backslash\mathbb{R}$ and define $\mathcal{M}_\pm$ by* (2.46) *so that $\mathcal{M}_\pm$ satisfy the Riccati-type equation* (2.47). *Then, for a.e. $x \in \mathbb{R}$,*

$$\begin{aligned}\mathfrak{g}' &= -(\mathfrak{g}_1 + \mathfrak{g}_2), && (2.64)\\ \mathfrak{g}_1' &= -(\mathcal{Q} - z\mathcal{I}_m)\mathfrak{g} - \mathfrak{h} && (2.65)\\ &= (-\mathfrak{g}'' + \mathfrak{g}\mathcal{Q} - \mathcal{Q}\mathfrak{g})/2, && (2.66)\\ \mathfrak{g}_2' &= -\mathfrak{g}(\mathcal{Q} - z\mathcal{I}_m) - \mathfrak{h} && (2.67)\\ &= (-\mathfrak{g}'' + \mathcal{Q}\mathfrak{g} - \mathfrak{g}\mathcal{Q})/2, && (2.68)\\ \mathfrak{h}' &= -\mathfrak{g}_1(\mathcal{Q} - z\mathcal{I}_m) - (\mathcal{Q} - z\mathcal{I}_m)\mathfrak{g}_2, && (2.69)\\ \mathfrak{h} &= [\mathfrak{g}'' - \mathfrak{g}(\mathcal{Q} - z\mathcal{I}_m) - (\mathcal{Q} - z\mathcal{I}_m)\mathfrak{g}]/2 && (2.70)\end{aligned}$$

if $Q \in L^1_{\mathrm{loc}}(\mathbb{R})^{m\times m}$, *and*

$$\mathfrak{g}_1'' = -2(Q - z\mathcal{I}_m)\mathfrak{g}' - Q'\mathfrak{g} + \mathfrak{g}_1 Q - Q\mathfrak{g}_1, \tag{2.71}$$

$$\mathfrak{g}_2'' = -2\mathfrak{g}'(Q - z\mathcal{I}_m) - \mathfrak{g}Q' + Q\mathfrak{g}_2 - \mathfrak{g}_2 Q \tag{2.72}$$

if in addition $Q' \in L^1_{\mathrm{loc}}(\mathbb{R})^{m\times m}$.

Proof. (2.47) rewritten in terms of $\mathfrak{g}$, $\mathfrak{g}_1$, $\mathfrak{g}_2$ yields

$$\begin{aligned}&\pm (1/2)\mathfrak{g}^{-1}\mathfrak{g}'\mathfrak{g}^{-1} + \mathfrak{g}^{-1}\mathfrak{g}'\mathfrak{g}^{-1}\mathfrak{g}_2 - \mathfrak{g}^{-1}\mathfrak{g}_2' + (1/4)\mathfrak{g}^{-2} + \mathfrak{g}^{-1}\mathfrak{g}_2\mathfrak{g}^{-1}\mathfrak{g}_2\\ &\pm (1/2)\mathfrak{g}^{-2}\mathfrak{g}_2 \pm (1/2)\mathfrak{g}^{-1}\mathfrak{g}_2\mathfrak{g}^{-1} = Q - z\mathcal{I}_m.\end{aligned} \tag{2.73}$$

Taking the difference of the two equations in (2.73) yields (2.64). Adding the two equations in (2.73) and using (2.57), (2.60), (2.61), and (2.64) yields (2.65) and (2.67). Combining (2.65), (2.67), and (2.64) implies (2.70). Inserting (2.70) into (2.65) and (2.67) yields (2.66) and (2.68). (2.69) follows from differentiating $\mathfrak{h} = \mathfrak{g}_1^2\mathfrak{g}^{-1} - (1/4)\mathfrak{g}^{-1}$, inserting $\mathfrak{g}'$ and $\mathfrak{g}_1'$ from (2.64) and (2.65), and making repeated use of the identities (2.57), (2.61). Finally, (2.71) (resp. (2.72)) follows from differentiating (2.65) (resp. (2.67)) inserting (2.69) for $\mathfrak{h}'$. Alternatively, (2.67)–(2.72) follow directly from (2.65)–(2.71) using (2.56). □

3. Polynomial pencils of matrices in a nutshell

Since self-adjoint polynomial pencils of matrices play a role in our principal section 4, we briefly review some of the corresponding definitions and basic results, mainly following the monograph of Markus [71] and papers by Markus and Matsaev [72], [73]. While all results below are discussed for operator pencils by Markus and Matsaev, we will only quote them in the matrix context, for simplicity.

Given $m \in \mathbb{N}$, we denote by

$$\mathcal{A}(z) = \sum_{k=0}^{n} A_k z^k, \quad A_k \in \mathbb{C}^{m\times m}, \, 1 \le k \le n, \, z \in \mathbb{C}, \tag{3.1}$$

a polynomial pencil of $m \times m$ matrices (in short, a pencil) in the following. $\mathcal{A}$ is called of degree $n \in \mathbb{N}_0$ if $A_n \neq 0$ and monic if $A_n = \mathcal{I}_m$.

Definition 3.1. *Let* $\mathcal{A}$ *be a pencil of the type* (3.1).

(i) *The spectrum of* $\mathcal{A}$, *denoted by* spec($\mathcal{A}$), *is defined by*

$$\mathrm{spec}(\mathcal{A}) = \{z \in \mathbb{C} \,|\, \mathcal{A}(z) \textit{ is not invertible}\}. \tag{3.2}$$

$z_0 \in \mathbb{C}$ *is called an eigenvalue of* $\mathcal{A}$ *if* $\mathcal{A}(z_0)f_0 = 0$ *has a solution* $f_0 \in \mathbb{C}^m\backslash\{0\}$.

(ii) *A monic pencil* $\mathcal{C}$ *is called a (right) divisor of* $\mathcal{A}$ *if*

$$\mathcal{A}(z) = \mathcal{B}(z)\mathcal{C}(z), \quad z \in \mathbb{C} \tag{3.3}$$

for some pencil $\mathcal{B}$. *If in addition* spec($\mathcal{B}$) ∩ spec($\mathcal{C}$) = ∅, *then* $\mathcal{C}$ *is called a (right) spectral divisor of* $\mathcal{A}$.

(iii) $\mathcal{Z} \in \mathbb{C}^{m\times m}$ *is called a (matrix) root of the pencil* $\mathcal{A}$ *if*

$$\mathcal{A}(\mathcal{Z}) = 0, \tag{3.4}$$

where $\mathcal{A}(\mathcal{Z})$ *is defined as*

$$\mathcal{A}(\mathcal{Z}) = \sum_{k=0}^{n} \mathcal{A}_k \mathcal{Z}^k. \tag{3.5}$$

$\mathcal{Z} \in \mathbb{C}^{m\times m}$ *is called a (matrix) spectral root of the pencil* $\mathcal{A}$ *if* $(z\mathcal{I}_m - \mathcal{Z})$ *is a spectral divisor of* $\mathcal{A}$.

(iv) *The pencil* $\mathcal{A}$ *is called self-adjoint if* $\mathcal{A}_k = \mathcal{A}_k^*$ *for all* $1 \le k \le n$ *(i.e.,* $\mathcal{A}(\overline{z})^* = \mathcal{A}(z)$ *for all* $z \in \mathbb{C}$*).*

(v) *A self-adjoint pencil* $\mathcal{A}$ *is called weakly hyperbolic if* $\mathcal{A}_n > 0$ *and for all* $f \in \mathbb{C}^m \backslash \{0\}$*, the roots of the polynomial* $(f, \mathcal{A}(\cdot)f)_{\mathbb{C}^m}$ *are real. If in addition all these zeros are distinct, the pencil* $\mathcal{A}$ *is called hyperbolic.*

(vi) *Let* $\mathcal{A}$ *be a weakly hyperbolic pencil and denote by* $\{p_j(\mathcal{A}, f)\}_{1\le j\le n}$,

$$p_j(\mathcal{A}, f) \le p_{j+1}(\mathcal{A}, f),\ 1 \le j \le n-1,\ f \in \mathbb{C}^m \backslash \{0\}, \tag{3.6}$$

the roots of the polynomial $(f, \mathcal{A}(\cdot)f)_{\mathbb{C}^m}$ *ordered in magnitude. The range of the roots* $p_j(\mathcal{A}, f)$, $f \in \mathbb{C}^m \backslash \{0\}$ *is denoted by* $\Delta_j(\mathcal{A})$ *and called the* j*th root zone of* $\mathcal{A}$.

(vii) *A hyperbolic pencil* $\mathcal{A}$ *is called strongly hyperbolic if* $\overline{\Delta_j(\mathcal{A})}$ *and* $\overline{\Delta_k(\mathcal{A})}$ *are mutually disjoint for* $j \ne k$, $1 \le j, k \le n$.

Moreover, the Vandermonde matrix corresponding to a collection $\{\mathcal{Z}_1, \dots, \mathcal{Z}_n\} \subset \mathbb{C}^{m\times m}$ is defined by

$$\mathcal{V}(\mathcal{Z}_1, \dots, \mathcal{Z}_n) = \begin{pmatrix} \mathcal{I}_m & \mathcal{I}_m & \dots & \mathcal{I}_m \\ \mathcal{Z}_1 & \mathcal{Z}_2 & \dots & \mathcal{Z}_n \\ \vdots & \vdots & \ddots & \vdots \\ \mathcal{Z}_1^{n-1} & \mathcal{Z}_2^{n-1} & \dots & \mathcal{Z}_n^{n-1} \end{pmatrix}. \tag{3.7}$$

Theorem 3.2 (Markus [71], Sect. 29). *Let* $\mathcal{A}$ *be a pencil of the type* (3.1).

(i) $(z\mathcal{I}_m - \mathcal{Z})$ *is a divisor of* $\mathcal{A}$ *if and only if* $\mathcal{A}(\mathcal{Z}) = 0$. *(This justifies the notation introduced in the last part of Definition* 3.1 *(ii).)*

(ii) *Let* $\mathcal{A}$ *be a monic pencil of degree* n *and* $\mathcal{Z}_1, \dots, \mathcal{Z}_n$ *spectral roots of* $\mathcal{A}$. *Then the following assertions are equivalent:*

(α) $\mathcal{V}(\mathcal{Z}_1, \dots, \mathcal{Z}_n)$ *is invertible.*

(β) $\operatorname{spec}(\mathcal{Z}_j) \cap \operatorname{spec}(\mathcal{Z}_k) = \emptyset$, $j \ne k$, $1 \le j, k \le n$.

(γ) $\operatorname{spec}(\mathcal{A}) = \bigcup_{j=1}^{n} \operatorname{spec}(\mathcal{Z}_j)$.

(iii) *If* $\mathcal{A}$ *is a self-adjoint pencil, then* $\operatorname{spec}(\mathcal{A})$ *is symmetric with respect to* $\mathbb{R}$.

Theorem 3.3 (Markus [71], Sect. 31). *Let $\mathcal{A}$ be a pencil of the type* (3.1).

(i) *If $\mathcal{A}$ is a weakly hyperbolic pencil, then* $\operatorname{spec}(\mathcal{A}) \subseteq \mathbb{R}$.

(ii) *The root zones $\Delta_j(\mathcal{A})$, $1 \le j \le n$, of a weakly hyperbolic pencil $\mathcal{A}$ are intervals (possibly degenerating to a point).*

(iii) *If $\mathcal{A}$ is a weakly hyperbolic pencil of order n with root zones $\{\Delta_j(\mathcal{A})\}_{1\le j\le n}$ and $\Delta_k(\mathcal{A}) = \{\lambda_k\}$ for some $k \in \{1, \dots, n\}$, then $\mathcal{A}(z) = (z - \lambda_k)\mathcal{B}(z)$, where $\mathcal{B}$ is a weakly hyperbolic pencil of order $n-1$, with root zones*

$$\Delta_1(\mathcal{A}), \dots, \Delta_{k-1}(\mathcal{A}), \Delta_{k+1}(\mathcal{A}), \dots, \Delta_n(\mathcal{A}). \tag{3.8}$$

(iv) *If $\mathcal{A}$ is a weakly hyperbolic pencil then $\Delta_j(\mathcal{A}) \cap \Delta_{j+1}(\mathcal{A})$, $1 \le j \le n-1$, consists of at most one point.*

(v) *If $\mathcal{A}$ is a hyperbolic pencil, then $\Delta_j(\mathcal{A}) \cap \Delta_k(\mathcal{A}) = \emptyset$, $j \ne k$, $1 \le j,k \le n$. Thus, a hyperbolic pencil is strongly hyperbolic if and only if*

$$\overline{\Delta_j(\mathcal{A})} \cap \overline{\Delta_{j+1}(\mathcal{A})} = \emptyset, \quad 1 \le j \le n-1. \tag{3.9}$$

(vi) *Suppose $\mathcal{A}$ is a self-adjoint pencil of degree n, $\mathcal{A}_n > 0$, and $\mathcal{A}(\lambda) \ne 0$ for all $\lambda \in \mathbb{R}$. Then $\mathcal{A}$ is a weakly hyperbolic pencil if and only if there exist numbers $\gamma_1 < \gamma_2 < \cdots < \gamma_{n-1}$ such that $(-1)^j \mathcal{A}(\gamma_j) \ge 0$, $1 \le j \le n-1$.*

(vii) *Suppose $\mathcal{A}$ is a self-adjoint pencil of degree n with $\mathcal{A}_n > 0$. Then $\mathcal{A}$ is a strongly hyperbolic pencil if and only if there exist numbers $\gamma_1 < \gamma_2 < \cdots < \gamma_{n-1}$ such that $(-1)^j \mathcal{A}(\gamma_j) > 0$, $1 \le j \le n-1$.*

Theorem 3.4 (Markus [71], Sect. 31). *Let $\mathcal{A}$ be a pencil of the type* (3.1).

(i) *Suppose $\mathcal{A}$ is a weakly hyperbolic pencil and $\overline{\Delta_{j_0-1}(\mathcal{A})} \cap \overline{\Delta_{j_0}(\mathcal{A})} = \overline{\Delta_{j_0}(\mathcal{A})} \cap \overline{\Delta_{j_0+1}(\mathcal{A})} = \emptyset$ for some $j_0 \in \{1, \dots, n\}$. Then $\mathcal{A}$ has a spectral root $\mathcal{Z}_{j_0}$ such that $\operatorname{spec}(\mathcal{A}) \cap \overline{\Delta_{j_0}(\mathcal{A})} = \operatorname{spec}(\mathcal{Z}_{j_0})$ and $\mathcal{Z}_{j_0}$ is similar to a self-adjoint matrix.*

(ii) *A strongly hyperbolic pencil $\mathcal{A}$ has n spectral roots $\{\mathcal{Z}_j\}_{1\le j\le n}$ such that $\operatorname{spec}(\mathcal{Z}_j) = \operatorname{spec}(\mathcal{A}) \cap \overline{\Delta_j(\mathcal{A})}$ and each $\mathcal{Z}_j$, $1 \le j \le n$, is similar to a self-adjoint matrix.*

Theorem 3.5 (Markus [71], Sect. 31, Markus and Matsaev [72], [73]). *Let $\mathcal{A}$ be a pencil of the type* (3.1).

(i) *A weakly hyperbolic monic pencil $\mathcal{A}$ of degree n is decomposable as*

$$\mathcal{A}(z) = (z\mathcal{I}_m - \mathcal{Y}_n)(z\mathcal{I}_m - \mathcal{Y}_{n-1}) \cdots (z\mathcal{I}_m - \mathcal{Y}_1), \tag{3.10}$$

with $\operatorname{spec}(\mathcal{Y}_j) \subset \overline{\Delta_j(\mathcal{A})}$, $1 \le j \le n$.

(ii) *Let $\mathcal{A}$ be a strongly hyperbolic monic pencil of degree n. Then $\mathcal{A}$ is decomposable as*

$$\mathcal{A}(z) = (z\mathcal{I}_m - \mathcal{Y}_n)(z\mathcal{I}_m - \mathcal{Y}_{n-1}) \cdots (z\mathcal{I}_m - \mathcal{Y}_1), \tag{3.11}$$

with $\operatorname{spec}(\mathcal{Y}_j) \subset \overline{\Delta_j(\mathcal{A})}$, $1 \le j \le n$. *Moreover, each $\mathcal{Y}_j$ is similar to a spectral root $\mathcal{Z}_j$ of $\mathcal{A}$ and hence,*

$$\operatorname{spec}(\mathcal{Y}_j) = \operatorname{spec}(\mathcal{Z}_j) = \operatorname{spec}(\mathcal{A}) \cap \overline{\Delta_j(\mathcal{A})}, \quad 1 \le j \le n. \tag{3.12}$$

4. A class of matrix-valued Schrödinger operators with prescribed finite-band spectra

This section is devoted to the construction of a class of matrix-valued Schrödinger operators with a prescribed finite-band spectrum of uniform maximum multiplicity, the principal result of this paper.

To begin our analysis we start with a useful result on (scalar) Herglotz functions. Even though the result is probably well known to experts, we provide an elementary proof for completeness.

Let

$$\{E_\ell\}_{0\le\ell\le 2n} \subseteq \mathbb{R},\ n\in\mathbb{N},\ \text{with } E_\ell < E_{\ell+1},\ 0\le\ell\le 2n-1, \tag{4.1}$$

and introduce the polynomial

$$R_{2n+1}(z) = \prod_{\ell=0}^{2n}(z-E_\ell), \quad z\in\mathbb{C}. \tag{4.2}$$

Moreover, we define the square root of R_{2n+1} by

$$R_{2n+1}(\lambda)^{1/2} = \lim_{\varepsilon\downarrow 0} R_{2n+1}(\lambda+i\varepsilon)^{1/2}, \quad \lambda\in\mathbb{R}, \tag{4.3}$$

and

$$R_{2n+1}(\lambda)^{1/2} = |R_{2n+1}(\lambda)^{1/2}| \begin{cases} (-1)^n i & \text{for } \lambda\in(-\infty,E_0),\\ (-1)^{n+j} i & \text{for } \lambda\in(E_{2j-1},E_{2j}),\ j=1,\dots,n,\\ (-1)^{n+j} & \text{for } \lambda\in(E_{2j},E_{2j+1}),\ j=0,\dots,n-1,\\ 1 & \text{for } \lambda\in(E_{2n},\infty), \end{cases}$$
$$\lambda\in\mathbb{R} \tag{4.4}$$

and analytically continue $R_{2n+1}^{1/2}$ from $\mathbb{R}$ to all of $\mathbb{C}\backslash\Sigma$, where Σ is defined by

$$\Sigma = \left\{\bigcup_{j=0}^{n-1}[E_{2j},E_{2j+1}]\right\} \cup [E_{2n},\infty). \tag{4.5}$$

In this context we also mention the useful formula

$$\overline{R_{2n+1}(\overline{z})^{1/2}} = -R_{2n+1}(z)^{1/2}, \quad z\in\mathbb{C}_+. \tag{4.6}$$

Theorem 4.1. *Let $z\in\mathbb{C}\backslash\Sigma$ and $n\in\mathbb{N}$. Define $R_{2n+1}^{1/2}$ as in (4.1)–(4.4) followed by an analytic continuation to $\mathbb{C}\backslash\Sigma$. Moreover let F_n and H_{n+1} be two monic polynomials of degree n and $n+1$, respectively. Then*

$$\frac{iF_n(z)}{R_{2n+1}(z)^{1/2}} \tag{4.7}$$

is a Herglotz function if and only if all zeros of F_n are real and there is precisely one zero in each of the intervals $[E_{2j-1},E_{2j}]$, $1\le j\le n$. Moreover, if $iR_{2n+1}^{-1/2}F_n$

is a Herglotz function, then it can be represented in the form

$$\frac{iF_n(z)}{R_{2n+1}(z)^{1/2}} = \frac{1}{\pi}\int_\Sigma \frac{F_n(\lambda)d\lambda}{R_{2n+1}(\lambda)^{1/2}}\frac{1}{\lambda - z}, \quad z \in \mathbb{C}\backslash\Sigma. \tag{4.8}$$

Similarly,

$$\frac{iH_{n+1}(z)}{R_{2n+1}(z)^{1/2}} \tag{4.9}$$

is a Herglotz function if and only if all zeros of H_{n+1} are real and there is precisely one zero in each of the intervals $(-\infty, E_0]$ and $[E_{2j-1}, E_{2j}]$, $1 \leq j \leq n$. Moreover, if $iR_{2n+1}^{-1/2}H_{n+1}$ is a Herglotz function, then it can be represented in the form

$$\begin{aligned}\frac{iH_{n+1}(z)}{R_{2n+1}(z)^{1/2}} &= \mathrm{Re}\left(\frac{iH_{n+1}(i)}{R_{2n+1}(i)^{1/2}}\right)\\ &\quad + \frac{1}{\pi}\int_\Sigma \frac{H_{n+1}(\lambda)d\lambda}{R_{2n+1}(\lambda)^{1/2}}\left(\frac{1}{\lambda - z} - \frac{\lambda}{1+\lambda^2}\right), \quad z \in \mathbb{C}\backslash\Sigma. \end{aligned}\tag{4.10}$$

Proof. We start with the case of $F_n(z)/R_{2n+1}(z)^{1/2}$ in (4.7). Consider a closed oriented contour $\Gamma_{R,\varepsilon}$ which consists of the clockwise oriented semicircle $C_\varepsilon = \{z \in \mathbb{C} \,|\, z = E_0 - \varepsilon\exp(-i\alpha), -\pi/2 \leq \alpha \leq \pi/2\}$ centered at E_0, the straight line $L_+ = \{z \in \mathbb{C}_+ \,|\, z = x + i\varepsilon, E_0 \leq x \leq E_0 + R\}$ (oriented from left to right), the following part of the counterclockwise oriented circle of radius $(R^2+\varepsilon^2)^{1/2}$ centered at E_0, $C_R = \{z \in \mathbb{C} \,|\, z = E_0 + (R^2+\varepsilon^2)^{1/2}\exp(i\beta), \arctan(\varepsilon/R) \leq \beta \leq 2\pi - \arctan(\varepsilon/R)\}$, and the straight line $L_- = \{z \in \mathbb{C}_- \,|\, z = x - i\varepsilon, E_0 \leq x \leq E_0 + R\}$ (oriented from right to left). Then, for $\varepsilon > 0$ small enough and $R > 0$ sufficiently large, one infers

$$\begin{aligned}\frac{iF_n(z)}{R_{2n+1}(z)^{1/2}} &= \frac{1}{2\pi i}\oint_{\Gamma_{R,\varepsilon}} \frac{1}{\zeta - z}\frac{iF_n(\zeta)}{R_{2n+1}(\zeta)^{1/2}}d\zeta\\ &\underset{\varepsilon\downarrow 0, R\uparrow\infty}{=} \frac{1}{\pi}\int_\Sigma \frac{1}{\lambda - z}\frac{F_n(\lambda)d\lambda}{R_{2n+1}(\lambda)^{1/2}}.\end{aligned}\tag{4.11}$$

Here we used (4.4) to compute the contributions of the contour integral along $[E_0, R]$ in the limit $\varepsilon \downarrow 0$ and note that the integral over C_R tends to zero as $R \uparrow 0$ since

$$\frac{F_n(\zeta)}{R_{2n+1}(\zeta)^{1/2}} \underset{\zeta\to\infty}{=} O(|\zeta|^{-1/2}). \tag{4.12}$$

Next, utilizing the fact that F_n is monic and using (4.4) again, one infers that $F_n(\lambda)d\lambda/R_{2n+1}(\lambda)^{1/2}$ represents a positive measure supported on Σ if and only if F_n has precisely one zero in each of the intervals $[E_{2j-1}, E_{2j}]$, $1 \leq j \leq n$. In other words,

$$\frac{F_n(\lambda)}{R_{2n+1}(\lambda)^{1/2}} \geq 0 \text{ on } \Sigma \tag{4.13}$$

if and only if F_n has precisely one zero in each of the intervals $[E_{2j-1}, E_{2j}]$, $1 \leq j \leq n$. The Herglotz representation theorem, Theorem 2.3, then finishes the proof of (4.8). The proof of (4.10) follows along similar lines taking into account the additional residues at $\pm i$ inside $\Gamma_{R,\varepsilon}$ which are responsible for the real part on the right-hand side of (4.10). $\square$

Theorem 4.1 can be improved by invoking ideas developed in the Appendix of [60] (cf. also [104]). We will pursue this further in [9].

Corollary 4.2. *Let $z \in \mathbb{C}\backslash\Sigma$ and $m, n \in \mathbb{N}$. Define $R_{2n+1}^{1/2}$ as in (4.1)–(4.4) followed by an analytic continuation to $\mathbb{C}\backslash\Sigma$. Moreover let $\mathcal{F}_n$ and $\mathcal{H}_{n+1}$ be two monic $m \times m$ matrix pencils of degree n and $n+1$, respectively. Then $(i/2)R_{2n+1}^{-1/2}\mathcal{F}_n$ is a Herglotz matrix if and only if the root zones $\Delta_j(\mathcal{F}_n)$ of $\mathcal{F}_n$ satisfy*

$$\Delta_j(\mathcal{F}_n) \subseteq [E_{2j-1}, E_{2j}], \quad 1 \le j \le n. \tag{4.14}$$

Analogously, $(i/2)R_{2n+1}^{-1/2}\mathcal{H}_{n+1}$ is a Herglotz matrix if and only if the root zones $\Delta_j(\mathcal{H}_{n+1})$ of $\mathcal{H}_{n+1}$ satisfy

$$\Delta_0(\mathcal{H}_{n+1}) \subset (-\infty, E_0], \quad \Delta_j(\mathcal{H}_{n+1}) \subseteq [E_{2j-1}, E_{2j}], \quad 1 \le j \le n. \tag{4.15}$$

If (4.14) (resp., (4.15)) holds, then $\mathcal{F}_n$ (resp., $\mathcal{H}_{n+1}$) is a strongly hyperbolic pencil.

Proof. We recall that $(i/2)R_{2n+1}^{-1/2}\mathcal{F}_n$ is an $m \times m$ Herglotz matrix if and only if $(f, (i/2)R_{2n+1}^{-1/2}\mathcal{F}_n f)_{\mathbb{C}^m}$ is a Herglotz function for all $f \in \mathbb{C}^m\backslash\{0\}$. Thus, it suffices to apply Theorem 4.1, identifying F_n and $(f, \mathcal{F}_n f)_{\mathbb{C}^m}$, to arrive at (4.14). The same argument applied to $\mathcal{H}_{n+1}$ yields (4.15). $\square$

Next, we define the following $2m \times 2m$ matrix $\mathcal{M}_\Sigma(z, x_0)$ which will turn out to be the underlying Weyl–Titchmarsh matrix associated with a class of $m \times m$ matrix-valued Schrödinger operators with prescribed finite-band spectra. We introduce, for fixed $x_0 \in \mathbb{R}$,

$$\begin{aligned} \mathcal{M}_\Sigma(z, x_0) &= \big(\mathcal{M}_{\Sigma,p,q}(z, x_0)\big)_{1 \le p,q \le 2} \\ &= \frac{i}{2R_{2n+1}(z)^{1/2}} \begin{pmatrix} \mathcal{H}_{n+1,\Sigma}(z, x_0) & -\mathcal{G}_{2,n-1,\Sigma}(z, x_0) \\ -\mathcal{G}_{1,n-1,\Sigma}(z, x_0) & \mathcal{F}_{n,\Sigma}(z, x_0) \end{pmatrix}, \quad z \in \mathbb{C}\backslash\Sigma. \end{aligned} \tag{4.16}$$

Here $R_{2n+1}(z)^{1/2}$ is defined as in (4.1)–(4.4) followed by analytic continuation into $\mathbb{C}\backslash\Sigma$ and the polynomial matrix pencils $\mathcal{F}_{n,\Sigma}$, $\mathcal{G}_{1,n-1,\Sigma}$, $\mathcal{G}_{2,n-1,\Sigma}$, and $\mathcal{H}_{n+1,\Sigma}$ are introduced as follows:

(i) $\mathcal{F}_{n,\Sigma}(\cdot, x_0)$ is an $m \times m$ monic matrix pencil of degree n, that is, $\mathcal{F}_{n,\Sigma}(\cdot, x_0)$ is of the type

$$\mathcal{F}_{n,\Sigma}(z, x_0) = \sum_{\ell=0}^{n} \mathcal{F}_{n-\ell,\Sigma}(x_0) z^\ell, \quad \mathcal{F}_{0,\Sigma}(x_0) = \mathcal{I}_m, \ z \in \mathbb{C} \tag{4.17}$$

and

$$\frac{i}{2R_{2n+1}^{1/2}} \mathcal{F}_{n,\Sigma}(\cdot, x_0) \text{ is assumed to be an } m \times m \text{ Herglotz matrix.} \tag{4.18}$$

Hence $\mathcal{F}_{n,\Sigma}(\cdot, x_0)$ is a self-adjoint (in fact, strongly hyperbolic) pencil,

$$\mathcal{F}_{n,\Sigma}(\overline{z}, x_0)^* = \mathcal{F}_{n,\Sigma}(z, x_0), \quad z \in \mathbb{C}. \tag{4.19}$$

In addition, $(i/2)R_{2n+1}^{-1/2}\mathcal{F}_{n,\Sigma}$ and $2iR_{2n+1}^{1/2}\mathcal{F}_{n,\Sigma}^{-1}$ admit the Herglotz representations

$$\frac{i}{2R_{2n+1}(z)^{1/2}}\mathcal{F}_{n,\Sigma}(z,x_0) = \frac{1}{2\pi}\int_\Sigma \frac{d\lambda}{R_{2n+1}(\lambda)^{1/2}}\mathcal{F}_{n,\Sigma}(\lambda,x_0)\frac{1}{\lambda - z}, \quad z\in\mathbb{C}\backslash\Sigma, \tag{4.20}$$

$$\begin{aligned} &iR_{2n+1}(z)^{1/2}\mathcal{F}_{n,\Sigma}(z,x_0)^{-1} \\ &= \frac{1}{\pi}\int_\Sigma d\lambda\, R_{2n+1}(\lambda)^{1/2}\mathcal{F}_{n,\Sigma}(\lambda,x_0)^{-1}\left(\frac{1}{\lambda - z} - \frac{\lambda}{1+\lambda^2}\right) \\ &\quad + \Gamma_{\Sigma,0}(x_0) - \sum_{k=1}^N (z-\mu_k(x_0))^{-1}\Gamma_{\Sigma,k}(x_0), \\ &\qquad z\in\mathbb{C}\backslash\{\Sigma\cup\{\mu_k(x_0)\}_{1\le k\le N}\}, \end{aligned} \tag{4.21}$$

where

$$\begin{aligned} &\Gamma_{\Sigma,0}(x_0) = \Gamma_{\Sigma,0}(x_0)^* \in \mathbb{C}^{m\times m}, \quad 0\le \Gamma_{\Sigma,k}(x_0)\in\mathbb{C}^{m\times m},\ 1\le k\le N, \\ &\sum_{k=1}^N \operatorname{rank}(\Gamma_{\Sigma,k}(x_0)) \le mn, \quad \mu_k(x_0)\in\bigcup_{j=1}^n [E_{2j-1},E_{2j}], \quad 1\le k\le N. \end{aligned} \tag{4.22}$$

In fact, there are precisely m numbers $\mu_k(x_0)$ in $[E_{2j-1},E_{2j}]$ for each $1\le j\le n$, counting multiplicity (they are the points z where $\mathcal{F}_{n,\Sigma}(z,x_0)$ is not invertible).

(ii) Given these facts we now define

$$\mathcal{G}_{1,n-1,\Sigma}(z,x_0) = \left(\sum_{k=1}^N \frac{\varepsilon_k(x_0)}{z-\mu_k(x_0)}\Gamma_{\Sigma,k}(x_0)\right)\mathcal{F}_{n,\Sigma}(z,x_0), \tag{4.23}$$

$$\mathcal{G}_{2,n-1,\Sigma}(z,x_0) = \mathcal{F}_{n,\Sigma}(z,x_0)\left(\sum_{k=1}^N \frac{\varepsilon_k(x_0)}{z-\mu_k(x_0)}\Gamma_{\Sigma,k}(x_0)\right), \tag{4.24}$$

$$\varepsilon_k(x_0)\in\{1,-1\},\ 1\le k\le N, \quad z\in\mathbb{C}\backslash\{\mu_k(x_0)\}_{1\le k\le N}, \tag{4.25}$$

and

$$\begin{aligned} \mathcal{H}_{n+1,\Sigma}(z,x_0) &= R_{2n+1}(z)\mathcal{F}_{n,\Sigma}(z,x_0)^{-1} \\ &+ \left(\sum_{k=1}^N \frac{\varepsilon_k(x_0)}{z-\mu_k(x_0)}\Gamma_{\Sigma,k}(x_0)\right)\mathcal{F}_{n,\Sigma}(z,x_0)\left(\sum_{\ell=1}^N \frac{\varepsilon_\ell(x_0)}{z-\mu_\ell(x_0)}\Gamma_{\Sigma,\ell}(x_0)\right), \\ &\qquad z\in\mathbb{C}\backslash\{\mu_k(x_0)\}_{1\le k\le N}. \end{aligned} \tag{4.26}$$

Lemma 4.3. *Let $z\in\mathbb{C}\backslash\{\mu_k(x_0)\}_{1\le k\le N}$. $\mathcal{G}_{p,n-1,\Sigma}(\cdot,x_0)$, $p=1,2$, are $m\times m$ polynomial matrix pencils of equal degree at most $n-1$ and $\mathcal{H}_{n+1,\Sigma}(\cdot,x_0)$ is a self-adjoint*

$m\times m$ monic matrix pencil of degree $n+1$. Moreover, the following identities hold.

$$\mathcal{G}_{2,n-1,\Sigma}(\overline{z},x_0)^* = \mathcal{G}_{1,n-1,\Sigma}(z,x_0), \tag{4.27}$$

$$\mathcal{F}_{n,\Sigma}(z,x_0)\mathcal{G}_{1,n-1,\Sigma}(z,x_0) = \mathcal{G}_{2,n-1,\Sigma}(z,x_0)\mathcal{F}_{n,\Sigma}(z,x_0), \tag{4.28}$$

$$\mathcal{H}_{n+1,\Sigma}(z,x_0)\mathcal{G}_{2,n-1,\Sigma}(z,x_0) = \mathcal{G}_{1,n-1,\Sigma}(z,x_0)\mathcal{H}_{n+1,\Sigma}(z,x_0), \tag{4.29}$$

$$\mathcal{F}_{n,\Sigma}(z,x_0)\mathcal{H}_{n+1,\Sigma}(z,x_0) - \mathcal{G}_{2,n-1,\Sigma}(z,x_0)^2 = R_{2n+1}(z)\mathcal{I}_m, \tag{4.30}$$

$$\mathcal{H}_{n+1,\Sigma}(z,x_0)\mathcal{F}_{n,\Sigma}(z,x_0) - \mathcal{G}_{1,n-1,\Sigma}(z,x_0)^2 = R_{2n+1}(z)\mathcal{I}_m. \tag{4.31}$$

Proof. The identities (4.28)–(4.31) are obvious from (4.23)–(4.26). Similarly, (4.27) is clear from (4.23)–(4.25) and (4.19). By (4.21), one infers

$$\lim_{z\to\mu_k(x_0)} \Gamma_{\Sigma,k}(x_0)\mathcal{F}_{n,\Sigma}(z,x_0) = 0 = \lim_{z\to\mu_k(x_0)} \mathcal{F}_{n,\Sigma}(z,x_0)\Gamma_{\Sigma,k}(x_0) \tag{4.32}$$

and hence $\mathcal{G}_{p,n-1,\Sigma}(\cdot,x_0)$, $p=1,2$, are polynomial matrix pencils of degree at most $n-1$. By (4.27),

$$\deg(\mathcal{G}_{1,n-1,\Sigma}(\cdot,x_0)) = \deg(\mathcal{G}_{2,n-1,\Sigma}(\cdot,x_0)) \leq n-1. \tag{4.33}$$

Next, using (4.21) again, one notes that

$$iR_{2n+1}(\mu_k(x_0))^{1/2}\Gamma_{\Sigma,k}(x_0) = -\Gamma_{\Sigma,k}(x_0)[(d/dz)\mathcal{F}_{n,\Sigma}(\mu_k(x_0))]\Gamma_{\Sigma,k}(x_0), \quad 1\leq k\leq N \tag{4.34}$$

and thus, combining (4.26) and (4.34),

$$\begin{aligned}&\operatorname{res}_{z=\mu_k(x_0)}\mathcal{H}_{n+1,\Sigma}(z,x_0) \\ &= iR_{2n+1}(\mu_k(x_0))^{1/2}\Gamma_{\Sigma,k}(x_0) + \Gamma_{\Sigma,k}(x_0)[(d/dz)\mathcal{F}_{n,\Sigma}(\mu_k(x_0))]\Gamma_{\Sigma,k}(x_0) = 0, \\ &\qquad 1\leq k\leq N.\end{aligned} \tag{4.35}$$

Hence, $\mathcal{H}_{n+1,\Sigma}$ is indeed a polynomial matrix pencil of degree $n+1$. □

In fact, $\mathcal{H}_{n+1,\Sigma}$ is a strongly hyperbolic pencil as shown in Theorem 4.8.

In the following it will be convenient to use the following set of assumptions.

Hypothesis 4.4. *Let $m,n\in\mathbb{N}$. Define R_{2n+1} as in (4.1), (4.2) and $R_{2n+1}^{1/2}$ as in (4.3), (4.4) followed by an analytic continuation to $\mathbb{C}\backslash\Sigma$, with Σ introduced in (4.5). Moreover, let the polynomial $m\times m$ matrix pencils $\mathcal{F}_{n,\Sigma}(\cdot,x_0)$, $\mathcal{G}_{1,n-1,\Sigma}(\cdot,x_0)$, $\mathcal{G}_{2,n-1,\Sigma}(\cdot,x_0)$, and $\mathcal{H}_{n+1,\Sigma}(\cdot,x_0)$ be defined as in (4.17), (4.18), (4.23)–(4.26).*

Next, we introduce

$$\begin{aligned}&\mathcal{M}_{\pm,\Sigma}(z,x_0) \\ &= \pm iR_{2n+1}(z)^{1/2}\mathcal{F}_{n,\Sigma}(z,x_0)^{-1} - \mathcal{G}_{1,n-1,\Sigma}(z,x_0)\mathcal{F}_{n,\Sigma}(z,x_0)^{-1} &\text{(4.36a)}\\ &= \pm iR_{2n+1}(z)^{1/2}\mathcal{F}_{n,\Sigma}(z,x_0)^{-1} - \mathcal{F}_{n,\Sigma}(z,x_0)^{-1}\mathcal{G}_{2,n-1,\Sigma}(z,x_0), &\text{(4.36b)}\\ &\qquad z\in\mathbb{C}\backslash\{\Sigma\cup\{\mu_k(x_0)\}_{1\leq k\leq N}\}\end{aligned}$$

and

$$\mathcal{N}_{\pm,\Sigma}(z,x_0) = \mathcal{M}_{-,\Sigma}(z,x_0)\pm\mathcal{M}_{+,\Sigma}(z,x_0), \quad z\in\mathbb{C}\backslash\{\Sigma\cup\{\mu_k(x_0)\}_{1\leq k\leq N}\}. \tag{4.37}$$

We also introduce the open interior Σ^o of Σ defined by $\Sigma^o = \bigcup_{j=0}^{n-1}(E_{2j}, E_{2j+1}) \cup (E_{2n}, \infty)$. Then one verifies the following fundamental facts.

Theorem 4.5. *Assume Hypothesis 4.4 and let* $z \in \mathbb{C}\backslash\{\Sigma \cup \{\mu_k(x_0)\}_{1\le k\le N}\}$. *Moreover, introduce the* $2m \times 2m$ *matrix* $\mathcal{M}_\Sigma(\cdot, x_0)$ *as in* (4.16), (4.37) *and the* $m \times m$ *matrices* $\mathcal{M}_{\pm,\Sigma}(\cdot, x_0)$ *as in* (4.36). *Then,*

(i) $\pm\mathcal{M}_{\pm,\Sigma}(\cdot, x_0)$ *are* $m \times m$ *Herglotz matrices with representations*

$$\pm\mathcal{M}_{\pm,\Sigma}(z,x_0) = \frac{1}{\pi}\int_\Sigma d\lambda\, R_{2n+1}(\lambda)^{1/2}\mathcal{F}_{n,\Sigma}(\lambda,x_0)^{-1}\Big(\frac{1}{\lambda - z} - \frac{\lambda}{1+\lambda^2}\Big)$$
$$+\Gamma_{\Sigma,0}(x_0) - \sum_{k=1}^N \frac{1 \pm \varepsilon_k(x_0)}{z - \mu_k(x_0)}\Gamma_{\Sigma,k}(x_0), \tag{4.38}$$
$$z \in \mathbb{C}\backslash\{\Sigma \cup \{\mu_k(x_0)\}_{1\le k\le N}\}.$$

Moreover, $\mathcal{M}_{\pm,\Sigma}(\cdot, x_0)$ *are the half-line* $\mathcal{M}$*-matrices associated with self-adjoint Schrödinger operators* $H^D_{\pm,x_0,\Sigma}$ *in* $L^2([x_0, \pm\infty))^m$, *with a Dirichlet boundary condition at the point* x_0 *and an* $m \times m$ *matrix-valued potential* $\mathcal{Q}_\Sigma$ *satisfying*

$$\mathcal{Q}_\Sigma = \mathcal{Q}_\Sigma^* \in L^1_{\text{loc}}(\mathbb{R})^{m\times m} \cap C^\infty(\mathbb{R}\backslash\{x_0\})^{m\times m}, \tag{4.39}$$

given by

$$H^D_{\pm,x_0,\Sigma} = -\mathcal{I}_m\frac{d^2}{dx^2} + \mathcal{Q}_\Sigma, \tag{4.40}$$
$$\text{dom}(H^D_{\pm,x_0,\Sigma}) = \{g \in L^2((x_0, \pm\infty))^m \mid g, g' \in AC([x_0, c])^m \text{ for all } c \gtrless x_0;$$
$$\lim_{\varepsilon\downarrow 0} g(x_0 \pm \varepsilon) = 0;\ (-g'' + \mathcal{Q}_\Sigma g) \in L^2((x_0, \pm\infty))^m\}.$$

(ii) *The differential expression* $\mathcal{L}_\Sigma = -\mathcal{I}_m\frac{d^2}{dx^2} + \mathcal{Q}_\Sigma$ *is in the limit point case at* $\pm\infty$.

(iii) *The matrix* $\mathcal{M}_\Sigma(\cdot, x_0)$, *defined by*

$$\mathcal{M}_\Sigma(z, x_0) = \big(\mathcal{M}_{\Sigma,p,q}(z,x_0)\big)_{1\le p,q\le 2} \tag{4.41}$$
$$= \begin{pmatrix} \mathcal{M}_{\pm,\Sigma}(z,x_0)\mathcal{N}_{-,\Sigma}(z,x_0)^{-1}\mathcal{M}_{\mp,\Sigma}(z,x_0) & \mathcal{N}_{-,\Sigma}(z,x_0)^{-1}\mathcal{N}_{+,\Sigma}(z,x_0)/2 \\ \mathcal{N}_{+,\Sigma}(z,x_0)\mathcal{N}_{-,\Sigma}(z,x_0)^{-1}/2 & \mathcal{N}_{-,\Sigma}(z,x_0)^{-1} \end{pmatrix},$$

is a $2m \times 2m$ *Herglotz matrix admitting a representation of the type* (2.43), *with absolutely continuous measure* $\Omega_\Sigma(\cdot, x)$ *given by*

$$d\Omega_\Sigma(\lambda, x_0) = \begin{cases} \frac{1}{2\pi R_{2n+1}(\lambda)^{1/2}}\begin{pmatrix} \mathcal{H}_{n+1,\Sigma}(\lambda,x_0) & -\mathcal{G}_{2,n-1,\Sigma}(\lambda,x_0) \\ -\mathcal{G}_{1,n-1,\Sigma}(\lambda,x_0) & \mathcal{F}_{n,\Sigma}(\lambda,x_0)\end{pmatrix} d\lambda, & \lambda \in \Sigma^o, \\ 0, & \lambda \in \mathbb{R}\backslash\Sigma. \end{cases} \tag{4.42}$$

In addition, $\mathcal{M}_\Sigma(\cdot, x_0)$ is the Weyl–Titchmarsh $\mathcal{M}$-matrix associated with the self-adjoint Schrödinger operator H_Σ in $L^2(\mathbb{R})^m$ defined by

$$H_\Sigma = -\mathcal{I}_m \frac{d^2}{dx^2} + \mathcal{Q}_\Sigma, \tag{4.43}$$
$$\operatorname{dom}(H_\Sigma) = \{g \in L^2(\mathbb{R})^m \mid g, g' \in \mathrm{AC}_{\mathrm{loc}}(\mathbb{R})^m;\ (-g'' + \mathcal{Q}_\Sigma g) \in L^2(\mathbb{R})^m\}.$$

(iv) *H_Σ has purely absolutely continuous spectrum Σ,*

$$\operatorname{spec}(H_\Sigma) = \operatorname{spec}_{ac}(H_\Sigma) = \Sigma, \quad \operatorname{spec}_{\mathrm{p}}(H_{\Sigma_n}) = \operatorname{spec}_{\mathrm{sc}}(H_{\Sigma_n}) = \emptyset, \tag{4.44}$$

with $\operatorname{spec}(H_{\Sigma_n})$ *of uniform spectral multiplicity* $2m$.

Proof. The representations (4.38) for $\mathcal{M}_{\pm,\Sigma}$ immediately follow from combining (4.21), (4.23), and (4.36a). These representations also prove that $\pm\mathcal{M}_{\pm,\Sigma}(\cdot, x_0)$ are $m \times m$ Herglotz matrices (cf. Theorem 2.3). Combining (4.38) with Theorem 2.4, taking into account Lemma 8.3.2 in [64], then yields the properties stated for $\mathcal{Q}_\Sigma$. That $\mathcal{L}_\Sigma = -\mathcal{I}_m \frac{d^2}{dx^2} + \mathcal{Q}_\Sigma$ is in the limit point case at $\pm\infty$ can be proved in analogy to Wienholtz's proof [108] of a result originally due to Povzner [83], reproduced as Theorem 35 in [40, p. 58]. The corresponding details will be presented in [21]. Equation (4.41) follows from (4.16), (4.23)–(4.26), and (4.36). Relation (4.44) follows from the explicit formula (4.42) of the spectral measure. In particular, the support property $\operatorname{supp}(\Omega_\Sigma) = \Sigma$ of the measure Ω_Σ in (4.42) proves $\operatorname{spec}(H_\Sigma) = \Sigma$, etc. The uniform maximum spectral multiplicity $2m$ then follows from the fact that $\operatorname{rank}(d\Omega_\Sigma/d\lambda) = 2m$ on the interior Σ^o of Σ. □

At this point we cannot yet infer continuity of $\mathcal{Q}_\Sigma$ at the boundary point x_0. We will subsequently return to this issue in Theorem 4.8.

In the following we will apply these facts to our concrete class of matrix-valued Schrödinger operators discussed in Theorem 4.5. In order to find the corresponding Weyl–Titchmarsh matrices $\mathcal{M}_{\pm,\Sigma}(z, x)$, we need some preparations. We denote by $\psi_{\pm,\Sigma}(z, x, x_0)$ the Weyl solutions (2.31) associated with $\mathcal{Q}_\Sigma$, that is,

$$\psi_{\pm,\Sigma}(z, x, x_0) = \theta_\Sigma(z, x, x_0) + \phi_\Sigma(z, x, x_0)\mathcal{M}_{\pm,\Sigma}(z, x_0), \quad z \in \mathbb{C}\backslash\Sigma, \tag{4.45}$$

where, in obvious notation, $\theta_\Sigma(z, x, x_0)$, $\phi_\Sigma(z, x, x_0)$ denote the fundamental system (2.4) corresponding to $\mathcal{Q}_\Sigma$. Then straightforward computations of the right-hand sides of (2.46)–(2.49) (taking into account (2.16), (2.17), (4.28), and (4.31)) yield

$$\mathcal{M}_\Sigma(z, x) = \big(\mathcal{M}_{\Sigma,p,q}(z, x)\big)_{1 \le p,q \le 2} \tag{4.46}$$
$$= \frac{i}{2R_{2n+1}(z)^{1/2}} \begin{pmatrix} \mathcal{H}_{n+1,\Sigma}(z, x) & -\mathcal{G}_{2,n-1,\Sigma}(z, x) \\ -\mathcal{G}_{1,n-1,\Sigma}(z, x) & \mathcal{F}_{n,\Sigma}(z, x) \end{pmatrix}, \quad z \in \mathbb{C}\backslash\Sigma,$$

where we abbreviated

$$\begin{aligned}\mathcal{F}_{n,\Sigma}(z,x) &= \theta_\Sigma(z,x,x_0)\mathcal{F}_{n,\Sigma}(z,x_0)\theta_\Sigma(\overline{z},x,x_0)^* \\ &\quad + \phi_\Sigma(z,x,x_0)\mathcal{H}_{n+1,\Sigma}(z,x_0)\phi_\Sigma(\overline{z},x,x_0)^* \\ &\quad - \phi_\Sigma(z,x,x_0)\mathcal{G}_{1,n-1,\Sigma}(z,x_0)\theta_\Sigma(\overline{z},x,x_0)^* \\ &\quad - \theta_\Sigma(z,x,x_0)\mathcal{G}_{2,n-1,\Sigma}(z,x_0)\phi_\Sigma(\overline{z},x,x_0)^*, \qquad (4.47)\\ \mathcal{G}_{1,n-1,\Sigma}(z,x) &= -\theta'_\Sigma(z,x,x_0)\mathcal{F}_{n,\Sigma}(z,x_0)\theta_\Sigma(\overline{z},x,x_0)^* \\ &\quad - \phi'_\Sigma(z,x,x_0)\mathcal{H}_{n+1,\Sigma}(z,x_0)\phi_\Sigma(\overline{z},x,x_0)^* \\ &\quad + \phi'_\Sigma(z,x,x_0)\mathcal{G}_{1,n-1,\Sigma}(z,x_0)\theta_\Sigma(\overline{z},x,x_0)^* \\ &\quad + \theta'_\Sigma(z,x,x_0)\mathcal{G}_{2,n-1,\Sigma}(z,x_0)\phi_\Sigma(\overline{z},x,x_0)^*, \qquad (4.48)\\ \mathcal{G}_{2,n-1,\Sigma}(z,x) &= -\theta_\Sigma(z,x,x_0)\mathcal{F}_{n,\Sigma}(z,x_0)\theta'_\Sigma(\overline{z},x,x_0)^* \\ &\quad - \phi_\Sigma(z,x,x_0)\mathcal{H}_{n+1,\Sigma}(z,x_0)\phi'_\Sigma(\overline{z},x,x_0)^* \\ &\quad + \phi_\Sigma(z,x,x_0)\mathcal{G}_{1,n-1,\Sigma}(z,x_0)\theta'_\Sigma(\overline{z},x,x_0)^* \\ &\quad + \theta_\Sigma(z,x,x_0)\mathcal{G}_{2,n-1,\Sigma}(z,x_0)\phi'_\Sigma(\overline{z},x,x_0)^*, \qquad (4.49)\\ \mathcal{H}_{n+1,\Sigma}(z,x) &= \theta'_\Sigma(z,x,x_0)\mathcal{F}_{n,\Sigma}(z,x_0)\theta'_\Sigma(\overline{z},x,x_0)^* \\ &\quad + \phi'_\Sigma(z,x,x_0)\mathcal{H}_{n+1,\Sigma}(z,x_0)\phi'_\Sigma(\overline{z},x,x_0)^* \\ &\quad - \phi'_\Sigma(z,x,x_0)\mathcal{G}_{1,n-1,\Sigma}(z,x_0)\theta'_\Sigma(\overline{z},x,x_0)^* \\ &\quad - \theta'_\Sigma(z,x,x_0)\mathcal{G}_{2,n-1,\Sigma}(z,x_0)\phi'_\Sigma(\overline{z},x,x_0)^*. \qquad (4.50)\end{aligned}$$

Considerations of this type can be found in [64, Sect. 8.2] in the special scalar case $m=1$ and in the matrix context $m \in \mathbb{N}$ in [100, Sect. 9.4].

Differentiating (4.47)–(4.50) with respect to x (taking into account that $\theta'' = (\mathcal{Q}_\Sigma - z\mathcal{I}_m)\theta$, $\phi'' = (\mathcal{Q}_\Sigma - z)\phi$) then yields (4.51)–(4.58) below. Alternatively, these results directly follow from Lemma 2.9 identifying $\mathfrak{g}$ and $(i/2)R_{2n+1}^{-1/2}\mathcal{F}_n$, $\mathfrak{g}_p$ and $(i/2)R_{2n+1}^{-1/2}\mathcal{G}_{p,n-1}$, $p=1,2$, and $\mathfrak{h}$ and $(i/2)R_{2n+1}^{-1/2}\mathcal{H}_{n+1}$, respectively.

Lemma 4.6. *Assume Hypothesis* 4.4 *and let* $(z,x) \in \mathbb{C} \times \mathbb{R}$. *Then*

$$\begin{aligned}\mathcal{F}'_{n,\Sigma} &= -(\mathcal{G}_{1,n-1,\Sigma} + \mathcal{G}_{2,n-1,\Sigma}), \qquad (4.51)\\ \mathcal{G}'_{1,n-1,\Sigma} &= -(\mathcal{Q}_\Sigma - z\mathcal{I}_m)\mathcal{F}_{n,\Sigma} - \mathcal{H}_{n+1,\Sigma} \qquad (4.52)\\ &= (-\mathcal{F}''_{n,\Sigma} + \mathcal{F}_{n,\Sigma}\mathcal{Q}_\Sigma - \mathcal{Q}_\Sigma\mathcal{F}_{n,\Sigma})/2, \qquad (4.53)\\ \mathcal{G}''_{1,n-1,\Sigma} &= -2(\mathcal{Q}_\Sigma - z\mathcal{I}_m)\mathcal{F}'_{n,\Sigma} - \mathcal{Q}'_\Sigma\mathcal{F}_{n,\Sigma} + \mathcal{G}_{1,n-1,\Sigma}\mathcal{Q}_\Sigma - \mathcal{Q}_\Sigma\mathcal{G}_{1,n-1,\Sigma}, \qquad (4.54)\\ \mathcal{G}'_{2,n-1,\Sigma} &= -\mathcal{F}_{n,\Sigma}(\mathcal{Q}_\Sigma - z\mathcal{I}_m) - \mathcal{H}_{n+1,\Sigma} \qquad (4.55)\\ &= (-\mathcal{F}''_{n,\Sigma} + \mathcal{Q}_\Sigma\mathcal{F}_{n,\Sigma} - \mathcal{F}_{n,\Sigma}\mathcal{Q}_\Sigma)/2, \qquad (4.56)\\ \mathcal{G}''_{2,n-1,\Sigma} &= -2\mathcal{F}'_{n,\Sigma}(\mathcal{Q}_\Sigma - z\mathcal{I}_m) - \mathcal{F}_{n,\Sigma}\mathcal{Q}'_\Sigma + \mathcal{Q}_\Sigma\mathcal{G}_{2,n-1,\Sigma} - \mathcal{G}_{2,n-1,\Sigma}\mathcal{Q}_\Sigma, \qquad (4.57)\\ \mathcal{H}'_{n+1,\Sigma} &= -\mathcal{G}_{1,n-1,\Sigma}(\mathcal{Q}_\Sigma - z\mathcal{I}_m) - (\mathcal{Q}_\Sigma - z\mathcal{I}_m)\mathcal{G}_{2,n-1,\Sigma}, \qquad (4.58)\\ \mathcal{H}_{n+1,\Sigma} &= [\mathcal{F}''_{n,\Sigma} - \mathcal{F}_{n,\Sigma}(\mathcal{Q}_\Sigma - z\mathcal{I}_m) - (\mathcal{Q}_\Sigma - z\mathcal{I}_m)\mathcal{F}_{n,\Sigma}]/2. \qquad (4.59)\end{aligned}$$

In particular, one also verifies the following facts from (2.56)–(2.63).

Lemma 4.7. *Assume Hypothesis* 4.4 *and let* $(z,x) \in \mathbb{C}\times\mathbb{R}$. *Then*

$$\mathcal{F}_{n,\Sigma}(\overline{z},x)^* = \mathcal{F}_{n,\Sigma}(z,x), \quad \mathcal{H}_{n+1,\Sigma}(\overline{z},x)^* = \mathcal{H}_{n+1,\Sigma}(z,x),$$
$$\mathcal{G}_{2,n-1,\Sigma}(\overline{z},x)^* = \mathcal{G}_{1,n-1,\Sigma}(z,x), \tag{4.60}$$
$$\mathcal{F}_{n,\Sigma}(z,x)\mathcal{G}_{1,n-1,\Sigma}(z,x) = \mathcal{G}_{2,n-1,\Sigma}(z,x)\mathcal{F}_{n,\Sigma}(z,x), \tag{4.61}$$
$$\mathcal{H}_{n+1,\Sigma}(z,x)\mathcal{G}_{2,n-1,\Sigma}(z,x) = \mathcal{G}_{1,n-1,\Sigma}(z,x)\mathcal{H}_{n+1,\Sigma}(z,x), \tag{4.62}$$
$$\mathcal{H}_{n+1,\Sigma}(z,x)\mathcal{F}_{n,\Sigma}(z,x) - \mathcal{G}_{1,n-1,\Sigma}(z,x)^2 = R_{2n+1}(z)\mathcal{I}_m, \tag{4.63}$$
$$\mathcal{F}_{n,\Sigma}(z,x)\mathcal{H}_{n+1,\Sigma}(z,x) - \mathcal{G}_{2,n-1,\Sigma}(z,x)^2 = R_{2n+1}(z)\mathcal{I}_m. \tag{4.64}$$

Proof. (4.60) is clear from (4.27), the fact that $\mathcal{F}_{n,\Sigma}(\cdot,x_0)$ and $\mathcal{H}_{n+1,\Sigma}(\cdot,x_0)$ are self-adjoint $m\times m$ matrix pencils, (4.48), and (4.49). Similarly, (4.61)–(4.64) follow from elementary (but somewhat tedious) calculations directly from (4.47)–(4.50), invoking (2.10)–(2.17) and (4.28)–(4.31) repeatedly. □

Combining (2.50)–(2.54) and (4.46) then yields

$$\begin{aligned}
&\mathcal{M}_{\pm,\Sigma}(z,x)\\
&= \pm i R_{2n+1}(z)^{1/2}\mathcal{F}_{n,\Sigma}(z,x)^{-1} - \mathcal{G}_{1,n-1,\Sigma}(z,x)\mathcal{F}_{n,\Sigma}(z,x)^{-1} && (4.65\text{a})\\
&= \pm i R_{2n+1}(z)^{1/2}\mathcal{F}_{n,\Sigma}(z,x)^{-1} - \mathcal{F}_{n,\Sigma}(z,x)^{-1}\mathcal{G}_{2,n-1,\Sigma}(z,x), && (4.65\text{b})\\
&\qquad z\in\mathbb{C}\backslash\mathbb{R}.
\end{aligned}$$

One observes that for each $x\in\mathbb{R}$, $\mathcal{M}_{+,\Sigma}(\cdot,x)$ is the analytic continuation of $\mathcal{M}_{-,\Sigma}(\cdot,x)$ through the set Σ, and vice versa,

$$\lim_{\varepsilon\downarrow 0}\mathcal{M}_{+,\Sigma}(\lambda+i\varepsilon,x) = \lim_{\varepsilon\downarrow 0}\mathcal{M}_{-,\Sigma}(\lambda-i\varepsilon,x), \tag{4.66}$$
$$\lambda \in \bigcup_{j=0}^{n-1}(E_{2j},E_{2j+1})\cup(E_{2n},\infty),\ x\in\mathbb{R}.$$

In other words, for each $x\in\mathbb{R}$, $\mathcal{M}_{+,\Sigma}(\cdot,x)$ and $\mathcal{M}_{-,\Sigma}(\cdot,x)$ are the two branches of an analytic matrix-valued function $\mathcal{M}_\Sigma(\cdot,x)$ on the two-sheeted Riemann surface of $R_{2n+1}^{1/2}$. Thus, the corresponding potential $\mathcal{Q}_\Sigma$ is *reflectionless* in the sense discussed in [22], [39], and [55].

Thus, one obtains the following results.

Theorem 4.8. *Assume Hypothesis* 4.4 *and let* $z\in\mathbb{C}\backslash\mathbb{R}$ *and* $x\in\mathbb{R}$. *Then*

(i) $\mathcal{M}_{\pm,\Sigma}(z,\cdot)$ *in* (4.65) *satisfy the matrix-valued Riccati-type equation*

$$\mathcal{M}'_{\pm,\Sigma}(z,x) + \mathcal{M}_{\pm,\Sigma}(z,x)^2 = \mathcal{Q}_\Sigma(x) - z\mathcal{I}_m, \quad x\in\mathbb{R},\ z\in\mathbb{C}\backslash\mathbb{R}. \tag{4.67}$$

Moreover, $\mathcal{M}_{\pm,\Sigma}(z,x)$ *in* (4.65) *are the* $m\times m$ *Weyl–Titchmarsh matrices associated with* $H^D_{\pm,x,\Sigma}$ *on the half-lines* $[x,\pm\infty)$ *and thus for each* $x\in\mathbb{R}$, $\mathcal{M}_\Sigma(z,x)$ *in* (4.46) *is a* $2m\times 2m$ *Weyl–Titchmarsh matrix associated with*

H_Σ on $\mathbb{R}$. In particular, $\mathcal{M}_\Sigma(\cdot,x_0)$ is a $2m\times 2m$ Herglotz matrix of H_Σ admitting a representation of the type (2.43), with absolutely continuous measure $\Omega_\Sigma(\cdot,x)$ given by

$$d\Omega_\Sigma(\lambda,x)=\begin{cases}\frac{1}{2\pi R_{2n+1}(\lambda)^{1/2}}\begin{pmatrix}\mathcal{H}_{n+1,\Sigma}(\lambda,x) & -\mathcal{G}_{2,n-1,\Sigma}(\lambda,x)\\ -\mathcal{G}_{1,n-1,\Sigma}(\lambda,x) & \mathcal{F}_{n,\Sigma}(\lambda,x)\end{pmatrix}d\lambda, & \lambda\in\Sigma^o,\\ 0, & \lambda\in\mathbb{R}\backslash\Sigma.\end{cases} \tag{4.68}$$

(ii) *$\mathcal{F}_{n,\Sigma}(\cdot,x)$ and $\mathcal{H}_{n+1,\Sigma}(\cdot,x)$ are strongly hyperbolic (and hence self-adjoint) $m\times m$ monic matrix pencils of degree n and $n+1$, respectively, and $\mathcal{G}_{p,n-1,\Sigma}(\cdot,x)$, $p=1,2$, are $m\times m$ matrix pencils of degree $n-1$.*

(iii) *$\mathcal{Q}_\Sigma\in C^\infty(\mathbb{R})^{m\times m}$.*

Proof. (4.67) is clear from Lemma 4.6 and (4.65). Since $\mathcal{Q}_\Sigma\in C^\infty(\mathbb{R})^{m\times m}$, the initial value problems

$$\mathcal{M}_\pm'(z,x)+\mathcal{M}_\pm(z,x)^2=\mathcal{Q}_\Sigma(x)-z\mathcal{I}_m,\quad x\in\mathbb{R},\ z\in\mathbb{C}\backslash\mathbb{R}, \tag{4.69a}$$
$$\mathcal{M}_\pm(z,x_0)=\mathcal{M}_{\pm,\Sigma}(z,x_0), \tag{4.69b}$$

with $\mathcal{M}_{\pm,\Sigma}(z,x_0)$ given by (4.36), has a unique solution. Since at $x=x_0$ this solution coincides with the Weyl–Titchmarsh M-matrix $\mathcal{M}_{\pm,\Sigma}(z,x)|_{x=x_0}$ in (4.65) (using the initial condition (2.5) in (4.47)–(4.50)), $\mathcal{M}_{\pm,\Sigma}(z,x)$ represents the Weyl–Titchmarsh matrices associated with $H^D_{\pm,\Sigma}$ on the half-lines $[x,\pm\infty)$. This proves part (i). By the known leading asymptotic behavior (2.33) of $\mathcal{M}_{\pm,\Sigma}(\cdot,x)$ (valid for each $x_0\in\mathbb{R}$, see [20]) and that of the diagonal Green's matrix $\mathcal{G}_\Sigma(\cdot,x,x)=\mathcal{M}_{\Sigma,2,2}(\cdot,x)$ of H_Σ as $|z|\to\infty$, $\mathcal{F}_{n,\Sigma}(\cdot,x)$ and $\mathcal{H}_{n+1,\Sigma}(\cdot,x)$ are monic matrix pencils of degree n and $n+1$, respectively, and $\mathcal{G}_{p,n-1,\Sigma}(\cdot,x)$, $p=1,2$, are $m\times m$ matrix pencils of degree $n-1$. Since the diagonal blocks of each Herglotz matrix are also Herglotz matrices, one concludes that $(i/2)R_{2n+1}^{-1/2}\mathcal{F}_{n,\Sigma}$ and $(i/2)R_{2n+1}^{-1/2}\mathcal{H}_{n+1,\Sigma}$ are Herglotz matrices. By Corollary 4.2 this then proves that $\mathcal{F}_{n,\Sigma}$ and $\mathcal{H}_{n+1,\Sigma}$ are strongly hyperbolic pencils and hence item (ii) holds. As in Theorem 4.5, $\mathcal{Q}_\Sigma\in C^\infty((-\infty,x)\cup(x,\infty))^{m\times m}$. Since $x\in\mathbb{R}$ is arbitrary, this proves (iii). □

It should be emphasized that the construction of $\mathcal{Q}_\Sigma$ along the lines of Section 4 in the scalar case $m=1$ is due to Levitan [61] (see also [62], [63], [64, Ch. 8], [66]).

5. Trace formulas and connections with the stationary matrix KdV hierarchy

In this section we introduce the stationary matrix Korteweg–de Vries (KdV) hierarchy and show that the class of finite-band potentials $\mathcal{Q}_\Sigma$ constructed in Section 4 satisfies some (and hence infinitely many) equations of the stationary KdV equations. We also introduce trace formulas for KdV invariants.

In order to extend the recursive approach constructing KdV Lax pairs in the scalar (Abelian) context to the present matrix-valued (non-Abelian) setting, we focus on an efficient approach introduced by Dubrovin [27] (in the scalar case

$m = 1$). Recalling (2.50)–(2.63) and Lemma 2.9, we state the following matrix-version of Dubrovin's generating function approach to higher-order Lax pairs.

We start by introducing the following hypothesis.

Hypothesis 5.1. *Fix $m \in \mathbb{N}$, suppose $\mathcal{Q} = \mathcal{Q}^* \in C^\infty(\mathbb{R})^{m\times m}$ and introduce the differential expression*

$$\mathcal{L} = -\mathcal{I}_m \frac{d^2}{dx^2} + \mathcal{Q}, \quad x \in \mathbb{R}. \tag{5.1}$$

Suppose $\mathcal{L}$ is in the limit point case at $\pm\infty$ and introduce the corresponding self-adjoint operator H in $L^2(\mathbb{R})^m$ by

$$\begin{aligned} &H = -\mathcal{I}_m \frac{d^2}{dx^2} + \mathcal{Q}, \\ &\mathrm{dom}(H) = \{g \in L^2(\mathbb{R})^m \mid g, g' \in \mathrm{AC}_{\mathrm{loc}}(\mathbb{R})^m;\ (-g'' + \mathcal{Q}g) \in L^2(\mathbb{R})^m\}. \end{aligned} \tag{5.2}$$

Given Hypothesis 5.1, we introduce the generating operator P_z by

$$\begin{aligned} &P_z = \Big(\mathfrak{g}(z,\cdot)\frac{d}{dx} + \mathfrak{g}_2(z,\cdot)\Big)(\mathcal{L} - z\mathcal{I}_m)^{-1}, \quad z \in \mathbb{C}\backslash\mathbb{R}, \\ &\mathrm{dom}(P_z) = \bigcup_{E\in\mathbb{C}\backslash\{\mathbb{R}\cup\{z\}\}} \ker(\mathcal{L} - E\mathcal{I}_m), \end{aligned} \tag{5.3}$$

where $\ker(\mathcal{L} - E\mathcal{I}_m)$ denotes the algebraic nullspace of $\mathcal{L} - E\mathcal{I}_m$ (as opposed to the functional analytic nullspace in $L^2(\mathbb{R})^m$) and $(\mathcal{L} - z\mathcal{I}_m)^{-1}$ acts in the obvious manner by

$$(\mathcal{L} - z\mathcal{I}_m)^{-1}\psi = (E - z)^{-1}\psi, \quad \psi \in \ker(\mathcal{L} - E\mathcal{I}_m). \tag{5.4}$$

The precise operator theoretic properties of P_z will be irrelevant in the following.

Lemma 5.2. *Assume Hypothesis 5.1 and let $z \in \mathbb{C}\backslash\mathbb{R}$, $x \in \mathbb{R}$. Then,*

$$\begin{aligned} &\Big[\Big(\mathfrak{g}(z,\cdot)\frac{d}{dx} + \mathfrak{g}_2(z,\cdot)\Big)(\mathcal{L} - z\mathcal{I}_m)^{-1}, \mathcal{L}\Big]\Big|_{\bigcup_{E\in\mathbb{C}\backslash\{\mathbb{R}\cup\{z\}\}} \ker(\mathcal{L}-E\mathcal{I}_m)} \\ &\quad = -2\mathfrak{g}'(z,\cdot)\Big|_{\bigcup_{E\in\mathbb{C}\backslash\{\mathbb{R}\cup\{z\}\}} \ker(\mathcal{L}-E\mathcal{I}_m)} \end{aligned} \tag{5.5}$$

Proof. Fix $\psi \in \ker(\mathcal{L} - E\mathcal{I}_m)$ for some $E \in \mathbb{C}\backslash\{\mathbb{R}\cup\{z\}\}$. Then one computes

$$\begin{aligned} &\Big[\Big(\mathfrak{g}(z,\cdot)\frac{d}{dx} + \mathfrak{g}_2(z,\cdot)\Big)(\mathcal{L} - z\mathcal{I}_m)^{-1}, \mathcal{L}\Big]\psi \\ &= (E - z)^{-1}\{[2\mathfrak{g}_2' + \mathfrak{g}'' + \mathfrak{g}\mathcal{Q} - \mathcal{Q}\mathfrak{g}]\psi' \\ &\quad [\mathfrak{g}_2'' + 2\mathfrak{g}'(\mathcal{Q} - z\mathcal{I}_m) + \mathfrak{g}\mathcal{Q}' + \mathfrak{g}_2\mathcal{Q} - \mathcal{Q}\mathfrak{g}_2 + 2(z - E)\mathfrak{g}']\psi\} \\ &= -2\mathfrak{g}'\psi, \end{aligned} \tag{5.6}$$

using (2.68) and (2.72). □

A variant of Dubrovin's idea of a generating operator for KdV Lax pairs was also used by Olmedilla, Martínez Alonso, and Guil [78]. Their approach, however, focuses on formal operator expansions and formal pseudo-differential operators. For a different approach we refer to [26, Ch. 15], [31].

Next, we recall that

$$\mathfrak{g}(z,x) = [\mathcal{M}_-(z,x) - \mathcal{M}_+(z,x)]^{-1}, \tag{5.7}$$

$$\mathfrak{g}_1(z,x) = \frac{1}{2}[\mathcal{M}_-(z,x) + \mathcal{M}_+(z,x)][\mathcal{M}_-(z,x) - \mathcal{M}_+(z,x)]^{-1}, \tag{5.8}$$

$$\mathfrak{g}_2(z,x) = \frac{1}{2}[\mathcal{M}_-(z,x) - \mathcal{M}_+(z,x)]^{-1}[\mathcal{M}_-(z,x) + \mathcal{M}_+(z,x)], \tag{5.9}$$

$$\mathfrak{h}(z,x) = \mathcal{M}_\pm(z,x)[\mathcal{M}_-(z,x) - \mathcal{M}_+(z,x)]^{-1}\mathcal{M}_\mp(z,x), \tag{5.10}$$

and note that by Theorem 2.7 the right-hand sides of (5.7) and (5.9) admit asymptotic expansions in cones avoiding the spectrum of H. In particular, one thus obtains the asymptotic expansions

$$\mathfrak{g}(z,x) \underset{\substack{|z|\to\infty \\ z\in C_\varepsilon}}{=} \frac{i}{2z^{1/2}} \sum_{k=0}^{\infty} \widehat{\mathfrak{R}}_k(x) z^{-k}, \quad \widehat{\mathfrak{R}}_0(x) = \mathcal{I}_m, \tag{5.11}$$

$$\mathfrak{g}_p(z,x) \underset{\substack{|z|\to\infty \\ z\in C_\varepsilon}}{=} \frac{i}{2z^{1/2}} \sum_{k=0}^{\infty} \widehat{\mathfrak{G}}_{p,k}(x) z^{-k}, \quad \widehat{\mathfrak{G}}_{p,0}(x) = 0, \quad p = 1,2, \tag{5.12}$$

$$\mathfrak{h}(z,x) \underset{\substack{|z|\to\infty \\ z\in C_\varepsilon}}{=} \frac{iz^{1/2}}{2} \sum_{k=0}^{\infty} \widehat{\mathfrak{H}}_k(x) z^{-k}, \quad \widehat{\mathfrak{H}}_0(x) = \mathcal{I}_m, \tag{5.13}$$

for some coefficients $\widehat{\mathfrak{R}}_k$, $\widehat{\mathfrak{G}}_{p,k}$, $p = 1,2$, and $\widehat{\mathfrak{H}}_k$, which are universal differential polynomials in $\mathcal{Q}$ by Remark 2.8 (i). Explicitly, one obtains

$$\widehat{\mathfrak{R}}_0 = \mathcal{I}_m, \quad \widehat{\mathfrak{R}}_1 = \tfrac{1}{2}\mathcal{Q}, \quad \widehat{\mathfrak{R}}_2 = -\tfrac{1}{8}\mathcal{Q}'' + \tfrac{3}{8}\mathcal{Q}^2, \tag{5.14}$$

$$\widehat{\mathfrak{G}}_{1,0} = -\tfrac{1}{4}\mathcal{Q}', \quad \widehat{\mathfrak{G}}_{1,1} = \tfrac{1}{16}\mathcal{Q}''' - \tfrac{1}{8}(\mathcal{Q}^2)' - \tfrac{1}{8}\mathcal{Q}'\mathcal{Q}, \tag{5.15}$$

$$\widehat{\mathfrak{G}}_{2,0} = -\tfrac{1}{4}\mathcal{Q}', \quad \widehat{\mathfrak{G}}_{2,1} = \tfrac{1}{16}\mathcal{Q}''' - \tfrac{1}{8}(\mathcal{Q}^2)' - \tfrac{1}{8}\mathcal{Q}\mathcal{Q}', \tag{5.16}$$

$$\widehat{\mathfrak{H}}_0 = \mathcal{I}_m, \quad \widehat{\mathfrak{H}}_1 = -\tfrac{1}{2}\mathcal{Q}, \quad \widehat{\mathfrak{H}}_2 = \tfrac{1}{8}\mathcal{Q}'' - \tfrac{1}{8}\mathcal{Q}^2, \tag{5.17}$$

etc.

Motivated by Lemma 5.2, we now introduce the $m \times m$ matrix-valued differential expressions $\widehat{\mathcal{P}}_{2k+1}$ by

$$\widehat{\mathcal{P}}_{2k+1} = \sum_{\ell=0}^{k} \Big(\widehat{\mathfrak{R}}_\ell \frac{d}{dx} + \widehat{\mathfrak{G}}_{2,\ell}\Big)\mathcal{L}^{k-\ell}, \quad k \in \mathbb{N}_0. \tag{5.18}$$

In analogy to Lemma 5.2 one then obtains the following result.

Lemma 5.3. *Assume Hypothesis* 5.1 *and let* $z \in \mathbb{C}\backslash\mathbb{R}$, $x \in \mathbb{R}$. *Then,*

$$[\widehat{\mathcal{P}}_{2k+1}, \mathcal{L}] = 2\widehat{\mathfrak{R}}_{k+1}', \quad k \in \mathbb{N}_0. \tag{5.19}$$

Proof. Assuming $\psi \in \ker(\mathcal{L} - z\mathcal{I}_m)$ one computes

$$\begin{aligned}
[\widehat{\mathcal{P}}_{2k+1}, \mathcal{L}]\psi &= \sum_{\ell=0}^{k} z^{k-\ell}(2\widehat{\mathfrak{G}}'_{2,\ell} + \widehat{\mathfrak{R}}''_\ell + \widehat{\mathfrak{R}}_\ell \mathcal{Q} - \mathcal{Q}\widehat{\mathfrak{R}}_\ell)\psi' \\
&\quad + \sum_{\ell=0}^{k} z^{k-\ell}(\widehat{\mathfrak{G}}''_{2,\ell} + 2\widehat{\mathfrak{R}}'_\ell(\mathcal{Q} - z\mathcal{I}_m) + \widehat{\mathfrak{R}}_\ell \mathcal{Q}' + \widehat{\mathfrak{G}}_{2,\ell}\mathcal{Q} - \mathcal{Q}\widehat{\mathfrak{G}}_{2,\ell})\psi \\
&= \sum_{\ell=0}^{k} z^{k-\ell}(2\widehat{\mathfrak{G}}'_{2,\ell} + \widehat{\mathfrak{R}}''_\ell + \widehat{\mathfrak{R}}_\ell \mathcal{Q} - \mathcal{Q}\widehat{\mathfrak{R}}_\ell)\psi' - 2\widehat{\mathfrak{R}}'_0\psi \\
&\quad + \sum_{\ell=0}^{k} z^{k-\ell}(\widehat{\mathfrak{G}}''_{2,\ell} - 2\widehat{\mathfrak{R}}'_{\ell+1} + 2\widehat{\mathfrak{R}}'_\ell \mathcal{Q} + \widehat{\mathfrak{R}}_\ell \mathcal{Q}' + \widehat{\mathfrak{G}}_{2,\ell}\mathcal{Q} - \mathcal{Q}\widehat{\mathfrak{G}}_{2,\ell})\psi + 2\widehat{\mathfrak{R}}'_{k+1}\psi \\
&= 2\widehat{\mathfrak{R}}'_{k+1}\psi. \qquad (5.20)
\end{aligned}$$

Here we used $\widehat{\mathfrak{R}}'_0 = 0$ and

$$2\widehat{\mathfrak{G}}'_{2,\ell} + \widehat{\mathfrak{R}}''_\ell + \widehat{\mathfrak{R}}_\ell \mathcal{Q} - \mathcal{Q}\widehat{\mathfrak{R}}_\ell = 0, \quad \ell \in \mathbb{N}_0, \qquad (5.21)$$

$$\widehat{\mathfrak{G}}''_{2,\ell} - 2\widehat{\mathfrak{R}}'_{\ell+1} + 2\widehat{\mathfrak{R}}'_\ell \mathcal{Q} + \widehat{\mathfrak{R}}_\ell \mathcal{Q}' + \widehat{\mathfrak{G}}_{2,\ell}\mathcal{Q} - \mathcal{Q}\widehat{\mathfrak{G}}_{2,\ell} = 0, \quad \ell \in \mathbb{N}_0, \qquad (5.22)$$

which follow from inserting the asymptotic expansion (5.11) and (5.12) into (2.68) and (2.72) (which is permitted by Theorem 2.7). Relation (5.20) implies (5.19) since $\widehat{\mathcal{P}}_{2k+1}$ and $\mathcal{L}$ are $m \times m$ matrix-valued differential expressions of finite-order while $\bigcup_{z\in\mathbb{C}} \ker(\mathcal{L} - z\mathcal{I}_m)$ is an infinite-dimensional space of $C^\infty(\mathbb{R})^m$-functions. □

Introducing

$$\mathcal{P}_{2k+1} = \sum_{\ell=0}^{k} c_{k-\ell}\widehat{\mathcal{P}}_{2\ell+1}, \quad k \in \mathbb{N}_0, \qquad (5.23)$$

where

$$\{c_\ell\}_{\ell=1,\dots,k} \subset \mathbb{C}, \quad c_0 = 1 \qquad (5.24)$$

denotes a set of constants, the pairs $(\mathcal{P}_{2k+1}, \mathcal{L})$, $k \in \mathbb{N}_0$, by definition, represent the Lax pairs of the (matrix-valued) KdV hierarchy. More precisely, varying $\{c_\ell\}_{\ell\in\mathbb{N}} \subset \mathbb{C}$, the set of evolution equations,

$$\frac{d}{dt}\mathcal{L} - [\mathcal{P}_{2k+1}, \mathcal{L}] = 0, \quad k \in \mathbb{N}_0, \qquad (5.25)$$

or equivalently, the set of equations,

$$\mathrm{KdV}_k(\mathcal{Q}) = \mathcal{Q}_t - 2\sum_{\ell=0}^{k} c_{k-\ell}\widehat{\mathfrak{R}}'_{\ell+1}(\mathcal{Q}, \dots) = 0, \quad k \in \mathbb{N}_0, \qquad (5.26)$$

represents the (matrix-valued) KdV hierarchy of evolution equations. The corresponding stationary KdV hierarchy, characterized by

$$\mathcal{Q}_t = 0, \text{ or equivalently, by } [\mathcal{P}_{2k+1}, \mathcal{L}] = 0, \quad k \in \mathbb{N}_0, \qquad (5.27)$$

is then given by

$$\text{s-KdV}_k(\mathcal{Q}) = -2\sum_{\ell=0}^{k} c_{k-\ell}\widehat{\mathfrak{R}}'_{\ell+1}(\mathcal{Q},\dots) = 0, \quad k \in \mathbb{N}_0. \tag{5.28}$$

Remark 5.4. *By Remark* 2.8, *each* $\widehat{\mathfrak{R}}_\ell$ *is a universal polynomial in* $\mathcal{Q}$ *and its* x-*derivatives and occasionally we slightly abuse notation and indicate this by writing* $\widehat{\mathfrak{R}}_\ell(\mathcal{Q},\dots)$ *for* $\widehat{\mathfrak{R}}_\ell(x)$, $\widehat{\mathfrak{R}}'_{\ell+1}(\mathcal{Q},\dots)$ *for* $\widehat{\mathfrak{R}}'_{\ell+1}(x)$, *etc. Explicit formulas for the differential polynomials* $\widehat{\mathfrak{R}}_\ell(\mathcal{Q},\dots)$ *were derived for instance, in* [8] *and* [82].

In order to make the connection with the finite-band formalism of Section 4 we now recall

$$\frac{i}{2R_{2n+1}(z)^{1/2}}\mathcal{F}_{n,\Sigma}(z,x) = [\mathcal{M}_{-,\Sigma}(z,x) - \mathcal{M}_{+,\Sigma}(z,x)]^{-1}, \tag{5.29}$$

$$\begin{aligned}&\frac{i}{2R_{2n+1}(z)^{1/2}}\mathcal{G}_{1,n-1,\Sigma}(z,x)\\ &\quad= \frac{1}{2}[\mathcal{M}_{-,\Sigma}(z,x) + \mathcal{M}_{+,\Sigma}(z,x)][\mathcal{M}_{-,\Sigma}(z,x) - \mathcal{M}_{+,\Sigma}(z,x)]^{-1},\end{aligned} \tag{5.30}$$

$$\begin{aligned}&\frac{i}{2R_{2n+1}(z)^{1/2}}\mathcal{G}_{2,n-1,\Sigma}(z,x)\\ &\quad= \frac{1}{2}[\mathcal{M}_{-,\Sigma}(z,x) - \mathcal{M}_{+,\Sigma}(z,x)]^{-1}[\mathcal{M}_{-,\Sigma}(z,x) + \mathcal{M}_{+,\Sigma}(z,x)],\end{aligned} \tag{5.31}$$

$$\begin{aligned}&\frac{i}{2R_{2n+1}(z)^{1/2}}\mathcal{H}_{n+1,\Sigma}(z,x)\\ &\quad= \mathcal{M}_{\pm,\Sigma}(z,x)[\mathcal{M}_{-,\Sigma}(z,x) - \mathcal{M}_{+,\Sigma}(z,x)]^{-1}\mathcal{M}_{\mp,\Sigma}(z,x),\end{aligned} \tag{5.32}$$

and note that (5.29)–(5.32) admit expansions convergent in a neighborhood of infinity. In particular,

$$\frac{1}{R_{2n+1}(z)^{1/2}}\mathcal{F}_{n,\Sigma}(z,x) \underset{\substack{|z|\to\infty\\ z\in C_\varepsilon}}{=} \frac{1}{z^{1/2}}\sum_{k=0}^{\infty}\widehat{\mathfrak{R}}_{k,\Sigma}(x)z^{-k}, \quad \widehat{\mathfrak{R}}_{0,\Sigma}(x) = \mathcal{I}_m, \tag{5.33}$$

$$\frac{1}{R_{2n+1}(z)^{1/2}}\mathcal{G}_{p,n-1,\Sigma}(z,x) \underset{\substack{|z|\to\infty\\ z\in C_\varepsilon}}{=} \frac{1}{z^{1/2}}\sum_{k=0}^{\infty}\widehat{\mathfrak{G}}_{p,k,\Sigma}(x)z^{-k}, \quad \widehat{\mathfrak{G}}_{p,0,\Sigma}(x) = 0, \ p = 1,2, \tag{5.34}$$

$$\frac{1}{R_{2n+1}(z)^{1/2}}\mathcal{H}_{n+1,\Sigma}(z,x) \underset{\substack{|z|\to\infty\\ z\in C_\varepsilon}}{=} \frac{1}{z^{1/2}}\sum_{k=0}^{\infty}\widehat{\mathfrak{H}}_{k,\Sigma}(x)z^{-k}, \quad \widehat{\mathfrak{H}}_{0,\Sigma}(x) = \mathcal{I}_m, \tag{5.35}$$

for $|z|$ sufficiently large. Here the coefficients $\widehat{\mathfrak{R}}_{k,\Sigma}$ and $\widehat{\mathfrak{G}}_{2,k,\Sigma}$ are the universal differential polynomials $\widehat{\mathfrak{R}}_k = \widehat{\mathfrak{R}}_k(\mathcal{Q}_\Sigma,\dots)$ and $\widehat{\mathfrak{G}}_{2,k} = \widehat{\mathfrak{G}}_{2,k}(\mathcal{Q}_\Sigma,\dots)$ in (5.11) and (5.12) (with $\mathcal{Q}$ replaced by $\mathcal{Q}_\Sigma$). We also recall

$$\mathcal{F}_{n,\Sigma}(z,x) = \sum_{\ell=0}^{n} \mathcal{F}_{n-\ell,\Sigma}(x) z^\ell, \quad \mathcal{F}_{0,\Sigma}(x) = \mathcal{I}_m, \tag{5.36}$$

$$\mathcal{G}_{p,n-1,\Sigma}(z,x) = \sum_{\ell=0}^{n-1} \mathcal{G}_{p,n-1-\ell,\Sigma}(x) z^\ell, \quad p=1,2, \tag{5.37}$$

$$\mathcal{H}_{n+1,\Sigma}(z,x) = \sum_{\ell=0}^{n} \mathcal{H}_{n+1-\ell,\Sigma}(x) z^\ell, \quad \mathcal{H}_{0,\Sigma}(x) = \mathcal{I}_m. \tag{5.38}$$

Since we seek the connection between the set of coefficients $\widehat{\mathfrak{K}}_{k,\Sigma}$, $\widehat{\mathfrak{G}}_{p,k,\Sigma}$, $\widehat{\mathfrak{H}}_{k,\Sigma}$ and $\mathcal{F}_{k,\Sigma}$, $\mathcal{G}_{p,k,\Sigma}$, $\mathcal{H}_{k,\Sigma}$, we next consider the following elementary expansions. Let

$$\eta \in \mathbb{C} \ \text{ such that } \ |\eta| < \min\{|E_0|^{-1}, \dots, |E_{2n}|^{-1}\}. \tag{5.39}$$

Then

$$\left(\prod_{\ell=0}^{2n} (1 - E_\ell \eta) \right)^{-1/2} = \sum_{k=0}^{\infty} \hat{c}_k(\underline{E}) \eta^k, \tag{5.40}$$

where

$$\hat{c}_0(\underline{E}) = 1,$$
$$\hat{c}_k(\underline{E}) = \sum_{\substack{j_0,\dots,j_{2n}=0 \\ j_0+\cdots+j_{2n}=k}}^{k} \frac{(2j_0)! \cdots (2j_{2n})!}{2^{2k}(j_0!)^2 \cdots (j_{2n}!)^2} E_0^{j_0} \cdots E_{2n}^{j_{2n}}, \quad k \in \mathbb{N}. \tag{5.41}$$

The first few coefficients explicitly read

$$\hat{c}_0(\underline{E}) = 1, \ \hat{c}_1(\underline{E}) = \frac{1}{2} \sum_{\ell=0}^{2n} E_\ell,$$
$$\hat{c}_2(\underline{E}) = \frac{1}{4} \sum_{\substack{\ell_1,\ell_2=0 \\ \ell_1<\ell_2}}^{2n} E_{\ell_1} E_{\ell_2} + \frac{3}{8} \sum_{\ell=0}^{2n} E_\ell^2, \quad \text{etc.} \tag{5.42}$$

Similarly, one has

$$\left(\prod_{\ell=0}^{2n} (1 - E_\ell \eta) \right)^{1/2} = \sum_{k=0}^{\infty} c_k(\underline{E}) \eta^k, \tag{5.43}$$

where

$$c_0(\underline{E}) = 1,$$
$$c_k(\underline{E}) = - \sum_{\substack{j_0,\dots,j_{2n}=0 \\ j_0+\cdots+j_{2n}=k}}^{k} \frac{(2j_0)! \cdots (2j_{2n})!}{2^{2k}(j_0!)^2 \cdots (j_{2n}!)^2 (2j_0-1) \cdots (2j_{2n}-1)} E_0^{j_0} \cdots E_{2n}^{j_{2n}},$$
$$k \in \mathbb{N}. \tag{5.44}$$

The first few coefficients explicitly are given by

$$c_0(\underline{E}) = 1, \quad c_1(\underline{E}) = -\frac{1}{2}\sum_{\ell=0}^{2n} E_\ell,$$
$$c_2(\underline{E}) = \frac{1}{4}\sum_{\substack{\ell_1,\ell_2=0\\ \ell_1<\ell_2}}^{2n} E_{\ell_1}E_{\ell_2} - \frac{1}{8}\sum_{\ell=0}^{2n} E_\ell^2, \quad \text{etc.} \tag{5.45}$$

Lemma 5.5. *Assume Hypothesis* 5.1 *and let* $x \in \mathbb{R}$. *Then,*

$$\mathcal{F}_{\ell,\Sigma}(x) = \sum_{k=0}^{\ell} c_{\ell-k}(\underline{E})\mathfrak{R}_{k,\Sigma}(x), \quad \ell = 0,\dots,n, \tag{5.46}$$

$$\widehat{\mathfrak{R}}_{\ell,\Sigma}(x) = \sum_{k=0}^{\ell} \hat{c}_{\ell-k}(\underline{E})\mathcal{F}_{k,\Sigma}(x), \quad \ell = 0,\dots,n, \tag{5.47}$$

$$\mathcal{G}_{p,\ell,\Sigma}(x) = \sum_{k=0}^{\ell} c_{\ell-k}(\underline{E})\widehat{\mathfrak{G}}_{p,k+1,\Sigma}(x), \quad \ell = 0,\dots,n-1, \ p = 1,2, \tag{5.48}$$

$$\widehat{\mathfrak{G}}_{p,\ell+1,\Sigma}(x) = \sum_{k=0}^{\ell} \hat{c}_{\ell-k}(\underline{E})\mathcal{G}_{p,k,\Sigma}(x), \quad \ell = 0,\dots,n-1, \ p = 1,2, \tag{5.49}$$

$$\mathcal{H}_{\ell,\Sigma}(x) = \sum_{k=0}^{\ell} c_{\ell-k}(\underline{E})\widehat{\mathfrak{H}}_{k,\Sigma}(x), \quad \ell = 0,\dots,n+1, \tag{5.50}$$

$$\widehat{\mathfrak{H}}_{\ell,\Sigma}(x) = \sum_{k=0}^{\ell} \hat{c}_{\ell-k}(\underline{E})\mathcal{H}_{k,\Sigma}(x), \quad \ell = 0,\dots,n+1. \tag{5.51}$$

Proof. It suffices to prove (5.46) and (5.47) and so we omit the analogous proofs of (5.48)–(5.51). Since for $|z|$ sufficiently large,

$$\begin{aligned} z^{-n}\mathcal{F}_{n,\Sigma}(z,x) &= \sum_{\ell=0}^{n} \mathcal{F}_{\ell,\Sigma}(x)z^{-\ell} \\ &= z^{-n-(1/2)}R_{2n+1}(z)^{1/2}\sum_{\ell=0}^{\infty}\widehat{\mathfrak{R}}_{\ell,\Sigma}(x)z^{-\ell} \\ &= \sum_{k=0}^{\infty} c_k(\underline{E})z^{-k}\sum_{\ell=0}^{\infty}\widehat{\mathfrak{R}}_{\ell,\Sigma}(x)z^{-\ell} \\ &= \sum_{\ell=0}^{\infty}\bigg(\sum_{k=0}^{\ell} c_{\ell-k}(\underline{E})\widehat{\mathfrak{R}}_{k,\Sigma}(x)\bigg)z^{-\ell} \end{aligned} \tag{5.52}$$

and hence (5.46). Equation (5.47) is then clear from (5.46) and

$$\sum_{\ell=0}^{k} \hat{c}_{k-\ell}(\underline{E})c_{\ell}(\underline{E}) = \delta_{k,0}, \quad k \in \mathbb{N}_0. \tag{5.53}$$

The latter follows from multiplying (5.40) and (5.43), comparing coefficients of η^k. □

Given these preliminaries we can now state the principal result of this section.

Theorem 5.6. *The self-adjoint finite-band potential $\mathcal{Q}_\Sigma \in C^\infty(\mathbb{R})^{m\times m}$, discussed in Theorems* 4.5 *and* 4.8, *is a stationary KdV solution satisfying*

$$\text{s-KdV}_n(\mathcal{Q}_\Sigma) = -2\sum_{\ell=0}^{n} c_{n-\ell}(\underline{E})\widehat{\mathfrak{R}}'_{\ell+1}(\mathcal{Q}_\Sigma, \dots) = 0, \tag{5.54}$$

with $c_\ell(\underline{E})$ defined in (5.44) *and $\widehat{\mathfrak{R}}_{\ell+1}$ the universal differential polynomials (with respect to $\mathcal{Q}$) in* (5.11).

Proof. Introducing the $m \times m$ matrix-valued differential expression $\mathcal{P}_{2k+1,\Sigma}$ by

$$\mathcal{P}_{2n+1,\Sigma} = \sum_{\ell=0}^{n} c_{n-\ell}(\underline{E})\widehat{\mathcal{P}}_{2\ell+1} = \sum_{\ell=0}^{n}\Big(\mathcal{F}_{n-\ell,\Sigma}(\cdot)\frac{d}{dx} + \mathcal{G}_{2,n-1-\ell,\Sigma}(\cdot)\Big)\mathcal{L}^\ell \tag{5.55}$$

(cf. (5.46), (5.48), and (5.23)), one computes for $\psi \in \ker(\mathcal{L} - z\mathcal{I}_m)$, $z \in \mathbb{C}$,

$$\mathcal{P}_{2n+1,\Sigma}\psi = \mathcal{F}_{n,\Sigma}(z,x)\psi' + \mathcal{G}_{2,n-1,\Sigma}(z,x)\psi, \tag{5.56}$$

and hence,

$$\begin{aligned}
[\mathcal{P}_{2n+1,\Sigma}, \mathcal{L}]\psi &= [2\mathcal{G}_{2,n-1,\Sigma} + \mathcal{F}''_{n,\Sigma}\mathcal{F}_{n,\Sigma}\mathcal{Q}_\Sigma - \mathcal{Q}_\Sigma\mathcal{F}_{n,\Sigma}]\psi' \qquad (5.57)\\
&+ [\mathcal{G}''_{2,n-1,\Sigma} + 2\mathcal{F}'_{n,\Sigma}(\mathcal{Q}_\Sigma - z\mathcal{I}_m) + \mathcal{F}_{n,\Sigma}\mathcal{Q}'_\Sigma + \mathcal{G}_{2,n-1,\Sigma}\mathcal{Q}_\Sigma - \mathcal{Q}_\Sigma\mathcal{G}_{2,n-1,\Sigma}]\psi = 0
\end{aligned}$$

by (4.56) and (4.57). Since $z \in \mathbb{C}$ is arbitrary, this implies

$$[\mathcal{P}_{2n+1,\Sigma}, \mathcal{L}] = 0, \tag{5.58}$$

completing the proof by (5.27), (5.28). □

Next, we turn to a discussion of trace formulas for the finite-band potential $\mathcal{Q}_\Sigma$ in terms of (matrix) roots of $\mathcal{F}_{n,\Sigma}$ and $\mathcal{H}_{n+1,\Sigma}$.

Theorem 5.7. *Let $(z,x) \in \mathbb{C}\times\mathbb{R}$ and assume $\mathcal{Q}_\Sigma$ to be the self-adjoint finite-band potential discussed in Theorems* 4.5 *and* 4.8. *In addition, let the monic self-adjoint matrix pencils $\mathcal{F}_{n,\Sigma}$ and $\mathcal{H}_{n+1,\Sigma}$ be given by* (4.47) *and* (4.50). *Then $\mathcal{F}_{n,\Sigma}(\cdot,x)$ and $\mathcal{H}_{n+1,\Sigma}(\cdot,x)$ are strongly hyperbolic pencils and hence admit the factorizations*

$$\mathcal{F}_{n,\Sigma}(z,x) = (z\mathcal{I}_m - \mathcal{U}_n(x))(z\mathcal{I}_m - \mathcal{U}_{n-1}(x))\cdots(z\mathcal{I}_m - \mathcal{U}_1(x)), \tag{5.59}$$

$$\mathcal{H}_{n+1,\Sigma}(z,x) = (z\mathcal{I}_m - \mathcal{V}_n(x))(z\mathcal{I}_m - \mathcal{V}_{n-1}(x))\cdots(z\mathcal{I}_m - \mathcal{V}_0(x)), \tag{5.60}$$

with

$$\operatorname{spec}(\mathcal{V}_0(x)) = \operatorname{spec}(\mathcal{H}_{n+1,\Sigma}) \cap \overline{\Delta_0(\mathcal{H}_{n+1,\Sigma})} \subset (-\infty, E_0], \tag{5.61}$$

$$\operatorname{spec}(\mathcal{U}_j(x)) = \operatorname{spec}(\mathcal{F}_{n,\Sigma}) \cap \overline{\Delta_j(\mathcal{F}_{n,\Sigma})} \subseteq [E_{2j-1}, E_{2j}], \quad 1 \le j \le n, \tag{5.62}$$

$$\operatorname{spec}(\mathcal{V}_j(x)) = \operatorname{spec}(\mathcal{H}_{n+1,\Sigma}) \cap \overline{\Delta_0(\mathcal{H}_{n+1,\Sigma})} \subseteq [E_{2j-1}, E_{2j}], \quad 1 \le j \le n. \tag{5.63}$$

Moreover, one obtains the sequence of trace formulas

$$\sum_{\substack{j_1,j_2,\dots,j_k=0\\ j_1<j_2<\cdots<j_k}}^{n} \mathcal{U}_{j_k}(x)\cdots\mathcal{U}_{j_2}(x)\mathcal{U}_{j_1}(x) = \sum_{\ell=0}^{k} c_{k-\ell}(\underline{E})\widehat{\mathfrak{R}}_{\ell,\Sigma}(x), \quad 1 \le k \le n, \tag{5.64}$$

$$0 = \sum_{\ell=0}^{k} c_{k-\ell}(\underline{E})\widehat{\mathfrak{R}}_{\ell,\Sigma}(x), \quad k \ge n+1, \tag{5.65}$$

$$\sum_{\substack{j_0,j_1,\dots,j_k=0\\ j_0<j_1<\cdots<j_k}}^{n} \mathcal{V}_{j_k}(x)\cdots\mathcal{V}_{j_1}(x)\mathcal{V}_{j_0}(x) = \sum_{\ell=0}^{k} c_{k-\ell}(\underline{E})\widehat{\mathfrak{H}}_{\ell,\Sigma}(x), \quad 1 \le k \le n, \tag{5.66}$$

$$0 = \sum_{\ell=0}^{k} c_{k-\ell}(\underline{E})\widehat{\mathfrak{H}}_{\ell,\Sigma}(x), \quad k \ge n+1. \tag{5.67}$$

In particular, in the special case $k=1$, $\mathcal{Q}_\Sigma$ satisfies the trace formulas

$$\mathcal{Q}_\Sigma(x) = \bigg(\sum_{\ell=0}^{2n} E_\ell\bigg)\mathcal{I}_m - 2\sum_{j=1}^{n}\mathcal{U}_j(x), \tag{5.68}$$

$$= -\bigg(\sum_{\ell=0}^{2n} E_\ell\bigg)\mathcal{I}_m + 2\sum_{k=0}^{n}\mathcal{V}_k(x). \tag{5.69}$$

In addition, one obtains

$$\mathcal{Q}_\Sigma^{(r)} \in C^\infty(\mathbb{R})^{m\times m} \cap L^\infty(\mathbb{R})^{m\times m} \text{ for all } r \in \mathbb{N}_0. \tag{5.70}$$

Proof. By (5.29) and (5.32), $i/R_{2n+1}^{1/2}\mathcal{F}_{n,\Sigma}(\cdot,x)$ and $i/R_{2n+1}^{-1/2}\mathcal{H}_{n+1,\Sigma}(\cdot,x)$ are Herglotz matrices and hence (4.14) and (4.15) apply. In particular, $\mathcal{F}_{n,\Sigma}(\cdot,x)$ and $\mathcal{H}_{n+1,\Sigma}(\cdot,x)$ are strongly hyperbolic pencils. Since both are monic, the factorizations (5.59) and (5.60), as well as (5.61)–(5.63), hold by Theorem 3.5 (ii). By (5.33) one infers

$$z^{-n}\mathcal{F}_{n,\Sigma}(z,x) = \sum_{k=0}^{n}\Bigg(\sum_{\substack{j_1,j_2,\dots,j_k=0\\ j_1<j_2<\cdots<j_k}}^{n} \mathcal{U}_{j_k}(x)\cdots\mathcal{U}_{j_2}(x)\mathcal{U}_{j_1}(x)\Bigg)z^{-k}$$

$$\underset{\substack{|z|\to\infty\\ z\in C_\varepsilon}}{=} z^{-n-(1/2)}R_{2n+1}(z)^{1/2}\sum_{k=0}^{\infty}\widehat{\mathfrak{R}}_{k,\Sigma}(x)z^{-k}$$

$$\underset{\substack{|z|\to\infty\\ z\in C_\varepsilon}}{=} \bigg(\prod_{\ell=0}^{2n}\big(1-(E_\ell/z)\big)\bigg)^{1/2}\sum_{k=0}^{\infty}\widehat{\mathfrak{R}}_{k,\Sigma}(x)z^{-k}$$

$$\underset{\substack{|z|\to\infty\\ z\in C_\varepsilon}}{=} \sum_{k=0}^{\infty}\bigg(\sum_{\ell=0}^{k}c_\ell(\underline{E})\widehat{\mathfrak{R}}_{k-\ell,\Sigma}(x)\bigg)z^{-k}. \tag{5.71}$$

Comparing coefficients z^{-k}, $k\in\mathbb{N}$, then yields (5.64) and (5.65). In the special case $k=1$ one infers

$$-\sum_{j=1}^{n}\mathcal{U}_j(x) = -\frac{1}{2}\bigg(\sum_{\ell=0}^{2n}E_\ell\bigg)\mathcal{I}_m+\widehat{\mathfrak{R}}_{1,\Sigma}(x) \tag{5.72}$$

and since $\widehat{\mathfrak{R}}_{1,\Sigma}=\mathcal{Q}_\Sigma/2$ by (5.14), the trace formula (5.68) for $\mathcal{Q}_\Sigma$ results. (5.66), (5.67), and (5.69) are proved analogously. By (5.62), $\mathcal{U}_j\in L^\infty(\mathbb{R})^{m\times m}$, $1\le j\le n$ and hence

$$\widehat{\mathfrak{R}}_k\in C^\infty(\mathbb{R})^{m\times m}\cap L^\infty(\mathbb{R})^{m\times m}, \quad k\in\mathbb{N}_0 \tag{5.73}$$

(since $\mathcal{Q}_\Sigma\in C^\infty(\mathbb{R})^{m\times m}$ by Theorem 4.8 (iii)). An analysis of the recursion relation (2.49) for $\mathcal{M}_{\pm,k}$ combined with (5.7), (5.11) then proves that $\widehat{\mathfrak{R}}_k$ is of the form

$$\widehat{\mathfrak{R}}_k = d_k\mathcal{Q}^{(2k-2)}+\mathcal{R}_k(\mathcal{Q}^{(2k-4)},\dots), \quad k\ge 2, \tag{5.74}$$

with $d_k\in\mathbb{R}$ appropriate constants and $\mathcal{R}_k$ abbreviating a differential polynomial in $\mathcal{Q}$ which contains $\mathcal{Q}^{(2k-4)}$ as the highest derivative of $\mathcal{Q}$. Hence one infers (5.70). □

The factorizations (5.59), (5.60), eigenvalue distributions (5.61)–(5.63), and trace formulas (5.64)–(5.69) are extensions of well-known formulas in the scalar case $m=1$ (see, e.g., [27], [30], [35], [37], [49], [66]).

Finally, the property $\mathcal{Q}_\Sigma\in C^\infty(\mathbb{R})^{m\times m}$ in Theorem 4.8 (iii) can be improved upon by using the system (4.51), (4.52), (4.55), and (4.58). In fact, writing

$$\mathcal{F}_{n,\Sigma}(z,x)=\sum_{\ell=0}^{n}\mathcal{F}_{n-\ell,\Sigma}(x)z^\ell, \quad \mathcal{F}_{0,\Sigma}(x)=\mathcal{I}_m, \tag{5.75}$$

$$\mathcal{G}_{p,n-1,\Sigma}(z,x)=\sum_{\ell=0}^{n-1}\mathcal{G}_{p,n-1-\ell,\Sigma}(x)z^\ell, \quad p=1,2, \tag{5.76}$$

$$\mathcal{H}_{n+1,\Sigma}(z,x)=\sum_{\ell=0}^{n+1}\mathcal{H}_{n+1-\ell,\Sigma}(x)z^\ell, \quad \mathcal{H}_{0,\Sigma}(x)=\mathcal{I}_m, \tag{5.77}$$

one obtains the following result.

Lemma 5.8. *Assume Hypothesis 4.4 and let* $(z,x) \in \mathbb{C} \times \mathbb{R}$. *Then the coefficients in* (5.75)–(5.77) *satisfy the autonomous nonlinear first-order system*

$$\mathcal{F}'_{\ell,\Sigma} = -(\mathcal{G}_{1,\ell-1,\Sigma} + \mathcal{G}_{2,\ell-1,\Sigma}), \quad 1 \leq \ell \leq n, \tag{5.78}$$

$$\mathcal{G}'_{1,\ell,\Sigma} = -(\mathcal{F}_{1,\Sigma} - \mathcal{H}_{1,\Sigma})\mathcal{F}_{\ell+1,\Sigma} + \mathcal{F}_{\ell+2,\Sigma} - \mathcal{H}_{\ell+2,\Sigma}, \quad 0 \leq \ell \leq n-1, \tag{5.79}$$

$$\mathcal{G}'_{2,\ell,\Sigma} = -\mathcal{F}_{\ell+1,\Sigma}(\mathcal{F}_{1,\Sigma} - \mathcal{H}_{1,\Sigma}) + \mathcal{F}_{\ell+2,\Sigma} - \mathcal{H}_{\ell+2,\Sigma}, \quad 0 \leq \ell \leq n-1, \tag{5.80}$$

$$\mathcal{H}'_{\ell,\Sigma} = \mathcal{G}_{1,\ell-1,\Sigma} + \mathcal{G}_{2,\ell-1,\Sigma} - \mathcal{G}_{1,\ell-2,\Sigma}(\mathcal{F}_{1,\Sigma} - \mathcal{H}_{1,\Sigma}) - (\mathcal{F}_{1,\Sigma} - \mathcal{H}_{1,\Sigma})\mathcal{G}_{2,\ell-2,\Sigma}, \quad 1 \leq \ell \leq n+1, \tag{5.81}$$

$$\mathcal{F}_{n+1,\Sigma} = 0, \quad \mathcal{G}_{p,n,\Sigma} = \mathcal{G}_{p,-1,\Sigma} = 0, \quad p = 1,2. \tag{5.82}$$

Moreover, $\mathcal{F}_{\ell,\Sigma}$, $0 \leq \ell \leq n$, $\mathcal{G}_{p,\ell,\Sigma}$, $0 \leq \ell \leq n-1$, $p = 1,2$, $\mathcal{H}_{\ell,\Sigma}$, $0 \leq \ell \leq n+1$, *and hence* $\mathcal{Q}_\Sigma$, *are all analytic in an open neighborhood containing the real axis.*

Proof. Inserting (5.75)–(5.77) into (4.51), (4.52), (4.55), and (4.58) yields

$$\mathcal{F}'_{\ell,\Sigma} = -(\mathcal{G}_{1,\ell-1,\Sigma} + \mathcal{G}_{2,\ell-1,\Sigma}), \quad 1 \leq \ell \leq n, \tag{5.83}$$

$$\mathcal{G}'_{1,\ell,\Sigma} = -\mathcal{Q}_\Sigma \mathcal{F}_{\ell+1,\Sigma} + \mathcal{F}_{\ell+2,\Sigma} - \mathcal{H}_{\ell+2,\Sigma}, \quad 0 \leq \ell \leq n-1, \tag{5.84}$$

$$\mathcal{G}'_{2,\ell,\Sigma} = -\mathcal{F}_{\ell+1,\Sigma}\mathcal{Q}_\Sigma + \mathcal{F}_{\ell+2,\Sigma} - \mathcal{H}_{\ell+2,\Sigma}, \quad 0 \leq \ell \leq n-1, \tag{5.85}$$

$$\mathcal{H}'_{\ell,\Sigma} = \mathcal{G}_{1,\ell-1,\Sigma} + \mathcal{G}_{2,\ell-1,\Sigma} - (\mathcal{G}_{1,\ell-2,\Sigma}\mathcal{Q}_\Sigma + \mathcal{Q}_\Sigma \mathcal{G}_{2,\ell-2,\Sigma}), \quad 1 \leq \ell \leq n+1, \tag{5.86}$$

$$\mathcal{F}_{n+1,\Sigma} = 0, \quad \mathcal{G}_{p,n,\Sigma} = \mathcal{G}_{p,-1,\Sigma} = 0, \quad p = 1,2, \tag{5.87}$$

$$\mathcal{Q}_\Sigma = \mathcal{F}_{1,\Sigma} - \mathcal{H}_{1,\Sigma}. \tag{5.88}$$

Insertion of (5.88) into (5.84)–(5.86) yields the autonomous nonlinear first-order system (5.78)–(5.82). Given the initial conditions

$$\begin{aligned} &\mathcal{F}_{\ell,\Sigma}(x_0), \quad 1 \leq \ell \leq n, \\ &\mathcal{G}_{p,\ell,\Sigma}(x_0), \quad p = 1,2, \ 0 \leq \ell \leq n-1, \\ &\mathcal{H}_{\ell,\Sigma}(x_0), \quad 1 \leq \ell \leq n+1, \end{aligned} \tag{5.89}$$

determined by (5.75)–(5.77) and the half-line Weyl-Titchmarsh matrices $\mathcal{M}_\pm(z,x_0)$ in (4.36), the maximal interval of existence of the solution of the autonomous system (5.78)–(5.82), (5.89) is all of $[0,\infty)$ and $(-\infty,0]$, and hence all of $\mathbb{R}$ (cf. [41, p. 18]), applying (5.70). Thus, one recovers the $C^\infty(\mathbb{R})^{m\times m}$-property $\mathcal{F}_\ell$, $\mathcal{G}_{p,\ell}$, $p = 1,2$, $\mathcal{H}_\ell$, and hence that of $\mathcal{Q}_\Sigma$. Moreover, since the Picard iterations are convergent in sufficiently small circles in $\mathbb{C}$ centered around each $x \in \mathbb{R}$ (cf. [43, Sect. 2.3]), the unique solution obtained by these Picard iterations is analytic in each of these circles. □

We conclude with a remark that puts the construction of $\mathcal{Q}_\Sigma$ in Section 4 into proper perspective.

Remark 5.9. *The simplest examples of potentials $\mathcal{Q}_\Sigma$ described in Theorem* 4.8 *are of the type*

$$\mathcal{Q}_\Sigma(x) = \operatorname{diag}(q_{1,\Sigma}(x), \dots, q_{m,\Sigma}(x)), \tag{5.90}$$

where $q_{j,\Sigma}$, $1 \le j \le m$, are isospectral algebro-geometric finite-band potentials associated with scalar Schrödinger operators in $L^2(\mathbb{R})$ with spectrum Σ. The next simplest and closely related set of such examples for $\mathcal{Q}_\Sigma$ then will be of the type

$$\mathcal{Q}_\Sigma(x) = \mathcal{U} \operatorname{diag}(q_{1,\Sigma}(x), \dots, q_{m,\Sigma}(x))\mathcal{U}^{-1}, \tag{5.91}$$

where $\mathcal{U} \in \mathbb{C}^{m\times m}$ is a unitary $m \times m$ matrix independent of $x \in \mathbb{R}$. At first sight one might think that perhaps all matrix potentials $\mathcal{Q}_\Sigma$ are of the form (5.91)*. That this is certainly not the case will be argued next. Indeed, assuming that* (5.91) *holds for some unitary $m \times m$ matrix $\mathcal{U}$ independent of x, one infers that*

$$\mathcal{Q}_\Sigma^{(r)}(x) = \mathcal{U} \operatorname{diag}(q_{1,\Sigma}^{(r)}(x), \dots, q_{m,\Sigma}^{(r)}(x))\mathcal{U}^{-1} \text{ for all } r \in \mathbb{N}_0. \tag{5.92}$$

Since by Remark 2.8 *(i) and Lemma* 5.5 *all coefficients of $\mathcal{F}_{n,\Sigma}$, $\mathcal{G}_{p,n-1,\Sigma}$, $p = 1, 2$, and $\mathcal{H}_{n+1,\Sigma}$ are differential polynomials with respect to $\mathcal{Q}_\Sigma$,* (5.92) *implies*

$$\mathcal{U}\mathcal{F}_{n,\Sigma}(z,x)\mathcal{U}^{-1} = \operatorname{diag}(F_{n,\Sigma,1}(z,x), \dots, F_{n,\Sigma,m}(z,x)), \tag{5.93}$$

$$\mathcal{U}\mathcal{G}_{p,n-1,\Sigma}(z,x)\mathcal{U}^{-1} = \operatorname{diag}(G_{p,n-1,\Sigma,1}(z,x), \dots, G_{p,n-1,\Sigma,m}(z,x)), \quad p = 1, 2, \tag{5.94}$$

$$\mathcal{U}\mathcal{H}_{n+1,\Sigma}(z,x)\mathcal{U}^{-1} = \operatorname{diag}(H_{n+1,\Sigma,1}(z,x), \dots, H_{n+1,\Sigma,m}(z,x)), \tag{5.95}$$

$$\mathcal{U}\mathcal{M}_{\pm,\Sigma}(z,x)\mathcal{U}^{-1} = \operatorname{diag}(m_{\pm,\Sigma,1}(z,x), \dots, m_{\pm,\Sigma,m}(z,x)). \tag{5.96}$$

Consequently, one obtains for all $z, z' \in \mathbb{C}_+$, $x, x' \in \mathbb{R}$,

$$[\mathcal{Q}_\Sigma^{(r)}(x), \mathcal{Q}_\Sigma^{(r)}(x')] = 0, \quad r \in \mathbb{N}_0, \tag{5.97}$$

$$[\mathcal{F}_{n,\Sigma}(z,x), \mathcal{F}_{n,\Sigma}(z',x')] = 0, \tag{5.98}$$

$$[\mathcal{G}_{p,n-1,\Sigma}(z,x), \mathcal{G}_{p,n-1,\Sigma}(z',x')] = 0, \quad p = 1, 2, \tag{5.99}$$

$$[\mathcal{H}_{n+1,\Sigma}(z,x), \mathcal{H}_{n+1,\Sigma}(z',x')] = 0, \tag{5.100}$$

$$[\mathcal{M}_{\pm,\Sigma}(z,x), \mathcal{M}_{\pm,\Sigma}(z',x')] = 0. \tag{5.101}$$

In particular,

$$[\mathcal{F}_{n,\Sigma}(z,x_0), \mathcal{F}_{n,\Sigma}(z',x_0)] = 0 \text{ for all } z, z' \in \mathbb{C}_+. \tag{5.102}$$

Thus, whenever

$$[\mathcal{F}_{n,\Sigma}(z,x_0), \mathcal{F}_{n,\Sigma}(z',x_0)] \neq 0 \text{ for some } z, z' \in \mathbb{C}_+ \tag{5.103}$$

(which can easily be arranged for $n \ge 2$), (5.91) *cannot hold for a unitary $m \times m$ matrix $\mathcal{U}$ independent of x.*

Additional results, including extensions of Borg's and Hochstadt's theorems in the special cases $n = 0, 1$, respectively, will appear in [9].

Acknowledgements. We thank Robert Carlson, Konstantin Makarov, Mark Malamud, Fedor Rofe-Beketov, and Barry Simon for many helpful discussions on the material presented in this paper. In particular, we are grateful to Konstantin Makarov for a critical reading of this manuscript.

References

[1] Z.S. Agranovich and V.A. Marchenko, *The Inverse Problem of Scattering Theory*, Gordon and Breach, New York, 1963.

[2] D. Alpay and I. Gohberg, *Inverse spectral problem for differential operators with rational scattering matrix functions*, J. Diff. Eq. **118**, 1–19 (1995).

[3] D. Alpay and I. Gohberg, *Inverse problem for Sturm-Liouville operators with rational reflection coefficient*, Integral Equations Operator Theory **30**, 317–325 (1998).

[4] D.Z. Arov and H. Dym, *J-inner matrix functions, interpolation and inverse problems for canonical systems, I: foundations*, Integral Equations Operator Theory **29**, 373–454 (1997).

[5] D.Z. Arov and H. Dym, *J-inner matrix functions, interpolation and inverse problems for canonical systems, II: the inverse monodromy problem*, Integral Equations Operator Theory **36**, 11–70 (2000).

[6] D.Z. Arov and H. Dym, *J-inner matrix functions, interpolation and inverse problems for canonical systems, III: more on the inverse monodromy problem*, Integral Equations Operator Theory **36**, 127–181 (2000).

[7] N. Asano and Y. Kato, *Algebraic and Spectral Methods for Nonlinear Wave Equations*, Longman, New York, 1990.

[8] I. Avramidi and R. Schimming, *A new explicit expression for the Korteweg–de Vries hierarchy*, Math. Nachr. **219**, 45–64 (2000).

[9] E.D. Belokolos, F. Gesztesy, K.A. Makarov, and L.A. Sakhnovich, *Matrix-valued generalizations of the theorems of Borg and Hochstadt*, to appear in *Recent Advances in Evolution Equations*, G. Ruiz Goldstein, R. Nagel, and S. Romanelli (eds.), Lecture Notes in Pure and Applied Mathematics, Marcel Dekker, New York.

[10] R. Benguria and M. Loss, *A simple proof of a theorem of Laptev and Weidl*, Math. Res. Lett. **7**, 195–203 (2000).

[11] F. Calogero and A. Degasperis, *Nonlinear evolution equations solvable by the inverse spectral transfor. II*, Nuovo Cim. **39B**, 1–54 (1977).

[12] R. Carlson, *Large eigenvalues and trace formulas for matrix Sturm-Liouville problems*, SIAM J. Math. Anal. **30**, 949–962 (1999).

[13] R. Carlson, *Compactness of Floquet isospectral sets for the matrix Hill's equation*, Proc. Amer. Math. Soc. **128**, 2933–2941 (2000).

[14] R. Carlson, *Eigenvalue estimates and trace formulas for the matrix Hill's equation*, J. Diff. Eq. **167**, 211–244 (2000).

[15] R. Carlson, *An inverse problem for the matrix Schrödinger equation*, preprint, 2001.

[16] R. Carmona and J. Lacroix, *Spectral Theory of Random Schrödinger operators*, Birkhäuser, Boston, MA, 1990.

[17] H.-H. Chern, *On the construction of isospectral vectorial Sturm-Liouville differential equations*, preprint, 1998.

[18] H.-H. Chern and C-L. Shen, *On the n-dimensional Ambarzumyan's theorem*, Inverse Problems **13**, 15–18 (1997).

[19] S.L. Clark, *Asymptotic behavior of the Titchmarsh–Weyl coefficient for a coupled second order system*, in "*Ordinary and Delay Differential Equations*", J. Wiener and J.K. Hale (eds.), Longman, New York, 1992, 24–28.

[20] S. Clark and F. Gesztesy, *Weyl–Titchmarsh M-function asymptotics for matrix-valued Schrödinger operators*, Proc. London Math. Soc. **82**, 701–724 (2001).

[21] S. Clark and F. Gesztesy, *On Povzner–Wienholtz-type Self-Adjointness Results for Matrix-Valued Sturm–Liouville Operators*, Proc. Roy. Soc. Edinburgh **A** (to appear).

[22] S. Clark, F. Gesztesy, H. Holden, and B.M. Levitan, *Borg-type theorems for matrix-valued Schrödinger and Dirac operators*, J. Diff. Eqs. **167**, 181–210 (2000).

[23] S. Clark and D. Hinton, *A Liapunov inequality for linear Hamiltonian systems*, Math. Inequ. Appl. **1**, 201–209 (1998).

[24] V.I. Derguzov, *The spectrum of Hamilton's operator with periodic coefficients*, Vestnik Leningrad Univ. Math. **12**, 280–285 (1980).

[25] B. Desprès, *The Borg theorem for the vectorial Hill's equation*, Inverse Probl. **11**, 97–121 (1995).

[26] L.A. Dickey, *Soliton Equations and Hamiltonian Systems*, World Scientific, Singapore, 1991.

[27] B.A. Dubrovin, *Periodic problems for the Korteweg-de Vries equation in the class of finite-gap potentials*, Funct. Anal. Appl. **9**, 215–223 (1975).

[28] B.A. Dubrovin, *Completely integrable Hamiltonian Systems associated with Matrix operators and Abelian varieties*, Funct. Anal. Appl. **11**, 265–277 (1977).

[29] B.A. Dubrovin, *Matrix finite-zone operators*, Revs. Sci. Tech. **23**, 20–50 (1983).

[30] B.A. Dubrovin, V.B. Matveev, and S.P. Novikov, *Non-linear equations of Korteweg-de Vries type, finite-zone linear operators, and Abelian varieties*, Russian Math. Surv. **31:1**, 59–146 (1976).

[31] I.M. Gel'fand and L.A. Dikii, *The resolvent and Hamiltonian systems*, Funct. Anal. Appl. **11**, 93–105 (1977).

[32] I.M. Gelfand and B.M. Levitan, *On the determination of a differential equation from its spectral function*, Izv. Akad. Nauk SSR. Ser. Mat. **15**, 309–360 (1951) (Russian); English transl. in Amer. Math. Soc. Transl. Ser. 2 **1**, 253–304 (1955).

[33] I.M. Gelfand and V.B. Lidskij, *On the structure of the regions of stability of linear canonical systems of differential equations with periodic coefficients*, in "Izrail M. Gelfand, Collected Papers", Vol. I, S.G. Gindikin, V.W. Guillemin, A.A. Kirillov, B. Kostant, S. Sternberg (eds.), Springer, Berlin, 1987, 466–504. (Amer. Math. Soc. Transl. (2) **8**, 143–181 (1958).)

[34] F. Gesztesy and H. Holden, *On trace formulas for Schrödinger-type operators*, in Multiparticle Quantum Scattering with Applications to Nuclear, Atomic and Molecular Physics, D.G. Truhlar and B. Simon (eds.), Springer, New York, 1997, 121–145.

[35] F. Gesztesy, H. Holden, B. Simon, and Z. Zhao, *Higher order trace relations for Schrödinger operators*, Revs. Math. Phys. **7**, 893–922 (1995).

[36] F. Gesztesy, A. Kiselev, and K.A. Makarov, *Uniqueness Results for Matrix-Valued Schrödinger, Jacobi, and Dirac-Type Operators*, Math. Nachr. **239–240**, 103–145 (2002).

[37] F. Gesztesy, R. Ratnaseelan, and G. Teschl, *The KdV hierarchy and associated trace formulas*, in proceedings of the *International Conference on Applications of Operator Theory*, I. Gohberg, P. Lancaster, P.N. Shivakumar (eds.), Operator Theory: Advances and Applications, Vol. 87, Birkhäuser, 1996, 125–163.

[38] F. Gesztesy and B. Simon, *On local Borg-Marchenko uniqueness results*, Commun. Math. Phys. **211**, 273–287 (2000).

[39] F. Gesztesy and E. Tsekanovskii, *On matrix-valued Herglotz functions*, Math. Nachr. **218**, 61–138 (2000).

[40] I.M. Glazman, *Direct Methods of Qualitative Spectral Analysis of Singular Differential Operators*, Israel Program for Scientific Translations, Jerusalem, 1965.

[41] J.K. Hale, *Ordinary Differential Equations*, 2nd ed., Krieger, Malabar, Fl, 1980.

[42] E. Hille, *Lectures on Ordinary Differential Equations*, Addison-Wesley, Reading, 1969.

[43] E. Hille, *Ordinary Differential Equations in the Complex Domain*, Dover, New York, 1997.

[44] D.B. Hinton and J.K. Shaw, *On Titchmarsh–Weyl $M(\lambda)$-functions for linear Hamiltonian systems*, J. Diff. Eqs. **40**, 316–342 (1981).

[45] D.B. Hinton and J.K. Shaw, *On the spectrum of a singular Hamiltonian system*, Quaest. Math. **5**, 29–81 (1982).

[46] D.B. Hinton and J.K. Shaw, *Hamiltonian systems of limit point or limit circle type with both endpoints singular*, J. Diff. Eqs. **50**, 444–464 (1983).

[47] D.B. Hinton and J.K. Shaw, *On boundary value problems for Hamiltonian systems with two singular points*, SIAM J. Math. Anal. **15**, 272–286 (1984).

[48] D.B. Hinton and J.K. Shaw, *On the spectrum of a singular Hamiltonian system, II*, Quaest. Math. **10**, 1–48 (1986).

[49] A.R. Its and V.B. Matveev, *Schrödinger operators with finite-gap spectrum and N-soliton solutions of the Korteweg-de Vries equation*, Theoret. Math. Phys. **23**, 343–355 (1975).

[50] M. Jodeit and B.M. Levitan, *Isospectral vector-valued Sturm-Liouville problems*, Lett. Math. Phys. **43**, 117–122 (1998).

[51] M. Jodeit and B.M. Levitan, *The isospectrality problem for some vector-valued Sturm-Liouville boundary problems*, Russ. J. Math. Phys **6**, 375–393 (1999).

[52] R.A. Johnson, *m-Functions and Floquet exponents for linear differential systems*, Ann. Mat. Pura Appl., Ser. 4, **147**, 211–248 (1987).

[53] R. Johnson, S. Novo, and R. Obaya, *Ergodic properties and Weyl M-functions for random linear Hamiltonian systems*, Proc. Roy. Soc. Edinburgh **130A**, 1045–1079 (2000).

[54] V.I. Kogan and F.S. Rofe-Beketov, *On square-integrable solutions of symmetric systems of differential equations of arbitrary order*, Proc. Roy. Soc. Edinburgh **74A**, 1–40 (1974).

[55] S. Kotani and B. Simon, *Stochastic Schrödinger operators and Jacobi matrices on the strip*, Commun. Math. Phys. **119**, 403–429 (1988).

[56] A.M. Krall, *$M(\lambda)$ theory for singular Hamiltonian systems with one singular point*, SIAM J. Math. Anal. **20**, 664–700 (1989).

[57] A.M. Krall, *$M(\lambda)$ theory for singular Hamiltonian systems with two singular points*, SIAM J. Math. Anal. **20**, 701–715 (1989).

[58] M.G. Krein, *Foundations of the theory of λ-zones of stability of a canonical system of linear differential equations with periodic coefficients*, Amer. Math. Soc. Transl. (2) **120**, 1–70 (1983).

[59] M.G. Krein, *On tests for stable boundedness of solutions of periodic canonical systems*, Amer. Math. Soc. Transl. (2) **120**, 71–110 (1983).

[60] M.G. Krein and A.A. Nudelman, *The Markov Moment Problem and Extremal Problems*, Amer. Math. Soc., Providence, 1977.

[61] B.M. Levitan, *On the solvability of the inverse Sturm-Liouville problem on the entire line*, Sov. Math. Dokl. **18**, 597–600 (1977).

[62] B.M. Levitan, *On the solvability of the inverse Sturm–Liouville problem on the whole axis in the case of an infinite number of lacunas in the spectrum*, Sov. Math. Dokl. **18**. 964–967 (1977).

[63] B.M. Levitan, *An inverse problem for the Sturm–Liouville operator in the case of finitezone and infinite-zone potentials*, Trans. Moscow Math. Soc. 1984, No. 1, 1–34.

[64] B.M. Levitan, *Inverse Sturm-Liouville Problems*, VNU Science Press, Utrecht, 1987.

[65] B.M. Levitan and M.G. Gasymov, *Determination of a differential equation by two of its spectra*, Russ. Math. Surv. **19:2**, 1–63 (1964).

[66] B.M. Levitan and A.V. Savin, *The inverse problem on the half-line for finite zone potentials*, Moscow Univ. Math. Bull. **43**, No. 1, 27–34 (1988).

[67] M.M. Malamud, *Similarity of Volterra operators and related questions of the theory of differential equations of fractional order*, Trans. Moscow Math. Soc. **55**, 57–122 (1994).

[68] M.M. Malamud, *Uniqueness questions in inverse problems for systems of ordinary differential equations on a finite interval*, Trans. Moscow Math. Soc. **60**, 173–224 (1999).

[69] Yu.I. Manin, *Matrix solitons and bundles over curves with singularities*, Funct. Anal. Appl. **12**, 286–295 (1978).

[70] V.A. Marchenko, *Nonlinear Equations and Operator Algebras*, Reidel, Dordrecht, 1988.

[71] A.S. Markus, *Introduction to the Spectral Theory of Polynomial Operator Pencils*, Translations of Mathematical Monographs, Vol. 71, Amer. Math. Soc., Providence, RI, 1988.

[72] A.S. Markus and V.I. Matsaev, *Factorization of a weakly hyperbolic bundle*, Funct. Anal. Appl. **10**, 69–71 (1976).

[73] A.S. Markus and V.I. Matsaev, *On the spectral factorization of holomorphic operator-functions*, Sel. Math. Sov. **4**, 325–354 (1985).

[74] L. Martínez Alonso and E. Olmedilla, *Trace identities in the inverse scattering transform method associated with matrix Schrödinger operators*, J. Math. Phys. **23**, 2116–2121 (1982).

[75] M.A. Naimark, *Linear Differential Operators, Part II*, F. Ungar, New York, 1968.

[76] R.G. Newton and R. Jost, *The construction of potentials from the S-matrix for systems of differential equations*, Nuovo Cim. **1**, 590–622 (1955).

[77] E. Olmedilla, *Inverse scattering transform for general matrix Schrödinger operators and the related symplectic structure*, Inverse Problems **1**, 219–236 (1985).

[78] E. Olmedilla, L. Martínez Alonso, and F. Guil, *Infinite-dimensional Hamiltonian systems associated with matrix Schrödinger operators*, Nuovo Cim. **61 B**, 49–61 (1981).

[79] P.J. Olver and V.V. Sokolov, *Integrable evolution equations on assciative algebras*, Commun. Math. Phys. **193**, 245–268 (1998).

[80] S.A. Orlov, *Nested matrix disks analytically depending on a parameter, and theorems on the invariance of ranks of radii of limiting disks*, Math. USSR Izv. **10**, 565–613 (1976).

[81] V.G. Papanicolaou, *Trace formulas and the behavior of large eigenvalues*, SIAM J. Math. Anal. **26**, 218–237 (1995).

[82] I. Polterovich, *Heat kernel asymptotics for Laplace type operators and matrix KdV hierarchy*, preprint, 2000.

[83] A.Ya. Povzner, *The expansion of arbitrary functions in eigenfunctions of the operator $-\Delta u + cu$*, Mat. Sbornik **32**, 109–156 (1953) (Russian.) English translation in Amer. Math. Soc. Transl. (2) **60**, 1–49 (1967).

[84] F.S. Rofe-Beketov, *Expansions in eigenfunctions of infinite systems of differential equations in the non-self-adjoint and self-adjoint cases*, Mat. Sb. **51**, 293–342 (1960). (Russian.)

[85] F.S. Rofe-Beketov, *The spectrum of non-selfadjoint differential operators with periodic coefficients*, Sov. Math. Dokl. **4**, 1563–1566 (1963).

[86] F.S. Rofe-Beketov, *The spectral matrix and the inverse Sturm-Liouville problem on the axis* $(-\infty, \infty)$, Teor. Funktsiĭ Funkts. Analiz Prilozh. **4**, 189–197 (1967). (Russian.)

[87] F.S. Rofe-Beketov, *The inverse Sturm-Liouville problem for the spectral matrix on the whole axis and associated problems*, in *Integral Equations and Inverse Problems*, V. Petkov and R. Lazarov Eds.), Longman, New York, 1991, p. 234–238.

[88] F.S. Rofe-Beketov and A.M. Hol'kin, *On the connection between spectral and oscillation properties of the Sturm-Liouville matrix problem*, Math. USSR Sbornik **31**, 365–378 (1977).

[89] F.S. Rofe-Beketov and A.M. Kholkin, *Spectral Analysis of Differential Operators. Connection between Spectral and Oscillatory Properties*, Mariupol, 2001.

[90] A.L. Sakhnovich, *Asymptotic behavior of spectral functions of an S-node*, Sov. Math. (Iz. VUZ) **32:9**, 92–105 (1988).

[91] A.L. Sakhnovich, *Spectral functions of a canonical system of order* $2n$, Math. USSR Sbornik **71**, 355–369 (1992).

[92] A. Sakhnovich, *Iterated Backlund-Darboux transform for canonical systems*, J. Funct. Anal. **144**, 359–370 (1997).

[93] A. Sakhnovich, *Canonical systems and transfer matrix-functions*, Proc. Amer. Math. Soc. **125**, 1451–1455 (1997).

[94] L.A. Sakhnovich, *Evolution of spectral data and nonlinear equations*, Ukrain. Math. J. **40**, 459–461 (1988).

[95] L.A. Sakhnovich, *Inverse problems for equations systems*, in *Matrix and Operator Valued Functions: The Vladimir Petrovich Potapov Memorial Volume*, I. Gohberg and L.A. Sakhnovich (eds.), Operator Theory: Advances and Applications, Vol. 72, Birkhäuser, Basel, 1994, 202–211.

[96] L.A. Sakhnovich, *Method of operator identities and problems of analysis*, St. Petersburg Math. J. **5**, 1–69 (1994).

[97] L.A. Sakhnovich, *Spectral problem on half-axis*, Methods Funct. Anal. Topology **2**, 128–140 (1996).

[98] L.A. Sakhnovich, *Interpolation Theory and its Applications*, Kluwer, Dordrecht, 1997.

[99] L.A. Sakhnovich, *Spectral analysis of a class of canonical differential systems*, St. Petersburg Math. J. **10**, 147–158 (1999).

[100] L.A. Sakhnovich, *Spectral Theory of Canonical Differential Systems. Method of Operator Identities*, Operator Theory: Advances and Applications, Vol. 107, Birkhäuser, Basel, 1999.

[101] C.-L. Shen, *Some eigenvalue problems for the vectorial Hill's equation*, Inverse Probl. **16**, 749–783 (2000).

[102] C.-L. Shen, *Some inverse spectral problems for vectorial Sturm-Liouville equations*, Inverse Probl. **17**, 1253–1294 (2001).

[103] C.-L. Shen and C.-T. Shieh, *Two inverse eigenvalue problems for vectorial Sturm-Liouville equations*, Inverse Probl. **14**, 1331–1343 (1998).

[104] M. Sodin and P. Yuditskii, *Almost periodic Sturm-Liouville operators with Cantor homogeneous spectrum*, Comment. Math. Helvetici **70**, 639–658 (1995).

[105] C. Thurlow, *A generalisation of the inverse spectral theorem of Levitan and Gasymov*, Proc. Roy. Soc. Edinburgh **84A**, 185–196 (1979).

[106] I. Trooshin, *Asymptotics for the spectral and Weyl functions of the operator-valued Sturm-Liouville problem*, in "*Inverse Problems and Related Topics*", G. Nakamura, S. Saitoh, J.K. Seo, and M. Yamamoto (eds.), Chapman & Hall/CRC, Res. Notes Math. **419**, Boca Raton, FL, 2000, 189–208.

[107] M. Wadati and T. Kamijo, *On the extension of inverse scattering method*, Progr. Theoret. Phys. **52**, 397–414 (1974).

[108] E. Wienholtz, *Halbbeschränkte partielle Differentialoperatoren zweiter Ordnung vom elliptischen Typus*, Math. Ann., **135**, 50–80 (1958).

[109] V.A. Yakubovich, *Nonoscillation of linear periodic Hamiltonian equations, and related topics*, St. Petersburg Math. J. **3**, 1165–1188 (1992).

[110] V.A. Yakubovich and V.M. Starzhinskii, *Linear Differential Equations with Periodic Coefficients 1*, Wiley, New York, 1975.

[111] V.A. Yakubovich and V.M. Starzhinskii, *Linear Differential Equations with Periodic Coefficients 2*, Wiley, New York, 1975.

Fritz Gesztesy
Department of Mathematics
University of Missouri
Columbia, MO 65211, USA
e-mail: fritz@math.missouri.edu
URL: http://www.math.missouri.edu/people/fgesztesy.html

Lev A. Sakhnovich
735 Crawford Ave.
Brooklyn, NY 11223, USA
e-mail: Lev.Sakhnovich@verizon.net

Operator Theory:
Advances and Applications, Vol. 143, 255–272

Laplace Transform Asymptotics, Bergman Kernels and Composition Operators

Thomas L. Kriete

Abstract. We determine the asymptotics at infinity and zero of the Laplace transforms of fast weights on a half-line. This yields the exact growth rate of the norm of the kernel function in a weighted Bergman space associated with a rotation-invariant fast weight on the unit disk. Applications to composition operators on such spaces are discussed.

1. Introduction

Let $G(r)$, $0 < r < 1$, be a positive, continuous integrable function that ultimately decreases to zero as $r \to 1$. We consider the weighted Bergman space A^2_G, consisting of all functions f analytic on the unit disk $\mathbb{D} = \{z : |z| < 1\}$ such that

$$\int_{\mathbb{D}} |f(z)|^2 G(|z|) dA(z) < \infty;$$

here dA is Lebesgue area measure on $\mathbb{D}$. On equipping A^2_G with the norm $\|f\|$ defined as the square root of the above integral, we obtain a reproducing kernel Hilbert space with kernel function $\{k_w : w \in \mathbb{D}\}$ given by

$$k_w(z) = \frac{1}{2\pi} \sum_{n=0}^{\infty} \frac{1}{p_n} (z\overline{w})^n,$$

where $p_n = \int_0^1 r^{2n+1} G(r) dr$. Each k_w lies in A^2_G and satisfies $f(w) = \langle f, k_w \rangle$ for all f in A^2_G. Consequently, $k_w(z) = \langle k_w, k_z \rangle$ and $\|k_w\|^2 = k_w(w) = \|k_{|w|}\|^2$ for all z and w in $\mathbb{D}$.

When $G(r)$ is one of the *standard weights* $(1-r)^\alpha$ or $\left(\log \frac{1}{r}\right)^\alpha$ with $\alpha > 0$, it is well known that

$$\|k_r\|^2 \sim \frac{\alpha+1}{4\pi} \frac{1}{(1-r)^{\alpha+2}},$$

2000 *Mathematics Subject Classification.* Primary: 44A15, 41A60, 46E22, 47B33.
Key words and phrases. Laplace transform asymptotics, reproducing kernel, weighted Bergman space, composition operator.

as $r \to 1$. (Here and throughout we interpose $\sim$ between two quantities to signify that their ratio tends to unity as the indicated limit is taken.) In this article, we are concerned with the asymptotics as $r \to 1$ of

$$\|k_r\|^2 = \frac{1}{2\pi} \sum_{n=0}^{\infty} \frac{1}{p_n} r^{2n}, \tag{1.1}$$

when $G(r)$ is a *fast weight*, that is, when for each $\alpha > 0$, $G(r)/(1-r)^\alpha$ is decreasing for r near 1. We write our weight as

$$G(r) = e^{-v\left(\log \frac{1}{r}\right)}, \tag{1.2}$$

and analyze, rather than the sum (1.1), a Laplace transform with the same asymptotics:

$$\|k_{e^{-t}}\|^2 \sim \frac{1}{4\pi} \int_{x_0}^{\infty} \frac{1}{P(x)} e^{-xt} dx \text{ as } t \to 0 \tag{1.3}$$

(see [K2, Lemma 2]), where

$$P(x) = \int_{r_0}^{1} r^{x-1} G(r) dr = \int_0^{t_0} e^{-v(t)-xt} dt,$$

x_0 and t_0 are arbitrary positive numbers, and $r_0 = e^{-t_0}$. Note that the asymptotics of $P(x)$ as $x \to \infty$ are independent of the choice of t_0 since the dominant part of the integral occurs at small t. Similarly, the growth of the integral in (1.3) as $t \to 0$ depends only on the size of $1/P(x)$ for x large, so the choice of x_0 is immaterial.

Consider first $P(x)$. We assume that v is of class C^3 on $(0, t_0]$ and that

(i) $v'(t) < 0,\ v''(t) > 0,\ v'''(t) < 0$ for $0 < t \leq t_0$.
(ii) $v(t) \to +\infty$ as $t \to 0$.

The Legendre-Fenschel transform of v is the function

$$L(x) = \inf_{0<t<t_0} (v(t) + xt).$$

For large enough positive x, $L(x)$ is differentiable to the same degree as v and the infimum on the right is attained at $t = L'(x)$. Moreover, for small t, $v(t)$ is recaptured as the inverse Legendre-Fenschel transform

$$v(t) = \sup_{x>x_0} (L(x) - xt).$$

We have the inverse function relations

$$L'(-v'(t)) = t,\ t \text{ small}; \qquad -v'(L'(x)) = x,\ x \text{ large}.$$

A good reference for these facts is Koosis [Ko]. It is useful to consider the example of the standard weights $G(r) = \left(\log \frac{1}{r}\right)^\alpha$, $\alpha > 0$. Here $v(t) = \alpha \log \frac{1}{t}$ and

$$L(x) = \alpha \log x - \alpha \log \alpha + \alpha.$$

Whatever the choice of $t_0 > 0$, we have

$$P(x) \sim \int_0^{\infty} t^\alpha e^{-xt} dt = \frac{\Gamma(\alpha+1)}{x^{\alpha+1}} = \frac{\Gamma(\alpha+1)}{\left(\frac{\alpha}{e}\right)^\alpha \sqrt{\alpha}} \sqrt{-L''(x)}\, e^{-L(x)}$$

as $x \to \infty$. Though the constant factor on the right varies with α, by "taking $\alpha \to \infty$," we can read off what should be the correct constant factor for fast weights, by Stirling's formula:

$$\frac{\Gamma(\alpha+1)}{\left(\frac{\alpha}{e}\right)^{\alpha}\sqrt{\alpha}} \sim \sqrt{2\pi} \text{ as } \alpha \to \infty.$$

To encode our requirement that $G(r)$ is fast, that is, that for each $\alpha > 0$, $e^{-v(t)}/t^{\alpha}$ is ultimately decreasing to zero as $t \to 0$, we require that the quantity

$$\sqrt{v''(t)}\,\frac{d}{dt}\,\frac{1}{v''(t)} = -\,\frac{v'''(t)}{v''(t)^{3/2}}$$

tends to zero as $t \to 0$. The reader can easily verify that

- If $\sqrt{v''(t)}\,\frac{d}{dt}\,\frac{1}{v''(t)} \to 0$ as $t \to 0$, then for each $\alpha > 0$, $e^{-v(t)}/t^{\alpha}$ ultimately decreases to zero as $t \to 0$, so that the corresponding weight $G(r)$ is fast.
- If $\sqrt{v''(t)}\,\frac{d}{dt}\,\frac{1}{v''(t)}$ is bounded away from zero as $t \to 0$, then for small t we have $e^{-v(t)} \geq t^{\beta}$ for some $\beta > 0$, that is, $G(r) \geq (1-r)^{\beta}$ for r near 1. In this case, $G(r)$ is not fast.

Theorem 1. *Let $v(t)$, $0 < t \leq t_0$, be of class C^3 and satisfy (i) and (ii) above. Assume in addition that*

$$\sqrt{v''(t)}\,\frac{d}{dt}\,\frac{1}{v''(t)} \to 0 \text{ as } t \to 0. \tag{1.4}$$

Then

$$P(x) \sim \sqrt{2\pi}\,\sqrt{-L''(x)}\;e^{-L(x)} \quad \text{as } x \to \infty.$$

To handle the Laplace transform in (1.3) and thus the kernel function itself, we need a result dual to Theorem 1. Let $N(x)$, $x \geq x_0$, be a function of class C^3 such that

(i)′ $N'(x) > 0$, $N''(x) < 0$, $N'''(x) > 0$ for $x \geq x_0$.
(ii)′ $N(x) \to \infty$ as $x \to \infty$.
(iii)′ $N'(x) \to 0$ as $x \to \infty$.

Note that condition (iii)′ tells us that $N(x)/x \to 0$ as $x \to \infty$, so that the Laplace transform

$$Q(t) = \int_{x_0}^{\infty} e^{N(x)-xt}dx$$

is defined for all $t > 0$. The particular value of x_0 is immaterial to the asymptotics of $Q(t)$ as $t \to 0$ since the dominant part of the integral comes from large x. We consider the inverse Legendre-Fenschel transform u of N given by

$$u(t) = \sup_{x > x_0} (N(x) - xt). \tag{1.5}$$

For small t, the function $u(t)$ is differentiable to the same degree as N, and the supremum is attained at the point $x = -u'(t)$ in (x_0, ∞). Moreover, N' and $-u'$ are inverse functions:

$$N'(-u'(t)) = t, \ t \text{ small}; \qquad -u'(N'(x)) = x, \ x \text{ large}.$$

The function N can be recovered from u as a (forward) Legendre-Fenschel transform:

$$N(x) = \inf_{0<t<t_0} (u(t) + tx),$$

for large x and suitable $t_0 > 0$. Again, see [Ko].

Theorem 2. *Let $N(x)$, $x \geq x_0$, be of class C^3 and satisfy (i)', (ii)', and (iii)' above. Assume in addition that*

$$\sqrt{-N''(x)}\,\frac{d}{dx}\,\frac{1}{N''(x)} \to 0 \text{ as } x \to \infty. \tag{1.6}$$

Then

$$Q(t) \sim \sqrt{2\pi}\,\sqrt{u''(t)}\,e^{u(t)} \text{ as } t \to 0.$$

Note: The hypothesis (1.6) says that as $x \to \infty$, $e^{N(x)} \to \infty$ faster than any power x^α, $\alpha > 0$.

Theorem 2 yields a version of a result of Hayman [H, Corollary III] about MacLaurin series having suitably regular coefficients. The proof involves replacing the series by an integral as in [K2, Lemma 2].

Corollary 1. *Let $\{a_n\}_{n=0}^\infty$ be a real sequence such that for large n, $a_n = e^{N(n)}$, where $N(x)$ is a function satisfying the hypothesis of Theorem 2. Let u be the inverse Legendre-Fenschel transform of N as in (1.5). Then*

$$\sum_{n=0}^\infty a_n r^n \sim \sqrt{2\pi}\,\sqrt{u''\left(\log\frac{1}{r}\right)}\,e^{u(\log\frac{1}{r})},$$

as $r \nearrow 1$.

We wish to combine Theorems 1 and 2 to determine the asymptotics of the Laplace transform in (1.3) in terms of the function v. Let us call the function $v(t)$, $0 < t \leq t_0$, *admissible* if v is of class C^5 and satisfies

- $v(t) \to +\infty$ as $t \to 0$,
- $v' < 0$, $v'' > 0$, $v''' < 0$ and $v^{(4)} > 0$ on $(0, t_0]$,
- $\frac{1}{v''(t)}\left\{1 + \frac{1}{2}\,\frac{d^2}{dt^2}\,\frac{1}{v''(t)}\right\}$ is positive and strictly increasing on $(0, t_0]$.

Given an admissible v as above, we say that v is *admissibly fast* provided it satisfies in addition

(a) $\sqrt{v''(t)}\,\frac{d}{dt}\,\frac{1}{v''(t)} \to 0$ as $t \to 0$,

(b) $\frac{d^2}{dt^2}\,\frac{1}{v''(t)} \to 0$ as $t \to 0$,

(c) $\frac{1}{\sqrt{v''(t)}}\,\frac{d^3}{dt^3}\,\frac{1}{v''(t)} \to 0$ as $t \to 0$.

Finally, we call a weight $G(r)$, $0 < r < 1$, admissible or admissibly fast if its restriction to some interval $[r_0, 1)$ has the form (1.2), where $v(t)$, $0 < t \leq t_0$, has that property with $r_0 = e^{-t_0}$.

Some observations about these definitions are in order. First, condition (a) is the "rate of growth" part of the hypothesis on v in Theorem 1. Moreover, each of the conditions (a), (b), (c) says more or less the same thing: $1/v''(t)$ tends to zero with t faster than t^2, the rate at which it tends to zero when G is standard. In fact, if

$$\frac{d^2}{dt^2}\frac{1}{v''(t)} \quad \text{and} \quad \frac{1}{v''(t)}\frac{\frac{d^3}{dt^3}\frac{1}{v''(t)}}{\frac{d}{dt}\frac{1}{v''(t)}}$$

have limiting values in the extended real numbers as $t \to 0$, then (a) $\Longleftrightarrow$ (b) $\Rightarrow$ (c), as one can check by two applications of L'Hospital's rule.

Theorem 3. *Let $v(t)$, $0 < t \leq t_0$, be admissibly fast. We put*

$$P(x) = \int_0^{t_0} e^{-v(t)-xt}\,dt,$$

and let $x_0 > 0$. Then

$$\int_{x_0}^{\infty} \frac{1}{P(x)}\,e^{-xt}\,dx \sim v''(t)e^{v(t)} \quad \text{as } t \to 0.$$

Theorem 3 and the relation (1.3) yield an immediate corollary.

Corollary 2. *Let G be admissibly fast with*

$$G(r) = e^{-v\left(\log\frac{1}{r}\right)}, \quad r_0 \leq r < 1.$$

Let $\{k_w : w \in \mathbb{D}\}$ be the kernel function for A_G^2. Then

$$\|k_r\|^2 \sim \frac{1}{4\pi}\,v''\left(\log\frac{1}{r}\right)\frac{1}{G(r)} \quad \text{as } r \nearrow 1.$$

Let us apply the above to complete part of the study of composition operators on A_G^2 begun by MacCluer and the author in [KM1]. We fix an analytic map $\phi : \mathbb{D} \to \mathbb{D}$ and define the composition operator C_ϕ by $C_\phi f = f \circ \phi$, where f is any function on $\mathbb{D}$. We want to know when C_ϕ acts as a bounded operator from A_G^2 to itself. A sufficient condition presented in [KM1] was shown in [K2] to be necessary for fast weights in a certain range ("quick weights"); a related result is in [K1]. As pointed out in [K2], the key to necessity is to show that the kernel function for A_G^2 satisfies

$$\|k_r\|^2 \sim \frac{\beta(r)}{G(r)} \quad \text{as } r \nearrow 1, \tag{1.7}$$

where $\beta(r)$ is monotone increasing for r near 1, a relation established in [K2] for quick weights. From Corollary 2 we can now see that (1.7) holds for any admissibly fast weight; simply put

$$\beta(r) = \frac{1}{4\pi}\,v''\left(\log\frac{1}{r}\right),$$

and use the hypothesis that $v'''(t) < 0$ for small t. The complete boundedness theorem for admissibly fast weights then follows. The proof, depending as it does on [KM1] and (1.7), is exactly as in [K2]. We write $M(r)$ for the maximum value of $|\phi(re^{i\theta})|$, $0 \le \theta < 2\pi$.

Theorem 4. *Let G be admissibly fast and suppose $\phi : \mathbb{D} \to \mathbb{D}$ is analytic. Then the following are equivalent.*

(i) *C_ϕ is a bounded operator on A^2_G;*

(ii) $\limsup_{r\to 1} \dfrac{G(r)}{G(M(r))} < \infty$;

(iii) $\sup_{w\in\mathbb{D}} \dfrac{\|k_{\phi(w)}\|}{\|k_w\|} < \infty$.

Moreover, if the above conditions hold, then the essential norm $\|C_\phi\|_e$ of C_ϕ acting on A^2_G satisfies

$$\limsup_{r\to 1} \frac{G(r)}{G(M(r))} \le \|C_\phi\|_e^2 \le c\, \frac{1+|\phi(0)|}{1-|\phi(0)|} \limsup_{r\to 1} \frac{G(r)}{G(M(r))},$$

where $c > 0$ depends only on G.

Theorem 4 allows us to extend the comparison theorem of [K2] to all admissibly fast weights; the proof is as in [K2]. We write our weight in the form $G = e^{-h}$ so that $h(r) = v\left(\log \frac{1}{r}\right)$.

Theorem 5 (Comparison of weights). *Let $G_1 = e^{-h_1}$, $G_2 = e^{-h_2}$ be admissibly fast weights and suppose that C_ϕ is a bounded composition operator on $A^2_{G_1}$. If $h_2'(r)/h_1'(r)$ remains bounded as $r \to 1$, then C_ϕ also acts boundedly on $A^2_{G_2}$.*

In our final result, "almost equivalence" of conditions (i) and (ii) was established quite generally in [KM2] and was improved to exact equivalence for quick weights in [K2]. Let us record that exact equivalence in fact holds for any admissibly fast weight; the proof, based on [KM2], is as in [K2] but now using Theorem 4 above.

Theorem 6. *Let $G = e^{-h}$ be an admissibly fast weight. Then the following are equivalent:*

(i) *Every bounded composition operator on A^2_G is either compact or unitary,*

(ii) $\limsup_{r\to 1} (1-r)^3 h'(r) = +\infty$.

Other authors have treated various versions of the questions considered here. Trent [T] determined the asymptotics of $\|k_r\|$ and the moments $\{p_n\}$ for the weights

$$G(r) = (1-r)^b \exp\left\{-\frac{c}{1-r}\right\}, \quad b \text{ real}, \ c > 0.$$

Lin and Rochberg [LR] considered not necessarily rotation-invariant weights $G(z) = e^{-h(z)}$, showing via function-theoretic and $\bar{\partial}$-methods that under suitable hypotheses on G, there exist positive constants c_1 and c_2 with

$$c_1 \leq \frac{\|k_z\|^2 G(z)}{\Delta h(z)} \leq c_2, \quad z \in \mathbb{D},$$

where Δh is the Laplacian of h. More recently Holland and Rochberg [HR1] have given analogues of Theorems 1, 2, 3 and Corollary 2 above for a large class of rotation-invariant weights on the complex plane for which a Gaussian is the prototype. Applications of these results to operator theory on the associated generalized Segal-Bargmann spaces are presented in [HR2].

On the composition operator side recent papers by Jones [J] and Böttger [B] uncover new information. Let us write $\mathcal{C}(G)$ for the collection of all analytic self-maps of $\mathbb{D}$ for which C_ϕ acts as a bounded operator on A^2_G. According to Theorem 4, if $G_1 = e^{-h_1}$ and $G_2 = e^{-h_2}$ are admissibly fast weights for which $h_2'(r)/h_1'(r)$ remains bounded as $r \to 1$, then $\mathcal{C}(G_1) \subset \mathcal{C}(G_2)$. Jones [J] has shown that if $h_2'(r)/h_1'(r) \to 0$ as $r \to 1$, then the containment $\mathcal{C}(G_1) \subset \mathcal{C}(G_2)$ is proper, a fact which has an interesting interpretation in terms of maximum modulus functions. In [B] Böttger studies composition operators on spaces A^2_G for which the weight G is *regularly slow* (that is, G is non-fast and satisfies a mild regularity condition). Böttger shows that any C_ϕ acts boundedly on such a space. On the question of whether C_ϕ is a compact operator on A^2_G, he finds a dichotomy. For certain G decaying to zero more slowly than any standard weight, compactness of C_ϕ on A^2_G is determined by a counting function in a manner analogous to Shapiro's theorem [S] on the Hardy space. However, if G decays at least as rapidly as some standard weight, the pure angular derivative criterion of MacCluer and Shapiro [MS] exactly suffices.

The papers [T], [HR1] and the present article carry out various implementations of "Laplace's method," in which one seeks to exploit the quadratic nature near its maximum of the exponent in an exponential integral. Holland and Rochberg approximate the exponent by its second Taylor polynomial and break the integral in pieces to estimate the error. Our approach here is to first change variables, producing an exact Gaussian factor in the integrand, and then to do all error estimates on the derivative of the corresponding inverse function.

The present paper can be considered a sequel to [K2], but in the author's opinion, a sequel that is best read first with a copy of [K2] close at hand. For parts of the proof of Theorem 1 we refer to [K2] where the result was established for the smaller class of quick weights. Accordingly, our notation throughout matches that of [K2] with this exception: we now, as noted above, write L for the Legendre-Fenschel transform of v, rather than M as in [K2], so as to avoid confusion with the maximum modulus function of ϕ.

The author thanks Barbara MacCluer, Loren Pitt, Richard Rochberg, Larry Thomas, and Tavan Trent for valuable comments on this work.

2. Proofs of Theorems 1, 2, and 3

Curiously, our proofs of Theorems 1 and 2 are formally almost identical, though the functions therein have quite different shapes and the asymptotics are taken at infinity and zero, respectively. A special case of Theorem 1 was presented in [K2], and we here refer to the definitive parts of that proof while superceding the rest. Accordingly, to set the pattern clearly we start with the proof of Theorem 2 which we present here in its entirety.

Proof of Theorem 2. We want to determine the asymptotics of

$$Q(t) = \int_{x_0}^{\infty} e^{N(x)-xt} dx$$

as $t \to 0$. Recall that

$$u(t) = \sup_{x > x_0} (N(x) - xt).$$

We note that $u(t)$ tends to $+\infty$ as $t \to 0$, as does $-u'(t)$; the latter holds because $-u'(N'(x)) = x$ for large x and $N'(x) \to 0$ as $x \to \infty$. In what follows we assume that t is small enough to ensure $N'(-u'(t)) = t$ so that $-N''(-u'(t)) = 1/u''(t)$.

We define

$$\Gamma_t(x) = (xt - N(x) + u(t))^{\frac{1}{2}}$$

and note that

$$Q(t) = e^{u(t)} \int_{x_0}^{\infty} e^{-\Gamma_t(x)^2} dx.$$

Putting $y = \Gamma_t(x)$ we observe that

$$\Gamma_t'(x) = \frac{t - N'(x)}{2(xt - N(x) + u(t))^{\frac{1}{2}}} = \frac{t - N'(x)}{2y}, \tag{2.1}$$

provided $\Gamma_t(x) \neq 0$. Moreover, Γ_t is decreasing on $[x_0, -u'(t)]$, increasing on $[-u'(t), \infty)$, and takes its minimum value of zero when $x = -u'(t)$, that is, when $N'(x) = t$. On applying the Cauchy mean value theorem to the quotient obtained by squaring the right side of (2.1), we see that there exists c between x and $-u'(t)$ with $\Gamma_t'(x)^2 = -N''(c)/2$, that is

$$\Gamma_t'(x) = \left\{ \begin{array}{ll} -\sqrt{-\dfrac{N''(c)}{2}} & \text{if } x_0 < x < -u'(t) \\ \sqrt{-\dfrac{N''(c)}{2}} & \text{if } x > -u'(t) \end{array} \right\}. \tag{2.2}$$

Though $\Gamma_t(x)$ is not differentiable at $x = -u'(t)$, it is clear from (2.2) that it does have left- and right-hand derivatives at $-u'(t)$ given by

$$D_-\Gamma_t(-u'(t)) = -\sqrt{-\frac{N''(-u'(t))}{2}} \text{ and } D_+\Gamma_t(-u'(t)) = \sqrt{-\frac{N''(-u'(t))}{2}},$$

respectively. Moreover, (2.1) and (2.2) combine to give

$$\Gamma_t''(x) = -\frac{N''(x)}{y} - \frac{(t-N'(x))^2}{4y^3} = \frac{N''(c)-N''(x)}{2y}, \tag{2.3}$$

when $x \neq -u'(t)$. Since N'' is increasing, Γ_t is convex on $[x_0, -u'(t)]$ and concave on $[-u'(t), \infty)$.

Let us write

$$\int_{x_0}^{\infty} e^{-\Gamma_t(x)^2}\,dx = \int_{x_0}^{-u'(t)} + \int_{-u'(t)}^{\infty} \stackrel{\text{def}}{=} a_1(t) + a_2(t),$$

and consider the asymptotics of first a_1 and then a_2. We know that Γ_t maps $[x_0, -u'(t)]$ onto $[0, \Gamma_t(x_0)]$ in a decreasing manner; we let $\eta_t : [0, \Gamma_t(x_0)] \to [x_0, -u'(t)]$ denote its inverse function and write $y = \Gamma_t(x)$. Then

$$a_1(t) = \int_{x_0}^{-u'(t)} e^{-\Gamma_t(x)^2}\,dx = -\int_0^{\Gamma_t(x_0)} e^{-y^2}\eta_t'(y)\,dy.$$

Since $-N''$ is decreasing, whenever $x < -u'(t)$ we have

$$-\Gamma_t'(x) = \sqrt{\frac{-N''(c)}{2}} \geq \sqrt{\frac{-N''(-u'(t))}{2}} = \frac{1}{\sqrt{2u''(t)}},$$

where we have again used the relation $-N''(-u'(t)) = 1/u''(t)$. Thus

$$-\eta_t'(y) = -1/\Gamma_t'(x) \leq \sqrt{2u''(t)}. \tag{2.4}$$

Next we obtain a lower bound on $-\eta_t'(y)$. For $x < c < -u'(t)$ with c as above,

$$-\Gamma_t'(x) = \sqrt{-\frac{N''(c)}{2}} \leq \sqrt{-\frac{N''(x)}{2}},$$

whence

$$-\eta_t'(y) \geq \sqrt{\frac{2}{-N''(x)}} = \sqrt{\frac{2}{-N''(\eta_t(y))}}. \tag{2.5}$$

Since Γ_t is convex on $(x_0, -u'(t))$, η_t is convex as well. Thus $\eta_t(y) \geq \eta_t(0) + \eta_t'(0)y$. Clearly we have $\eta_t(0) = -u'(t)$ and

$$\eta_t'(0) = \frac{1}{D_-\Gamma_t(-u'(t))} = -\sqrt{\frac{2}{-N''(-u'(t))}} = -\sqrt{2u''(t)},$$

so that

$$\eta_t(y) \geq -u'(t) - \sqrt{2u''(t)}y.$$

We invoke the mean value theorem, applied to $-1/N''$, together with the fact that N'' is increasing, to see that there exists z with $-u'(t) - \sqrt{2u''(t)}y < z < -u'(t)$

such that

$$\frac{N''(-u'(t))}{N''(\eta_t(y))} \geq \frac{-N''(-u'(t))}{-N''(-u'(t)-\sqrt{2u''(t)}y)}$$

$$= 1-\sqrt{2}\frac{N'''(z)}{N''(z)^2}\sqrt{-N''(-u'(t))}y$$

$$\geq 1-\sqrt{2}\frac{N'''(z)}{(-N''(z))^{\frac{3}{2}}}\, y.$$

On combining this with (2.4), (2.5) and the relation $-N''(-u'(t)) = \frac{1}{u''(t)}$, we see that

$$1 \geq \frac{-\eta_t'(y)}{\sqrt{2u''(t)}} \geq \sqrt{\left(1-\sqrt{2}\frac{N'''(z)}{(-N''(z))^{\frac{3}{2}}}\, y\right)_+}$$

where $(v)_+$ denotes $\max\{v, 0\}$.

Recall that $-u'(t) \to +\infty$ as $t \to 0$. It follows from (1.6) that $x^2 N''(x) \to -\infty$ as $x \to \infty$, which in turn implies that for fixed positive y, $-u'(t) - \sqrt{2u''(t)}y$, and thus z, tends to infinity as $t \to 0$. By our hypothesis (1.6), $N'''(z)/(-N''(z))^{3/2} \to 0$ as $t \to 0$, and so the dominated convergence theorem tells us that

$$\lim_{t\to 0}\frac{a_1(t)}{\sqrt{2u''(t)}} = -\lim_{t\to 0}\int_0^{\Gamma_t(x_0)} e^{-y^2}\frac{\eta_t'(y)}{\sqrt{2u''(t)}}\,dy = \int_0^\infty e^{-y^2}dy = \frac{\sqrt{\pi}}{2}.$$

Now we turn to $a_2(t)$. We know that Γ_t maps $[-u'(t), \infty)$ onto $[0, \infty)$ in an increasing and concave manner. We let $\nu_t : [0, \infty) \to [-u'(t), \infty)$ denote the inverse function, which is increasing and convex. Then

$$a_2(t) = \int_{-u'(t)}^\infty e^{-\Gamma_t(x)^2}dx = \int_0^\infty e^{-y^2}\nu_t'(y)dy.$$

Using the mean value theorem we write

$$\nu_t'(y) = \nu_t'(0) + \nu_t''(z)y, \tag{2.6}$$

where $0 < z < y$ and z depends on both t and y. First note that

$$\nu_t'(0) = \frac{1}{D_+\Gamma_t(-u'(t))} = \sqrt{\frac{2}{-N''(-u'(t))}} = \sqrt{2u''(t)}.$$

We put $w = -u'(t)$ so that $-N''(w) = 1/u''(t)$, and we see from (2.6) that

$$\frac{a_2(t)}{\sqrt{2u''(t)}} = \int_0^\infty e^{-y^2}\left[1+\sqrt{\frac{-N''(w)}{2}}\,\nu_t''(z)y\right]dy.$$

Inasmuch as ν_t is convex, it is only necessary to estimate ν_t'' from above.

Since Γ_t and ν_t are inverse functions,

$$\nu_t''(y) = -\frac{1}{\Gamma_t'(x)^3}\Gamma_t''(x).$$

Then, using (2.2), (2.1), and (2.3), and writing $w = -u'(t)$, or equivalently $t = N'(w)$, we find

$$\begin{aligned}\nu_t''(y) &= -\frac{1}{\Gamma_t'(x)^3}\frac{N''(c)-N''(x)}{2y}\\ &= -\frac{1}{\Gamma_t'(x)^2}\cdot\frac{N''(c)-N''(x)}{N'(w)-N'(x)}\\ &= -\frac{2}{N''(c)}\frac{N''(x)-N''(c)}{N'(w)-N'(x)}.\end{aligned}$$

Keeping in mind that $-N''$ is positive and decreasing, we have

$$\nu_t''(y) \le -\frac{2}{N''(w)}\frac{N''(x)-N''(w)}{N'(w)-N'(x)} = -\frac{2}{N''(w)}\frac{N'''(d)}{-N''(d)},$$

for some d with $w < d < x$, the last equality following by the Cauchy mean value theorem. Now, in the above inequality for $\nu_t''(y)$, replace y by z and x by r where $z = \Gamma_t(r)$ to obtain

$$\sqrt{\frac{-N''(w)}{2}}\,\nu_t''(z) \le \sqrt{\frac{2}{-N''(w)}}\,\frac{N'''(d)}{-N''(d)} \le \sqrt{2}\,\frac{N'''(d)}{(-N''(d))^{3/2}},$$

where now $w < d < r$. As $t \to 0$, we have $d > w = -u'(t) \to \infty$, whence $N'''(d)/(-N''(d))^{3/2} \to 0$ uniformly in y by our hypothesis on N. Thus

$$\lim_{t\to 0}\frac{a_2(t)}{\sqrt{2u''(t)}} = \lim_{t\to 0}\int_0^\infty e^{-y^2}\left[1+\sqrt{\frac{-N''(w)}{2}}\,\nu_t''(z)y\right]dy = \int_0^\infty e^{-y^2}dy = \frac{\sqrt{\pi}}{2}.$$

On combining our results for $a_1(t)$ and $a_2(t)$ we have

$$Q(t) = [a_1(t)+a_2(t)]e^{u(t)} \sim \sqrt{2\pi}\,\sqrt{u''(t)}\,e^{u(t)}, \qquad \text{as } t\to 0. \qquad \square$$

Proof of Theorem 1. We study the behavior of

$$P(x) = \int_0^{t_0} e^{-v(t)-xt}dt$$

as $x \to \infty$. Recall that

$$L(x) = \inf_{0<t<t_0}(v(t)+xt).$$

We observe that $L(x)$ tends to ∞ as $x\to\infty$, while $L'(x)\to 0$. The latter follows because $L'(-v'(t)) = t$ for small t. In what follows we assume that x is large enough to guarantee that $-v'(L'(x)) = x$. Then the infimum defining $L(x)$ is attained when $-v'(t) = x$, or equivalently, $t = L'(x)$. We put

$$\Phi_x(t) = (xt + v(t) - L(x))^{\frac{1}{2}},$$

so that

$$P(x) = e^{-L(x)} \int_0^{t_0} e^{-\Phi_x(t)^2} dt.$$

Let us write

$$\int_0^{t_0} e^{-\Phi_x(t)^2} dt = \int_0^{L'(x)} + \int_{L'(x)}^{t_0} \stackrel{\text{def}}{=} b_1(x) + b_2(x).$$

The asymptotic behavior of $b_1(x)$ and an effective lower bound for $b_2(x)$ were worked out in [K2]; the relevant material begins with the last paragraph on p. 80 and ends with Eqn. (6) on p. 82.

First we review $b_1(x)$. Let $\sigma_x : [0, \infty) \to (0, L'(x)]$ be the inverse function of the restriction of Φ_x to $(0, L'(x)]$, so that $\sigma_x(y) = t$ if $\Phi_x(t) = y$. Since said restriction is decreasing, a change of variables gives

$$b_1(x) = -\int_0^{\infty} e^{-y^2} \sigma_x'(y) dy.$$

From [K2, p. 80–82] we have the following upper and lower estimates on the positive function $-\sigma_x'(y)$:

$$\begin{aligned} 1 \;&\geq\; \frac{-\sigma_x'(y)}{\sqrt{-2L''(x)}} \qquad (2.7) \\ &\geq\; \sqrt{\frac{v''(L'(x))}{v''(\sigma_x(y))}} \\ &\geq\; \sqrt{\left(1 + \sqrt{2}\,\frac{\sqrt{v''(L'(x))}\,v'''(z)}{v''(z)^2}\, y\right)_+} \\ &\geq\; \sqrt{\left(1 + \sqrt{2}\,\frac{v'''(z)}{v''(z)^{\frac{3}{2}}}\, y\right)_+}, \end{aligned}$$

where z, depending on x and y, satisfies $L'(x) - \sqrt{\frac{2}{v''(L'(x))}}\, y < z < L'(x)$, and $(u)_+$ denotes $\max\{u, 0\}$; the last inequality in (2.7) follows since v'' is decreasing. The reader will note the analogy with a dual inequality in the proof of Theorem 2. On letting $x \to \infty$, we have $L'(x) \to 0$, whence $z \to 0$ and, by our hypothesis on v, $v'''(z)/v''(z)^{\frac{3}{2}} \to 0$. On integrating the inequality (2.7) over $[0, \infty)$, we find by the dominated convergence theorem that

$$\lim_{x\to\infty} \frac{b_1(x)}{\sqrt{-2L''(x)}} = -\lim_{x\to\infty} \int_0^{\infty} e^{-y^2} \frac{\sigma_x'(y)}{\sqrt{-2L''(x)}} dy = \int_0^{\infty} e^{-y^2} dy = \frac{\sqrt{\pi}}{2}.$$

We next consider the asymptotics of $b_2(x)$. For large enough x, $L'(x) < t_0$ and the function Φ_x maps $[L'(x), t_0]$ in an increasing manner onto $[0, \Phi_x(t_0)]$. Let

$\tau_x : [0, \Phi_x(t_0)] \to [L'(x), t_0]$ denote its inverse function, so that

$$b_2(x) = \int_0^{\Phi_x(t_0)} e^{-y^2} \tau_x'(y) dy.$$

For each t with $L'(x) < t < t_0$ there exists c between $L'(x)$ and t with

$$\Phi_x(t) = \sqrt{\frac{v''(c)}{2}}, \tag{2.8}$$

see the last paragraph on p. 80 and the middle of p. 82 in [K2]. We put $s = L'(x)$ so that $x = -v'(s)$ (this inverse function relation holds for x large enough, which we assume throughout). Differentiation gives $v''(s) = -1/L''(x)$. If $0 < y < \Phi_x(t_0)$, there exists z between 0 and y (depending on x and y) so that $\tau_x'(y) = \tau_x'(0) + \tau_x''(z)y$. On letting t approach $s = L'(x)$ from the right in (2.8) and using the chain rule we find

$$\tau_x'(0) = \frac{1}{\Phi_x'(s)} = \sqrt{\frac{2}{v''(s)}}.$$

Thus

$$\frac{b_2(x)}{\sqrt{-2L''(x)}} = \sqrt{\frac{v''(s)}{2}}\, b_2(x) = \int_0^{\Phi_x(t_0)} e^{-y^2} \left[1 + \sqrt{\frac{v''(s)}{2}} \tau_x''(z) y \right] dy.$$

Now Φ_x is increasing and concave on $[L'(x), t_0]$, so τ_x is convex and $\tau_x''(z) \geq 0$. We will show that there exists $M > 0$ such that for large x and all y,

$$\sqrt{v''(s)} \tau_x''(z) \leq M, \tag{2.9}$$

and

$$\sqrt{v''(s)} \tau_x''(z) \to 0 \text{ as } x \to \infty, \text{ pointwise in } y. \tag{2.10}$$

Since $\Phi_x(t_0) \to \infty$ as $x \to \infty$, (2.9) and (2.10) plus the dominated convergence theorem will imply that

$$\lim_{x \to \infty} \frac{b_2(x)}{\sqrt{-2L''(x)}} = \int_0^\infty e^{-y^2} dy = \frac{\sqrt{\pi}}{2},$$

and we will be done.

Since τ_x and Φ_x are inverse functions, we have $\tau_x''(y) = -\Phi_x''(t)/\Phi_x(t)^3$, with $y = \Phi_x(t)$. Further, from [K2, p. 80–81],

$$\Phi_x'(t) = \frac{v'(t) + x}{2y} = \frac{v'(t) - v'(s)}{2y} \quad \text{and} \quad \Phi_x''(t) = \frac{v''(t) - v''(c)}{2y}$$

where c is as in (2.8). It follows from these equations that

$$\begin{aligned}\tau_x''(y) &= \frac{1}{\Phi_x'(t)^2}\,\frac{v''(c)-v''(t)}{v'(t)-v'(s)} \\ &= \frac{2}{v''(c)}\,\frac{v''(c)-v''(t)}{v'(t)-v'(s)} \\ &\leq \frac{2}{v''(s)}\,\frac{v''(s)-v''(t)}{v'(t)-v'(s)} \\ &= \frac{2}{v''(s)}\left(-\frac{v'''(d)}{v''(d)}\right),\end{aligned}$$

where we have used the fact that v'' is decreasing plus the Cauchy mean value theorem to produce the indicated d with $s < d < t$. Now use the above estimate on τ_x'' with y replaced by z and t replaced by the unique r with $s < r < t$ and $\Phi_x(r) = z$. Our estimate now implies that

$$\sqrt{v''(s)}\;\tau_x''(z) \leq -2\,\frac{v'''(d)}{v''(d)^{\frac{3}{2}}}, \tag{2.11}$$

where now $s < d < r$ and we have again used the fact that v'' is decreasing. Moreover, $v'''/(v'')^{3/2}$ tends to zero at the origin and hence is bounded on $(0, t_0]$. Thus (2.9) follows from (2.11). For (2.10), it suffices by (2.11) to show that if we fix y and let $x \to \infty$, then $d \to 0$. To this end, note that

$$y^2 = \Phi_x(t)^2 = xt + v(t) - L(x) > xt - L(x),$$

so for fixed y,

$$d < r < t < \frac{y^2}{x} + \frac{L(x)}{x}.$$

However, L' is the inverse function of $-v'$, so that $L'(x)$, and thus $L(x)/x$, tends to zero as $x \to \infty$. □

Proof of Theorem 3. We assume that v satisfies all hypotheses of our theorem. Theorem 1 tells us that

$$P(x) \sim \sqrt{2\pi}\,\sqrt{-L''(x)}e^{-L(x)} \text{ as } x \to 0. \tag{2.12}$$

We define $N(x)$ by setting $e^{-N(x)}$ equal to the right side of (2.12), that is

$$N(x) = -\log\sqrt{2\pi} - \frac{1}{2}\log(-L''(x)) + L(x).$$

We want to apply Theorem 2 to N, and so check that N satisfies its hypotheses. We calculate

$$N'(x) = -\frac{L'''(x)}{2L''(x)} + L'(x). \tag{2.13}$$

We know that $L'(-v'(t)) = t$, $L''(-v'(t)) = -1/v''(t)$ and thus $L'''(-v'(t)) = -\frac{v'''(t)}{v''(t)^3}$ is positive. Therefore $L'''(x) > 0$ and $N'(x) > L'(x)$ for large x. On plugging $-v'(t)$ into (2.13), we find

$$N'(-v'(t)) = t - \frac{v'''(t)}{2v''(t)^2} = t + \frac{1}{2}\frac{d}{dt}\frac{1}{v''(t)}, \tag{2.14}$$

which implies that $N'(x) \to 0$ as $x \to \infty$, that is, as $t \to 0$. Note that since $N'(-u'(t)) = t$, (2.14) expresses in some sense how u is a perturbation of v. Taking another derivative yields

$$N''(-v'(t)) = -\frac{1}{v''(t)}\left\{1 + \frac{1}{2}\frac{d^2}{dt^2}\frac{1}{v''(t)}\right\}.$$

Since $-v'(t)$ increases to ∞ as $t \to 0$, the admissibility hypothesis on v tells us that $N''(x) < 0$ and $N'''(x) > 0$ for large x. Another differentiation and division by $(-N''(-v'(t)))^{3/2}$ gives

$$\begin{aligned}\frac{N'''(-v'(t))}{(-N''(-v'(t)))^{3/2}} &= -\left(1 + \frac{1}{2}\frac{d^2}{dt^2}\frac{1}{v''(t)}\right)^{-\frac{1}{2}} \frac{v'''(t)}{v''(t)^{3/2}} \\ &\quad + \frac{1}{2}\left(1 + \frac{1}{2}\frac{d^2}{dt^2}\frac{1}{v''(t)}\right)^{-\frac{3}{2}} \frac{1}{\sqrt{v''(t)}}\frac{d^3}{dt^3}\frac{1}{v''(t)}.\end{aligned}$$

The hypotheses (a)–(c) on v tell us that this expression tends to zero with t, and so N satisfies all hypotheses of Theorem 2.

Now consider the inverse transform u of N given by (1.5). We have the following asymptotic equivalences:

$$\int_{x_0}^{\infty} \frac{1}{P(x)} e^{-xt} dx \sim \int_{x_0}^{\infty} e^{N(x)-xt} dx \sim \sqrt{2\pi}\,\sqrt{u''(t)} e^{u(t)} \tag{2.15}$$

as $t \to 0$. The first equivalence follows from (2.12) and the fact that x_0 can be taken as large as desired without affecting the asymptotics of the integrals; Theorem 2 yields the second.

Our remaining task is to relate $u(t)$ to $v(t)$. We know that for small t the supremum defining $u(t)$ as in (1.5) is attained at $x = -u'(t)$, so

$$\begin{aligned} u(t) &= N(-u'(t)) + tu'(t) \\ &\geq N(-v'(t)) + tv'(t) \\ &= -\log\sqrt{2\pi} + \frac{1}{2}\log v''(t) + L(-v'(t)) + tv'(t). \end{aligned}$$

Here we have used the relation $L''(-v'(t)) = -\frac{1}{v''(t)}$. Now $v(t) = L(-v'(t)) + tv'(t)$ and we have the lower bound

$$e^{u(t)} \geq \frac{1}{\sqrt{2\pi}} \sqrt{v''(t)}\, e^{v(t)}. \tag{2.16}$$

For an upper bound, start with the inequality $N'(x) > L'(x)$, x large, so that the respective inverse functions satisfy $-u'(t) > -v'(t)$ for small t. Since N' is decreasing, $L(-v'(t)) = v(t) - tv'(t)$ and (2.14) holds, we have

$$\begin{aligned} u(t) &= N(-u'(t)) + tu'(t) \qquad (2.17)\\ &\leq N(-v'(t)) + N'(-v'(t))(-u'(t) + v'(t)) + tu'(t)\\ &= -\log\sqrt{2\pi} + \frac{1}{2}\log v''(t) + L(-v'(t))\\ &\quad + \left(t - \frac{v'''(t)}{2v''(t)^2}\right)(-u'(t) + v'(t)) + tu'(t)\\ &= -\log\sqrt{2\pi} + \frac{1}{2}\log v''(t) + v(t) - \frac{v'''(t)}{2v''(t)^2}(-u'(t) + v'(t)). \end{aligned}$$

Now

$$t = N'(-u'(t)) = N'(-v'(t)) + N''(b)(-u'(t) + v'(t)), \tag{2.18}$$

for some b with $-v'(t) < b < -u'(t)$. Since $N''(b) \leq N''(-u'(t)) = -\frac{1}{u''(t)}$, we see from (2.14) and (2.18) that

$$-u'(t) + v'(t) \leq -\frac{v'''(t)}{2v''(t)^2}\, u''(t).$$

Substituting into (2.17) gives

$$e^{u(t)} \leq \frac{1}{\sqrt{2\pi}} \sqrt{v''(t)}\, e^{v(t) + \frac{v'''(t)^2}{4v''(t)^4} u''(t)}. \tag{2.19}$$

To finish the proof, we need only establish the following lemma.

Lemma. *$u''(t) \sim v''(t)$ as $t \to 0$.*

Let us check that the lemma does the job. First, the second summand in the exponent on the right in (2.19) must satisfy

$$\frac{v'''(t)^2}{4v''(t)^4}\, u''(t) \sim \frac{v'''(t)^2}{4v''(t)^3} \to 0$$

as $t \to 0$, by our hypothesis on v. Then (2.16) and (2.19) yield

$$e^{u(t)} \sim \frac{1}{\sqrt{2\pi}} \sqrt{v''(t)}\, e^{v(t)}$$

as $t \to 0$. The conclusion of Theorem 3 is then immediate from (2.15), by a second use of the lemma.

Proof of the Lemma. Since $-u'$ is the inverse function of N', applying it to both sides of the equation (2.14) gives $v'(t) = u'(t+w)$, where

$$w = -\frac{v'''(t)}{2v''(t)^2} = \frac{1}{2}\frac{d}{dt}\frac{1}{v''(t)} > 0.$$

Therefore,

$$v''(t) = u''(t+w)\left(1 + \frac{dw}{dt}\right),$$

while the mean value theorem gives $v''(t) = v''(t+w) - v'''(d)w$, where $t < d < t+w$. It follows that

$$1 - \frac{v'''(d)w}{v''(t+w)} = \frac{u''(t+w)}{v''(t+w)}\left(1 + \frac{dw}{dt}\right).$$

Since w and dw/dt tend to zero continuously with t, the conclusion will follow if we can show that

$$\frac{v'''(d)w}{v''(t+w)} \to 0 \text{ as } t \to 0. \tag{2.20}$$

But $v^{(4)} > 0$ and $t < d$, so on letting $t \to 0$ we find

$$\left|\frac{v'''(d)w}{v''(t)}\right| = \frac{(-v'''(d))(-v'''(t))}{2v''(t)^3} \leq \frac{v'''(t)^2}{2v''(t)^3} \to 0.$$

Moreover,

$$\frac{v''(t+w)}{v''(t)} = 1 + \frac{v'''(d)w}{v''(t)},$$

so $v''(t+w) \sim v''(t)$ as $t \to 0$. Thus (2.20) holds, as desired. □

References

[B] Ulrich Böttger, Composition operators on small weighted Bergman spaces, preprint, 1999.

[H] W.K. Hayman, A generalization of Stirling's formula, *Jour. Reine Angew. Math.* 196 (1956), 67–95.

[HR1] Finbarr Holland and Richard Rochberg, Bergman kernel asymptotics for generalized Fock spaces, *Jour. d'Analyse Math.* 83 (2001), 207–242.

[HR2] ——————, Bergman kernels and Hankel forms on generalized Fock spaces, Proc. Third Conference on Function Spaces, Edwardsville, IL 1998, K. Jarosz, ed., *Contemporary Mathematics*, Vol. 232, American Math. Soc., Providence, 1999, 189–200.

[J] Matthew M. Jones, Bounded composition operators on weighted Bergman spaces, *Jour. Math. Anal. Appl.* 256 (2001), 650–667.

[Ko] Paul Koosis, *The Logarithmic Integral*, Vol. I, Cambridge University Press, Cambridge, 1988.

[K1] Thomas L. Kriete, Kernel function estimates and norms of composition operators, 1994. (The results in this preprint are presented in Chapter 5 of *Composition Operators on Spaces of Analytic Functions*, by C. Cowen and B. MacCluer, CRC Press, Boca Raton, 1995.)

[K2] ________, Kernel functions and composition operators in weighted Bergman spaces, *Studies on Composition Operators*, F. Jafari, *et al.*, eds., Proc. Rocky Mtn. Math. Consortium, U. of Wyoming, 1996, Contemporary Mathematics, Vol. 213, Amer. Math. Soc., Providence, 1998, 73–91.

[KM1] T.L. Kriete and B.D. MacCluer, Composition operators on large weighted Bergman spaces, *Indiana Univ. Math. J.* 41 (1992), 755–788.

[KM2] ________, A rigidity theorem for composition operators on certain Bergman spaces, *Mich. Math. J.* 45 (1995), 379–386.

[LR] Peng Lin and Richard Rochberg, Trace ideal criteria for Toeplitz and Hankel operators on the weighted Bergman spaces with exponential type weights, *Pac. Jour. Math.* 173 (1996), 127–146.

[S] Joel Shapiro, The essential norm of a composition operator, *Annals of Math.* 125 (1987), 375–404.

[T] Tavan T. Trent, A measure inequality, preprint, 1986.

Thomas L. Kriete
Department of Mathematics
Kerchof Hall
University of Virginia
P. O. Box 400137
Charlottesville, VA 22904-4137, USA
e-mail: tlk8q@virginia.edu

Operator Theory:
Advances and Applications, Vol. 143, 273–302

On the Structure of Self-similar Systems: A Hilbert Space Approach

Mamadou Mboup

Abstract. This paper investigates the structural properties of linear self-similar systems, using an invariant subspace approach. The self-similar property is interpreted in terms of invariance of the corresponding transfer function space to a given transformation in a Hilbert space, in a same way as the time invariance property for linear systems is related to the shift-invariance of the Hardy spaces. The transformation in question is exactly that defining the de Branges homogeneous spaces. We show that any de Branges homogeneous space of order $\nu \geq -\frac{1}{2}$ belongs to the Paley-Wiener space so that each element of such a space may be viewed as the transfer function of some linear self-similar system of parameter ν. The explicit form of the corresponding impulse response, which is shown to be described by a hyperbolic partial differential equation, is given. Finally, we emphasize on the infinite dimension nature of self-similar systems through an abstract state space description.

1. Introduction

Signals presenting the properties of long memory and self-similarity appear in various fields of physics and engineering science such as hydrology, electronic, music, road traffic and others. The long memory (or long dependence), is related to a slow decrease (of hyperbolic type) of the correlation of the signals whereas the self-similarity is related to a form of invariance in scale.

In particular, traffic measurements taken over several years and under various conditions, on a high quality LAN Ethernet network have highlighted the presence of self-similarity and long memory in telecommunications networks [19]. These characteristics may have a significant effect on the queuing delays. Unfortunately, the analysis and synthesis of such behaviors are still incomplete and this, undoubtedly, explains why these signals are not taken into account in, for example, current high-speed networks congestion control schemes.

2000 *Mathematics Subject Classification.* 46E22, 47A15, 93A30, 30D10, 26A33.
Key words and phrases. Self-similar systems, reproducing kernel Hilbert space, invariant subspace, entire function, homogeneous space, infinite dimension.

Although the two properties of long memory and self-similarity are coupled for most of the time, this paper stresses the second property and more precisely on the self-similar nature of certain class of continuous-time linear systems.

From an input-output point of view, a given system $\mathcal{S}$ will be termed self-similar with parameter ν, if for a causal input $u(t)$, the relation

$$\mathcal{S}\{u(t)\} = \int_0^{\infty} K(t,\lambda)u(\lambda)d\lambda = y(t), \Longrightarrow \mathcal{S}\{u(\alpha t)\} = \alpha^{-\nu} y(\alpha t) \tag{1}$$

holds for any positive real number α. One common approach to previously presented causal self-similar models relies on the choice of a convolutive kernel of the form $K(t,\lambda) = D^{\nu}(t-\lambda)$, where

$$D^{\nu}(t) = \frac{1}{\Gamma(\nu)} t_+^{\nu-1} \tag{2}$$

The corresponding output then appears as the Riemann-Liouville [32] fractional integration of order $\nu > 0$ of the input. An important feature of these models is that they can be realized in an appropriate state space from which reliable finite-dimensional approximation schemes have been derived by [13], taking advantage of the diffusive approach of fractional integration developed by Audounnet and Montseny [3].

In the stochastic setting, the most popular model is undoubtedly the fractional Brownian motion of Mandelbrot and Van Ness [21]. A number of models derived from the fractional Brownian motion have been proposed by several authors (see, *e.g.*, [6], [12], [16], [20], [27], [8]). For example, Barnes and Allen's model, which turns out to be a causal and simplified version (without stationary increments) of the fractional Brownian motion, yields a process X_t in terms of fractional integration

$$X_t = \frac{1}{\Gamma(\nu)} \int_0^t (t-\lambda)^{\nu-1} dB(\lambda), \tag{3}$$

where $B(\lambda)$ is a standard Brownian motion. We may mention other forms for representing the fractional Brownian motion as, for example, that based on wavelet decomposition (see, *e.g.*, [2], [23] and the references therein).

Generally speaking, the bulk of the models for describing the self-similarity and long memory behaviors are approached directly in the time domain, in terms of fractional integration. By way of an exception, we may cite the models derived from the generalized Euler-Cauchy system and that proposed by Yazıcı and Kashyap [33], on the basis of a new concept of stationarity.

The intent of this paper is not the synthesis, in the engineering point of view, of the self-similarity and long memory behaviors. Rather, this paper investigates the frequency domain structural properties of linear self-similar systems, using an invariant subspace approach. The self-similar property is interpreted in terms of invariance of the corresponding transfer function space to a certain transformation in a Hilbert space, in a same way as the time-invariance property for linear systems is related to the shift-invariance of the Hardy spaces. The transformation in question is exactly that defining the de Branges homogeneous spaces [22]. We show that any de Branges homogeneous space of order $\nu \geq -\frac{1}{2}$ belongs to the Paley-Wiener space so that each element of such a space may be viewed as the transfer function of some linear self-similar system of parameter ν. The explicit form of the corresponding impulse response, which is shown to be described by a hyperbolic partial differential equation, is given. Finally, we emphasize on the infinite dimension nature of self-similar systems through an abstract state space description.

The starting point of our study is the self-similar model presented in [33]. So, we begin in Section 2 by a review of the basic definitions pertaining to self-similarity, in the light of [33]. In Section 3, we introduce a (fractional) Hardy space which seems to be a natural transfer function space corresponding to the model proposed by [33]. From this Hardy space, we present, in Section 4, a numerical example of a linear self-similar model related to the model of [33]. An important feature of the above-mentioned Hardy space is its invariance to a transformation of the form $F(s) \to \alpha^{1+\nu}F(\alpha s)$, $\alpha > 0$. We use this invariance to define the self-similar behavior, in the frequency domain. This approach leads us to the de Branges theory of Hilbert spaces of entire functions. In Section 5, we start with a brief exposition of some elements of this theory (which to our knowledge, is not well known in signal and system theory) that allow us to derive a class of linear self-similar models. Finally concluding remarks are given in Section 6.

2. Basic definitions

Take a linear time-invariant (LTI) system f, with finite energy (*i.e.*, $f \in L_2(\mathbb{R}, dt)$). Denote by g the image of f under the logarithmic distortion of the time axis: $g(t) = f(\ln t)$, $t > 0$. With this distortion, the time-invariance property of f,

$$\int_{-\infty}^{\infty} |f(t)|^2 dt = \int_{-\infty}^{\infty} |f(t+\tau)|^2 dt < \infty, \quad \tau \in \mathbb{R}, \tag{4}$$

reverberates into g as:

$$\int_{0}^{\infty} |g(t)|^2 d\ln t = \int_{0}^{\infty} |g(\lambda t)|^2 d\ln t < \infty, \tag{5}$$

where $\ln\lambda = \tau \in \mathbb{R}$. From this relation, we have

Definition 2.1 (scale invariance). A continuous time-linear system g is said to be scale-invariant (LSI) and finite energy if (5) holds for any $\lambda > 0$.

Let now f be a LSI system with finite energy and set $g(t) = t^{-H}f(t)$ for $-\infty < H < \infty$. Again, reverberating the scale-invariance property of f into g yields:

$$\int_{0}^{\infty} t^{2H}|g(t)|^2 d\ln t = \lambda^{2H}\int_{0}^{\infty} t^{2H}|g(\lambda t)|^2 d\ln t < \infty \tag{6}$$

which shows that the energy of g is proportional to the scale factor, under a scale transformation. This introduces

Definition 2.2 (Self-similar). A continuous time-linear system g is said to be self-similar (LSS) and finite energy if (6) holds for any $\lambda > 0$.

From an input-output point of view, the kernel $K(t,\lambda)$ associated with such a system as in (1) must satisfy the invariance relation

$$K(t,\lambda) = \alpha^{1-H} K(\alpha t, \alpha\lambda); \quad \forall\, \alpha, \lambda, t \geq 0. \tag{7}$$

These definitions are given in a nice exposition of self-similar system theory by Yazıcı and Kashyap [33]. They also introduced the function

$$h_H(t) = t^{-H} K(t,1), \tag{8}$$

obtained from (7) by setting $\alpha = 1/\lambda$. The output of the LSS system $\mathcal{S}$ to an arbitrary self-similar input $u(t)$ thus reads as

$$\begin{aligned} y(t) &= t^H \int_0^\infty h_H\left(\frac{t}{\lambda}\right) u(\lambda) d\ln\lambda \\ &= t^H \int_0^\infty h_H(\lambda) u\left(\frac{t}{\lambda}\right) d\ln\lambda \quad t > 0 \\ &\triangleq t^H \tilde{y}(t). \end{aligned} \tag{9}$$

If we denote by $\tilde{\delta}(\cdot)$ the image of the unit impulse, $\delta(\cdot)$, under the logarithmic distortion, namely as:

$$\tilde{\delta}(\frac{t}{\lambda}) = \begin{cases} 1 & t = \lambda \\ 0 & t \neq \lambda \end{cases}$$

then the impulse response in the logarithmic domain reads as:

$$\mathcal{S}\{\tilde{\delta}(t)\} = t^H h_H(t). \tag{10}$$

Following [33], the function $h_H(t)$ in (8) is thus referred to as the *pseudo-impulse response* of the system.

The next section will investigate the connection of this model for self-similar systems with the class of autoregressive fractional integrated moving average (ARFIMA) models, using a frequency domain approach. Indeed, we shall show that the LSS model presented by Yazıcı and Kashyap [33], although stemming from a different approach, based on a new concept of stationarity, turns out to belong to the continuous time version of the well-known ARFIMA class. We shall also show that this model may be interpreted in terms of fractional differencing directly in the frequency domain.

3. Fractional Hardy space

To begin, assume that the output $y(t)$ in (9) admits a Laplace transform, say $\hat{y}(s)$, which reads as:

$$\hat{y}(s) = \int_0^\infty t^H \tilde{y}(t) e^{-st} dt \tag{11a}$$

$$= \int_0^\infty \lambda^H u(\lambda) \int_0^\infty t^H h_H(t) e^{-\lambda st} dt d\lambda = \int_0^\infty \lambda^H u(\lambda) H_H(\lambda s) d\lambda \tag{11b}$$

$$= \int_0^\infty \lambda^H h_H(\lambda) \int_0^\infty t^H u(t) e^{-\lambda st} dt d\lambda = \int_0^\infty \lambda^H h_H(\lambda) U_H(\lambda s) d\lambda \tag{11c}$$

where $H_H(s)$ is the integral transform of $h_H(t)$ defined by

$$H_H(s) = \int_0^\infty t^H h_H(t) e^{-st} dt$$

and likewise for $U_H(s)$. Now, it appears that this integral transform defines a Hardy space [22] [31, Chap. V]. The mathematical framework of Hardy spaces, which underlies the bulk of frequency domain linear system theory, will then provide a clear interpretation of the input-output relations in eqs. (11b)–(11c).

We proceed to introduce a mathematical framework in which LSS systems act very naturally.

Definition 3.1. Let $\nu > -1$ be a real number. We define by $\mathcal{H}_2^\nu$ the set of functions $F(s)$, analytic in the right half-plane and of the form

$$F(s) = \int_0^\infty t^{\nu/2} e^{-st} f(t) dt, \tag{12}$$

where $f(t) \in L_2(0, \infty)$.

$\mathcal{H}_2^\nu$ is a Hardy space that we shall denote as *fractional* ν-Hardy space and we shall call the function $F(s)$, the ν-transform of $f(t)$. For $\nu = 0$, we recover the classical Hardy space $\mathcal{H}_2$ of the right half-plane and the ν-transform reduces to the Laplace transform.

3.1. Frequency domain fractional differ-integral model

According to the above definition, the functions $\hat{y}(s)$, $H_H(s)$ and $U_H(s)$ appearing in (11) belong to the fractional ν-Hardy space $\mathcal{H}_2^\nu$, for $2_H = \nu > -1$, provided that $\tilde{y}(t)$, $h_H(t)$ and $u(t)$ are in $L_2(0, \infty)$, respectively. Therefore, these conditions are sufficient for the output $y(t)$ to be Laplace transformable.

The integral in (12) defines an operator which acts on Laplace transforms. Now, according to Nishimoto [26], this operator represents the fractional differ-integral (differentiation or integration is considered for $\nu > 0$ or $\nu < 0$ respectively) of order ν of the Laplace transform $\int_0^\infty e^{-st} f(t) dt$. An interpretation of the LSS model is thus obtained in terms of fractional differ-integration, in the frequency domain.

When the input is set to the (pseudo) unit impulse, $u(t) = \tilde{\delta}(t)$, the response of the system reads, in the frequency domain, as:

$$\hat{y}(s) = \int_0^\infty t^{\nu/2} e^{-st} \int_0^\infty \tilde{\delta}(\frac{t}{\lambda}) h_H(\lambda) d \ln \lambda dt = H_H(s), \tag{13}$$

showing that the function $H_H(s)$, which is also the Laplace transform of (10), may be interpreted in terms of "transfer function" of the system although this notion does not have its classical meaning here. Instead, the frequency domain output of the system is obtained here by a (continuous) linear combination[1] of scaled and dilated versions of $H_H(s)$, where the coefficients of the linear combination are given directly by the time domain input signal (see eq. (11b)). Equivalently, this frequency domain output appears, from eq. (11c), as a (continuous) linear combination of scaled and dilated versions of the fractional derivative (or integral) of the Laplace transform of the input signal, where the coefficients are given by the values of the pseudo impulse response.

This fractional differ-integral interpretation will be made more explicit, from the structure and properties of the fractional ν-Hardy space $\mathcal{H}_2^\nu$.

3.2. ARFIMA models

This paragraph investigates the properties of $\mathcal{H}_2^\nu$ which will be shown to represent a mathematical framework, as natural for the class of ARFIMA systems, as $\mathcal{H}_2$ is for the class of ARMA systems. In particular, we shall establish that each element of $\mathcal{H}_2^\nu$ is the transfer function of a continuous-time ARFIMA system. This conclusion derives directly from the structure and properties of the fractional Hardy space, that we study below.

Property 3.2. *The set $\mathcal{H}_2^\nu$ given in Definition* 3.1 *is a reproducing kernel Hilbert space,*

1. *with reproducing kernel*

$$\mathcal{K}(\xi, s) = \Gamma(1+\nu)(\bar{\xi} + s)^{-1-\nu} \tag{14}$$

2. *and norm*

$$\|F(s)\|_\nu^2 = \int_0^\infty |f(t)|^2 dt \tag{15}$$

Proof. This property follows from a direct application of

Theorem 3.3 (see [31]). *Let Ω be an arbitrary set of complex numbers and $h(p,t)$ a function defined on $\Omega \times T$ such that $h(p,t) \in L_2(T)$ for all $p \in \Omega$. The set $\mathcal{H}$, of functions $F(p)$ defined on Ω by an integral operator of the form*

$$F(p) = \int_T h(p,t) f(t) dt,$$

[1]Note that the integrals in eqs. (11b)–(11c) converge as they define the Laplace transform of a function of $L_2(0,\infty)$. To be more precise, they converge as $\lim_{a\to 0 b\to\infty} \int_a^b$.

where $f(t) \in L_2(T)$, is a reproducing kernel Hilbert space which admits the reproducing kernel

$$R(q,p) = \int_T h(p,t)\overline{h(q,t)}dt.$$

Moreover, the integral transform $f(t) \to F(p)$ is isometric between $L_2(T)$ and $\mathcal{H}$ if and only if the family $\{h(p,t), p \in \Omega\}$ is complete in $L_2(T)$.

Since $t^{\nu/2}e^{-st}$, which represents the function $h(s,t)$ above, is an element of $L_2(0,\infty)$ for $\Re s > 0$, we obtain Part 1 of the above property by setting

$$\mathcal{K}(\xi, s) = \int_0^\infty t^\nu e^{-(s+\bar{\xi})t}dt = \Gamma(1+\nu)(s+\bar{\xi})^{-1-\nu},$$

where $\Gamma(\cdot)$ denotes Euler's function.

On the other hand, observe that an element $f(t)$ of $L_2(0,\infty)$ is orthogonal to the family $\{t^{\nu/2}e^{-st}; s = \sigma + i\lambda \text{ with } \sigma > 0\}$ if and only if the Fourier transform of $t^{\nu/2}e^{-\sigma t}f(t)$, vanishes for all $\sigma > 0$. Now, this holds if and only if $f(t)$ vanishes identically. Part 2 then follows and this completes the proof. □

Property 3.4. *Let $\nu > -1$ be a real number and set $\nu' = \frac{\nu-1}{2}$. If $F(s)$ is in $\mathcal{H}_2^\nu$ then, for any $\alpha > 0$, $\alpha^{\nu'+1}F(\alpha s)$ also belongs to $\mathcal{H}_2^\nu$ and this expression has the same norm as $F(s)$.*

By this property, it appears that the mathematical framework of fractional Hardy spaces allows the concept of self-similarity to be expressed, in the frequency domain, in terms similar than that in the time domain.

Proof. This follows directly by noting that if $F(s)$ is the image of $f(t) \in L_2(0,\infty)$ under (12) then $\alpha^{\nu'+1}F(\alpha s)$ is precisely the image of $\frac{1}{\sqrt{\alpha}}f(t/\alpha)$. Now, $\frac{1}{\sqrt{\alpha}}f(t/\alpha)$ is also in $L_2(0,\infty)$ and has the same norm as $f(t)$. The transformation $F(s) \to \alpha^{\nu'+1}F(\alpha s)$, with $\alpha > 0$, then maps $\mathcal{H}_2^\nu$ isometrically onto itself. □

Property 3.5. *The family of functions $\{\Phi_n(s)\}_{n=0}^\infty$ defined by*

$$\Phi_n(s) = \sqrt{\frac{\Gamma(1+n+\nu)}{n!}}\left(\frac{1-s}{1+s}\right)^n\left(\frac{\sqrt{2}}{1+s}\right)^{1+\nu}, \quad n = 0,1,2,\ldots, \tag{16}$$

forms a complete orthonormal set in $\mathcal{H}_2^\nu$.

Proof. Since $\mathcal{H}_2^\nu$ is separable, it suffices to verify that the function $\sum_n \Phi_n(s)\overline{\Phi_n(\xi)}$ coincides with the reproducing kernel (14). To this end, set $u = (1-s)/(1+s)$

and $v = (1-\xi)/(1+\xi)$. Then we have

$$\begin{aligned}
\sum_{n=0}^{\infty} \Phi_n(s)\overline{\Phi_n(\xi)} &= \left[\frac{(1+s)(1+\bar{\xi})}{2}\right]^{-1-\nu} \sum_{n=0}^{\infty} \frac{\Gamma(1+n+\nu)}{n!} \left[\frac{(1-s)(1-\bar{\xi})}{(1+s)(1+\bar{\xi})}\right]^n \\
&= \left[\frac{(1+s)(1+\bar{\xi})}{2}\right]^{-1-\nu} \Gamma(1+\nu) \sum_{n=0}^{\infty} \frac{(1+\nu)\cdots(n+\nu)}{n!} (u\bar{v})^n \\
&= \left[\frac{(1+s)(1+\bar{\xi})}{2}\right]^{-1-\nu} \Gamma(1+\nu)(1-u\bar{v})^{-1-\nu} \\
&= \left[\frac{(1+s)(1+\bar{\xi})}{2}\right]^{-1-\nu} \Gamma(1+\nu) \left[\frac{2(\bar{\xi}+s)}{(1+s)(1+\bar{\xi})}\right]^{-1-\nu} \\
&= \Gamma(1+\nu)(\bar{\xi}+s)^{-1-\nu} = \mathcal{K}(\xi, s),
\end{aligned}$$

and this completes the proof. □

Recall that the functions $\{\varphi_n(s) \triangleq \pi^{-1/2}(1-s)^n/(1+s)^{n+1}\}_{n=0}^{\infty}$ form a complete orthonormal set in $\mathcal{H}_2$ of the right half-plane (see, *e.g.*, [28]). Therefore, a direct consequence of this property is that every element $F(s)$ of $\mathcal{H}_2^{\nu}$ admits a factorization of the form

$$F(s) = (1+s)^{-\nu} R_{\nu}(s), \tag{17}$$

for some $R_{\nu}(s)$ expressible in the form

$$R_{\nu}(s) = \sum_{n=0}^{\infty} a_n \gamma_{\nu}(n) \varphi_n(s), \tag{18}$$

with

$$\gamma_{\nu}(n) \triangleq \sqrt{\frac{2^{1+\nu}\pi\Gamma(1+n+\nu)}{n!}} \text{ and } \sum_n |a_n|^2 = \|F(s)\|_{\nu}^2 < \infty.$$

For negative values of ν, *i.e.*, $-1 < \nu < 0$, $\{\gamma_{\nu}(n)^2\}$ is a strictly decreasing sequence satisfying $\frac{1}{\pi} 2^{-(1+\nu)} \gamma_{\nu}(n)^2 < 1$, for all $n > 1$. In this setting, one may readily check that the factor $R_{\nu}(s)$ belongs to $\mathcal{H}_2$. Therefore, we deduce that each element of $\mathcal{H}_2^{\nu}$ corresponds to the transfer function of a continuous-time ARFIMA system.

Now, $\hat{y}(s)$ from (11) is in $\mathcal{H}_2^{2H}$. Indeed, $\hat{y}(s)$ appears, by property (3.4), as a linear combination of elements of $\mathcal{H}_2^{2H}$. The LSS model (9), with $-\frac{1}{2} < H < 0$, proposed by Yazıcı and Kashyap [33] then turns out to belong to the class of ARFIMA systems though it is obtained by a quite different approach.

The case when $\nu > 0$ is somewhat different, for in this case, $\gamma_{\nu}(n)^2 > 1$ for all n and as such, $R_{\nu}(s)$ does not belong to $\mathcal{H}_2$ in general. The space generated by the set of functions of the form (18) may, however, be given explicitly if the factorization (17) is expressed in the unit disc. Indeed, under the conformal mapping between the Hardy spaces of the right half-plane and of the unit disc $\mathbb{D} = \{z \in \mathbb{C}; |z| < 1\}$, the orthonormal functions $\Phi_n(s)$, $n = 0, 1, 2, \ldots$ are transformed to

$$\Psi_n(z) = \sqrt{\frac{2^{1-\nu}\pi\Gamma(1+n+\nu)}{n!}}\, z^n(1+z)^\nu, \quad n = 0, 1, 2, \ldots, \tag{19}$$

and the image of the reproducing kernel $\mathcal{K}(\xi, s)$ reads as:

$$\mathcal{R}(\omega, z) = 2^{1-\nu}\pi\Gamma(1+\nu)\frac{(1+z)^\nu(1+\bar{\omega})^\nu}{(1-\bar{\omega}z)^{1+\nu}} \triangleq \overline{\phi(\omega)}\kappa(\omega, z)\phi(z), \tag{20}$$

where $\phi(z) \triangleq \sqrt{2^{1-\nu}\pi\Gamma(1+\nu)}\,(1+z)^\nu$ and $\kappa(\omega, z) \triangleq (1-\bar{\omega}z)^{-1-\nu}$. For $\nu > 0$, the function $\kappa(\omega, z)$ is the reproducing kernel for the space $\mathcal{H}$ composed of all functions $G(z)$ analytic on $\mathbb{D}$ with finite norm [31]

$$\|G(z)\|_{\mathcal{H}}^2 = \iint_{\mathbb{D}} |G(z)|^2 \frac{dx\, dy}{(1-|z|^2)^{1-\nu}} < \infty.$$

Thus, as in the half-plane, every element $F(z)$ of $\mathcal{H}_2^\nu(\mathbb{D})$ admits a factorization of the form

$$F(z) = \phi(z)R_\nu(z) = \sqrt{2^{1-\nu}\pi\Gamma(1+\nu)}\,(1+z)^\nu R_\nu(z), \tag{21}$$

for some $R_\nu(z)$ in $\mathcal{H}$.

Remark 3.6. Note that for positive ν, one may verify, after some algebra, that the norm defined in (15) also reads as

$$\|F(s)\|_\nu^2 = \int_0^\infty |f(t)|^2 dt = \frac{1}{2^{1-\nu}\pi\Gamma(\nu)}\int_0^\infty\int_{-\infty}^\infty |F(s)|^2 a^{\nu-1} db\, da\,; \quad s = a+ib,\ a > 0. \tag{22}$$

If $G(z)$ is the image of $F(s) \in \mathcal{H}_2^\nu$ under the conformal map *i.e.*, $G(z) = F\left(\frac{1-z}{1+z}\right)$ then, using the change of variable $s = \frac{1-z}{1+z}$ in (22), we have

$$\|F(s)\|_\nu^2 = \frac{1}{2^{1-\nu}\pi\Gamma(\nu)}\iint_{\mathbb{D}}\left|\frac{2G(z)}{(1+z)^\nu}\right|^2 \frac{dx\, dy}{(1-|z|^2)^{1-\nu}}\,; \quad z = x+iy,\ |z| < 1, \tag{23}$$

which shows that $2(1+z)^{-\nu}G(z)$ belongs to $\mathcal{H}_2^\nu(\mathbb{D})$ and has the same norm as $F(s)$. Therefore to obtain an isomorphic map which is also isometric, one should divide all functions in $\mathcal{H}_2^\nu(\mathbb{D})$ by the factor $(1+z)^\nu/2$, in much the same way as $\mathcal{H}_2$ of the unit disc is obtained from $\mathcal{H}_2$ of the right half-plane (see, *e.g.*, [15]).

On the other hand, observe that the function $\mathcal{Q}(\omega, z) = \ln\kappa(\omega, z)$ is the reproducing kernel for the Dirichlet space $\mathcal{D}$ [31], [4] of all analytic functions $f(z)$ on $\mathbb{D}$, normalized by $f(0) = 0$, with finite Dirichlet integral

$$\iint_{\mathbb{D}} |f'(z)|^2\, dx\, dy < \infty.$$

With this observation and with the relation

$$\kappa(\omega, z) = e^{\mathcal{Q}(\omega, z)} = 1 + \mathcal{Q}(\omega, z) + \mathcal{Q}(\omega, z)^2/2 + \cdots + \mathcal{Q}(\omega, z)^k/k! + \cdots,$$

we deduce that every element of $\mathcal{H}_2^\nu(\mathbb{D})$ is of the form given in (21) for some $R_\nu(z)$ in

$$\mathcal{H} = \mathbb{C} \oplus \mathcal{D} \oplus [\mathcal{D} \otimes \mathcal{D}] \oplus \cdots \oplus [\mathcal{D} \otimes \cdots \otimes \mathcal{D}]_k \oplus \cdots , \tag{24}$$

where $[\mathcal{D}\otimes \cdots \otimes \mathcal{D}]_k$ denotes the restriction, to the diagonal set $\mathbb{D}\times\mathbb{D}$ of $(\mathbb{D}\times\mathbb{D})\times(\mathbb{D}\times\mathbb{D})$, of the k-fold tensor product of $\mathcal{D}$ by itself or simply, the Hilbert space whose reproducing kernel is $\mathcal{Q}(\omega, z)^k$ (see, *e.g.*, [31] or [4]).

4. Example of LSS model

We have seen that the self-similar nature of the system described by the input-output relation (1) stemmed from the invariance property of the kernel $K(t,\lambda)$ given in (7). Using a frequency domain approach, the previous section provided an interpretation of self-similarity in terms of the invariance of $\mathcal{H}_2^\nu$ to the transformation $F(s) \mapsto \alpha^{\nu'+1}F(\alpha s)$. Exploiting this interpretation, we propose in this section, a simple and straightforward example of self-similar systems along with a numerical computation.

This example follows directly through the model proposed by [33], with a specific choice for the kernel $K(t,\lambda)$. Indeed, setting in (1) $K(t,\lambda) = F(t+i\lambda)$ for some $F(s = t+i\lambda) \in \mathcal{H}_2^\nu$, one immediately observes that the invariance of $\mathcal{H}_2^\nu$ to the transformation $F(s) \mapsto \alpha^{\nu'+1}F(\alpha s)$ translates in terms of self-similarity of the corresponding system $\mathcal{S}$, which input-output relation reads as:

$$\begin{aligned}\mathcal{S}\{u(t)\} \triangleq y_\nu(t) &= \int_0^\infty K(t,\lambda)u(\lambda)d\lambda = \int_0^\infty F(t+i\lambda)u(\lambda)d\lambda \\ &= \int_0^\infty\int_0^\infty \tau^{\nu/2}e^{-t\tau}f(\tau)e^{-i\lambda\tau}u(\lambda)d\tau\, d\lambda.\end{aligned} \tag{25}$$

Using Fubini's theorem, we obtain

$$y_\nu(t) = \int_0^\infty \tau^{\nu/2}e^{-t\tau}f(\tau)\hat{u}(\tau)d\tau, \tag{26}$$

where $\hat{u}(\tau)$ denotes the Fourier transform of the causal input $u(t)$.

Remark 4.1. The action of the above system $\mathcal{S}\{\cdot\}$ on its input $u(\cdot)$ may be decomposed in two steps:

1. The first step consists in a linear filtering. The function $f(\cdot)$ is interpreted as the restriction, to the positive frequency axis, of the Fourier transform of the impulse response of the filter. In what follows, we denote this impulse response by $\check{f}$. Thus, $y(t) = (\check{f}\star u)(t)$ represents the time response of the filter to the input $u(t)$ and $\hat{y}(\tau) = f(\tau)\hat{u}(\tau)$ the corresponding Fourier transform at the frequency τ.
2. In the second step, this output is transformed by an integral operator. It is this operator which confers the self-similar property on the system.

We may note that the output, $y_\nu(t)$, of the overall system is the restriction to the positive real axis of $y_\nu(s = t + i\lambda)$, the ν-transform of $\hat{y}(\tau)$. Now it is clear that $\hat{y}(\tau)$ belongs to $L_2(0,\infty)$ as long as the input $u(t)$ does. Consequently, $y_\nu(t)$ is an element of $\mathcal{H}_2^\nu$ and then, $\alpha^{\nu/2}y_\nu(\alpha t)$ is, by Property 3.4, the ν-transform of $1/\alpha\hat{y}(\tau/\alpha)$. The self-similar property

$$\mathcal{S}\{u(t)\} = y_\nu(t) \Longrightarrow \mathcal{S}\{u(\alpha t)\} = \alpha^{\nu/2}y_\nu(\alpha t),$$

therefore holds for the system $\mathcal{S}$ described by (25)–(26).

In the frequency domain, we have:

Lemma 4.2. *For $-1 < \nu < 1$, $y_\nu(t)$ admits a Laplace transform, say $Y_\nu(s)$, given by*

$$Y_\nu(s) = \int_0^\infty \frac{\sigma^{\nu/2}\hat{y}(\sigma)}{s+\sigma}d\sigma, \tag{27}$$

and $Y_\nu(s)$ is devoid of poles in the open right half-plane.

Remark 4.3. This expression of $Y_\nu(s)$ may be viewed as an infinite set of first order low-pass filters in parallel, in which the set of cut-off frequencies (represented by the variable σ) forms a continuum. A $1/f^\beta$ behavior is therefore obtained, as illustrated in the figures presented below.

Proof. The expression of $Y_\nu(s)$ given in equation (27), as the Laplace transform of $y_\nu(t)$, is obtained directly from Fubini's theorem.

For the remaining of the proof, let us consider first the case $0 < \nu < 1$. We shall show that, in this case, $y_\nu(t)$ is in $L_2(0,+\infty)$. To see this recall that the norm of $y_\nu(s)$ is given by

$$\|y_\nu(s)\|_\nu^2 = \frac{1}{2^{1-\nu}\pi\Gamma(\nu)}\int_0^\infty\int_{-\infty}^\infty |y_\nu(t+i\lambda)|^2 t^{\nu-1}d\lambda\, dt < \infty,$$

so that, due to the analyticity of $y_\nu(s)$, we have

$$\int_0^1 t^{\nu-1}|y_\nu(t)|^2 \leq \int_0^\infty t^{\nu-1}|y_\nu(t)|^2 < \infty.$$

The Lebesgue dominated convergence theorem then shows that $y_\nu(t)$ is in $L_2(0,1)$.

Next, because $\mathcal{H}_2^\nu$ is a reproducing kernel Hilbert space, the reproducing kernel property

$$F(\xi) = \langle F(s), \mathcal{K}(\xi,s)\rangle_\nu, \tag{28}$$

where $\langle\cdot,\cdot\rangle_\nu$ stands for the scalar product in $\mathcal{H}_2^\nu$, holds for every function $F \in \mathcal{H}_2^\nu$ and every point ξ in the right half-plane. In particular, for the function $\mathcal{K}(\xi,\cdot)$ (which belongs to $\mathcal{H}_2^\nu$ for all $\Re(\xi) > 0$), this reproducing property yields the identity

$$\mathcal{K}(\xi,\xi) = \langle\mathcal{K}(\xi,s),\mathcal{K}(\xi,s)\rangle_\nu = \|\mathcal{K}(\xi,s)\|_\nu^2. \tag{29}$$

From this identity and the Cauchy-Schwartz inequality, we immediately deduce that

$$|y_\nu(t)|^2 \leq \|y_\nu\|_\nu^2\,\Gamma(1+\nu)\,(2t)^{-1-\nu}. \tag{30}$$

This inequality allows one to deduce that $y_\nu(t) \in L_2(1,\infty)$, which completes the proof that $y_\nu(t) \in L_2(0,\infty)$.

It now remains to handle the case $-1 < \nu < 0$. In this case, we already know that $\mathcal{H}_2^\nu \subset \mathcal{H}_2$ but $y_\nu(t)$ may present a singularity as t approaches 0. To show that $Y_\nu(s)$ does not have any unstable pole, it is sufficient to prove that there exists $0 < \alpha < 1$ such that $t^\alpha y_\nu(t) \to 0$ as $t \to 0^+$. One may readily check that this holds true for, *e.g.*, $\alpha = 1/2$, by multiplying both sides of inequality (30) by t. □

For numerical computation purposes, observe that the integral equation (26), defining the output $y_\nu(t)$, is in a form suitable for Gaussian quadrature approximation corresponding to the generalized Laguerre orthonormal polynomials, $L_k^{\nu/2}(\tau)$. A numerical approximation of this output may then be computed via

$$y_\nu(t) = t^{-1-\nu/2} \int_0^\infty \tau^{\nu/2} e^{-\tau}\, \hat{y}\left(\frac{\tau}{t}\right) d\tau, \qquad (31a)$$

$$\approx t^{-1-\nu/2} \sum_{j=1}^{n} w_j\, \hat{y}\left(\frac{\tau_j}{t}\right), \qquad (31b)$$

where the abscissas τ_j are the zeros of $L_n^{\nu/2}(\tau)$, and the weights w_j are given by

$$w_j = \frac{\langle \tau^{\nu/4} e^{-\tau/2} L_{n-1}^{\nu/2}(\tau), \tau^{\nu/4} e^{-\tau/2} L_{n-1}^{\nu/2}(\tau)\rangle}{L_{n-1}^{\nu/2}(\tau_j)\, L_n'^{\nu/2}(\tau_j)}.$$

A routine for the computation of these abscissas and weights can be found in, *e.g.*, Numerical Recipes [29].

In the numerical example presented below, the input signal was set to $u(t) = t^2 e^{-t^2}$ and the filter f to $f(\tau) = \frac{i\tau+2}{(i\tau+3)(i\tau+5-2i)}$.

The time response of the system $\mathcal{S}$, computed via (31b) with $n = 10$ is displayed in Figure 1, with a logarithmic scale for the time axis. Figures 2 and 3 show the Bode plots corresponding to the intermediate output $y(t)$ and to the overall output $y_\nu(t)$ respectively. Observe the long dependence behavior of $y_\nu(t)$ in opposition to the very fast decay of the spectrum of the intermediate output $y(t)$.

Before closing this example, we may point out the following

Remark 4.4. The model presented in this example may be related to the concept of "diffusive representation" of a pseudo-differential operator, initially introduced by Montseny, Audounet and Mbodje [25]. Roughly speaking, the diffusive representation of the operator $u \to h \star u$ is a function $\mu : \mathbb{R}_+ \to \mathbb{C}$ such that $h(t)$ is the restriction to the positive real axis of the Laplace transform of μ. The Laplace transform of $h(t)$ is called the symbol of the operator [24].

With these settings, it appears from (26) and (27) that $Y_\nu(s)$ may be interpreted as the symbol of the operator $v \to y_\nu \star v$ whose diffusive representation is given by $\mu(\sigma) = \sigma^{\nu/2}\hat{y}(\sigma)$.

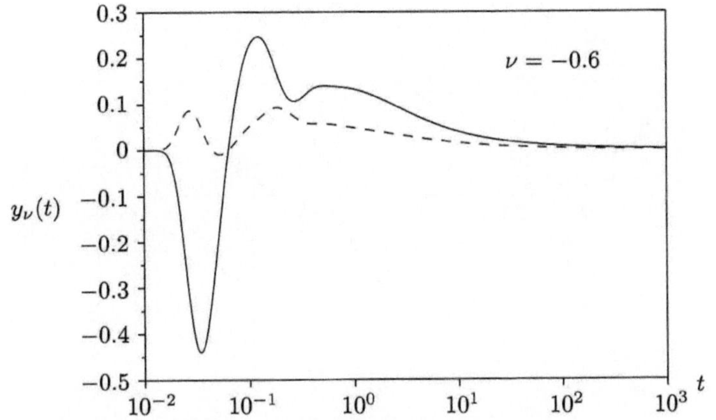

FIGURE 1. Time response of system $\mathcal{S}$: solid and dashed lines correspond to the real and imaginary parts, respectively

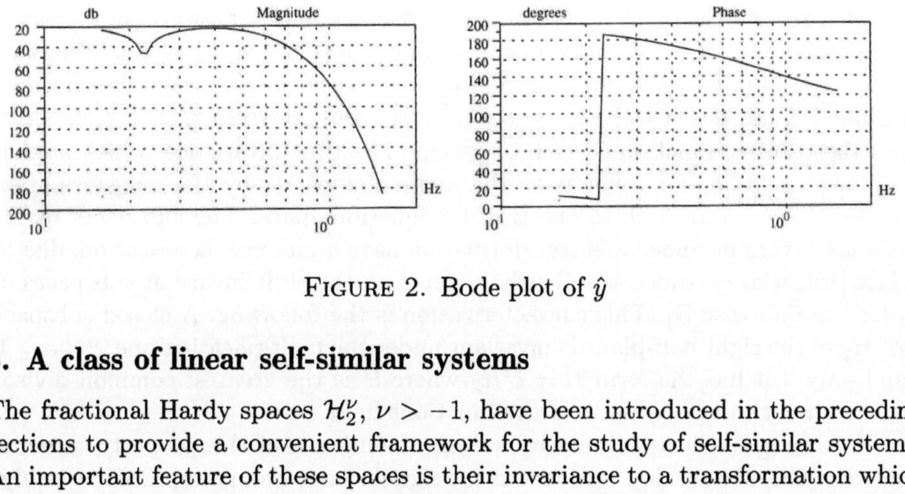

FIGURE 2. Bode plot of $\hat{y}$

5. A class of linear self-similar systems

The fractional Hardy spaces $\mathcal{H}_2^\nu$, $\nu > -1$, have been introduced in the preceding sections to provide a convenient framework for the study of self-similar systems. An important feature of these spaces is their invariance to a transformation which amounts to a weighted shift in scale. We have attached to this invariance an interpretation in terms of self-similarity.

In order to fix the scope of this section, let us make a brief review of the basic facts concerning $\mathcal{H}_2$. The starting point is the well-known Paley-Wiener theorem (see, *e.g.*, [15]) which is this:

Theorem 5.1 (Paley-Wiener I). *A complex-valued function F in the right half-plane belongs to $\mathcal{H}_2$ if and only if F has the form*

$$F(s) = \frac{1}{2\pi} \int_0^\infty f(t) e^{-st}\, dt \tag{32}$$

for some $f \in L_2(0, \infty)$. This representation is unique.

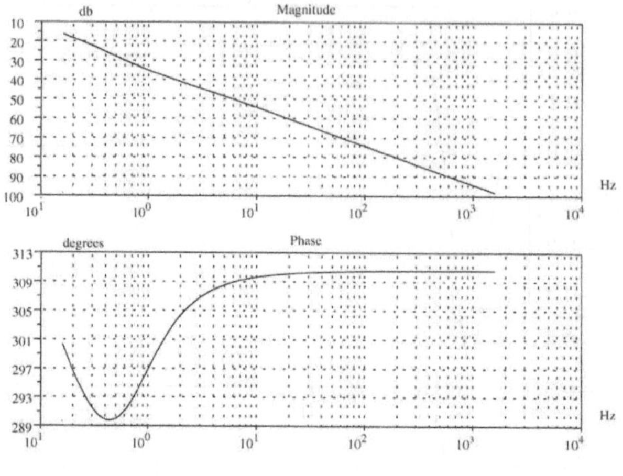

FIGURE 3. Bode plot of the output of system $\mathcal{S}$ in the frequency range $w_{\min} = 1$, $w_{\max} = 10^4$

This theorem is on the basis of the Fourier-Laplace transform and as such, it constitutes one of the foundations of signal processing and linear system theory. Indeed, as is well known, every causal and exponentially stable linear time-invariant system has a transfer function in $\mathcal{H}_2$ and conversely, each element of $\mathcal{H}_2$ is the transfer function of such a system. The time-invariance, which applies in the signal space, corresponds to the invariance of $\mathcal{H}_2$ by the transformation $F(s) \to e^{-\alpha s}F(s)$, $\alpha \geq 0$, in the transfer function space. The subspaces of $\mathcal{H}_2$ which are invariant under this transformation have a nice characterization, due to P. Lax [18], who extended the Beurling result on the shift invariant subspaces of $\mathcal{H}_2$ for the unit disc [7]. This characterization is the following. A closed subspace $\mathcal{H}$ of $\mathcal{H}_2$ of the right half-plane is invariant under the multiplication of $e^{-\alpha s}$, $\alpha \geq 0$ if and only if it has the form $\mathcal{H} = U\mathcal{H}_2$ where U is the greatest common divisor of all the inner (all-pass) parts of the functions in $\mathcal{H}$.

Returning to the self-similarity, we may mention that this property amounts to a (weighted) shift scale-invariance in the signal space. From our preceding analysis, it seems that this invariance translates into the invariance of the corresponding "transfer function" space, under a transformation of the form $F(s) \to \alpha^r F(\alpha s)$, with $r > 0$. Yet, the fractional Hardy classes $\mathcal{H}_2^\nu$, $\nu > -1$ have supplied us with an example of such space. Recall from (17) and (21) that $\mathcal{H}_2^\nu$ admits a factorization of the form $\mathcal{H}_2^\nu = U\mathcal{H}$ where, in the right half-plane and for negative ν, $U = (1+s)^{-\nu}$ with $\mathcal{H}$ being a subspace of $\mathcal{H}_2$ and in the unit disc and for positive ν, $U = (1+z)^\nu$ with $\mathcal{H}$ given by (24). At this point, it is worth mentioning the similarity between the close connection of the time-invariance property and the Beurling-Lax factorization for shift-invariant subspaces, on the one hand and the close connection of the time domain self-similar property and the factorization of $\mathcal{H}_2^\nu$, on the other hand.

Now, our next task will be to draw the general structure of the transfer function space for linear self-similar systems, on the basis of an invariant subspace concept.

To proceed, our program which, basically, reduces to mimicking the linear time-invariant case is the following.

First, we present a generalization of the Paley-Wiener spaces. This generalization is given by the de Branges spaces. The de Branges theory of Hilbert spaces of entire functions completely characterizes the subspaces for which a transformation of the above-mentioned form is an isometry. These spaces are called homogeneous. These facts are briefly exposed in Subsection 5.1.

Second, we attach to de Branges's homogeneous spaces the usual meaning of a transfer function space. Then, we turn back to the signal space, to express the input-output relationship for linear self-similar systems. This is the subject of Subsection 5.2.

5.1. de Branges spaces

Let us start with the second form of the Paley-Wiener theorem (see, *e.g.*, [30]) which is this.

Theorem 5.2 (Paley-Wiener II). *if $f(t) \in L_2(-\infty,\infty)$ and vanishes outside the interval $[-T,T]$ then the integral*

$$F(z) = \frac{1}{2\pi}\int_{-\infty}^{\infty} e^{-izt} f(t)dt \tag{33}$$

is absolutely convergent and defines an entire function $F(z)$ of exponential type[2] *less than or equal to T. The function $F(z)$ is square integrable on the real line.*

Conversely, any entire function $F(z)$ of exponential type at most T and square integrable in the real line is the Fourier-Laplace transform of some function f in $L_2(-\infty,\infty)$ with support on the interval $[-T,T]$.

The Hilbert space theory of entire functions is due to L. de Branges [22] who developed a generalization of the Fourier analysis at the end of the sixties. Starting with the Paley-Wiener theory which characterizes entire functions of exponential type and square integrable on the real line, the author introduced the theory of the spaces $\mathcal{H}\{E\}$, generated from an entire function $E(z)$ which is subject to

$$|E(x-iy)| < |E(x+iy)| \ ; \quad y > 0. \tag{34}$$

This function plays the same role for the spaces $\mathcal{H}\{E\}$ as does the exponential function for the Paley-Wiener spaces.

If $F(z)$ is analytic in the upper half-plane, we set $F^*(z) = \overline{F(\bar{z})}$ for its extension in the lower half-plane. The upper half-plane will be subsequently denoted by $\mathbb{C}_+$. Then we have

[2] An entire function $F(z)$ is said to be of exponential type τ, if $|F(z)| \leq Ce^{\tau|z|}$ for some constant C.

Theorem 5.3 (Definition). *Let $E(z) = A(z) - iB(z)$ be an entire function where $A(z)$ and $B(z)$ are entire functions, real for z real. Suppose that $E(z)$ satisfies the constraint* (34).

Then the set $\mathcal{H}\{E\}$ of all entire functions $F(z)$ such that

1. *$F(z)/E(z)$ and $F^*(z)/E(z)$ are analytic in $\mathbb{C}_+$ and of non-positive* ***mean type***[3]

2. $$\int_{-\infty}^{\infty} \left|\frac{F(t)}{E(t)}\right|^2 dt < \infty \tag{35}$$

is a reproducing kernel Hilbert space of entire functions, called the de Branges space. The norm of $F \in \mathcal{H}\{E\}$ is defined by the left-hand side of (35) *and the reproducing kernel associated to $\mathcal{H}\{E\}$ is given by*

$$\mathcal{K}(\omega, z) = \frac{E(z)E^*(\bar{\omega}) - E^*(z)E(\bar{\omega})}{2\pi i(\bar{\omega} - z)} = \frac{A(z)B(\bar{\omega}) - B(z)A(\bar{\omega})}{\pi(\bar{\omega} - z)} \tag{36}$$

In particular, for $E(z) = e^{-iaz}$, $a > 0$, the de Branges space $\mathcal{H}\{e^{-iaz}\}$ reduces to the Paley-Wiener space of entire functions of exponential type at most a.

We proceed to present the successive stages of the de Branges generalization of the Paley-Wiener theory. The first two theorems below introduce the concept of a one-parameter family of de Branges's spaces which is the analog of the family $E(a, z) = e^{-iaz}$. The third theorem is the proper generalization of the Fourier-Laplace transform in the Paley-Wiener space.

5.1.1. A HILBERT SPACE-VALUED FUNCTION We now present a construction of an ascending chain of $\mathcal{H}\{E\}$ spaces which gives rise to a Hilbert space-valued function of one parameter $t > 0$. The construction uses a matrix-valued function defined hereafter.

Let

$$m(\lambda) = \begin{bmatrix} \alpha(\lambda) & \beta(\lambda) \\ \beta(\lambda) & \gamma(\lambda) \end{bmatrix} \tag{37}$$

be a matrix-valued function of $\lambda > 0$, where $\alpha(\lambda), \beta(\lambda)$ and $\gamma(\lambda)$ are absolutely continuous real functions subject to

C-1 $\quad \alpha'(\lambda) \geq 0\,; \quad \gamma'(\lambda) \geq 0\,; \quad [\beta'(\lambda)]^2 \leq \alpha'(\lambda)\gamma'(\lambda)$ *a.e* for $\lambda > 0$.

These conditions imply that $m'(\lambda)$ (which is to be understood as the matrix obtained by taking the derivative of each entry of $m(\lambda)$) is non-negative definite or

[3]The notion of mean type is a generalization of that of the exponential type to meromorphic functions which are the quotient of two bounded analytic functions. Such functions are said to be of bounded type or bounded characteristic. If $G(z)$ is such a function then it admits a Nevanlinna factorization which is of the form $G(z) = B(z)e^(- ihz)e^(Q(z))$ where $B(z)$ is a Blaschke product and $Q(z)$ is analytic in $\mathbb{C}_+$. The number h which appears in this factorization is called the mean type of G. It is given by a formula analogous to that of the exponential type: $h = \limsup_{y\to\infty} y^{-1} \ln|G(iy)|$

equivalently that $m(\lambda)$ is non-decreasing.

$$C\text{-}2 \qquad \alpha(0) = \lim_{\lambda \downarrow 0} \alpha(\lambda) < \infty \quad \text{and} \quad \int_0^1 [\alpha(\lambda) - \alpha(0)] d\gamma(\lambda) < \infty$$

Definition 5.4. A real number b is said to be singular with respect to $m(\lambda)$ if it belongs to an open interval (a, c) in which there exists a single function $h(\lambda)$ and some constants k_1, k_2 and k_3 with $k_3^2 = k_1 k_2$, such that *a.e.*,

$$k_1 \alpha'(\lambda) = k_2 \gamma'(\lambda) = k_3 \beta'(\lambda) = h(\lambda).$$

Otherwise, b will be termed regular with respect to $m(\lambda)$.

Theorem 5.5. *If $m(\lambda)$ is as defined above, with $\alpha(\lambda) > \alpha(0)$, $\lambda > 0$ then there exists a unique family $E(\lambda, z) = A(\lambda, z) - iB(\lambda, z), \lambda \geq 0$ of entire functions such that*

1. *for each complex number z, $E(\lambda, z)$ or equivalently $X(\lambda, z) \triangleq \begin{bmatrix} A(\lambda, z) \\ B(\lambda, z) \end{bmatrix}$, is the solution of the Cauchy problem*
$$\begin{cases} \frac{d}{d\lambda} X(\lambda, z) = zJm'(\lambda) X(\lambda, z) \\ X(0, z) = \begin{bmatrix} 1 \\ 0 \end{bmatrix} \end{cases} \tag{38}$$
where J is the signature matrix $J = \begin{bmatrix} 0 & -1 \\ 1 & 0 \end{bmatrix}$,
and $E(\lambda, z)$ satisfies $\lim_{\lambda \to 0} E(\lambda, z) e^{\beta(\lambda) z} = 1$.
2. *for each fixed $\lambda > 0$, $\mathcal{H}\{E(\lambda, z)\}$ defines a de Branges's space, $E(\lambda, z)$ has no real zeros and $E(\lambda, 0) = 1$. If $0 < \lambda_1 < \lambda_2$ are regular points with respect to $m(\lambda)$, then $\mathcal{H}\{E(\lambda_1, z)\}$ is contained isometrically in $\mathcal{H}\{E(\lambda_2, z)\}$.*

Theorem 5.6. *If, in addition to the situation of the last theorem, $m(\lambda)$ satisfies*

$$\lim_{\lambda \to \infty} [\alpha(\lambda) + \gamma(\lambda)] = \infty, \tag{C-3}$$

then there exists a unique non-negative measure μ on the Borel sets of the real line such that for each $a > 0$, regular with respect to $m(\lambda)$, $\mathcal{H}\{E(a, z)\}$ is contained isometrically in $L_2(d\mu)$. The union of such $\mathcal{H}\{E(a, z)\}$, where a is regular with respect to $m(\lambda)$, is dense in $L_2(d\mu)$.

5.1.2. GENERALIZED FOURIER TRANSFORM One of the essential parts, if not the corner-stone, of the de Branges theory is the development of an integral representation for $\mathcal{H}\{E\}$ spaces and, incidentally, the Parseval equality in these spaces. The guide line of this development is exposed now. Let $L_2(\mathbb{R}^+, dm)$ be the Hilbert space of equivalent classes of $\mathbb{C}^{2\times 1}$-valued functions $f(\lambda) = \begin{bmatrix} f_1(\lambda) \\ f_2(\lambda) \end{bmatrix}$ of $\lambda \geq 0$ with norm

$$\|f\|_{dm}^2 = \int_0^\infty [\bar{f}_1(\lambda)\ \bar{f}_2(\lambda)]\, dm(\lambda) \begin{bmatrix} f_1(\lambda) \\ f_2(\lambda) \end{bmatrix} < \infty. \tag{39}$$

By the equivalence of two elements f and g of $L_2(\mathbb{R}^+, dm)$, in an interval $[a,\ b]$ we mean that $\int_a^b \{f(t) - g(t)\}^T dm(t) \{\overline{f(t) - g(t)}\} = 0$. We may remark that $L_2(\mathbb{R}^+, dm)$ represents the space of finite energy self-similar signals as introduced by Yazıcı and Kashyap [33] and given in Definition 2.2. The closed subspace of $L_2(\mathbb{R}^+, dm)$, composed of the elements f which are equivalent to a constant in any interval containing only singular points with respect to $m(\lambda)$, is denoted by $L_2^0(dm)$.

Theorem 5.7. *Let $a > 0$ be regular with respect to $m(\lambda)$. For each $f \in L_2^0(dm)$ and equivalent to zero outside of $[0,\ a]$, the integral equation*

$$F(a, z) = \frac{1}{\pi} \int_0^a f^T(\lambda) dm(\lambda) X(\lambda, z) \tag{40}$$

defines an entire function $F(a, z)$ which belongs to $\mathcal{H}\{E(a, z)\}$ and the Parseval equality

$$\|F(a, \cdot)\|^2_{\mathcal{H}\{E\}} = \int |F(a, t)|^2 d\mu(t) = \frac{1}{\pi} \int_0^a f^T(\lambda) dm(\lambda) \bar{f}(\lambda) = \|f\|^2_{dm} \tag{41}$$

holds. Conversely, every element of $\mathcal{H}\{E(a, z)\}$ is the integral transform (40) of some f in $L_2^0(dm)$.

For any $f \in L_2^0(dm)$, the function $F(x)$, defined on the real line by

$$F(x) = \frac{1}{\pi} \int_0^\infty f^T(\lambda) dm(\lambda) X(\lambda, x), \tag{42}$$

where the integral is to be understood as $\lim_{c \to \infty} \int_0^c$, exists in the metric of $L_2(d\mu)$ and (41) holds. Every element of $L_2(d\mu)$ is of this form.

5.2. Homogeneous spaces

Let us now give the definition of homogeneous spaces in the theory of de Branges

Definition 5.8. Let $\nu > -1$ be a real number. A de Branges space $\mathcal{H}\{E(z)\}$ is said to be homogeneous of order ν, if it is isometrically invariant to the transformation $F(z) \to \alpha^{1+\nu} F(\alpha z),\ 0 < \alpha < 1$.

Observe that this transformation is exactly that we have been dealing with in the study of the fractional Hardy spaces $\mathcal{H}_2^\nu$. Therefore, and now on, we shall define

Definition 5.9. A linear ν-self-similar system as a linear system whose transfer function belongs to a homogeneous space of order ν.

The de Branges theory provides a complete characterization of structure of homogeneous spaces. This structure is now given below. But, before entering into the details, let us remark that the structure of a given de Branges space is determined from the matrix-valued function $m(\lambda)$ introduced in the previous section. We thus have

Theorem 5.10. *Let $\nu > -1$ be a real number. If the matrix-valued function $m(\lambda)$, defined in* (37), *is chosen such as*

$$\alpha(\lambda) = \frac{\lambda^{2\nu+2}}{2\nu+2}; \ \beta(\lambda) = 0 \ \text{and} \ \gamma(\lambda) = -\frac{\lambda^{-2\nu}}{2\nu}, \tag{43}$$

then any point $a > 0$ is regular with respect to $m(\lambda)$ and the conditions of theorems 5.5 and 5.6 are satisfied. Let $\{E(\lambda, z)\}_{\lambda>0}$ be the corresponding family of entire functions. Then,

1. *for each $\lambda > 0$, $\mathcal{H}\{E(\lambda, z)\}$ is homogeneous of order ν,*
2. *if $F(z) \in \mathcal{H}\{E(b, z)\}$ and if $0 < a \leq b$ then $F(\frac{a}{b}z)$ is in $\mathcal{H}\{E(a, z)\}$,*
3. *for $\lambda > 0$, the measure μ is given by $\mu(\lambda) = \mu(1)\lambda^{2\nu+2}$ and $\mu(-\lambda) = \mu(-1)\lambda^{2\nu+2}$,*
4. *any homogeneous space of order ν, $\mathcal{H}\{E(z)\}$, such that $E(0) \neq 0$ is isometrically equal to $\mathcal{H}\{E(\lambda, z)\}$ for some fixed $\lambda > 0$.*

A consequence of this choice of $m(\lambda)$ is that $L_2^0(dm)$ simply reduces to $L_2(dm)$ because there is no singular point with respect to $m(\lambda)$.

In the sequel, we shall denote by $\mathcal{H}_\lambda^\nu$ the de Branges homogeneous space of order ν, $\mathcal{H}\{E(\lambda, z)\}$, defined in Theorems (5.5) and (5.10).

5.3. Linear time-invariant self-similar systems

We now exhibit the structure, according to our point of view, of linear self-similar systems directly in the signal space. Recall that the self-similarity is to be understood here in the sense of Definition 5.9.

Our approach is this. First we show that the usual concept of transfer function does apply to self-similar systems. This then allows us to introduce a linear time-invariant model which presents a self-similar behavior when observed through different scales. The model so obtained, will thus depend on two parameters: a time parameter and a scale parameter.

5.3.1. A TRANSFER FUNCTION INTERPRETATION We begin with

Lemma 5.11. *Let $m(\lambda)$ be chosen as in* (43). *Then the entire functions $A(\lambda, z)$ and $B(\lambda, z)$ that solve the Cauchy problem* (38) *are given by*

$$A(\lambda, z) = 2^\nu\,\Gamma(1+\nu)\,(\lambda z)^{-\nu}\mathcal{J}_\nu(\lambda z) \tag{44}$$

$$B(\lambda, z) = 2^\nu\,\Gamma(1+\nu)\,\lambda^{2\nu+1}(\lambda z)^{-\nu}\mathcal{J}_{\nu+1}(\lambda z) \tag{45}$$

where $\mathcal{J}_\eta(x) = \sum_{n=0}^{\infty}(-1)^n \frac{\left(\frac{x}{2}\right)^{\eta+2n}}{n!\Gamma(\eta+n+1)}$ is the Bessel function of the first kind, of order η.

Proof. Since $A(\lambda, z)$ and $B(\lambda, z)$ are entire functions of z for each fixed λ, the solution $X(\lambda, z)$ of (38) is a vector-valued entire function. Hence it has the form

$$X(\lambda, z) = \sum_{k=0}^{\infty} X_k(\lambda) z^k,$$

where $X_k(\lambda)$ is a vector-valued continuous function of λ for each k. Differentiating both sides of this equation with respect to λ and equating with (38) shows that the sequence of functions $\{X_k(\lambda)\}_{k\geq 0}$ must satisfy the recurrence relation

$$\frac{d}{d\lambda}X_k(\lambda) = Jm'(\lambda)X_{k-1}(\lambda), \quad k = 1, 2, \ldots$$

with $X_0(\lambda) = X(\lambda, 0) = \begin{bmatrix} 1 \\ 0 \end{bmatrix}$. One may directly check, by induction, that for any integer $n \geq 0$, we have

$$X_{2n}(\lambda) = \begin{bmatrix} \frac{(-1)^n (\lambda)^{2n}}{n! 2^{2n} (\nu+1)(\nu+2)\cdots(\nu+n)} \\ 0 \end{bmatrix} \text{ and } X_{2n+1}(\lambda) = \begin{bmatrix} 0 \\ \frac{(-1)^n (\lambda)^{2\nu+2n+2}}{n! 2^{2n+1} (\nu+1)(\nu+2)\cdots(\nu+n+1)} \end{bmatrix}$$

Combining this with the identity

$$\Gamma(\nu + n) = (\nu + n)(\nu + n - 1) \cdots (\nu + 1)\Gamma(\nu + 1),$$

we obtain the result of the lemma. □

Remark 5.12. Recall that by Theorem 5.7 every element $F(x)$ of $L_2(d\mu)$ admits the integral representation given in equation (42) for some $f = \begin{bmatrix} f_1 \\ f_2 \end{bmatrix} \in L_2(dm)$. In the present context, every such element can be written in the form $F(x) = F_1(x) + F_2(x)$ where

$$F_1(x) = \frac{2^\nu \, \Gamma(\nu + 1) x^{-\nu-1/2}}{\pi} \int_0^\infty \lambda^{\nu+\frac{1}{2}} f_1(\lambda) \, \mathcal{J}_\nu(\lambda x) \sqrt{(\lambda x)} \, d\lambda \tag{46a}$$

$$F_2(x) = \frac{2^\nu \, \Gamma(\nu + 1) x^{-\nu-1/2}}{\pi} \int_0^\infty \lambda^{-\nu-\frac{1}{2}} f_2(\lambda) \, \mathcal{J}_{\nu+1}(\lambda x) \sqrt{(\lambda x)} \, d\lambda \tag{46b}$$

Now, observe that the integrals appearing in these expressions define the Hankel transforms of orders ν and $\nu+1$ of the $L_2(d\lambda)$ functions $\lambda^{\nu+\frac{1}{2}} f_1(\lambda)$ and $\lambda^{-\nu-\frac{1}{2}} f_2(\lambda)$ respectively. Consequently, one may describe $L_2(d\mu)$ in terms of Hankel transforms of elements of $L_2(d\lambda)$.

To each element of $f(\lambda) = \begin{bmatrix} f_1(\lambda) \\ f_2(\lambda) \end{bmatrix} \in L_2(dm)$ we associate the scalar-valued function $\varphi(\lambda)$ defined by

$$\varphi(\lambda) = \lambda^{\nu+\frac{1}{2}} f_1(\lambda) - i\lambda^{-\nu-\frac{1}{2}} f_2(\lambda) \tag{47a}$$

$$\text{and } \varphi(-\lambda) = \lambda^{\nu+\frac{1}{2}} f_1(\lambda) + i\lambda^{-\nu-\frac{1}{2}} f_2(\lambda). \tag{47b}$$

As one may readily verify, the original function f can be recovered from its correspondence φ by

$$f_1(\lambda) = \lambda^{-\nu-\frac{1}{2}} \frac{\varphi(\lambda) + \varphi(-\lambda)}{2} \tag{48a}$$

$$\text{and } f_2(\lambda) = \lambda^{\nu+\frac{1}{2}} \frac{\varphi(\lambda) - \varphi(-\lambda)}{2i} \tag{48b}$$

The function φ is then perfectly defined and if f vanishes outside an interval $[a,\ b]$ then φ also vanishes outside $[a,\ b]$. Moreover, it is immediate to show that φ belongs to $L_2(d\lambda)$ as long as f belongs to $L_2(dm)$ and

$$\|\varphi\|_{L_2(d\lambda)} = \|f\|_{L_2(dm)}.$$

The correspondence introduced in (47) therefore defines an isometrically isomorphic map between $L_2(d\lambda)$ and $L_2(dm)$. This map is subsequently denoted as $\mathfrak{G}_\lambda$. Since for each fixed z, $X(\lambda, z)$ is in[4] $L_2(dm)$ as a function of λ, we can define its image under $\mathfrak{G}_\lambda$, that we denote by $\mathcal{E}(\lambda, z)$. This image is given by

$$\mathcal{E}(\lambda, z) \stackrel{\triangle}{=} \mathfrak{G}_\lambda X(\lambda, z) = \lambda^{\nu+\frac{1}{2}} A(\lambda, z) - i\lambda^{-\nu-\frac{1}{2}} B(\lambda, z). \tag{49}$$

Because $A(\lambda, z)$ and $B(\lambda, z)$ are real for z real, $\mathcal{E}(-\lambda, z)$ satisfies

$$\mathcal{E}(-\lambda, z) = \mathcal{E}^*(\lambda, z). \tag{50}$$

With these settings, we are now ready to reformulate Theorem 5.7 in our context, so as to mimick verbatim the Paley-Wiener theorem. This is achieved by

Theorem 5.13. *If $\varphi(x)$ is in $L_2(-\infty, \infty)$ and vanishes outside the interval $[-\lambda, \lambda]$ then the eigentransform $F(\lambda, z)$ defined by the integral*

$$F(\lambda, z) = \frac{1}{2\pi} \int_{-\infty}^{\infty} \mathcal{E}(x, z) \varphi(x) dx \tag{51}$$

is an entire function of exponential type at most λ, which belongs to $\mathcal{H}_\lambda^\nu$.

Conversely, any element $F(z)$ of exponential type at most λ and square integrable in the real line with respect to $d\mu$ is the eigentransform (51) *of some function $\varphi(x)$ in $L_2(-\infty, \infty)$ which vanishes outside the interval $[-\lambda, \lambda]$.*

Remark 5.14. Note the analogy of the function

$$\mathcal{E}(\lambda, z) = 2^\nu\, \Gamma(1+\nu)\, \lambda^{\nu+\frac{1}{2}} \{(\lambda z)^{-\nu} \mathcal{J}_\nu(\lambda z) - i(\lambda z)^{-\nu} \mathcal{J}_{\nu+1}(\lambda z)\}, \tag{52}$$

obtained from (49), (44) and (45), with the complex exponential function. In particular, when $\nu = -\frac{1}{2}$, $\mathcal{E}(\lambda, z)$ reduces to $e^{-i\lambda z}$, showing again that the Paley-Wiener spaces are de Branges homogeneous spaces of order $-\frac{1}{2}$.

Proof. As has been already remarked, this theorem is just a reformulation of Theorem 5.7 (see [22]) except for the exponential type property. This property may however be checked in various ways. For example, one may invoke a result due to Krein [17] stating that if an entire matrix function is J inner, then it is of exponential type (see also [11] where part of the de Branges theory is reviewed in the framework of J theory). The property then follows by applying this result to the monodromy matrix of the differential system (38). The J innerness of that

[4]Indeed the reproducing kernel of $\mathcal{H}\{E(\lambda, z)\}$ is the function $\mathcal{K}(\omega, z)$ obtained as the integral transform (40) of $X(\lambda, \bar{\omega})$

monodromy matrix stems from the positivity of the matrix $m'(\cdot)$. To see this, first recall that if $W(\lambda, z)$ is the monodromy matrix of (38), then we have

$$\frac{dW}{d\lambda} = zJm'(\lambda)W, \quad W(0,z) = I,$$

and every solution of (38) is of the form $X(\lambda, z) = W(\lambda, z)X(0, z)$. Next, let $\tilde{J} = iJ$ denote a signature matrix. Then we can derive from (38) that

$$\frac{d}{d\lambda}\{X^t(\lambda, z)\,\tilde{J}\,\overline{X(\lambda, z)}\} = -2\,\Im z\, X^t(\lambda, z)\, m'(\lambda)\,\overline{X(\lambda, z)}.$$

Finally, since $m'(\lambda)$ is positive definite, we deduce that $\frac{1}{\Im z}\{X^t(\lambda, z)\,\tilde{J}\,\overline{X(\lambda, z)}\}$ is a non-increasing function of λ so that we may infer the inequality

$$\frac{1}{\Im z}\langle \tilde{J}W(\lambda, z)X(0, z), W(\lambda, z)X(0, z)\rangle \leq \frac{1}{\Im z}\langle \tilde{J}X(0, z), X(0, z)\rangle \text{ for } \Im z \neq 0,$$

where equality holds for $\Im z = 0$. The monodromy matrix $W(\lambda, z)$ is therefore $\tilde{J}$ inner for each fixed $\lambda > 0$. □

Now that we have shown that each element of de Branges's homogeneous spaces is an entire function of exponential type, our next step is to ask whether such an element is square integrable on the real line. If so, then we meet the requirements of the Paley-Wiener theorem and consequently, we are able to identify every such element as the transfer function of some linear system.

Fortunately, the answer to this question is affirmative as soon as $\nu \geq \frac{1}{2}$, as given by

Theorem 5.15. *Let $F(\lambda, z)$ be an element of the de Branges homogeneous space as defined by* (51). *If $\nu \geq \frac{1}{2}$, then for any fixed $\lambda > 0$, we have*

$$\int_{-\infty}^{\infty} |F(\lambda, x)|^2\, dx < \infty.$$

Proof. If $F(\lambda, z)$ is an element of the de Branges homogeneous space then we know by Theorems 5.13 and 5.10 that it is of the form given by (51) and

$$\int_{-\infty}^{\infty} |F(\lambda, x)|^2 |x|^{2\nu+1} dx < \infty.$$

If $\nu \geq \frac{1}{2}$, $|x|^{2\nu+1} \geq 1$ for $|x| \geq 1$ so that we have by the Lebesgue dominated convergence theorem

$$\int_{|x|\geq 1} |F(\lambda, x)|^2 dx < \infty. \tag{53}$$

Using (51) and the Fubini theorem, we have the inequality

$$\int_{-1}^{1} |F(\lambda, x)|^2 dx \leq \frac{1}{4\pi^2}\int_{-\lambda}^{\lambda} |\varphi(\alpha)|^2 \int_{-1}^{1} |\mathcal{E}(\alpha, x)|^2 dx\, d\alpha.$$

Next, we show that the last integral of the second member of this inequality is bounded. This integral is given by

$$\int_{-1}^{1} |\mathcal{E}(\alpha, x)|^2 dx = \Gamma(\nu+1)\,|2\alpha|^{2\nu+1} \int_{0}^{1} |(\alpha x)^{-\nu}|^2 \{|\mathcal{J}_\nu(\alpha x)|^2 + |\mathcal{J}_{\nu+1}(\alpha x)|^2\} dx$$

where we observe that $(\lambda z)^{-\nu}\mathcal{J}_\nu(\lambda z)$ and $(\lambda z)^{-\nu}\mathcal{J}_{\nu+1}(\lambda z)$ are real for z real and that these functions are even and odd respectively. Now, using the following bound for the Bessel functions (see, *e.g.*, [1, Chapter 9.1.62])

$$|\mathcal{J}_\nu(z)| \leq \frac{2^\nu |z|^{-\nu} e^{|\Im z|}}{\Gamma(\nu+1)}, \text{ for } \quad \nu \geq -\frac{1}{2},$$

we deduce that

$$\int_{-1}^{1} |\mathcal{E}(\alpha, x)|^2 dx \leq 2|\alpha|^{2\nu+1} + \frac{2|\alpha|^{2\nu+3}}{3(\nu+1)^2}.$$

This then allows us to derive the bound

$$\int_{-1}^{1} |F(\lambda, x)|^2 dx \leq |\lambda|^{2\nu+1} \frac{\lambda^2 + 3(\nu+1)^2}{6\pi^2(\nu+1)^2} \int_{-\lambda}^{\lambda} |\varphi(\alpha)|^2 d\alpha < \infty,$$

which, combined with (53), concludes the proof. □

Finally, we come to the conclusion that every element of the de Branges homogeneous space of order $\nu \geq -\frac{1}{2}$ is of exponential type and square integrable on the real line. Every such element is therefore, by the Paley-Wiener theorem, the transfer function of some linear time-invariant system.

5.3.2. A SCALE-TIME SELF-SIMILAR MODEL By Theorems 5.13 and 5.15, a full sense is attached to our Definition 5.9 of self-similarity: A linear self-similar system is a system whose transfer function is an element of de Branges's homogeneous space of order $\nu \geq -\frac{1}{2}$. This defines the self-similar property in the frequency domain. In this section, we exhibit the time domain structure of such systems by showing that any linear self-similar system with parameter $\nu \geq -\frac{1}{2}$ admits a time domain input-output relation of the form:

$$y(\lambda, t) = \{h(\lambda, \cdot) \star u\}(t), \tag{54}$$

where $u(t)$ and $y(\lambda, t)$ stand for the input and the output for the scale λ, respectively. The function $h(\lambda, t)$ thus represents the impulse response at time t and scale λ.

We begin by the following elementary lemma:

Lemma 5.16. *Let $\psi(\lambda, t)$ denotes the inverse fourier transform of $\mathcal{E}(\lambda, z)$ as in*

$$\mathcal{E}(\lambda, z) = \frac{1}{2\pi} \int_{-\infty}^{\infty} e^{-izt} \psi(\lambda, t) dt \tag{55}$$

and let $\begin{bmatrix} \psi_r(\lambda, t) \\ \psi_i(\lambda, t) \end{bmatrix}$ be the inverse Fourier transform of $X(\lambda, z)$. Then the function

$\psi(\lambda, t)$ in (55) is given by

$$\psi(\lambda, t) = \mathfrak{G}_\lambda \begin{bmatrix} \psi_r(\lambda, t) \\ \psi_i(\lambda, t) \end{bmatrix} \tag{56}$$

Moreover, $\psi_r(\lambda, t)$ and $\psi_i(\lambda, t)$ are real-valued functions and $\psi(\lambda, t)$ satisfies

$$\psi(\lambda, -t) = \psi(\lambda, t).$$

Proof. First note that equation (56) is a direct consequence of the relation $\mathcal{E}(\lambda, z) = \mathfrak{G}_\lambda X(\lambda, z)$. Next, since $X(\lambda, z)$ is real for z real, its inverse Fourier transform must be an even function so that we may infer $\psi(\lambda, -t) = \psi(\lambda, t)$. Finally, it is straightforward to verify that the identity (50) induces the equality $\overline{\psi(\lambda, t)} = \psi(-\lambda, -t) = \psi(-\lambda, t)$ which, combined with (56) shows that $\psi_r(\lambda, t)$ and $\psi_i(\lambda, t)$ are real-valued functions. □

Now, observe that one may rewrite the differential system (38) in terms of the function $\mathcal{E}(\lambda, z)$ as:

$$\frac{d}{d\lambda}\mathcal{E}(\lambda, z) = \frac{\nu + \frac{1}{2}}{\lambda}\, \mathcal{E}^*(\lambda, z) - iz\mathcal{E}(\lambda, z), \tag{57}$$

where $\mathcal{E}^*(\lambda, z)$ is as defined in (50). To proceed, let us introduce the functions $\mathcal{E}_+(\lambda, z)$ and $\mathcal{E}_-(\lambda, z)$ as the causal and anti-causal projections of $\mathcal{E}(\lambda, z)$, according to the decomposition of L_2 into orthogonal complement Hardy spaces. Expressing equation (57) in terms of these projections, yields the differential system

$$\frac{d}{d\lambda}\mathcal{E}_+(\lambda, z) = \frac{\nu + \frac{1}{2}}{\lambda}\, \mathcal{E}_-^*(\lambda, z) - iz\mathcal{E}_+(\lambda, z), \tag{58}$$

$$\frac{d}{d\lambda}\mathcal{E}_-(\lambda, z) = \frac{\nu + \frac{1}{2}}{\lambda}\, \mathcal{E}_+^*(\lambda, z) - iz\mathcal{E}_-(\lambda, z). \tag{59}$$

Because $F(\lambda, z)$, and consequently $\mathcal{E}(\lambda, z)$, is of exponential type at most λ, $\psi(\lambda, t)$ must be zero for $|t| \geq \lambda$. Therefore, the causal part of the above differential system may be put in the form:

$$\begin{aligned}\frac{d}{d\lambda}\mathcal{E}_+(\lambda,z) &= \frac{1}{2\pi}\int_0^\infty e^{-izt}\frac{\partial\psi(\lambda,t)}{\partial\lambda}dt\\ &= \frac{\nu+\frac{1}{2}}{\lambda}\frac{1}{2\pi}\int_0^\infty e^{-izt}\psi(-\lambda,t)dt + \frac{1}{2\pi}\int_0^\infty \left(e^{-izt}\right)'\psi(\lambda,t)dt\\ &= \frac{\nu+\frac{1}{2}}{\lambda}\frac{1}{2\pi}\int_0^\infty e^{-izt}\psi(-\lambda,t)dt - \frac{1}{2\pi}\int_0^\infty e^{-izt}\frac{\partial\psi(\lambda,t)}{\partial t}dt + \frac{1}{2\pi}\psi(\lambda,0).\end{aligned} \tag{60}$$

The differential system (38) (or equivalently (57)) thus reads in the time domain as the partial differential equation

$$\frac{\partial\psi(\lambda, t)}{\partial\lambda} = \frac{\nu + \frac{1}{2}}{\lambda}\psi(-\lambda, t) - \frac{\partial\psi(\lambda, t)}{\partial t}, \quad a.e. \tag{61}$$

with the initial condition

$$\psi(\lambda,0)=\int_0^\infty \mathcal{E}(\lambda,x)dx=\frac{\Gamma(\frac{1}{2})\,\Gamma(\nu+1)}{\Gamma(\nu+\frac{1}{2})}\lambda^{\nu-\frac{1}{2}}-i\lambda^{\nu-\frac{1}{2}}. \tag{62}$$

Recall from Lemma 5.16 that the function $\psi(\lambda,t)$ is given by

$$\psi(\lambda,t)=\lambda^{\nu+\frac{1}{2}}\psi_r(\lambda,t)-i\lambda^{-\nu-\frac{1}{2}}\psi_i(\lambda,t).$$

Now, it turns out that, this function $\psi(\lambda,t)$ solves (61) with the initial condition (62) if and only if the functions $\psi_r(\lambda,t)$ and $\psi_i(\lambda,t)$ are solutions of the partial differential equation

$$\frac{\partial\tilde{\psi}(\lambda,t)}{\partial\lambda}+\frac{\partial\tilde{\psi}(\lambda,t)}{\partial t}=0, \tag{63}$$

with the initial conditions

$$\tilde{\psi}(\lambda,0)=\begin{cases}\frac{\sqrt{\pi}\,\Gamma(\nu+1)}{\Gamma(\nu+\frac{1}{2})}\dfrac{1}{\lambda} & \text{for } \psi_r(\lambda,t);\\ -\lambda^{2\nu} & \text{for } \psi_i(\lambda,t).\end{cases} \tag{64}$$

On the other hand, we may readily verify that the homogeneous property of $\mathcal{E}(\lambda,z)$

$$\mathcal{E}(\lambda,\alpha z)=\alpha^{-\nu-\frac{1}{2}}\mathcal{E}(\alpha\lambda,z), \tag{65}$$

reverberates on $\psi(\lambda,t)$ as:

$$\psi(\lambda,t)=\alpha^{-\nu+\frac{1}{2}}\,\psi(\alpha\lambda,\alpha t), \tag{66}$$

for all $\alpha>0$. This, in turn, translates into $\psi_r(\lambda,t)$ and $\psi_i(\lambda,t)$ by

$$\psi_r(\lambda,t)=\alpha\psi_r(\alpha\lambda,\alpha t) \tag{67}$$

$$\psi_i(\lambda,t)=\alpha^{-2\nu}\psi_i(\alpha\lambda,\alpha t). \tag{68}$$

Since these relations are valid for all $\alpha>0$, we obtain by setting $\alpha=\frac{1}{\lambda}$,

$$\psi_r(\lambda,t)=\frac{1}{\lambda}g_r\left(\frac{t}{\lambda}\right) \tag{69}$$

$$\psi_i(\lambda,t)=\lambda^{2\nu}g_i\left(\frac{t}{\lambda}\right), \tag{70}$$

where $g_{\{r,i\}}(x)$ are the one variable functions defined by $g_{\{r,i\}}(x)\stackrel{\triangle}{=}\psi_{\{r,i\}}(1,x)$. If we express the partial differential equation (63) with the initial conditions (64) directly in terms of these functions, then we see that they must solve the Cauchy problems

$$g_r'(x)=\frac{1}{1-x}\,g_r(x);\quad g_r(0)=\frac{\sqrt{\pi}\,\Gamma(\nu+1)}{\Gamma(\nu+\frac{1}{2})}, \tag{71}$$

$$g_i'(x)=\frac{-2\nu}{1-x}\,g_i(x);\quad g_i(0)=-1. \tag{72}$$

Now, the solutions of these problems are easy to find and they are given by

$$g_r(x) = \frac{\sqrt{\pi}\,\Gamma(\nu+1)}{\Gamma(\nu+\frac{1}{2})}\frac{1}{1-x}\,; \tag{73}$$

$$g_i(x) = -(1-x)^{2\nu}. \tag{74}$$

Inverting the preceding steps, we obtain the expression of $\psi(\lambda, t)$, for $\lambda > 0$, as

$$\psi(\lambda,t) = \begin{cases} \frac{\sqrt{\pi}\,\Gamma(\nu+1)}{\Gamma(\nu+\frac{1}{2})}\dfrac{\lambda^{\nu+\frac{1}{2}}}{\lambda - t} - i\lambda^{-\nu-\frac{1}{2}}\,(\lambda-t)^{2\nu}, & \text{for } |t| < \lambda \\ 0 & \text{for } |t| \geq \lambda \end{cases} \tag{75}$$

with $\psi(-\lambda, t) = \overline{\psi(\lambda, t)}$. We can now summarize all the preceding developments by

Theorem 5.17. *Let $\nu \geq -\frac{1}{2}$ be a real number and let $F(\lambda, z)$, where λ is some fixed positive real number, be an element of $\mathcal{H}_\lambda^\nu$. Then $F(\lambda, z)$ is expressible in the form*

$$F(\lambda, z) = \frac{1}{2\pi}\int_{-\lambda}^{\lambda} \varphi(x) \int_{-\infty}^{\infty} \psi(x,t) e^{-itz}\,dt\,dx, \tag{76}$$

for some $\varphi \in L_2$, with support in $[-\lambda, \lambda]$, and where $\psi(x, t)$ is the function defined by (75).

Proof. Use equations (55) and (75) in Theorem 5.13. □

Corollary 5.18. *A linear self-similar system with parameter $\nu \geq -\frac{1}{2}$ has, at the scale $\lambda > 0$, an impulse response $h(\lambda, t)$ of the form*

$$\begin{aligned} h(\lambda,t) =& \frac{\sqrt{\pi}\,\Gamma(\nu+1)}{\Gamma(\nu+\frac{1}{2})}\int_t^{\lambda} \frac{x^{\nu+\frac{1}{2}}[\varphi(x)+\varphi(-x)]}{x-t}\,dx \\ & - i\int_t^{\lambda} x^{-\nu-\frac{1}{2}}[\varphi(x)-\varphi(-x)](x-t)^{2\nu}\,dx \end{aligned} \tag{77}$$

At each fixed scale λ, the transfer function of the system belongs to the de Branges homogeneous space $\mathcal{H}_\lambda^\nu$.

Proof. By the Fubini theorem, we can invert the order of integration in (76) to obtain

$$\begin{aligned} F(\lambda, z) &= \frac{1}{2\pi}\int_{-\infty}^{\infty}\left[\int_{-\lambda}^{0} \varphi(x)\,\psi(x,t)\,dx + \int_0^{\lambda} \varphi(x)\,\psi(x,t)\,dx\right] e^{-itz}dt \\ &= \frac{1}{2\pi}\int_{-\infty}^{\infty} \underbrace{\int_0^{\lambda}\left[\varphi(-x)\,\overline{\psi(x,t)} + \varphi(x)\,\psi(x,t)\right]dx}_{\triangleq h(\lambda,t)}\; e^{-itz}dt. \end{aligned}$$

Since $\psi(x, t)$ vanishes for $x < t$, we have

$$h(\lambda,t) = \int_t^{\lambda}\left[\varphi(-x)\,\overline{\psi(x,t)} + \varphi(x)\,\psi(x,t)\right]dx, \tag{78}$$

and the result follows by using the expression of $\psi(x, t)$ given in (75). □

5.3.3. STATE SPACE DESCRIPTION Linear self-similar systems belong to the class of distributed parameter systems and as such, they are clearly of infinite dimension. Note however that this appeared only implicitly through the impulse response description and even less in the transfer function space. Important enough, the generalization to infinite dimensions of the fundamental state-space concepts has long been studied [5] [14] and these concepts are by now well established for infinite-dimensional linear systems [9]. So, we give in this paragraph, a (A, B, C) description of self-similar systems in an abstract state-space form, on an infinite-dimensional Hilbert space $\mathfrak{L}$ (to be specified below) as the state space. The state operator A is the generator of a strongly continuous semigroup on $\mathfrak{L}$ (see, *e.g.*, [10] for the theory of semigroups of operators) and B and C are respectively the input and output operators.

To see what these operators look like, let us consider the input-output relation of a linear self-similar system, namely

$$y(\lambda, t) = h(\lambda, t) \star u(t) = \int_{-\lambda}^{\lambda} \varphi(x)\psi(x, t) \star u(t)dx \tag{79}$$

where we have used Fubini's theorem. Let $\xi(x, t)$ be defined by the convolution

$$\xi(x, t) = \psi(x, t) \star u(t) \tag{80}$$

If the input u is bounded, then $\xi(x, t)$ is, by (56), an element of $L_2([-\lambda, \lambda])$ as a function of x. Now, using (61), it follows that $\xi(x, t)$ satisfies the hyperbolic partial differential equation

$$\frac{\partial \xi(x, t)}{\partial t} = \frac{\nu + \frac{1}{2}}{x} \xi(-x, t) - \frac{\partial \xi(x, t)}{\partial x} + \psi(x, 0)u(t), \quad \text{for a.e. } x \tag{81}$$

Define by σ the argument sign reversal operator defined on $L_2([-\lambda, \lambda])$ by $\sigma f(x) = f(-x)$. Then, choosing $\mathfrak{L} = L_2([-\lambda, \lambda])$ as the state space and $\xi(\cdot, t)$ as the state, one can put the input-output relation (79) in an abstract state space form, as

$$\begin{aligned} \frac{d}{dt} \xi(\cdot, t) &= A\xi(\cdot, t) + Bu(t) \\ y(\lambda, t) &= C\xi(\cdot, t) \end{aligned} \tag{82}$$

where A is the partial differential operator defined by

$$A = \frac{\nu + \frac{1}{2}}{x} \sigma - \frac{\partial}{\partial x}$$

the output operator C is the linear functional defined by φ by

$$Cf = \int_{-\lambda}^{\lambda} \varphi(x) f(x)dx$$

and the input operator B is given by $B = \psi(x, 0)$.

6. Conclusion

In this paper, we give a frequency domain review of the model for linear self-similar systems, proposed by Yazıcı and Kashyap [33]. This review has allowed us to identify the fractional Hardy space $\mathcal{H}_2^\nu$, $\nu > -1$, as the corresponding transfer function space. An important feature of this space is its invariance to the transformation $F(s) \to \alpha^{1+\nu'} F(\alpha s)$, with $\nu' = \frac{\nu-1}{2}$. As the shift-invariance property for the classical Hardy spaces is intimately related to the time-invariance for linear systems, so the above-mentioned invariance of $\mathcal{H}_2^\nu$ may be thought as a basic property, which defines the self-similar nature of a linear system. Furthermore, note that the fractional Hardy space $\mathcal{H}_2^\nu$ admits a factorization in a form similar to the Beurling-Lax characterization of shift-invariant subspaces. Following this idea, we define a linear self-similar system as any system whose transfer function belongs to a subspace invariant to the above-mentioned transformation and vice versa.

Our definition of self-similarity has prompted us to investigate the general structure of invariant subspaces to the transformation we deal with. Whereas we are concerned with transfer function spaces, it was natural to expect these invariant subspaces to be related to the Paley-Wiener spaces. Then, we have given a brief exposition of de Branges's generalization of the Paley-Wiener theory. Fortunately, the de Branges theory of Hilbert spaces of entire functions completely characterizes the subspaces for which a transformation of the above-mentioned form is an isometry. These spaces are called homogeneous.

We have attached to de Branges's homogeneous spaces the usual meaning of a transfer function space. Then, returning back to the signal space (*i.e., the time domain*), we have given explicitly the structure of the impulse response corresponding to linear self-similar systems. These systems have also been described in an abstract state space, involving a hyperbolic partial differential operator as the state operator. The input and output operators have also been given explicitly. Finally, note that this description exposes the infinite-dimensional nature of self-similar systems more directly than does the impulse response description and much less the transfer function description.

References

[1] M. Abramowitz and I.A. Stegun, eds., *Handbook of mathematical functions*, Dover Publications, 1965.

[2] P. Abry, P. Gonçalvès, and P. Flandrin, *Wavelets, spectrum analysis and 1/f processes*, in Wavelets and statistics, Lecture notes in Statistics, 1995.

[3] J. Adounnet and G. Montseny, *Modèles non éréditaires pour l'analyse et le contrôle de systèmes à mémoire longue de type fractionnaire.* Prepublication, 1998.

[4] D. Alpay, *Algorithme de Schur, espaces à noyau reproduisant et théorie des systèmes*, vol. 6 of Panoramas et Synthèses, Société mathématique de France, 1998.

[5] J.S. BARAS AND P. DEWILDE, *Invariant subspace methods in linear multivariable-distributed systems and lumped-distributed network synthesis*, Proc. of the IEEE, 64 (1976), 160–178.

[6] J.A. BARNES AND D.W. ALLAN, *A statistical model of flicker noise*, Proceedings of the IEEE, 54 (1966), 176–178.

[7] A. BEURLING, *On two problems concerning linear transformations in Hilbert space*, Acta Mathematica, 81 (1949), 239–255.

[8] P. CARMONA, L. COUTIN, AND G. MONTSENY, *A diffusive Markovian representation of fractional Brownian motion with hurst parameter less than 1/2.* Prepublication, 1998.

[9] R.F. CUTAIN AND H.J. ZWART, *An introduction to infinite-dimensional linear systems theory*, Texts in Applied Mathematics, Springer-Verlag, 1995.

[10] N. DUNFORD AND J.T. SCHWARTZ, *Linear operators - Part I*, Interscience Publishers, New York, 1957.

[11] L. GOLINSKII AND I. MIKHAILOVA, *Hilbert spaces of entire functions as a J theory subject*, in Operator Theory: Advances and Applications, V. Potapov, ed., vol. 95, Birkhäuser Verlag, Basel, 1997, 205–251.

[12] C.W.J. GRANGER AND R. JOYEUX, *An introduction to long memory time series models and fractional differencing*, Journ. of times series analysis, 1 (1980), 15–29.

[13] D. HELESCHEWITZ AND D. MATIGNON, *Diffusive realisations of fractional integro-differential operators: structural analysis under approximation*, in Proc. IFAC'98, System, Structure and Control, vol. 2, Nantes, France, July 1998, 243–248.

[14] J.W. HELTON, *Systems with infinite-dimensional state space: the Hilbert space approach*, Proc. of the IEEE, 64 (1976), 145–160.

[15] K. HOFFMAN, *Banach Spaces of Analytic Functions*, Modern analysis, Prentice-Hill, N.J., 1962.

[16] J.R.M. HOSKING, *Fractional differencing*, Biometrika, 68 (1981), 165–176.

[17] M.G. KREIN, *On the theory of entire matrix-functions of exponential type*, in Operator Theory: Advances and Applications, V. Potapov, ed., vol. 95, Birkhäuser Verlag, Basel, 1997, 361–371.

[18] P. LAX, *Translation invariant subspaces*, Acta Mathematica, 101 (1959).

[19] W. LELAND, M. TAQQU, W. WILLINGER, AND D. WILSON, *On the self-similar nature of Ethernet traffic (extended version)*, IEEE/ACM Trans. Networking, (1994), 1–15.

[20] O. MAGRÉ AND M. GUGLIELMI, *Modeling and analysis of fractional Brownian motions*, Chaos, Solitons and Fractals, 8 (1997), 377–388.

[21] B.B. MANDELBROT AND J.W.V. NESS, *Fractional Brownian motions, fractional noises and applications*, SIAM Review, 10 (1968), 422–437.

[22] L. DE BRANGES, *Espaces Hilbertiens de Fonctions Entières*, Masson, Paris, 1972.

[23] Y. MEYER, F. SELLAN, AND M. TAQQU, *Wavelets, generalised white noise and fractional integration: the synthesis of fractional Brownian motion.* Prepublication, 1997.

[24] G. Montseny, *Diffusive representation of pseudo-differential time-operators*, in ESAIM: Proceedings, vol. 5, SMF, 1998, 159–175.
URL: http://www.emath.fr/proc/vol.5/.

[25] G. Montseny, J. Audounet, and B. Mbodje, *Optimal models of fractional integrators and application to systems with fading memory*, in IEEE System, Man and Cybernetics Conference, Le Touquet, France, October 1993, 65–70.

[26] K. Nishimoto, *Fractional derivative and integrals (Part I)*, J. Coll. Eng. Nihon Uni., Japan, 1976.

[27] E. Noret, *Contribution à l'étude des signaux auto-similaires et à mémoire longue au moyen de systèmes linéaires non-stationnaires*, PhD thesis, Université de Nantes, 1999.

[28] J.R. Partington, *An Introduction to Hankel operators*, Cambridge University Press, 1988.

[29] W.H. Press, S.A. Teukolsky, W.T. Vetterling, and B.P. Flannery, *Numerical Recipes in C*, Cambridge Univ. Press, 1997.

[30] W. Rudin, *Real and complex analysis*, McGraw-Hill, 1966.

[31] S. Saitoh, *Theory of reproducing kernels and its applications*, Pitman Research Notes in Mathematics, Longman Scientic & Technical, UK, 1988.

[32] S.G. Samko, A. Kilbas, and I. Marichev, *Fractional integrals and derivatives*, Gordon & Breach Science, 1993.

[33] B. Yazici and R.L. Kashyap, *A class of second-order stationary self-similar processes for $1/f$ phenomena*, IEEE Trans. on Signal Processing, 45 (1997), 396–410.

Mamadou Mboup
UFR de Mathématiques et Informatique – CRIP5
Université René Descartes-Paris V
45, rue des Saints-Pères
F-75270 Paris cedex 06, France
e-mail: mboup@math-info.univ-paris5.fr

Operator Theory:
Advances and Applications, Vol. 143, 303–312

Reproducing Kernels and a Family of Bounded Linear Operators

Saburou Saitoh

Abstract. We consider a generalization of the Pythagorean theorem with geometric meanings and from the generalization we were able to obtain a general and fundamental concept for the inversion of a family of bounded linear operators with continuous parameters on a Hilbert space into various Hilbert spaces. After reviewing the applications to linear mappings in the framework of Hilbert spaces of the general theory of reproducing kernels, we shall state the results for the case of operator versions. We shall consider solutions and generalized solutions of general bounded linear operator equations with continuous parameters on a Hilbert space.

We consider any positive matrix $K(p,q)$ on E; that is, for an abstract set E and for a complex-valued function $K(p,q)$ on $E \times E$, it satisfies that for any finite points $\{p_j\}$ of E and for any complex numbers $\{C_j\}$,

$$\sum_j \sum_{j'} C_j \overline{C_{j'}} K(p_{j'}, p_j) \geq 0.$$

Then, by the fundamental theorem by Moore–Aronszajn, we have:

Proposition 1 ([1]) *For any positive matrix $K(p,q)$ on E, there exists a uniquely determined functional Hilbert space H_K(RKHSH_K) comprising functions $\{f\}$ on E and admitting the reproducing kernel $K(p,q)$ satisfying and characterized by*

$$K(\cdot, q) \in H_K \text{ for any } q \in E \tag{1}$$

and, for any $q \in E$ and for any $f \in H_K$

$$f(q) = (f(\cdot), K(\cdot, q))_{H_K}. \tag{2}$$

For some general properties for reproducing kernel Hilbert spaces and for various constructions of the RKHS H_K from a positive matrix $K(p,q)$, see the recent book [15] and its Chapter 2, Section 5, respectively.

2000 *Mathematics Subject Classification.* 44A05, 46E22.
Key words and phrases. Operator equations, reproducing kernels, transforms.

1. Connections with linear mappings

Let us connect linear mappings in the framework of Hilbert spaces with reproducing kernels ([8]).

For an abstract set E and for any Hilbert (possibly finite-dimensional) space $\mathcal{H}$, we shall consider an $\mathcal{H}$-valued function $\boldsymbol{h}$ on E

$$\boldsymbol{h}: \quad E \longrightarrow \mathcal{H} \tag{3}$$

and the linear mapping for $\mathcal{H}$

$$f(p) = (\boldsymbol{f}, \boldsymbol{h}(p))_{\mathcal{H}} \quad \text{for} \quad \boldsymbol{f} \in \mathcal{H} \tag{4}$$

into a linear space comprising functions on E. For this linear mapping (4), we form the positive matrix $K(p,q)$ on E defined by

$$K(p,q) = (\boldsymbol{h}(q), \boldsymbol{h}(p))_{\mathcal{H}} \quad \text{on} \quad E \times E. \tag{5}$$

Then, we have the following fundamental results:

(I) For the RKHS H_K admitting the reproducing kernel $K(p,q)$ defined by (5), the images $\{f(p)\}$ by (4) for $\mathcal{H}$ are characterized as the members of the RKHS H_K.

(II) In general, we have the inequality in (4)

$$\|f\|_{H_K} \leq \|\boldsymbol{f}\|_{\mathcal{H}}, \tag{6}$$

however, for any $f \in H_K$ there exists a uniquely determined $\boldsymbol{f}^* \in \mathcal{H}$ satisfying

$$f(p) = (\boldsymbol{f}^*, \boldsymbol{h}(p))_{\mathcal{H}} \quad \text{on } E \tag{7}$$

and

$$\|f\|_{H_K} = \|\boldsymbol{f}^*\|_{\mathcal{H}}. \tag{8}$$

In (6), the isometry holds if and only if $\{\boldsymbol{h}(p); p \in E\}$ is complete in $\mathcal{H}$.

(III) We can obtain the inversion formula for (7) in the form

$$f \longrightarrow \boldsymbol{f}^*, \tag{9}$$

by using the RKHS H_K.

However, this inversion formula will depend on, case by case, the realizations of the RKHS H_K.

(IV) Conversely, if we have an isometric mapping $\tilde{L}$ from a RKHS H_K admitting a reproducing kernel $K(p,q)$ on E onto a Hilbert space $\mathcal{H}$, then the mapping is linear and its isometric inversion $\tilde{L}^{-1}$ is represented in the form (4). Here, the Hilbert space $\mathcal{H}$-valued function $\boldsymbol{h}$ satisfying (3) and (4) is given by

$$\boldsymbol{h}(p) = \tilde{L}K(\cdot, p) \text{ on } E \tag{10}$$

and, then $\{\boldsymbol{h}(p); p \in E\}$ is complete in $\mathcal{H}$.

When (4) is isometrical, sometimes we can use the isometric mapping for a realization of the RKHS H_K, conversely – that is, if the inverse L^{-1} of the linear mapping (4) is known, then we have $\|f\|_{H_K} = \|L^{-1}f\|_{\mathcal{H}}$.

We shall state some general applications of the results (I)∼(IV) to several wide subjects and their basic references:

(1) Linear mappings ([8],[12]).
The fact that the image spaces of linear mappings in the framework of Hilbert spaces are characterized as reproducing kernel Hilbert spaces defined by (5) is the most important one in the general theory of reproducing kernels. Therefore, the fact will mean that the theory of reproducing kernels is fundamental and a general concept in mathematics. To look for the characterization of the image space is a starting point when we consider the linear equation (4). (II) gives a generalization of the Pythagorean theorem (see also [7]) and means that in the general linear mapping (4) there exists essentially an isometric identity between the input and the output. (III) gives a generalized (natural) inverse (solution) of the linear mapping (equation) (4). (IV) gives a general method determining and constructing the linear system from an isometric relation between outputs and inputs by using the reproducing kernel in the output space.

(2) Linear mappings among smooth functions ([19]).
We considered linear mappings in the framework of Hilbert spaces, however, we can consider linear mappings in the framework of Hilbert spaces comprising smooth functions, similarly. Conversely, reproducing kernel Hilbert spaces are considered as the images of some Hilbert spaces by considering some decomposed representations (5) of the reproducing kernels. Such decomposition is, in general, possible. This idea is important in [19] and also in the following items (6) and (7).

(3) Nonharmonic linear mappings ([9]).
If the linear system vectors $\boldsymbol{h}(p)$ move in a small way (perturbation of the linear system) in the Hilbert space $\mathcal{H}$, then we cannot calculate the related positive matrix (5), however, we can discuss the inversion formula and an isometric identity of the linear mapping. The prototype result is the Paley-Wiener theorem on nonharmonic Fourier series.

(4) Various norm inequalities ([9],[13]).
Relations among positive matrices correspond to those of the associated reproducing kernel Hilbert spaces, by the minimum principle. So, we can derive various norm inequalities among reproducing kernel Hilbert spaces. We were able to derive many beautiful norm inequalities.

(5) Nonlinear mappings ([13],[16]).
In a very general nonlinear mapping of a reproducing kernel Hilbert space, we can look for a natural reproducing kernel Hilbert space containing the image space and furthermore, we can derive a natural norm inequality in the nonlinear mapping. How to catch nonlinearity in connection with linearity? It seems that the theory of reproducing kernels gives a fundamental and interesting answer for this question.

(6) Linear integral equations ([20]).

(7) Linear differential equations with variable coefficients ([20]).
In linear integro-differential equations with general variable coefficients, we can discuss the existence and construction of the solutions, if the solutions exist. This method is called a backward transformation method and by reducing the equations to Fredholm integral equations of the first kind -(4)- and we can discuss the classical solutions, in very general linear equations.

(8) Approximation theory ([3],[2]).
Reproducing kernel Hilbert spaces are very nice function spaces, because the point evaluations are continuous. Then, the reproducing kernels are a fundamental tool in the related approximation theory.

(9) Representations of inverse functions ([14]).
For any mapping, we discussed the problem of representing its inverse in term of the direct mapping and we derived a unified method for this problem. As a simple example, we can represent the Taylor coefficients of the inverse of the Riemann mapping function on the unit disc on the complex plane in terms of the Riemann mapping function. This fact was important in the representation of analytic functions in terms of local data in ([23],[24]).

(10) Various operators among Hilbert spaces ([17]).
Among various abstract Hilbert spaces, we can introduce various operators of sum, product, integral and derivative by using the linear mapping (4) or very general nonlinear mappings. The prototype operator is convolution and we discussed it from a wide and general viewpoint with concrete examples.

(11) Sampling theorems ([15], Chapter 4, Section 2; [5]).
The Whittaker-Kotel'nikov-Shannon sampling theorem may be interpretated by (I) and (II), very well and we can discuss the truncation errors in the sampling theory. J.R. Higgins [5] established a fully general theory for [15].

(12) Interpolation problems of Pick-Nevanlinna type ([9],[10]).
General and abstract theory of interpolation problems of Pick-Nevanlinna type may be discussed by using the general theory of reproducing kernels.

(13) Analytic extension formulas and their applications ([21],[11]).
We were able to obtain various analytic extension formulas and their applications from various isometric identities (II). For their applications to nonlinear partial differential equations, see the survey article by N. Hayashi [4].

In this survey article, we shall present also new results on

(14) Inversions of a family of bounded linear operators on a Hilbert space into various Hilbert spaces, which are generalizations of [22] and [7].

Furthermore, in connection with Kaczmarz's Method for a finite number of bounded linear operator equations on a Hilbert space, we shall give our generalized solutions for general operator equations with continuous parameters.

2. Operator versions

We shall give operator versions of the fundamental theory (I)~(IV) which may be expected to have many concrete applications. In particular, for full generalizations of the Pythagorean theorem with geometric meanings, see [7]. Some special versions were given in [22].

For an abstract set $\mathbf{\Lambda}$, we shall consider an operator-valued function L_λ on $\mathbf{\Lambda}$,

$$\lambda \longrightarrow L_\lambda \tag{11}$$

where L_λ are bounded linear operators from a Hilbert space $\mathcal{H}$ into various Hilbert spaces $\mathcal{H}_\lambda$,

$$L_\lambda : \mathcal{H} \longrightarrow \mathcal{H}_\lambda. \tag{12}$$

In particular, we are interested in the inversion formula

$$L_\lambda x \longrightarrow x, \quad x \in \mathcal{H}. \tag{13}$$

Here, we consider $\{L_\lambda x; \lambda \in \Lambda\}$ as information obtained from x and we wish to determine x from the information. However, the information $L_\lambda x$ belong to various Hilbert spaces $\mathcal{H}_\lambda$, and so, in order to unify the information in a sense, we shall take fixed elements $\mathbf{b}_{\lambda,\omega} \in \mathcal{H}_\lambda$ and consider the linear mapping from $\mathcal{H}$

$$\begin{aligned} X_{\mathbf{b}}(\lambda,\omega) &= (L_\lambda x, \mathbf{b}_{\lambda,\omega})_{\mathcal{H}_\lambda} \\ &= (x, L_\lambda^* \mathbf{b}_{\lambda,\omega})_{\mathcal{H}}, \quad x \in \mathcal{H} \end{aligned} \tag{14}$$

into a linear space comprising functions on $\mathbf{\Lambda} \times \mathbf{\Omega}$. For the information $L_\lambda x$, we shall consider $X_{\mathbf{b}}(\lambda,\omega)$ as observations (measurements, in fact) for x depending on λ and ω. For this linear mapping (14), we form the positive matrix $K_{\mathbf{b}}(\lambda,\omega;\lambda',\omega')$ on $\mathbf{\Lambda} \times \mathbf{\Omega}$ defined by

$$\begin{aligned} K_{\mathbf{b}}(\lambda,\omega;\lambda',\omega') &= (L_{\lambda'}^* \mathbf{b}_{\lambda',\omega'}, L_\lambda^* \mathbf{b}_{\lambda,\omega})_{\mathcal{H}} \\ &= (L_\lambda L_{\lambda'}^* \mathbf{b}_{\lambda',\omega'}, \mathbf{b}_{\lambda,\omega})_{\mathcal{H}_\lambda} \quad \text{on} \quad \mathbf{\Lambda} \times \mathbf{\Omega}. \end{aligned} \tag{15}$$

Then, as in (I)~(IV), we have the following fundamental results:

(I′) For the RKHS $H_{K_{\mathbf{b}}}$ admitting the reproducing kernel $K_{\mathbf{b}}(\lambda,\omega;\lambda',\omega')$ defined by (15), the images $\{X_{\mathbf{b}}(\lambda,\omega)\}$ by (14) for $\mathcal{H}$ are characterized as the members of the RKHS $H_{K_{\mathbf{b}}}$.

(II′) In general, we have the inequality in (14)

$$\|X_{\mathbf{b}}\|_{H_{K_{\mathbf{b}}}} \leq \|x\|_{\mathcal{H}}, \tag{16}$$

however, for any $X_{\mathbf{b}} \in H_{K_{\mathbf{b}}}$ there exists a uniquely determined $x' \in \mathcal{H}$ satisfying

$$X_{\mathbf{b}}(\lambda,\omega) = (x', L_\lambda^* \mathbf{b}_{\lambda,\omega})_{\mathcal{H}} \quad \text{on} \quad \mathbf{\Lambda} \times \mathbf{\Omega} \tag{17}$$

and

$$\|X_{\mathbf{b}}\|_{H_{K_{\mathbf{b}}}} = \|x'\|_{\mathcal{H}}. \tag{18}$$

In (16), the isometry holds if and only if $\{L_\lambda^* \mathbf{b}_{\lambda,\omega}; (\lambda,\omega) \in \mathbf{\Lambda} \times \mathbf{\Omega}\}$ is complete in $\mathcal{H}$.

(III′) We can obtain the inversion formula for (14) and so, for the mapping (13) as in (III), in the form

$$L_\lambda x \longrightarrow (L_\lambda x, \mathbf{b}_{\lambda,\omega})_{\mathcal{H}_\lambda} = X_{\mathbf{b}}(\lambda,\omega) \longrightarrow x', \tag{19}$$

by using the RKHS $H_{K_{\mathbf{b}}}$.

(IV′) Conversely, if we have an isometric mapping $\tilde{L}$ from a RKHS $H_{K_{\mathbf{b}}}$ admitting a reproducing kernel $K_{\mathbf{b}}(\lambda,\omega;\lambda',\omega')$ on $\mathbf{\Lambda}\times\mathbf{\Omega}$ in the form (15) using bounded linear operators L_λ and fixed vectors $\mathbf{b}_{\lambda,\omega}$ onto a Hilbert space $\mathcal{H}$, then the mapping $\tilde{L}$ is linear and the isometric inversion $\tilde{L}^{-1}$ is represented in the form (14) by using

$$L_\lambda^* \mathbf{b}_{\lambda,\omega} = \tilde{L} K_{\mathbf{b}}(\cdot,\cdot;\lambda,\omega) \text{ on } \mathbf{\Lambda}\times\mathbf{\Omega}. \tag{20}$$

Further, then $\{L_\lambda^* \mathbf{b}_{\lambda,\omega}; (\lambda,\omega) \in \mathbf{\Lambda}\times\mathbf{\Omega}\}$ is complete in $\mathcal{H}$.

The author obtained the above concept for the operator versions from a generalization of the Pythagorean theorem in the following way:

Let $x \in \mathbf{R}^n$ and $\{e_j\}_{j=1}^n$ be linearly independent unit vectors. We consider the linear mappings

$$L : x \longrightarrow \{x - (x, e_j)e_j\}_{j=1}^n \tag{21}$$

from $\mathbf{R}^n$ into $\mathbf{R}^n$. Then we wish to establish an isometric identity and inversion formula in the operators. Recall the Pythagorean theorem for $n = 2$. By our operator versions, we can establish the desired results.

Note that in (21), for $n \geq 3$ if we consider

$$\{\|x - (x, e_j)e_j\|\}_{j=1}^n \tag{22}$$

as scalar-valued mappings, then the mappings are not linear. So, we must consider the operator-valued mappings (21) in order to obtain isometric mappings in the framework of Hilbert spaces.

We found the book [6] and we see that some related equations were considered as in the following way ([6], pages 128–157):

Let $H, H_j; j = 1, 2, \ldots, p$ be Hilbert spaces and let

$$R_j : H \longrightarrow H_j, \quad j = 1, 2, \ldots, p \tag{23}$$

be linear continuous maps from H onto H_j. Let $g_j \in H_j$ be given. Then, consider the problem to compute $f \in H$ such that

$$R_j f = g_j, \quad j = 1, 2, \ldots, p. \tag{24}$$

These equations are very important in the theory of computerized tomography by discretization. The typical method is Kaczmarz's Method based on an iterative method by using the orthogonal projections P_j in H onto the affine subspaces $R_j f = g_j$.

Our direct solutions for (24) seem that the result is stable for the sake of the use (24) as data, because we use (19) which is given by the inner product.

In general, in equations (24) we have noises and errors for the data g_j and so, in those cases the equations do, in general, not have solutions. So, we will consider a more general solution which is called a generalized solution (inverse) in the next section.

3. Generalized solutions and best approximations

In order to represent our generalized solutions explicitly, we shall consider bounded linear operators on a reproducing kernel Hilbert space. So, we consider the Hilbert space H_K on E stated in Section 1. We consider H as H_K in Section 3. We assume that the direct integral

$$\mathbf{H} = \int_{\mathbf{\Lambda}}^{\oplus} \mathbf{H}_\lambda d\mu(\lambda) \tag{25}$$

of the Hilbert spaces $\mathbf{H}_\lambda$ on $\mathbf{\Lambda}$ converges with a σ finite positive measure $d\mu$ on $\mathbf{\Lambda}$. We assume that the bounded linear operators L_λ in (12) are bounded on H_K into $\mathbf{H}$ in the sense:

$$\int_{\mathbf{\Lambda}} \| L_\lambda f \|_{\mathbf{H}_\lambda}^2 \, d\mu(\lambda) \leq M \| f \|_{H_K}^2 \tag{26}$$

for some constant $M \geq 0$.

In this setting, we consider the extremal problem:

$$\inf_{f \in H_K} \int_{\mathbf{\Lambda}} \| L_\lambda f - g(\lambda) \|_{\mathbf{H}_\lambda}^2 \, d\mu(\lambda), \tag{27}$$

which gives a generalized solution for the equations

$$L_\lambda f = g(\lambda) \quad \text{on} \quad H_K \quad \text{and} \quad \mathbf{H}_\lambda. \tag{28}$$

We shall write the operators $\{L_\lambda\}$ as L from H_K into $\mathbf{H}$ in the sense (26). Let L^* be the adjoint operator of L from $\mathbf{H}$ into H_K. We form the positive matrix

$$k(p,q) = (L^*LK(\cdot,q), L^*LK(\cdot,p))_{H_K} \quad \text{on} \quad E \times E. \tag{29}$$

Then, we obtain

Theorem 1 *For a function $g \in \mathbf{H}$, there exists a function $\tilde{f}$ in H_K such that*

$$\inf_{f \in H_K} \int_{\mathbf{\Lambda}} \| L_\lambda f - g(\lambda) \|_{\mathbf{H}_\lambda}^2 \, d\mu(\lambda) = \int_{\mathbf{\Lambda}} \| L_\lambda \tilde{f} - g(\lambda) \|_{\mathbf{H}_\lambda}^2 \, d\mu(\lambda) \tag{30}$$

if and only if, for the RKHS H_k

$$L^*g \in H_k. \tag{31}$$

Furthermore, if there exist the best approximations $\tilde{f}$ satisfying (30), then there exists a unique extremal function $\check{f}$ with the minimum norm in H_K, and this function is expressible in the form

$$\check{f}(p) = (L^*g, L^*LK(\cdot,p))_{H_k} \quad \text{on} \quad E. \tag{32}$$

In this theorem, note that

$$(L^*g)(p) = (L^*g, K(\cdot,p))_{H_K} = (g, LK(\cdot,p))_{\mathbf{H}}; \tag{33}$$

that is, the adjoint operator L^* is expressible in terms of $g, L, K(\cdot,p)$ and $\mathbf{H}$.

For a proof of this theorem, we can apply the argument in [3].

As a simple example, we shall consider the space H_K on $[0,\infty)$ for $K(x,y) = \min\{x,y\}$. This space is composed of all absolutely continuous real-valued functions on $[0,\infty)$ and $f(0) = 0$ equipped with the norm

$$\|f\|_{H_K}^2 = \int_0^\infty f'(x)^2 dx.$$

As a space $\mathbf{H}$ we consider the space $L_2((0,\infty), e^{-\lambda}d\lambda)$ and a bounded linear operator L:

$$L_\lambda f = \int_0^\lambda f(\xi)d\xi \tag{34}$$

from H_K into $\mathbf{H}$. Then, the adjoint operator L^* from $\mathbf{H}$ into H_K is given by

$$(L^*g)(x) = \frac{1}{2}\int_0^x g(\lambda)\lambda^2 e^{-\lambda}d\lambda + \int_x^\infty g(\lambda)[x\lambda - \frac{1}{2}x^2]e^{-\lambda}d\lambda$$

and we can discuss the problem, for any $g \in \mathbf{H}$,

$$\inf_{f\in H_K} \int_0^\infty |\int_0^\lambda f(\xi)d\xi - g(\lambda)|^2 e^{-\lambda}d\lambda. \tag{35}$$

We can give a complete solution for the problem. We would like to discuss those concrete problems in separate papers.

We are interested in some concrete results for typical problems such as generalized solutions for ordinary differential equations in connection with reproducing kernels, Green's functions and the related completeness in (32).

Acknowledgments

The author wishes to express his deep thanks Professor Daniel Alpay for his kind invitation to this special volume.

References

[1] N. Aronszajn. Theory of reproducing kernels. *Trans. Amer. Math. Soc.*, **68** (1950), 337–404.

[2] D.-W. Byun and S. Saitoh. Approximation by the solutions of the heat equation. *J. Approximation Theory*, **78** (1994), 226–238.

[3] D.-W. Byun and S. Saitoh. Best approximation in reproducing kernel Hilbert spaces. *Proc. of the 2th International Colloquium on Numerical Analysis, VSP-Holland*, (1994), 55–61.

[4] N. Hayashi. Analytic function spaces and their applications to nonlinear evolution equations. *Analytic Extension Formulas and their Applications*, (2001), Kluwer Academic Publishers, 59–86.

[5] J.R. Higgins. A sampling principle associated with Saitoh's fundamental theory of linear transformations. *Analytic Extension Formulas and their Applications*, (2001), Kluwer Academic Publishers, 73–86.

[6] F. Natterer. The Mathematics of Computerized Tomography, SIAM, *In Applied Mathematics* 32 (2001), Philadelphia.

[7] Th.M. Rassias and S. Saitoh. The Pythagorean theorem and linear mappings. *PanAmerican Math. J.* 12(2002), 1–10.

[8] S. Saitoh. Hilbert spaces induced by Hilbert space valued functions. *Proc. Amer. Math. Soc.*, **89** (1983), 74–78.

[9] S. Saitoh. Theory of Reproducing Kernels and its Applications. *Pitman Research Notes in Mathematics Series*, **189** (1988), Longman Scientific & Technical, UK.

[10] S. Saitoh. Interpolation problems of Pick-Nevanlinna type. *Pitman Research Notes in Mathematics Series*, **212** (1989), 253–262.

[11] S. Saitoh. Representations of the norms in Bergman-Selberg spaces on strips and half-planes. *Complex Variables*, **19** (1992), 231–241.

[12] S. Saitoh. One approach to some general integral transforms and its applications. *Integral Transforms and Special Functions*, **3** (1995), 49–84.

[13] S. Saitoh. Natural norm inequalities in nonlinear transforms. *General Inequalities* **7**, (1997), 39–52. Birkhäuser Verlag, Basel, Boston.

[14] S. Saitoh. Representations of inverse functions. *Proc. Amer. Math. Soc.*, **125** (1997), 3633–3639.

[15] S. Saitoh. Integral Transforms, Reproducing Kernels and their Applications. *Pitman Research Notes in Mathematics Series*, **369** (1997). Addison Wesley Longman, UK.

[16] S. Saitoh. Nonlinear transforms and analyticity of functions. *Nonlinear Mathematical Analysis and Applications*, (1998), 223–234. Hadronic Press, Palm Harbor.

[17] S. Saitoh. Various operators in Hilbert space induced by transforms. *International J. of Applied Math.*, **1** (1999), 111–126.

[18] S. Saitoh. Applications of the general theory of reproducing kernels. *Reproducing Kernels and their Applications*, (1999), Kluwer Academic Publishers, 165–188.

[19] S. Saitoh and M. Yamamoto. Integral transforms involving smooth functions. *Reproducing Kernels and their Applications*, (1999), Kluwer Academic Publishers, 149–164.

[20] S. Saitoh. Linear integro-differential equations and the theory of reproducing kernels. *Volterra Equations and Applications.* C. Corduneanu and I.W. Sandberg (eds.), Gordon and Breach Science Publishers (2000), Amsterdam.

[21] S. Saitoh. Analytic extension formulas, integral transforms and reproducing kernels. *Analytic Extension Formulas and their Applications*, (2001), Kluwer Academic Publishers, 207–232.

[22] S. Saitoh. Applications of the reproducing kernel theory to inverse problems. *Comm. Korean Math. Soc.*, **16** (2001), 371–383.

[23] S. Saitoh and M. Mori. Representations of analytic functions in terms of local values by means of the Riemann mapping function. *Complex Variables*, **45** (2001), 387–393.

[24] S. Saitoh. Principle of telethoscope. *Functional-Analytic and Complex Methods, their Interaction and Applications to Partial Differential Equations – Proceedings of the International Graz Workshop*, Graz, Austria, 12–16 February 2001. World Scientific (2001), 101–117

Saburou Saitoh
Department of Mathematics
Faculty of Engineering
Gunma University
Kiryu 376-8515, Japan
e-mail: ssaitoh@math.sci.gunma-u.ac.jp

Operator Theory:
Advances and Applications, Vol. 143, 313–331

Multipliers in the Reproducing Kernel Hilbert Space, Subnormality and Noncommutative Complex Analysis

Franciszek Hugon Szafraniec

Abstract. Operators of multiplication in the reproducing kernel Hilbert space (cf. footnote [4], p. 315) and their subnormality as a background for noncommutative complex analysis is to be furnished. By the way we proclaim subnormality as a missed cell in the quantization scheme.

Based on [24], our belief is in that (bounded or unbounded) *subnormal* operators have to be put in the table below instead of **?** which, in turn, replaces (somehow enigmatic in the context) "operators" declared to be the quantization correspondent for complex variable; for recent evidence of the latter see [6] and [7].

real variable	selfadjoint operators
complex variable	**?**
⋮	⋮

Further support comes from [15] and this is why we start in this paper with pulling out the results therefrom to the general context of the reproducing kernel Hilbert space. A look at Theorem 4 of [15], the main result therein, makes it evident that most of the items are independent of specific character of the space. Thus atomizing it enables us to perform more systematic study of operators of multiplication, their adjoints as well as these abstract operators which commute in a weak sense with those. The aforesaid commutativity is what Rochberg and Weaver [15] propose to be the quantized analogue of the Cauchy-Riemann equations on the plane. Though they consider the one-dimensional Segal-Bargmann space their philosophy extends to other cases as well: the reproducing kernel Hilbert space

2000 *Mathematics Subject Classification.* Primary 47B20, 47B32, 47A57; Secondary 30C40, 32A25, 33D15.
Key words and phrases. Reproducing kernel Hilbert space; multiplier and operator of multiplication; integrability of RKHS; subnormal system of operators; joint cyclicity; q-calculus.
The research resulting in this paper was initiated when the author was assisted with a support of the KBN, grant # 2 P03A 00417.

may serve well as a settlement for noncommutative complex analysis in other domains (while the multiplication operator looks rather to be solid under different occasions, its adjoint, the 'differential' one, may take contrasting forms). Anyway, a favourable circumstance, if it happens, is the presence of cyclic subnormal operators in the space in question, or, in other words, to have the RKHS *integrable*; the latter is intimately related to *subnormality* of multiplication operators.

The present exposition is rather oriented toward unbounded operators; for instance, the question of integrability we are going to stand up here is invisible in the 'bounded' case to some extend; here it is uppermost along with a pretty high level of difficulty in making any progress. The 'unbounded' case mostly corresponds to the situation when the underlying set (domain) is unbounded in a sense and in fact there is no unique treatment of this kind of RKHS and any observation in this matter, even this which may look like a rather apparent fact, is worthy of being recorded as well.

Multipliers or rather operators of multiplication

Let X be a set. Suppose we are given a couple $(\mathcal{H}, K)$ where $\mathcal{H}$ is a Hilbert space of complex functions on X (with inner product denoted by $\langle \cdot, - \rangle$) and K is a complex function on $X \times X$. The function K is called a *reproducing kernel of* $\mathcal{H}$ and the space $\mathcal{H}$ a *reproducing kernel Hilbert space* (with respect to K) if

$$\begin{aligned} &K_x \in \mathcal{H}, \\ &f(x) = \langle f, K_x \rangle, \end{aligned} \qquad x \in X,\ f \in \mathcal{H}, \tag{1}$$

where $K_x \stackrel{\text{df}}{=} K(\cdot, x)$, sometimes a bit confusingly called a kernel function. We refer to $(\mathcal{H}, K)$ as a *RKHS couple on* X (for anything more in this matter see [23]). Let us remind that, given an orthonormal basis $\boldsymbol{e} \stackrel{\text{df}}{=} \{e_\alpha\}_{\alpha \in \mathrm{A}}$ in $\mathcal{H}$ [1], we always have

$$K_x = \sum_{\alpha \in \mathrm{A}} \overline{e_\alpha(x)} e_\alpha, \quad x \in X; \tag{2}$$

convergent in $\mathcal{H}$ [2] and, consequently,

$$K(x, y) = \sum_{\alpha \in \mathrm{A}} e_\alpha(x) \overline{e_\alpha(y)}, \quad x, y \in X \tag{3}$$

pointwise.

Proposition 1 (RKHS test). *A function* $f : X \mapsto \mathbb{C}$ *is in* $\mathcal{H}$ *if and only if there is* $C > 0$ *such that*

$$\Big|\sum_{i=0}^{N} f(x_i)\lambda_i\Big|^2 \le C^2 \sum_{i,j=1}^{N} K(x_i, x_j)\lambda_i\bar{\lambda}_j, \quad x_1, \ldots, x_N \in X,\ \lambda_1, \ldots \lambda_N \in \mathbb{C}. \tag{4}$$

[1] Notice that a reproducing kernel Hilbert space may be non-separable as well, *exemplum:* $\ell^2(\mathrm{A})$ with A uncountable.

[2] It means that $e_\alpha(x) = 0$ but a countable number of α's depending on x of course.

If this happens, then

$$\|f\| = \inf\{C;\ condition\ (4)\ holds\}.$$

The RKHS test is a kind of *interpolation* result (for the RKHS test and backward references look at [23] at least). An interesting facet of this is that a *quadratic form* argument (4) implies regularity (like analyticity) of members of $\mathcal{H}$, if relevant.

Let us notify the possible convergence interrelations.

Corollary 2. (a) *If $\|f_n - f\| \to 0$ as $n \to \infty$ then $\{f_n\}_{n=0}^{\infty} \subset \mathcal{H}$ converges to f pointwise on X (and uniformly on each subset of X on which $K(x,x)$ is bounded).*

(b) *If a sequence $\{f_n\}_{n=0}^{\infty} \subset \mathcal{H}$ converges to f pointwise on X and the sequence $\{\|f_n\|\}_{n=0}^{\infty}$ of norms is bounded, then f is in $\mathcal{H}$ and $\|f_n - f\| \to 0$ as $n \to \infty$.*

(c) *A net $\{f_\iota\}_{\iota\in I} \subset \mathcal{H}$ converges to f pointwise on X and $f \in \mathcal{H}$ if and only if it converges to f in the $\sigma(\mathcal{H}, \mathcal{D}_K)$ topology.*

Proof. Parts (a) and (c) come immediately from the reproducing kernel property. To show (b) use Proposition 1 and the Hahn-Banach Theorem. □

Let φ be a complex function on X. The (maximal) operator M_φ of multiplication by φ is defined as [3]

$$M_\varphi f \stackrel{\mathrm{df}}{=} \varphi f$$

for

$$f \in \mathcal{D}(M_\varphi) \stackrel{\mathrm{df}}{=} \{f \in \mathcal{H};\ \varphi f \in \mathcal{H}\}.$$

It follows immediately from the reproducing kernel property that M_φ is a closed operator. Moreover we have the following

Proposition 3. *The mapping $M_\varphi : \mathcal{D}(M_\varphi) \mapsto \mathcal{H}$ is continuous (bicontinuous if invertible) when in $\mathcal{D}(M_\varphi)$ the topology is $\sigma(\mathcal{D}(M_\varphi), \mathcal{D}_K)$ and in $\mathcal{H}$ the topology is $\sigma(\mathcal{H}, \mathcal{D}_K)$.*

Proof. For this notice that $\langle \varphi f, K_x\rangle = \varphi(x)\langle f, K_x\rangle$, $f \in \mathcal{D}(M_\varphi)$. □

Letting occasionally a function φ play more active role it is commonly called a multiplier if $\mathcal{D}(M_\varphi) = \mathcal{H}$. Holding this up we may say that a function φ is a *(densely defining) multiplier* if M_φ is a densely defined operator in $\mathcal{H}$; for a given (dense) $\mathcal{D}$ we say that φ is a *$\mathcal{D}$-multiplier* if $\mathcal{D} \subset \mathcal{D}(M_\varphi)$ (notation: $\mathcal{M}(\mathcal{D})$ for the set of all $\mathcal{D}$-multipliers[4]). Exemplifying this mention that the class $\mathcal{M}(\mathcal{D}_K)$ for the Bargmann-Segal space characterized by Newman and Shapiro is in [15] as Theorem 1. In general, by the RKHS test (cf. [23]), we get

[3] Let us remind $\mathcal{D}(A)$ and $\mathcal{R}(A)$ stand for the domain and range of A.

[4] The densely defined context makes it clear that one has to distinguish between multipliers (the functions) and operators of multiplication they define. Thus, for instance, the Newman-Shapiro theorem [14] says something on multipliers while the main result of [15], Theorem 4 deals with operators of multiplication. Now, referring to the content of the present paper, one may say its title is delusive. It is deliberatively so because we want to provoke attention of those who are a bit careless playing on words.

Proposition 4. *Suppose $\mathcal{S}$ is a subset of $\mathcal{H}$ such that $\mathcal{D} = \operatorname{lin}\mathcal{S}$. A function φ is in $\mathcal{M}(\mathcal{D})$ if and only if*

$$f \in \mathcal{S} \Rightarrow \text{there is } C > 0 \text{ such that } \Big|\sum_{i=1}^{n} \varphi(x_i)f(x_i)\xi_i\Big|^2 \leq C \sum_{i,j=1}^{n} K(x_i, x_j)\xi_i\bar{\xi}_j$$
$$\text{for any finite choice of } x_i\text{'s in } X \text{ and } \xi_i\text{'s in } \mathbb{C}. \quad (5)$$

From (5) we get a half of our generalization of the Newman-Shapiro characterization; the other half will be proved in the next section. To have it more flexible for further applications we make it a statement on individual vectors rather than one concerning the set $\mathcal{M}(\mathcal{D}_K)$.

Corollary 5. *If $\varphi \in \mathcal{M}(\mathcal{D}_K)$ then*

to every $y \in X$ there is C such that $|\varphi(x)| \leq C|K(x,y)|^{-1}K(x,x)^{1/2}$, $x \in X$.

Apparently $aM_\varphi + bM_\psi \subset M_{a\varphi+b\psi}$ and $M_\varphi M_\psi \subset M_{\varphi\psi}$. Reminding that $\varphi \in \mathcal{M}(\mathcal{D})$ means precisely $\mathcal{D} \subset \mathcal{D}(M_\varphi)$, we are very far from seeing in $\mathcal{M}(\mathcal{D})$ any algebraic structure.

Our basic linear subspaces of $\mathcal{H}$ are

$$\mathcal{D}_K \stackrel{\text{df}}{=} \operatorname{lin}\{K_x;\ x \in X\} \text{ and } \mathcal{D}_{\boldsymbol{e}} \stackrel{\text{df}}{=} \operatorname{lin}\{e_\alpha;\ \alpha \in \mathrm{A}\}.$$

Introduce one more object, R_φ with domain $\mathcal{D}_K$,

$$R_\varphi : \sum_i \xi_i K_{x_i} \mapsto \sum_i \xi_i \varphi(x_i) K_{x_i}.$$

In general R_φ is multivalued (or in other words, a linear relation) however we have the following [23]

Theorem 6. *If M_φ is densely defined then R_φ is a well-defined operator and $R_\varphi \subset M_\varphi^*$. Conversely, if R_φ is a well-defined operator then $R^* \subset M_\varphi$, hence M_φ is densely defined. If the above happens then $R_\varphi^* = M_\varphi$.*

Now we can declare

any M_φ is supposed to be densely defined whenever it appears,
any other (abstract) operator is densely defined and closable. (★)

Shorthand notation: set[5] $M_x \stackrel{\text{df}}{=} M_{K_x}$, $R_x \stackrel{\text{df}}{=} R_{K_x}$, $M_\alpha \stackrel{\text{df}}{=} M_{e_\alpha}$ and $R_\alpha \stackrel{\text{df}}{=} R_{e_\alpha}$ for any x and any α. Of course, these kinds of operators are subject to the convention (★) as well.

First we contribute to the question when the multiplication operators are bounded, see [9], Problem 53 (the converse will be discussed later).

Proposition 7. *If M_φ is a bounded operator on $\mathcal{H}$ then φ is a bounded function in a sense that $|\varphi(x)| \leq \|M_\varphi\|$ whenever $K(x,x) > 0$.*

[5] A formalist would require for this kind of definition the sets X and A to be disjoint.

Proof. Using the reproducing kernel property (1) we can write

$$|\varphi(x)K_x(x)| = |\langle M_\varphi K_x, K_x\rangle| \le \|M_\varphi\|\,\|K_x\|\,\|K_x\| = \|M_\varphi\|K(x,x)$$

which gives $|\varphi(x)| \le \|M_\varphi\|$. □

Now we present a number of facts which all together cover the aforesaid generalization of Theorem 4 of [15], the main result therein. Proving them we make multiple use of Theorem 6 without pointing it out explicitly.

Proposition 8. *Let A be an operator such that $\mathcal{D}_K \cup \mathcal{D}_{\boldsymbol{e}} \subset \mathcal{D}(A)$. Suppose*

$$\langle AK_x, h\rangle = \langle \sum_{\alpha\in\mathrm{A}} \overline{e_\alpha(x)} Ae_\alpha, h\rangle, \text{ for every } h \in \mathcal{R}(\bar{A}) \text{ and } x \in X. \tag{6}$$

Then $\mathcal{D}_K$ is a core of A if and only if so is $\mathcal{D}_{\boldsymbol{e}}$.

Proof. Suppose $\mathcal{D}_K$ is a core of A. Denote the graph inner product by $\langle\cdot,-\rangle_A$ (that is, $\langle f,g\rangle_A \stackrel{\text{df}}{=} \langle f,g\rangle + \langle \bar{A}f,\bar{A}g\rangle$, $f,g\in\mathcal{D}(\bar{A})$) and take $g\in\mathcal{D}(\bar{A})$ such that $\langle e_\alpha, g\rangle_A = 0$ for every α. Then, by (6),

$$\langle K_x, g\rangle_A = \langle \sum_{\alpha\in\mathrm{A}} \overline{e_\alpha(x)} e_\alpha, g\rangle_A = \sum_{\alpha\in\mathrm{A}} \overline{e_\alpha(x)}\langle e_\alpha, g\rangle_A = 0.$$

Consequently $g = 0$ which means $\mathcal{D}_{\boldsymbol{e}}$ is a core of A.

Suppose now $\mathcal{D}_{\boldsymbol{e}}$ is a core of A. Take $g\in\mathcal{D}(\bar{A})$ such that $\langle K_x, g\rangle_A = 0$ for every $x\in X$. Because (using (6) again) we have

$$\langle K_x, g\rangle_A = \sum_{\alpha\in\mathrm{A}} \overline{e_\alpha(x)}\langle e_\alpha, g\rangle_A,$$

we have $\sum_{\alpha\in\mathrm{A}}\langle g, e_\alpha\rangle_A e_\alpha = 0$ which, due to [23], implies that $\langle g, e_\alpha\rangle_A = 0$ for every α. Thus $g = 0$ and the proof is completed. □

Remark 9. For (6) to hold it is sufficient, and sometimes practical, to suppose that

$$\textit{to every } x \textit{ there is } C_x \textit{ which dominates the norms of all } \sum_\alpha^{\text{finite}} \overline{e_\alpha(x)} Ae_\alpha \tag{7}$$

and $\mathcal{D}_K \subset \mathcal{D}(A^*)$ (the latter is always for multiplication operators). Indeed, write

$$\Big(\sum_\alpha^{\text{finite}} \overline{e_\alpha(x)} Ae_\alpha\Big)(x) = \langle \sum_\alpha^{\text{finite}} \overline{e_\alpha(x)} e_\alpha, A^*K_x\rangle$$
$$\to \langle \sum_{\alpha\in\mathrm{A}} \overline{e_\alpha(x)} e_\alpha, A^*K_x\rangle = \Big(A\sum_{\alpha\in\mathrm{A}} \overline{e_\alpha(x)} e_\alpha\Big)(x).$$

Using (7) and applying Corollary 2, (b) we infer that (6) holds (the convergence involved in it is even that of the norm topology).

Remark. Proposition 8 is related to the example constructed in [5, Theorem]

Proposition 10. *Suppose A is an operator in $\mathcal{H}$ and φ is a complex function on X. Then the following conditions are equivalent*

(a) *$\mathcal{D}_K \subset \mathcal{D}(A^*)$ and $A^*K_x = \overline{\varphi(x)}K_x$, $x\in X$;*

(b) *φ is a densely defining multiplier and $\bar{A} \subset M_\varphi$.*

If the above equivalence happens then $\bar{A} = M_\varphi$ if and only if $\mathcal{D}_K$ is a core of A^*.

Proof. (a) means that $R_\varphi \subset A^*$. Theorem 6 gives immediately that M_φ is densely defined and that (b) holds. Again Theorem 6 and (b) give $R_\varphi \subset M_\varphi \subset A^*$ and this is (a).

$\mathcal{D}_K$ is a core of A^* if and only if $A^* = \bar{R}_\varphi$ and this happens if and only if (Theorem 6!) $\bar{A} = A^{**} = \bar{R}_\varphi^* = R_\varphi^* = M_\varphi$. □

For the next portion we need the weakest possible version of *weak* commutativity of two unbounded operators[6]: we say that A and B $(\mathcal{D}, \mathcal{E})$*-weakly commute* if

$$\mathcal{D} \subset \mathcal{D}(A) \cap \mathcal{D}(B), \quad \mathcal{E} \subset \mathcal{D}(A^*) \cap \mathcal{D}(B^*),$$
$$\langle Af, B^*g \rangle = \langle Bf, A^*g \rangle, \quad f \in \mathcal{D},\ g \in \mathcal{E}.$$

The following observation, which is trivial at least, is stated so as to localize it for further use.

Proposition 11. (α) *If A and B $(\mathcal{D}, \mathcal{E})$-weakly commute and $\mathcal{D}' \subset \mathcal{D}$, $\mathcal{E}' \subset \mathcal{E}$, then A and B $(\mathcal{D}', \mathcal{E}')$-weakly commute.*

(β) *If A and B $(\mathcal{D}, \mathcal{E})$-weakly commute then A^* and B^* $(\mathcal{E}, \mathcal{D})$-weakly commute. On the other hand, if A^* and B^* $(\mathcal{E}, \mathcal{D})$-weakly commute and $\mathcal{D} \subset \mathcal{D}(A) \cap \mathcal{D}(B)$ then A and B $(\mathcal{D}, \mathcal{E})$-weakly commute.*

(γ) *If A and B $(\mathcal{E}, \mathcal{D})$-weakly commute, B is a bounded operator on $\mathcal{H}$ and $\mathcal{E}$ is a core of A^* then, for $f \in \mathcal{D}$, $Bf \in \mathcal{D}(\bar{A})$ and $BAf = \bar{A}Bf$.*

Continuing the main course of this section we will often need that

$$1_X \in \mathcal{H}. \tag{8}$$

Proposition 12. *Suppose* (8) *holds and A is an operator in $\mathcal{H}$. Then the condition*

(c) *$1_X \in \mathcal{D}(\bar{A})$, $\mathcal{D}_K \subset \mathcal{D}(A^*)$ and the operators $\bar{A}$ and M_x $(\mathbb{C} \cdot 1_X, \mathcal{D}_K)$-weakly commute for every $x \in X$*

implies (a) *with $\varphi = \bar{A}1_X$.*

Proof. If (c) holds we can write

$$\begin{aligned}\langle A^*K_x, K_y \rangle &= \langle A^*K_x, M_y 1_X \rangle = \langle M_y^* K_x, \bar{A}1_X \rangle = \langle \overline{K_y(x)} K_x, \bar{A}1_X \rangle \\ &= \overline{K_y(x)} \langle K_x, \bar{A}1_X \rangle = \overline{K_y(x)(\bar{A}1_X)(x)} = (\bar{A}1_X)(x) \langle K_x, K_y \rangle \\ &= \langle (\bar{A}1_X)(x) K_x, K_y \rangle, \quad x, y \in X.\end{aligned}$$

This leads to (a) with $\varphi \overset{\text{df}}{=} \bar{A}1_X$. □

Proposition 13. *Suppose* (8) *holds and A is an operator in $\mathcal{H}$. Then the condition*

(d) *$1_X \in \mathcal{D}(\bar{A})$, $\mathcal{D}_K \subset \mathcal{D}(A^*)$ and the operators $\bar{A}$ and M_α $(\mathbb{C} \cdot 1_X, \mathcal{D}_K)$-weakly commute for every α*

implies (a) *with $\varphi = \bar{A}1_X$.*

[6] The custom definition does not go that far.

Proof. Suppose (d) holds. We show that

$$e_\alpha(x)(\bar{A}1_X)(x) = \langle e_\alpha, A^* K_x \rangle. \tag{9}$$

Indeed, the left-hand side of (9) is

$$e_\alpha(x)\langle \bar{A}1_X, K_x \rangle = \langle \bar{A}1_X, \overline{e_\alpha(x)K_x} \rangle = \langle \bar{A}1_X, M_\alpha^* K_x \rangle,$$

while for the right-hand side we have

$$\langle e_\alpha, A^* K_x \rangle = \langle M_\alpha 1_X, A^* K_x \rangle.$$

Comparing this with $(\mathbb{C}\cdot 1_X, \mathcal{D}_K)$-weak commutativity of A and M_α we get (9). Now multiplying (9) by $\overline{e_\alpha(y)}$ and summing up over α (and using (3) as well as (2) on the way) we conclude that

$$\begin{aligned}\langle K_y, \overline{(\bar{A}1_X)(x)}K_x \rangle &= (\bar{A}1_X)(x)\langle K_y, K_x \rangle = (\bar{A}1_X)(x) K(x,y) \\ &= (\bar{A}1_X)(x) \sum_\alpha e_\alpha(x)\overline{e_\alpha(y)} = \sum_\alpha \overline{e_\alpha(y)}\langle e_\alpha, A^* K_x \rangle \quad \text{(here (9))} \\ &= \langle K_y, A^* K_x \rangle\end{aligned}$$

and this gives (a) with $\varphi = \bar{A}1_X$. □

As a kind of common converse to Propositions 12 and 13 we have

Proposition 14. *For arbitrary φ and ψ the operators M_φ and M_ψ $(\mathcal{D}(M_\varphi) \cap \mathcal{D}(M_\psi), \mathcal{D}_K)$-weakly commute.*

Proof. Indeed, for any $x \in X$, and $f \in \mathcal{D}(M_\varphi) \cap \mathcal{D}(M_\psi)$,

$$\langle M_\varphi f, M_\psi^* K_x \rangle = \langle \varphi f, R_\psi K_x \rangle = \langle \varphi f, \overline{\psi(x)} K_x \rangle = \psi(x)\varphi(x) = \cdots = \langle M_\psi f, M_\varphi^* K_x \rangle.$$

□

Notice that as for the strong (pointwise) commutativity of two multiplication operators we have $M_\varphi M_\psi f = M_\psi M_\varphi f$ for $f \in \mathcal{D}(M_\varphi M_\psi) \cap \mathcal{D}(M_\psi M_\varphi)$ at most. On the other hand, however the domains of M_φ and M_ψ may differ[7], the adjoints M_φ^* and M_ψ^* pointwise commute on $\mathcal{D}_K$ (actually $\mathcal{D}_K$ is an invariant core of any M_φ^*; this is an important message Theorem 6 brings forth).

As we have already mentioned the above conditions (a), (b), (c) and (d) come out as a result of an acute analysis of Theorem 4 of [15]. They together constitute half of the conditions therein; the other half can be obtained by applying Proposition 11.

Theorem 15. *Suppose* (8) *holds and A is an operator in $\mathcal{H}$.*

- *Then the following conditions are equivalent*
  - (a) *there is a densely defining multiplier φ such that $\bar{A} \subset M_\varphi$;*
  - (b) $1_X \in \mathcal{D}(\bar{A})$, $\mathcal{D}_K \subset \mathcal{D}(A^*)$ *and the operators $\bar{A}$ and M_x $(\mathbb{C}\cdot 1_X, \mathcal{D}_K)$-weakly commute for every $x \in X$;*

[7] The example of the function g provided on p. 2685 of [15] in the section 'Counterexamples' shows they may differ by a finite-dimensional set of polynomials, for instance.

(c) $1_X \in \mathcal{D}(\bar{A})$, $\mathcal{D}_K \subset \mathcal{D}(A^*)$ *and the operators* A^* *and* R_x $(\mathcal{D}_K, \mathbb{C}\cdot 1_X)$-*weakly commute for every* $x \in X$;

(d) $\mathcal{D}_K \subset \mathcal{D}(A^*)$ *and* $A^*K_x = \overline{(\bar{A}1_X)(x)}K_x$;

(e) $1_X \in \mathcal{D}(\bar{A})$, $\mathcal{D}_K \subset \mathcal{D}(A^*)$ *and the operators* $\bar{A}$ *and* M_α $(\mathbb{C}\cdot 1_X, \mathcal{D}_K)$-*weakly commute for every* α;

(f) $1_X \in \mathcal{D}(\bar{A})$, $\mathcal{D}_K \subset \mathcal{D}(A^*)$ *and the operators* A^* *and* R_α $(\mathcal{D}_K, \mathbb{C}\cdot 1_X)$-*weakly commute for every* α.

▲ *If* φ *is a complex function on* X *then the following conditions are equivalent*

(g) $\mathcal{D}_K \subset \mathcal{D}(A^*)$ *and* $A^*K_x = \overline{\varphi(x)}K_x$;

(h) φ *is a densely defining multiplier and* $\bar{A} \subset M_\varphi$.

★ *If* $\mathcal{D}_K$ *is a core of* A^* *then* $\bar{A} = M_\varphi$ *in* (a) *and* (h) *above.*

Remark 16. Theorem 15 is the most penetrating refinement in the abstract circumstances of the main result of [15]. A careful reader may notice that the conditions (b), (c), (e) and (f) treated as necessary ones look weaker than those of [15]. This is not so because one can easily derive from them their Rochberg-Weaver versions due to the group property of the Segal-Bargmann kernel. Also the role played by '$\mathcal{D}_K$ is a core of A^*', which is a subject of some dispute in [15], now becomes clear.

Integrability of RKHS's and subnormality of multiplication operators

For not going too far assume X is a subset of $\mathbb{C}^d$, d a positive integer; this case is illustrative enough and it is what we want to call attention to in this paper. Call a RKHS couple $(\mathcal{H}, K)$ *integrable* if there is a positive Borel measure μ on $\mathbb{C}^d$ and an isometry, say V, of $\mathcal{H}$ into $\mathcal{L}^2(\mu)$, which preserves pointwise multiplication of functions. Notice that such a μ may not exist at all (cf. [22] for some discussion) or may not be unique, cf. [21]. Even if it is unique in a sense, the Hardy space case shows that it may happen that $X \cap \operatorname{supp}\mu = \varnothing$. This forces some caution resulted in taking into consideration the isometry V. On the other hand, the case considered in [21] and recounted here in © below tells us that though always $\operatorname{supp}\mu \subset X$, one of supports is nowhere dense while for the other $\operatorname{supp}\mu = X = \mathbb{C}$.

We require additional properties of isometry V like

- the polynomials are in $\mathcal{L}^2(\mu)$ and V maps 1_X to $1_{\operatorname{supp}\mu}$ and Z_X to $Z_{\operatorname{supp}\mu}$;
- for any $\mathcal{D}$ dense in $\mathcal{H}$, V extends to $V_\mathcal{D}$, say, which maps $\mathcal{M}(\mathcal{D})$ in a (pointwise) linear and multiplicative way to functions on $\operatorname{supp}\mu$.

The very first benefit of integrability is immediate and constitutes the converse to Corollary 5 under the presence of a representing measure. This is a generalization of the Newman-Shapiro theorem [14].

Theorem 17. *Suppose* $(\mathcal{H}, K)$ *is integrable with* V *being the inclusion map and with a measure* μ *for which there exists a family* Ψ *of functions on* X *such that*

(α) *to every* $y \in X$ *there is* $\psi \in \Psi$ *such that* $x \mapsto \psi(x)|K_y(x)|K(x,x)^{1/2}$ *is in* $\mathcal{L}^2(\mu)$;

(β) *to every* $\psi \in \Psi$ *there is* $y \in X$ *such that for every* $x \in X$ $|K_y(x)|^{-1} \le \psi(x)$.

Then the following conditions are equivalent

(i) *to every* $y \in X$ *there is* C *such that* $|\varphi(x)| \le C|K(x,y)|^{-1}K(x,x)^{1/2}$, $x \in X$;
(ii) *to every* $\psi \in \Psi$ *there is* C *such that* $|\varphi(x)| \le C\psi(x)K(x,x)^{1/2}$, $x \in X$;
(iii) φ *is in* $\mathcal{M}(\mathcal{D}_K)$.

Proof. For the proof of "(i) *implies* (ii)" use (β). For "(ii) *implies* (iii)" apply (α). "(iii) *implies* (i)" is in Corollary 5. □

The family Ψ for the Segal-Bargmann space, which makes the Newman-Shapiro proof work, is $\{\psi;\ \text{there exists } N > 0 \text{ such that } \psi(z) = \mathrm{e}^{-N|z|},\ z \in \mathbb{C}\}$. Then the condition (ii) of Theorem 17 becomes that of Newman-Shapiro.

Also Proposition 7 profits having its converse as

Proposition 18. *If* $V\varphi$ *is in* $\mathcal{L}^\infty(\mu)$ *then the operator* M_φ *is bounded and* $\|M_\varphi\| \le \|V\varphi\|_\infty$.

Integrability gives rise to study Toeplitz operators, in particular the operator R_φ may be thought of as one of those with the symbol $\bar{\varphi}$; it would be proper to chose the right measure for that as one may have many of them. Thus as we see integrability of RKHS couples brings forth additional capability; in fact, it was already used by Zaremba in [27] in his pioneering work (cf. [3] where the abstract theory of RKHS's is built up).

Now a definition of subnormality: with notation $\mathcal{D}(\boldsymbol{S}) = \mathcal{D}(S_1) \cap \cdots \cap \mathcal{D}(S_\kappa)$ for a finite system $\boldsymbol{S} = (S_1, \ldots, S_\kappa)$ of operators in a Hilbert space $\mathcal{H}$ we say [20] that $\boldsymbol{S}$ is *subnormal* if $\overline{\mathcal{D}(\boldsymbol{S})} = \mathcal{H}$ and there exists another Hilbert space $\mathcal{K} \supset \mathcal{H}$ (the inclusion being isometric) and a system $\boldsymbol{N} = (N_1, \ldots, N_\kappa)$ of spectrally commuting normal operators in $\mathcal{K}$ such that $S_i \subset N_i$ for $i = 1, \ldots, \kappa$ (then $\boldsymbol{N}$ is a *normal extension* of $\boldsymbol{S}$). A necessary condition for subnormality is at hand; in general it is not sufficient. In fact, there are not too many sufficient conditions available[8], especially for more than one operator.

Proposition 19. *Suppose the system* $\boldsymbol{S} = (S_1, \ldots, S_\kappa)$ *is subnormal and* $\mathcal{D}(\boldsymbol{S})$ *is invariant for any* S_i. *Then*

$$\sum_{|\boldsymbol{m}|,|\boldsymbol{n}|=0}^{N} \langle \boldsymbol{S}^{\boldsymbol{m}} f_{\boldsymbol{n}}, \boldsymbol{S}^{\boldsymbol{n}} f_{\boldsymbol{m}} \rangle \ge 0, \quad \{f_{\boldsymbol{m},\boldsymbol{n}}\}_{|\boldsymbol{m}|,|\boldsymbol{n}|=0}^{N} \subset \mathcal{D}(\boldsymbol{S}), \quad N = 0, 1, \ldots \tag{10}$$

Proof. Use the spectral representation of $\boldsymbol{S}$. □

Because spectrally commuting normal operators, by definition, have a joint spectral measure, intuitively one may expect some connection between integrability of a reproducing kernel Hilbert space and subnormality of operators of multiplication therein. One way is simple.

[8] For some recent results on subnormality of a single operator look at [19].

Proposition 20. *Suppose $(\mathcal{H}, K)$ is integrable. If $\varphi_1, \ldots, \varphi_\kappa$ are measurable and $\mathcal{D}(M_{\varphi_1}) \cap \cdots \cap \mathcal{D}(M_{\varphi_\kappa})$ is dense in $\mathcal{H}$, then the system $(M_{\varphi_1}, \ldots, M_{\varphi_\kappa})$ is subnormal.*

Proof. Because φ_i are measurable, for the maximal operators $N_{V\varphi_i}$ of multiplication by φ_i in $\mathcal{L}^2(\mu)$ the subspace $\mathcal{D}(N_{V\varphi_1}) \cap \cdots \cap \mathcal{D}(N_{V\varphi_\kappa})$ is dense in $\mathcal{L}^2(\mu)$ (imitate the argument in the proof of (4.1), p. 51 of [26]). The operators $N_{V\varphi_i}$ are normal ($N^*_{V\varphi_i} = N_{\overline{V\varphi_i}}$) and extend M_{φ_i}. The conclusion follows after some measure theoretic manipulation showing spectral commutativity of the operators $N_{V\varphi_i}$. □

The other way resides in what follows.

Proposition 21. *Suppose the system $(A_1, \ldots, A_d)$ of operators in $\mathcal{H}$ is subnormal and such that the subspace $\{p(A_1, \ldots, A_d)f;\ p \in \mathbb{C}[Z_1, \ldots, Z_d]\}$, with some f, is dense in $\mathcal{H}$[9]. Then $(\mathcal{H}, K)$ is integrable.*

Proof. If E is the spectral measure of a normal extension of $(A_1, \ldots, A_d)$, then with $\mu \stackrel{\mathrm{df}}{=} \langle E(\cdot)f, f\rangle$ we have

$$\|p(A_1, \ldots, A_d)f\|^2 = \int_{\mathbb{C}^d} |p|^2 \, \mathrm{d}\mu, \quad p \in \mathbb{C}[Z_1, \ldots, Z_d]$$

and this establishes the isometry V. □

Now we are going to make the kernels more specific. First our multi-index notation: for $\boldsymbol{n} = (n_1, \ldots, n_d) \in \mathbb{N}^d$[10] and $\boldsymbol{z} = (z_1, \ldots, z_d), \boldsymbol{w} = (w_1, \ldots, w_d) \in \mathbb{C}^d$ we have $\bar{\boldsymbol{z}} \stackrel{\mathrm{df}}{=} (\bar{z}_1, \ldots, \bar{z}_d)$, $\boldsymbol{z}\cdot\boldsymbol{w} \stackrel{\mathrm{df}}{=} (z_1 w_1, \ldots, z_d w_d)$ and $\boldsymbol{z}^{\boldsymbol{n}} \stackrel{\mathrm{df}}{=} z_1^{n_1} \cdots z_d^{n_d}$ (this allows us to have a very convenient equality $(\boldsymbol{z}\cdot\boldsymbol{w})^{\boldsymbol{n}} = \boldsymbol{z}^{\boldsymbol{n}}\boldsymbol{w}^{\boldsymbol{n}}$). The notation $\langle \boldsymbol{z}, \boldsymbol{w}\rangle \stackrel{\mathrm{df}}{=} z_1\bar{w}_1 + \cdots + z_d\bar{w}_d$ serves for the inner product in $\mathbb{C}^d$ as well, no confusion expected. Moreover, $|\boldsymbol{n}| \stackrel{\mathrm{df}}{=} n_1 + \cdots + n_d$ and $\mathfrak{s}_i\boldsymbol{n} \stackrel{\mathrm{df}}{=} (n_1, \ldots n_{i+1}, \ldots, n_d)$.

A fairly general method of a defining positive definite kernel whose corresponding reproducing kernel space is composed of holomorphic functions is to take a holomorphic function F, say, in a single variable in a domain containing the origin with Taylor coefficient at zero being non-negative and $P \in \mathbb{C}[Z_1, \ldots, Z_d, \bar{Z}_1, \ldots, \bar{Z}_d]$ which itself is a positive definite kernel. Then it follows from the Schur lemma that

$$K(\boldsymbol{z}, \boldsymbol{w}) \stackrel{\mathrm{df}}{=} F(P(\boldsymbol{z}, \bar{\boldsymbol{w}})) \tag{11}$$

is a positive definite kernel wherever it is defined (with the only case when an orthonormal basis of polynomials comes at once from (11) being that of $P(\boldsymbol{z}, \bar{\boldsymbol{w}}) = Q(\boldsymbol{z})\overline{Q(\boldsymbol{w})}$). The instances which follow are patterned after (11).

---

9 This is less than joint cyclicity.

10 In this paper $\mathbb{N} \stackrel{\mathrm{df}}{=} \{0, 1, \ldots\}$.

Given a d-sequence $\{\alpha_{\boldsymbol{n}}\}_{|\boldsymbol{n}|=0}^{\infty}$ and a sequence $\{\alpha_n\}_{n=0}^{\infty}$, both of non-negative numbers, set

$$K_{\mathrm{p}}(\boldsymbol{z},\boldsymbol{w}) \overset{\mathrm{df}}{=} \sum_{|\boldsymbol{n}|=0}^{\infty} \alpha_{\boldsymbol{n}} \boldsymbol{z}^{\boldsymbol{n}} \bar{\boldsymbol{w}}^{\boldsymbol{n}}, \quad \boldsymbol{z},\boldsymbol{w} \in \mathbb{C}^d, \tag{12}$$

and

$$K_{\mathrm{b}}(\boldsymbol{z},\boldsymbol{w}) \overset{\mathrm{df}}{=} \sum_{n=0}^{\infty} \alpha_n \langle \boldsymbol{z},\boldsymbol{w}\rangle^n, \quad \boldsymbol{z},\boldsymbol{w} \in \mathbb{C}^d. \tag{13}$$

For $d = 1$ both the kernels (12) and (13) coincide. Notice that the invariance property

$$K(\boldsymbol{z},\boldsymbol{w}) = K(A\boldsymbol{z}, A\boldsymbol{w}) \tag{14}$$

holds for K_{p} when the matrix A is unitary and diagonal while for K_{b} when this matrix is just unitary (in the case the corresponding RKHS couples are integrable the appropriate properties of the representing measures imply (14) but not the other way; in absence of uniqueness, a representing measure which is not rotationally invariant is constructed [13], in addition to those which are so, cf. © below).

Remark. Compare K_{p} of (12) with the product $K_1(z_1,w_1)\cdots K_d(z_d,w_d)$ of kernels K_i's defined by (12) as well.

THE POLYDISC RHKS Let a d-sequence $\{\alpha_{\boldsymbol{n}}\}_{|\boldsymbol{n}|=0}^{\infty}$ be such that there is a polydisc[11] in which the series of $\sum_{|\boldsymbol{n}|=0}^{\infty} \alpha_{\boldsymbol{n}} \boldsymbol{z}^{\boldsymbol{n}}$ converges. The kernel K_{p} is apparently positive definite. Denote by $\mathcal{H}_{\mathrm{p}}$ the corresponding Hilbert space. It follows from [23] that the monomials $e_{\boldsymbol{n}} \overset{\mathrm{df}}{=} \sqrt{\alpha_{\boldsymbol{n}}} \boldsymbol{Z}^{\boldsymbol{n}}$ are in $\mathcal{H}_{\mathrm{p}}$ and form an orthonormal basis in it. The members of $\mathcal{H}_d$ are holomorphic in respective domain and the polynomials in $\boldsymbol{Z}$ are dense. Set

$$M_i \overset{\mathrm{df}}{=} M_{Z_i},\ i = 1,\dots,d,\ \boldsymbol{M} \overset{\mathrm{df}}{=} (M_1,\dots,M_d),\ \boldsymbol{M}^{\boldsymbol{n}} \overset{\mathrm{df}}{=} M_1^{n_1}\cdots M_d^{n_d},\ \boldsymbol{n} \in \mathbb{N}^d,$$

the latter is in accordance with the multi-index notation above. Thus $M_n = \boldsymbol{M}^{\mathfrak{s}_n \mathbf{0}}$ (warning $\boldsymbol{M}^{\mathbf{1}} \neq \boldsymbol{M}$). Referring to our notation so far we have $M_{\boldsymbol{n}} = \sqrt{\alpha_{\boldsymbol{n}}} \boldsymbol{M}^{\boldsymbol{n}} = \sqrt{\alpha_{\boldsymbol{n}}} M_{\boldsymbol{Z}}^{\boldsymbol{n}}$ (again M_n may differ from $M_{\boldsymbol{n}}$, the latter corresponds to M_α in the previous section). Moreover, $e_{\boldsymbol{n}} = \sqrt{\alpha_{\boldsymbol{n}}} \boldsymbol{M}^{\boldsymbol{n}} \mathbf{1}$.

It is a matter of direct verification to find out that the polynomials from $\mathbb{C}[Z_1,\dots,Z_d]$, as members of $\mathcal{H}_{\mathrm{p}}$, form a core of any $\boldsymbol{M}^{\boldsymbol{n}}$ (in fact it is joint core of any finite collection of them). Moreover, it is easy to verify that the conditions of Remark 9 are met which implies that $\mathcal{D}_K$ is a core of those operators as well. Because $\mathfrak{s}_i \mathfrak{s}_j = \mathfrak{s}_j \mathfrak{s}_i$, the operators M_i commute.

[11] By a polydisc we mean here the set $\{z_1;\ |z_1| < r_1\} \times \cdots \times \{z_d;\ |z_d| < r_d\}$, where $0 < r_i \leq +\infty$.

Propositions 20 and 9 make up the following

Theorem 22. *$(\mathcal{H}_{\mathrm{p}}, K_{\mathrm{p}})$ is integrable if and only if $\boldsymbol{M}$ is subnormal if and only if $\{\alpha_{\boldsymbol{n}}^{-1}\}_{|\boldsymbol{n}|=0}^{\infty}$ is an n-dimensional Stjelties moment sequence, that is there is a measure μ on $\mathbb{R}_{+}^{d} \stackrel{\mathrm{df}}{=} [0, +\infty)^{d}$ such that*

$$\alpha_{\boldsymbol{n}}^{-1} = \int_{\mathbb{R}_{+}^{d}} \boldsymbol{x}^{\boldsymbol{n}} \mu(\mathrm{d}\boldsymbol{x}), \quad \boldsymbol{n} \in \mathbb{N}^{d}.$$

Proof. Suppose $\boldsymbol{S}$ is subnormal with the joint spectral measure E. Then

$$\alpha_{\boldsymbol{n}}^{-1} = \langle \boldsymbol{Z}^{\boldsymbol{n}}, \boldsymbol{Z}^{\boldsymbol{n}} \rangle = \langle \boldsymbol{M}^{\boldsymbol{n}}\mathbf{1}, \boldsymbol{M}^{\boldsymbol{n}}\mathbf{1} \rangle = \int_{\mathbb{C}^{d}} \boldsymbol{z}^{\boldsymbol{n}} \langle E(\mathrm{d}\boldsymbol{z})\mathbf{1}, \mathbf{1} \rangle.$$

The required measure μ comes from $\langle E(\mathrm{d}\boldsymbol{z})\mathbf{1}, \mathbf{1} \rangle$ via the mapping

$$\mathbb{C}^{d} \ni (z_1, \dots, z_d) \mapsto (|z_1|^2, \dots, |z_d|^2) \in \mathbb{R}_{+}^{d}.$$

Thus $\{\alpha_{\boldsymbol{n}}^{-1}\}_{|\boldsymbol{n}|=0}^{\infty}$ is an n-dimensional Stjelties moment sequence.

Suppose now $\{\alpha_{\boldsymbol{n}}^{-1}\}_{|\boldsymbol{n}|=0}^{\infty}$ is an n-dimensional Stjelties moment sequence and ν is its representing measure. Consider on $\mathbb{R}_{+}^{d} \times \mathbb{T}^{d}$ the measure $\nu \otimes m^{d}$ where m is the normalized Lebesgue measure on $\mathbb{T}$ and m^{d} its d-th measure product power. Consider the mapping

$$\mathbb{R}_{+}^{d} \times \mathbb{T}^{d} \ni (r_1, \dots, r_d, \mathrm{e}^{\mathrm{i}\,t_1}, \dots, \mathrm{e}^{\mathrm{i}\,t_d}) \mapsto (r_1\,\mathrm{e}^{\mathrm{i}\,t_1}, \dots, r_d\,\mathrm{e}^{\mathrm{i}\,t_d}) \in \mathbb{C}^{d}$$

and transpose the measure $\nu \otimes m^{d}$ from $\mathbb{R}_{+}^{d} \times \mathbb{T}^{d}$ to $\mathbb{C}^{d}$ so as to get the measure μ. It is a matter of direct verification to check that $\mathcal{L}^{2}(\mu)$ realizes integrability of $\mathcal{H}_{\mathrm{p}}$. □

As a step towards subnormality of $\boldsymbol{M}_{\boldsymbol{Z}}$ let us record the following fact

Proposition 23. *$\boldsymbol{M}_{\boldsymbol{Z}}$ satisfies* (10) *if and only if*

$$\sum_{\boldsymbol{i},\boldsymbol{j}} \alpha_{\boldsymbol{k}+\boldsymbol{i}+\boldsymbol{j}}^{-1} \lambda_{\boldsymbol{k}+\boldsymbol{i},\boldsymbol{i}} \bar{\lambda}_{\boldsymbol{k}+\boldsymbol{j},\boldsymbol{j}} \geq 0, \quad \boldsymbol{k} \in \mathbb{N}^{d}, \quad \{\lambda_{\boldsymbol{m},\boldsymbol{n}}\}_{\boldsymbol{m},\boldsymbol{n}} \subset \mathbb{C}^{d} \quad \textit{is finite.} \tag{15}$$

Proof. Indeed, due to $\langle \boldsymbol{Z}^{\boldsymbol{i}}, \boldsymbol{Z}^{\boldsymbol{j}} \rangle = \alpha_{\boldsymbol{i}}^{-1} \delta_{\boldsymbol{i},\boldsymbol{j}}$, we have

$$\sum_{|\boldsymbol{m}|,|\boldsymbol{n}|=0}^{N} \langle \boldsymbol{M}_{\boldsymbol{Z}}^{\boldsymbol{m}} p_{\boldsymbol{n}}, \boldsymbol{M}_{\boldsymbol{Z}}^{\boldsymbol{n}} p_{\boldsymbol{m}} \rangle = \sum_{\boldsymbol{k}} \sum_{\boldsymbol{i},\boldsymbol{j}} \alpha_{\boldsymbol{k}+\boldsymbol{i}+\boldsymbol{j}}^{-1} p_{\boldsymbol{k}+\boldsymbol{i},\boldsymbol{i}} \bar{p}_{\boldsymbol{k}+\boldsymbol{j},\boldsymbol{j}}$$

where $p_{\boldsymbol{n}} = \sum_{|\boldsymbol{i}|=0} p_{\boldsymbol{n},\boldsymbol{i}} \boldsymbol{Z}^{\boldsymbol{i}}$ and the undefined $p_{\boldsymbol{n},\boldsymbol{i}}$'s are 0. Now choosing proper $p_{\boldsymbol{m},\boldsymbol{n}}$ we can split the above as in (15). The other way is obvious. □

For the converse we have

Theorem 24. *In the case the maximal polydisc in which the kernel K_{p} converges is bounded the RKHS couple $(\mathcal{H}_{\mathrm{p}}, K_{\mathrm{p}})$ is integrable if and only if* (15) *holds.*

Proof. Because

$$\|M_i p\|^2 = \sum_{\boldsymbol{k}} \alpha_{\mathfrak{s}_i \boldsymbol{k}}^{-1} |p_{\boldsymbol{k}}|^2 = \sup_{\boldsymbol{k}} \alpha_{\mathfrak{s}_i \boldsymbol{k}}^{-1} \alpha_{\boldsymbol{k}} \sum_{\boldsymbol{k}} \alpha_{\boldsymbol{k}}^{-1} |p_{\boldsymbol{k}}|^2 \leq C \sum_{\boldsymbol{k}} \alpha_{\boldsymbol{k}}^{-1} |p_{\boldsymbol{k}}|^2 \leq \|p\|,$$

where $p = \sum_{\boldsymbol{k}} p_{\boldsymbol{k}} \boldsymbol{Z}^{\boldsymbol{k}}$, M_i are bounded operators. As the system $\boldsymbol{M}_{\boldsymbol{Z}}$ satisfies (10), it is subnormal (use [10]). □

Remark. In the case $d = 1$ (15) reduces to two, those for $k = 0$ and $k = 1$. Because they are precisely what for solvability of the one-dimensional Stjelties moment problem is needed, in view of Theorem 24, they give integrability of $(K_{\mathrm{p}}, \mathcal{H}_{\mathrm{p}})$, cf. [17]. For $d > 1$ this is not the case though a question of further reducibility of (15) may be still asked.

In general, in the unbounded case, density of some customary classes of $\mathcal{C}^\infty$ vectors (like bounded, analytic or quasianalytic), if properly used, may lead to subnormality, cf. [16] or for a single operator [17]. Due to the link established here we can get this way sufficient conditions for integrability (plus uniqueness of the representing measure). Below, as a kind of sample, we state a result which comes out from what is in [20].

Theorem 25. *The couple* $(K_{\mathrm{p}}, \mathcal{H}_{\mathrm{p}})$ *is integrable if there is an operator* A *with* $\mathcal{D}(A) = \mathbb{C}[Z_1, \dots, Z_d]$ *such that* (10) *is satisfied by the system* $(M_1, \dots, M_d, A)$ *as well as*

$$\sum_{\boldsymbol{k}} \alpha_{\mathfrak{s}_i \boldsymbol{k}}^{-1} |p_{\boldsymbol{k}}|^2 \leq c\Big(\sum_{\boldsymbol{k}} \alpha_{\boldsymbol{k}}^{-1} |p_{\boldsymbol{k}}|^2 + \|Ap\|^2\Big),$$

$$i = 1, 2, \dots, d, \quad p = \sum_{\boldsymbol{k}} p_{\boldsymbol{k}} Z^{\boldsymbol{k}}, \quad \text{some } c > 0 \tag{16}$$

and

$$\sum_{n=0}^{\infty} \|A^n p\| = +\infty, \quad p \in \mathbb{C}[Z_1, \dots, Z_d] \tag{17}$$

hold. If this happens the representing measure is unique.

In particular, if (15) *is satisfied and instead of* (16) *we have*

$$\alpha_{\mathfrak{s}_i \boldsymbol{k}}^{-1} \leq c(\alpha_{\boldsymbol{k}}^{-1} + \alpha_{\mathfrak{s}_1 \boldsymbol{k}}^{-1}), \quad i = 2, \dots, d, \quad \boldsymbol{k} \in \mathbb{N}^d, \quad \text{some } c > 0 \tag{18}$$

as well as instead of (17)

$$\sum_{n=0}^{\infty} \alpha_{\mathfrak{s}_1^n \boldsymbol{k}} = +\infty, \quad \boldsymbol{k} \in \mathbb{N}^d, \tag{19}$$

the same conclusion holds.

Proof. The first conclusion comes immediately from Proposition 39 of [20]. For the second conclusion notice that, after a rather elementary calculation, (16) and (17) imply (18) and (19), respectively. □

THE BALL RKHS This is when the kernel is given by (13). Because it can be rearranged (at least formally) as an (infinite) sum of monomials with coefficients express by α_n's and the binomial coefficients, everything what we have said for the polydisc case is applicable here provided there is a nontrivial polydisc of convergence of the resulted decomposition (12).

Two illustrative examples

Among a number of concrete kernels available the two which follow have the same flavour.

THE SEGAL-BARGMANN SPACE ([4] AND ... [8]) The reproducing kernel of this space is given by

$$K(\boldsymbol{z},\boldsymbol{w}) \stackrel{\mathrm{df}}{=} \sum_{|\boldsymbol{k}|=0}^{\infty} \frac{\boldsymbol{z}^{\boldsymbol{k}}\bar{\boldsymbol{w}}^{\boldsymbol{k}}}{\boldsymbol{k}!} = \mathrm{e}^{\langle \boldsymbol{z},\boldsymbol{w}\rangle}, \quad \boldsymbol{z},\boldsymbol{w} \in \mathbb{C}^d$$

where $\boldsymbol{k}! \stackrel{\mathrm{df}}{=} k_1! \cdots k_d!$.

From Theorem 15 we deduce the following generalization of Theorem 4 of [15].

Theorem 26. *Suppose A is an operator in the Segal-Bargmann space.*

▼ *Then the following conditions are equivalent*
- (a) *there is a densely defining multiplier φ such that $\bar{A} \subset M_\varphi$;*
- (b) *$1 \in \mathcal{D}(\bar{A})$, $\mathcal{D}_K \subset \mathcal{D}(A^*)$ and the operators $\bar{A}$ and $M_{\boldsymbol{z}}$ $(\mathbb{C},\mathcal{D}_K)$-weakly commute for every $\boldsymbol{z} \in \mathbb{C}^d$;*
- (c) *$1 \in \mathcal{D}(\bar{A})$, $\mathcal{D}_K \subset \mathcal{D}(A^*)$ and the operators A^* and $R_{\boldsymbol{z}}$ $(\mathcal{D}_K,\mathbb{C})$-weakly commute for every $\boldsymbol{z} \in \mathbb{C}^d$;*
- (d) *$\mathcal{D}_K \subset \mathcal{D}(A^*)$ and $A^*K_{\boldsymbol{z}} = \overline{(\bar{A}1)(\boldsymbol{z})}K_{\boldsymbol{z}}$;*
- (e) *$1 \in \mathcal{D}(\bar{A})$, $\mathcal{D}_K \subset \mathcal{D}(A^*)$ and the operators $\bar{A}$ and $M_{\boldsymbol{n}}$ $(\mathbb{C},\mathcal{D}_K)$-weakly commute for every $\boldsymbol{n} \in \mathbb{N}^d$;*
- (f) *$1 \in \mathcal{D}(\bar{A})$, $\mathcal{D}_K \subset \mathcal{D}(A^*)$ and the operators A^* and $R_{\boldsymbol{n}}$ $(\mathcal{D}_K,\mathbb{C})$-weakly commute for every $\boldsymbol{n} \in \mathbb{N}^d$.*

▲ *If φ is a complex function on X then the following conditions are equivalent*
- (g) *$\mathcal{D}_K \subset \mathcal{D}(A^*)$ and $A^*K_{\boldsymbol{z}} = \overline{\varphi(\boldsymbol{z})}K_{\boldsymbol{z}}$;*
- (h) *φ is a densely defining multiplier and $\bar{A} \subset M_\varphi$.*

★ *If $\mathcal{D}_K$ is a core of A^* then $\bar{A} = M_\varphi$ in (a) and (h) above.*

Remark 27. Theorem 26 is the crux of our response to the statement of Rochberg and Weaver. Besides the fact we are here in several variables, some parts of Theorem 26 can be developed further to show its widespread nature.

First of all notice conditions (b) and (c), as sufficient ones, make our statement stronger. As for necessity

$$\bar{A} \text{ and } M_{\boldsymbol{z}}\ (\mathbb{C},\mathcal{D}_K)\text{-weakly commute for every } \boldsymbol{z} \in \mathbb{C}^d \text{ if and only if they } (\mathcal{D}_K,\mathcal{D}_K)\text{-weakly commute for every } \boldsymbol{z} \in \mathbb{C}^d \quad (20)$$

and

$$A^* \text{ and } R_{\boldsymbol{z}}\ (\mathcal{D}_K, \mathbb{C})\text{-weakly commute for every } \boldsymbol{z} \in \mathcal{D}_K \text{ if and only if}$$
$$\text{they } (\mathcal{D}_K, \mathcal{D}_K)\text{-weakly commute for every } \boldsymbol{z} \in \mathbb{C}^d$$

provided

$$\mathcal{D}_K \subset \mathcal{D}(\bar{A}) \cap \mathcal{D}(A^*). \tag{21}$$

On the other hand, conditions (e) and (f) overlap their correspondents in [15], Theorem 4 and the bridge is paved by

$$\bar{A} \text{ and } M_{\boldsymbol{n}}\ (\mathbb{C}, \mathcal{D}_K)\text{-weakly commute for every } \boldsymbol{n} \in \mathbb{N}^d \text{ if and only if}$$
$$\bar{A} \text{ and } M_i\ (\mathcal{E}, \mathcal{E})\text{-weakly commute for every } i = 1, \ldots, d$$

as well as

$$A^* \text{ and } R_{\boldsymbol{n}}\ (\mathcal{D}_K, \mathbb{C})\text{-weakly commute for every } \boldsymbol{n} \in \mathbb{N}^d \text{ if and only if}$$
$$A^* \text{ and } R_i\ (\mathcal{E}, \mathcal{E})\text{-weakly commute for every } i = 1, \ldots, d$$

provided

$$\mathcal{E} \subset \mathcal{D}(\bar{A}) \cap \mathcal{D}(A^*)$$

where $\mathcal{E} \stackrel{\text{df}}{=} \{pK_{\boldsymbol{z}};\ p \in \mathbb{C}[Z_1, \ldots, Z_d],\ \boldsymbol{z} \in \mathbb{C}^d\}$.

Moreover, notice that (21) means precisely that φ in (a) and (g) is a $\mathcal{D}_K$-multiplier. All these comments relate our Theorem 26 to Theorem 4 of [15] (in the case of $d = 1$).

q-SPACES Here we assume $d = 1$. Some q-notions first: for $x \geq 0$ and $q \in \mathbb{C}$, $[x]_q \stackrel{\text{df}}{=} (1 - q^x)(1 - q)^{-1}$ if $q \neq 1$, $[x]_1 \stackrel{\text{df}}{=} x$ and $[0]_0 \stackrel{\text{df}}{=} 1$. If x is a positive integer they merge into $[x]_q = 1 + q \cdots + q^{x-1}$. The so-called q-shifted factorials are defined as

$$(a;q)_0 \stackrel{\text{df}}{=} 1, \quad (a;q)_k \stackrel{\text{df}}{=} (1-a)(1-aq)(1-aq^2)\cdots(1-aq^{k-1}), \quad k = 1, 2, 3, \ldots$$

For $[n]_q! \stackrel{\text{df}}{=} [1]_q \cdots [n]_q$ if $n > 0$ and $[0]_q! \stackrel{\text{df}}{=} 1$ we have $(q,q)_n = [n]_q!(1-q)^n$. Moreover[12],

$$(a;q)_\infty \stackrel{\text{df}}{=} \prod_{k=0}^{\infty} (1 - aq^k), \quad |q| < 1.$$

The q-exponential function is usually defined [12] in two ways[13]

$$e_q(z) \stackrel{\text{df}}{=} \sum_{k=0}^{\infty} \frac{1}{(q;q)_k} z^k, \quad z \in \omega_q, \quad \text{and} \quad E_q(z) \stackrel{\text{df}}{=} \sum_{k=0}^{\infty} \frac{q^{\binom{k}{2}}}{(q;q)_k} z^k, \quad z \in \omega_{q^{-1}} \quad q \neq 0.$$

[12] This is used to define the q-gamma function which in turn for non-negative integers coincides (for $|q| < 1$ of course) with the q-factorial $[n]_q!$.

[13] What they have in common is they remind the standard exponential function, they tend to it when $q \to 1$ and they satisfy the Cauchy functional equation in q-commuting variables. For $|q| < 1$ we have $e_q(z) = ((z,q)_\infty))^{-1}$ and $E_q(z) = (-z,q)_\infty$.

where

$$\omega_q \overset{\text{df}}{=} \begin{cases} \{z;\ |z|<1\} & \text{if } |q|<1, \\ \mathbb{C} & \text{otherwise .} \end{cases}$$

The point of note, which seems to be overlooked in the vast 'q-literature', is

$$e_q(z) = E_{q^{-1}}(-z), \quad z \in \omega_q, \quad q \neq 0 \tag{22}$$

and it is a consequence of

$$(q,q)_k = (-1)^k \frac{(q^{-1})^{\binom{k}{2}}}{(q^{-1};q^{-1})_k}, \quad k = 0,1,\ldots, \quad q \neq 0.$$

In fact this is the duality relation (22) which protects performing the reciprocation $q \longleftrightarrow q^{-1}$ on some occasions, though usually being out of sight.

The q-derivative operator D_q is defined as

$$D_q f(z) \overset{\text{df}}{=} \begin{cases} \frac{f(z)-f(qz)}{(1-q)z} & \text{if } z \neq 0 \\ f'(z) & \text{if } z = 0. \end{cases}$$

For the members of $\mathbb{C}[Z]$ it acts as $D_q Z^n = [n]_q Z^{n-1}$ and $D_q e_q(z) = e_q(z)$, $D_q E_q(z) = E_q(z)$.

In what follows we will try to monitor to what kind of Segal-Bargmann space the exponential functions may lead.

ⓐ The only kernel of the form (12) with respect to which multiplication by the independent variable and the q-derivative are (formally) adjoint each to the other is[14]

$$K(z,w) \overset{\text{df}}{=} \sum_{k=0}^{\infty} \frac{1}{[k]_q!} z^k \bar{w}^k = e_q((1-q)z\bar{w}) \quad \text{for } z, w \in |1-q|^{-1/2}\omega_q.$$

With $M \overset{\text{df}}{=} M_Z$ and $R \overset{\text{df}}{=} R_Z$ we can state[15] explicitly the results of the preceding section. First of all notice that, due to $R = M^* = D_q$, we have the commutation relation

$$M^*M - qMM^* = I$$

satisfied on $\mathbb{C}[Z]$; this is that of one of the q deformed harmonic oscillators (for others and their interrelations look at [11] for instance).

The case of $0 < q < 1$ was considered in [2]. The operator M is bounded with $\|M\| \leq (1-q)^{-1/2}$, satisfies (10) and consequently is subnormal. The couple $(K, \mathcal{H})$ is integrable with the (unique) representing measure[16] $\mu = G\,\mathrm{d}z$ ($z = x + \mathrm{i}\,y$,

---

[14] This is the reason why the corresponding space can be viewed as the most natural q-version of the Segal-Bargmann space.

[15] Notice $(1-q)^{1/2}\omega_q$ is not the unit (poly)disc but it does not matter too much.

[16] In [2] this is expressed in terms of q-integral.

$\mathrm{d}z = \mathrm{d}x\,\mathrm{d}y$) with

$$G(z) \stackrel{\mathrm{df}}{=} \begin{cases} \frac{1}{2\pi}[q^{-k}(1-q)]^{1/2} E_q(-q(1-q)|z|^2) & \text{if } |z| = [q^k(1-q)^{-1}]^{1/2}, \\ & k = 0, \pm 1, \pm 2, \ldots \\ 0 & \text{otherwise} \end{cases}$$

This interpolates in a sense between the Hardy space ($q \to 0+$) and the Segal-Bargmann one ($q \to 1-$).

The case $q > 1$ will be treated in ⓒ below.

ⓑ The kernel

$$K(z, w) \stackrel{\mathrm{df}}{=} E_q((1-q)z\bar{w}), \quad z, w \in |1-q|^{-1/2}\omega_q$$

enjoys the usefulness of (22) and this will be extended in ⓒ.

ⓒ Let $0 < q < 1$. Set

$$\alpha_n \stackrel{\mathrm{df}}{=} \frac{q^{n+\binom{n}{2}}(1-q)^n}{(q;q)_n}$$

and define K as in (12) by

$$K(z, w) \stackrel{\mathrm{df}}{=} \sum_{k=0}^{\infty} \alpha_n z^n \bar{w}^n, \quad z, w \in \mathbb{C}.$$

Then

$$K(z, w) = E_q(q(1-q)z\bar{w})$$

which is as in ⓑ with q replaced by $1 - q + q^2$.

Set $A_q \stackrel{\mathrm{df}}{=} (\int_0^\infty e_q(-(1-q)t)\,\mathrm{d}t)^{-1}$ and $B_q \stackrel{\mathrm{df}}{=} (\sum_{k=-\infty}^{+\infty} e_q(-(1-q)q^k))^{-1}$. Then the measures $\mu_1 = G_i\,\mathrm{d}z$, $i = 1, 2$, where

$$G_1(z) = \frac{1}{\pi} A_q e_q(-(1-q)|z|^2)$$

and

$$G_2(z) \stackrel{\mathrm{df}}{=} \begin{cases} \frac{1}{2\pi} B_q q^{\frac{1}{2}k} e_q(-(1-q)|z|^2) & \text{if } |z| = q^{\frac{1}{2}k},\ k = 0, \pm 1, \pm 2, \ldots \\ 0 & \text{otherwise,} \end{cases}$$

make the RKHS couple considered here integrable. Notice one may have more representing measures taking any in the closed (in the appropriate week sense) convex envelope of these two.

This is a special case of what was presented in [21] as an example of a subnormal operator which has no minimal normal extension of cyclic type, cf. [18]; it is based on [1]. A throughout discussion of the contents of this subsection will appear in [25].

References

[1] R. Askey, *Ramanujan's extension of the gamma and beta functions,* Amer. Math. Monthly, **87** (1980), 346–359.

[2] M. Arik, D.D. Coon, *Hilbert spaces of analytic functions and generalized coherent states,* J. Math. Phys., **17** (1976), 524–527.

[3] N. Aronszajn, *Theory of reproducing kernels,* Trans. Amer. Math. Soc., **68**(1950), 337–404.

[4] V. Bargmann, *On a Hilbert space of analytic functions and an associated integral transform,* Comm. Pure App. Math., **14** (1961), 187–214.

[5] A.A. Borichev, *The polynomial approximation property in Fock-type spaces,* Math. Scand., **82** (1998), 256–264.

[6] A. Connes, *Noncommutative geometry year 2000,* Geom. Funct. Anal., **2000**, Special Volume, Part II, 481–559; also circulating as arXiv:QA/0011193 preprint.

[7] ———, *A short survey of noncommutative geometry,* J. Math. Phys., **41** (2000), 3832–3866.

[8] B.C. Hall, Holomorphic methods in analysis and mathematical physics, Contemp. Math., **260** (2000), 1–59.

[9] P.R. Halmos, *A Hilbert space problem book,* van Nostrand, Princeton, NY, Toronto, London, **1967**.

[10] *T. Ito, On the commutative family of subnormal operators,* J. Fac Sci. Hokkaido Univ., **14** (1958), 1–15.

[11] A.U. Klimyk, W. Schempp, *Classical and quantum Heisenberg groups, their representations and applications,* Acta Appl. Math., **45** (1996), 143–194.

[12] R. Koekoek and R.F. Swarttouw, *The Askey-scheme of hypergeometric orthogonal polynomials and its q-analogue,* Delft University of Technology, Report of the Department of Technical Mathematics and Informatics no. 98-17, **1998**.

[13] I. Królak, *Measures connected with Bargmann's representation of the q-commutation relation for $q > 1$,* in Quantum Probability, Banach Center Publications, vol. **43**, 253–257, Institute of Mathematics, Polish Academy of Sciences, Warsaw, 1998.

[14] D.J. Newman and H.S. Shapiro, *A Hilbert space of entire functions related to the operational calculus,* manuscript.

[15] R. Rochberg and N. Weaver, *Noncommutative complex analysis and Bargmann-Segal multipliers,* Proc. Amer. Math. Soc., **129** (2001), 2679–2687.

[16] J. Stochel, F.H. Szafraniec, *On normal extensions of unbounded operators. I,* J. Operator Theory, **14** (1985), 31–55.

[17] ———, *On normal extensions of unbounded operators. II,* Acta Sci. Math. (Szeged), **53** (1989), 153–177.

[18] ———, *On normal extensions of unbounded operators. III. Spectral properties,* Publ. RIMS, Kyoto Univ., **25** (1989), 105–139.

[19] ———, *The complex moment problem and subnormality: a polar decomposition approach,* J. Funct. Anal., **159** (1998), 432–491.

[20] ———, *Domination of unbounded operators and commutativity,* J. Math. Soc. Japan, **55** (2003) (to appear).

[21] F.H. Szafraniec, *A RKHS of entire functions and its multiplication operator*. An explicit example, Operator Theory Adv. Appl., **43** (1990), 309–312.

[22] ———, *A (little) step towards orthogonality of analytic polynomials*, J. Comput. Appl. Math., **49** (1993), 255–261.

[23] ———, *The reproducing kernel Hilbert space and its multiplication operators*, Operator Theory: Adv. Appl., **114** (2000), 253–263.

[24] ———, *Subnormality in the quantum harmonic oscillator*, Comm. Math. Phys., **210** (2000), 323–334.

[25] ———, *Operators of the q-oscillator*, (work in progress).

[26] J. Weidmann, *Linear operators in Hilbert spaces*, Springer Verlag, New York - Heidelberg - Berlin, **1980**.

[27] S. Zaremba, *L'équation biharmonique et une classe remarquable de fonctions fondamentales harmoniques*, Bulletin International de l'Académie des Sciences de Cracovie, **1907**, 147–196.

Franciszek Hugon Szafraniec
Instytut Matematyki
Uniwersytet Jagielloński
ul. Reymonta 4
PL-30059 Kraków
e-mail: `fhszafra@im.uj.edu.pl`

Operator Theory:
Advances and Applications, Vol. 143, 333–344

Existence of Unitary Dilations as a Moment Problem

F.-H. Vasilescu

Abstract. A new approach to the existence of unitary dilations for tuples of commuting Hilbert space contractions, via moment problems on the torus, is initiated.

1. Introduction

Let $\mathcal{H}$ be a complex Hilbert space and let $\mathcal{L}(H)$ be the algebra of all bounded linear operators, acting in $\mathcal{H}$. We recall that a commuting multioperator $T = (T_1, \ldots, T_n)$ from $\mathcal{L}(H)^n$ ($n \geq 1$ a fixed integer) is said to have a *unitary dilation* if there exist a Hilbert space $\mathcal{K} \supset H$ and a multioperator $U = (U_1, \ldots, U_n)$ from $\mathcal{L}(K)^n$ consisting of commuting unitary operators such that $T^\alpha = PU^\alpha|\mathcal{H}$ for all $\alpha \in \mathbf{Z}_+^n$ (here we use the standard multi-index notation), where P is the orthogonal projection of $\mathcal{K}$ onto $\mathcal{H}$. The problem to find a unitary dilation for a commuting multioperator (more generally, for a commuting family of Hilbert space operators) has been firstly raised by B. Sz.-Nagy (see [20], [21]), who, besides his well-known solution valid for a single contraction, also gave a first partial solution to the several variable problem.

An obvious necessary condition for the existence of unitary dilations is that $T_1, \ldots, T_n$ be contractions in $\mathcal{H}$. This is also a sufficient condition for $n = 1$, as shown by B. Sz.-Nagy [20], and for $n = 2$, as shown by T. Ando [1], which are already classical results. A counterexample, due to Parrott [14], shows that this is no longer true for $n \geq 3$.

As far as we know, in spite of the existence of some partial solutions, in particular the existence of a regular unitary dilation for some multioperators satisfying certain positivity conditions (see, for instance, [19], I.9), there is no general "explicit" characterization of those commuting multioperators having a unitary dilation. Recently, the problem has been converted into one concerning the extension of positive definite functions defined on a subset of a $*$-semigroup (see [18]).

The aim of this work is to initiate a new approach to the existence of a unitary dilation for commuting multioperators consisting of contractions, via some special

2000 *Mathematics Subject Classification.* 44A60, 47A20.
Key words and phrases. Moment problems, unitary dilations.

solutions of the moment problems on the torus $\mathbf{T}^n$. We take the opportunity to discuss some moment problems in the more general context of the semi-algebraic compact sets (defined by polynomial inequalities).

Definition. *Let $T = (T_1, \dots, T_n)$ be a commuting multioperator from $\mathcal{L}(H)^n$. A map $\Gamma : \mathbf{Z}^n \to \mathcal{L}(H)$ is said to be a $\mathbf{Z}^n$-extension of T if*

$$\Gamma(\alpha) = T^\alpha,\ \alpha \in \mathbf{Z}^n_+, \qquad \text{and} \qquad \Gamma(-\alpha) = \Gamma(\alpha)^*,\ \alpha \in \mathbf{Z}^n.$$

Before stating our main result, we introduce some notation. Let $\mathcal{T}_a$ denote the set all analytic polynomials of the form

$$z^\theta (z+e)^\alpha (z-e)^\beta (z+ie)^\xi (z-ie)^\eta,\ z \in \mathbf{C}^n,\ \theta, \alpha, \beta, \xi, \eta \in \mathbf{Z}^n_+,$$

where $e = (1, \dots, 1) \in \mathbf{Z}^n_+$. Let also $\mathcal{C}_a$ be the set all finite n-sequences of complex numbers $\{c_\alpha\}_{\alpha \ge 0}$ such that $p(z) = \sum_{\alpha \ge 0} c_\alpha z^\alpha$ for some $p \in \mathcal{T}_a$.

Theorem A. *Let $T = (T_1, \dots, T_n)$ be a commuting multioperator from $\mathcal{L}(H)^n$. The following conditions are equivalent:*

(1) *The multioperator T has a unitary dilation.*
(2) *There exists a $\mathbf{Z}^n$-extension Γ of T such that $\sum_{\alpha,\beta \ge 0} \langle \Gamma(\alpha - \beta) x_\alpha, x_\beta \rangle \ge 0$ for all finite collections of vectors $\{x_\alpha\}_{\alpha \ge 0}$ from $\mathcal{H}$.*
(3) *There exists a $\mathbf{Z}^n$-extension Γ of T such that $\sum_{\alpha,\beta \ge 0} c_\alpha \bar{c}_\beta \Gamma(\alpha - \beta) \ge 0$ for all collections of complex numbers $\{c_\alpha\}_{\alpha \ge 0} \in \mathcal{C}_a$.*

The equivalence (1) $\Leftrightarrow$ (2) is well known (see, for instance, [SzFo], Theorem I.7.1), via the classical Naimark dilation theorem. Condition (2) clearly implies condition (3). To prove the implication (3) $\Rightarrow$ (1) in Theorem A, we may apply the following Theorem B.

Of course, Theorem A is not an explicit solution for the existence of unitary dilations as long as some of the "parameters" $(\Gamma(\alpha))_{\alpha \in \mathbf{Z}^n}$ are not fairly determined. Nevertheless, condition (3) looks simpler than the positive definiteness from (2), and it might lead to explicit solutions.

Theorem B. *Let $\Theta = (\Theta_\alpha)_{\alpha \in \mathbf{Z}^n}$ be a sequence of bounded operators acting on $\mathcal{H}$, such that $\Theta^*_\alpha = \Theta_{-\alpha}$ for all $\alpha \in \mathbf{Z}^n$, and $\Theta_0 = 1$. There exists an operator-valued positive measure F_Θ on $\mathbf{T}^n$ such that*

$$\Theta_\alpha = \int_{\mathbf{T}^n} z^\alpha \, dF_\Theta(z),\ \alpha \in \mathbf{Z}^n, \qquad \text{if and only if} \qquad \sum_{\alpha,\beta \in \mathbf{Z}^n_+} c_\alpha \bar{c}_\beta \Theta_{\alpha - \beta} \ge 0$$

for all collections of complex numbers $\{c_\alpha\}_{\alpha \ge 0} \in \mathcal{C}_a$.

Theorem B is an operator version of the classical trigonometric moment problem in several variables. This statement does not seem to follow directly from the corresponding "full" (scalar or operator) version. We shall give a proof of Theorem B in the more general context of the moment problems on semi-algebraic compact sets.

2. Moments on semi-algebraic compact sets

For the convenience of the reader, we present the framework and some necessary statements from [22]. Then we shall discuss certain consequences of those results.

Let $t = (t_1, \dots, t_n)$ denote the variable in the real Euclidean space $\mathbf{R}^n$, and let $P(\mathbf{R}^n)$ be the algebra of all real polynomial functions in $t_1, \dots, t_n$. If $\alpha = (\alpha_1, \dots, \alpha_n) \in \mathbf{Z}_+^n$ is an arbitrary multi-index, we set $t^\alpha = t_1^{\alpha_1} \dots t_n^{\alpha_n}$.

Let $\gamma = (\gamma_\alpha)_{\alpha \in \mathbf{Z}_+^n}$ $(\gamma_0 > 0)$ be an n-sequence of real numbers. We set

$$L_\gamma(t^\alpha) = \gamma_\alpha, \ \alpha \in \mathbf{Z}_+^n, \tag{2.1}$$

and extend L_γ to $P(\mathbf{R}^n)$ by linearity.

Let $K \subset \mathbf{R}^n$ be a closed set. The n-sequence γ is said to be a *K-moment sequence* if there exists a positive Borel measure μ on K such that $t^\alpha \in L^1(\mu)$ and $\gamma_\alpha = \int_K t^\alpha d\mu(t)$ for all $\alpha \in \mathbf{Z}_+^n$. When such a measure μ exists, it is called a *representing measure* of the sequence γ.

Let $\mathcal{P} = \{p_1, \dots, p_m\}$ be a finite family in $P(\mathbf{R}^n)$, and let

$$K_\mathcal{P} = \{s \in \mathbf{R}^n; p_j(s) \geq 0, j = 1, \dots, m\}. \tag{2.2}$$

A closed subset $K \subset \mathbf{R}^n$ will be called in this text *semi-algebraic* if there exists a family $\mathcal{P}$ in $P(\mathbf{R}^n)$ such that $K = K_\mathcal{P}$.

Remark 2.1. *Let $\mathcal{P} = \{p_1, \dots, p_m\}$ be a finite family in $P(\mathbf{R}^n)$. Suppose that $K = K_\mathcal{P}$ is compact. We attach to the family $\mathcal{P}$ two families $\hat{\mathcal{P}}$ and $\Delta_\mathcal{P}$, constructed in the following way. As we clearly have $m_j = \sup_{t \in K} p_j(t) < \infty$, we set $\hat{p}_j(t) = m_j^{-1} p_j(t)$, $t \in \mathbf{R}^n$, if $m_j > 0$, and $\hat{p}_j = p_j$ if $m_j = 0$, $j = 1, \dots, m$.*

We define $\hat{\mathcal{P}} = \{0, 1, \hat{p}_1, \dots, \hat{p}_m\}$. Note that $K = K_\mathcal{P} = K_{\hat{\mathcal{P}}}$, and that $0 \leq \hat{p}(t) \leq 1$ for all $t \in K$ and $\hat{p} \in \hat{\mathcal{P}}$.

Let $\Delta_\mathcal{P}$ be the set of all products of the form

$$q_1 \cdots q_k (1 - r_1) \cdots (1 - r_l)$$

for polynomials $q_1, \dots, q_k, r_1, \dots, r_l \in \hat{\mathcal{P}}$ and integers $k, l \geq 1$.

We clearly have $p|K \geq 0$ for all $p \in \Delta_\mathcal{P}$. Note also that $\Delta_\mathcal{P}$ is explicitly constructed in terms of $\mathcal{P}$.

One can prove the following assertion (practically contained in the proof of Theorem 2.3 from [22]).

Theorem 2.2. *Let $\mathcal{P} = \{p_0 = 1, p_1, \dots, p_m\}$ be a finite family in $P(\mathbf{R}^n)$. Suppose that $K = K_\mathcal{P}$ is compact, and that the family $\mathcal{P}$ generates the algebra $P(\mathbf{R}^n)$.*

An n-sequence of real numbers $\gamma = (\gamma_\alpha)_{\alpha \in \mathbf{Z}_+^n}$ $(\gamma_0 > 0)$ is a K-moment sequence if and only if the linear form L_γ is nonnegative on the set $\Delta_\mathcal{P}$.

In addition, the representing measure of a K-moment sequence is uniquely determined.

Remark 2.3. *An alternate proof of Theorem 2.2 was given in* [16]. *Another proof can be derived from the representation theorem for real algebras given in* [6] *(see also* [9] *and* [10]*; see* [15] *for general representation theorems, leading to solutions to various moment problems, and* [4], [7],[17] *for other connexions). We are indebted to E. Becker and A. Prestel for some discussions concerning the actual state-of-the-art.*

A better localization of the support of the representing measure of a K-moment sequence is given by the following (see Theorem 2.9 from [22]; see also [2], [12] etc.).

Theorem 2.4. *Let $K = K_{\mathcal{P}}$ be a semi-algebraic compact set, let $\gamma = (\gamma_\alpha)_{\alpha \in \mathbf{Z}_+^n}$ ($\gamma_0 > 0$) be a K-moment sequence, and let μ be the representing measure of γ. Assume that there exists an $r \in P(\mathbf{R}^n)$ such that $L_\gamma(rp) \geq 0$ for all $p \in \Delta_{\mathcal{P}}$. Then*

$$\operatorname{supp}(\mu) \subset \{s \in K; r(s) \geq 0\}.$$

If $L_\gamma(rp) = 0$ for some $r \in P(K)$ and for all $p \in \Delta_{\mathcal{P}}$, then

$$\operatorname{supp}(\mu) \subset \{s \in K; r(s) = 0\}.$$

The methods developed in [22] also allow an approach to some so-called "complex moment problems", that is, moment problems in the complex Euclidean space $\mathbf{C}^n$. Identifying the space $\mathbf{C}^n$ with the space $\mathbf{R}^{2n}$, we may derive easily some useful assertions, stated only in terms of complex variables. We shall briefly present the necessary changes for this transfer of information.

If $z = (z_1, \dots, z_n)$ is the variable in the complex Euclidean space $\mathbf{C}^n$, we denote by $P(\mathbf{C}^n)$ the algebra of all complex polynomial functions in $z_1, \dots, z_n$, $\bar{z}_1, \dots, \bar{z}_n$. Let also $\Re P(\mathbf{C}^n)$ be the subalgebra of all real-valued polynomials from $P(\mathbf{C}^n)$.

Let $\gamma = (\gamma_{\alpha,\beta})_{\alpha,\beta \in \mathbf{Z}_+^n}$ be an $2n$-sequence of complex numbers. We set

$$L_\gamma(z^\alpha \bar{z}^\beta) = \gamma_{\alpha,\beta},\ \alpha, \beta \in \mathbf{Z}_+^n, \tag{2.3}$$

and extend L_γ to $P(\mathbf{C}^n)$ by linearity.

The concepts of K-*moment sequence* ($K \subset \mathbf{C}^n$ a closed subset) is defined as for real sequences.

Let $\mathcal{P} = \{p_1, \dots, p_m\}$ be a finite family in $\Re P(\mathbf{C}^n)$. We set $K_{\mathcal{P}} = \{w \in \mathbf{C}^n; p_j(w, \bar{w}) \geq 0, j = 1, \dots, m\}$. Suppose that $K = K_{\mathcal{P}}$ is compact. As in Remark 2.1, we attach to the family $\mathcal{P}$ a family $\hat{\mathcal{P}}$ constructed in a similar way. Specifically, if $m_j = \sup_{w \in K} p_j(w, \bar{w}) < \infty$, we set $\hat{p}_j(w, \bar{w}) = m_j^{-1} p_j(w, \bar{w}), t \in \mathbf{R}^n$, if $m_j > 0$, and $\hat{p}_j = p_j$ if $m_j = 0$, $j = 1, \dots, m$.

Define $\hat{\mathcal{P}} = \{0, 1, \hat{p}_1, \dots, \hat{p}_m,\}$. We have $K = K_{\mathcal{P}} = K_{\hat{\mathcal{P}}}$ and $0 \leq \hat{p}(w, \bar{w}) \leq 1$ for all $w \in K$ and $\hat{p} \in \hat{\mathcal{P}}$. We also denote by $\Delta_{\mathcal{P}}$ the set of all products of the form

$$q_1 \cdots q_k (1 - r_1) \cdots (1 - r_l)$$

for polynomials $q_1, \dots, q_k, r_1, \dots, r_l \in \hat{\mathcal{P}}$, and integers $k, l \geq 1$.

Theorems 2.2 and 2.4 lead to the following assertion (which is not explicitly mentioned in [22]).

Theorem 2.5. *Let $\mathcal{P} = \{p_0 = 1, p_1, \dots, p_m\}$ be a finite family in $\Re P(\mathbf{C}^n)$. Suppose that $K = K_{\mathcal{P}}$ is compact and that the family $\mathcal{P}$ generates the algebra $\Re P(\mathbf{C}^n)$*

(i) *A $2n$-sequence of complex numbers $\gamma = (\gamma_{\alpha,\beta})_{\alpha,\beta \in \mathbf{Z}_+^n}$ with $\bar{\gamma}_{\alpha,\beta} = \gamma_{\beta,\alpha}$ for all $\alpha, \beta \in \mathbf{Z}_+^n$ and $\gamma_{0,0} > 0$ is a K-moment sequence if and only if the linear form L_γ is nonnegative on the set $\Delta_{\mathcal{P}}$.*

(ii) *Assume that $\gamma = (\gamma_{\alpha,\beta})_{\alpha,\beta \in \mathbf{Z}_+^n}$ $(\gamma_{0,0} > 0)$ is a K-moment sequence, and let μ be the uniquely determined representing measure of γ. Also assume that there exists a polynomial $r \in \Re P(\mathbf{C}^n)$ such that $L_\gamma(rp) \geq 0$ for all $p \in \Delta_{\mathcal{P}}$. Then*

$$\operatorname{supp}(\mu) \subset \{w \in K; r(w, \bar{w}) \geq 0\}.$$

If $L_\gamma(rp) = 0$ for some $r \in P(\mathbf{C}^n)$ and for all $p \in \Delta_{\mathcal{P}}$, then

$$\operatorname{supp}(\mu) \subset \{w \in K; r(w, \bar{w}) = 0\}.$$

Proof. (i) Writing $z_j = t_j + is_j$, i.e., $t_j = \Re z_j$, $s_j = \Im z_j$, $j = 1, \dots, n$, we have a natural map, say θ, given by

$$\mathbf{C}^n \ni z = (z_1, \dots, z_n) \to (t_1, \dots, t_n, s_1, \dots, s_n) = (t, s) \in \mathbf{R}^n \times \mathbf{R}^n,$$

which allows the identification of $\mathbf{C}^n$ with $\mathbf{R}^{2n}$.

The hypothesis on γ implies the equality $L_\gamma(\bar{p}) = \overline{L_\gamma(p)}$ for all $p \in P(\mathbf{C}^n)$. Therefore, $L_\gamma(p)$ is a real number whenever $\bar{p} = p$. In particular, setting $\tilde{\gamma}_{\alpha,\beta} = L_\gamma(t^\alpha s^\beta)$, $\alpha, \beta \in \mathbf{Z}_+^n$, and $\tilde{\gamma} = (\tilde{\gamma}_{\alpha,\beta})_{\alpha,\beta \in \mathbf{Z}_+^n}$, we get a $2n$-sequence of real numbers with $\tilde{\gamma}_{0,0} > 0$. We have in fact $L_{\tilde{\gamma}}(q) = L_\gamma(q \circ \theta^{-1})$ for all $q \in P(\mathbf{R}^{2n})$.

If $\tilde{\mathcal{P}} = \{p \circ \theta^{-1}; p \in \mathcal{P}\}$, then $\Delta_{\tilde{\mathcal{P}}} = \{r \circ \theta^{-1}; r \in \Delta_{\mathcal{P}}\}$ and $\tilde{K} = \theta(K) = (\theta(K))_{\tilde{\mathcal{P}}}$. Therefore, $\tilde{\gamma}$ is a $\tilde{K}$-moment sequence if and only if γ is a K-moment sequence. Moreover, $\tilde{\mu}$ is a representing measure for $\tilde{\gamma}$ if and only if $\mu = \tilde{\mu} \circ \theta^{-1}$ is a representing measure for γ. Consequently, the assertion follows directly from Theorem 2.2.

(ii) Using the discussion from (i), we infer easily the assertion, as a consequence of Theorem 2.4. We only note that, without loss of generality, the polynomial $r \in P(\mathbf{C}^n)$ may be assumed to be a real-valued function. □

Example 2.6. *(i) We first present the complex version of Example* 2.11(2) *from* [22]. *Let $K = \{z \in \mathbf{C}^n; |z_1|^2 + \cdots + |z_n|^2 \leq 1\}$, i.e., the unit ball in $\mathbf{C}^n$. We consider the following polynomials:*

$$p_{1,j}(z, \bar{z}) = \frac{1 + \Re z_j}{2} = \frac{1}{4}(|z_j + 1|^2 + 1 - |z_j|^2),$$
$$p_{2,j}(z, \bar{z}) = \frac{1 + \Im z_j}{2} = \frac{1}{4}(|z_j + i|^2 + 1 - |z_j|^2),$$

$$p_{3,j}(z,\bar z)=\frac{1-\Re z_j}{2}=\frac{1}{4}(|z_j-1|^2+1-|z_j|^2),$$
$$p_{4,j}(z,\bar z)=\frac{1-\Im z_j}{2}=\frac{1}{4}(|z_j-i|^2+1-|z_j|^2),$$

for all $j=1,\dots,n$, *via the obvious identity* $2+z_j+\bar z_j=|z_j+1|^2+1-|z_j|^2$, *and*

$$p_5(z,\bar z)=1-|z_1|^2-\cdots-|z_n|^2,\ p_6(z,\bar z)=|z_1|^2+\cdots+|z_n|^2.$$

If $\mathcal{P}=\{1,p_{1,j},p_{2,j},p_5;j=1,\dots,n\}$, *then* $K=K_{\mathcal{P}}$ *and* $\mathcal{P}$ *generates the algebra* $\Re P(\mathbf{C}^n)$. *Note also that* $\hat{\mathcal{P}}=\{0,1,p_{1,j},p_{2,j},p_{3,j},p_{4,j},p_5,p_6;j=1,\dots,n\}$.
By virtue of Theorem 2.5, the 2n-sequence of complex numbers $\gamma=(\gamma_{\alpha,\beta})_{\alpha,\beta\in\mathbf{Z}_+^n}$, *with* $\bar\gamma_{\alpha,\beta}=\gamma_{\beta,\alpha}$ *for all* $\alpha,\beta\in\mathbf{Z}_+^n$ *and* $\gamma_{0,0}>0$, *is a K-moment sequence if and only if*

$$L_\gamma(p_{1,1}^{k_{1,1}}p_{1,2}^{k_{1,2}}\cdots p_{1,n}^{k_{1,n}}p_{2,1}^{k_{2,1}}\cdots p_{4,n}^{k_{4,n}}p_5^{k_5}p_6^{k_6})\ge 0, \tag{2.4}$$

for all integers $k_{1,1}\ge 0,\dots,k_{4,n}\ge 0,k_5\ge 0,k_6\ge 0$.

Condition (2.4) is implied by the condition

$$L_\gamma((1-|z_1|^2)^{\eta_1}\cdots(1-|z_n|^2)^{\eta_n})(1-\|z\|^2)^{\eta_{n+1}}|p(z)|^2)\ge 0 \tag{2.4$'$}$$

for all $\eta_1\ge 0,\dots,\eta_n\ge 0,\eta_{n+1}\ge 0$ *integers and p an analytic polynomial. Indeed, it is easily seen that each expression* $p_{1,1}^{k_{1,1}}p_{1,2}^{k_{1,2}}\cdots p_{4,n}^{k_{4,n}}p_5^{k_5}p_6^{k_6}$ *can be written as a linear combination with positive coefficients of expressions of the form*

$$(1-|z_1|^2)^{\eta_1}\cdots(1-|z_n|^2)^{\eta_n})(1-\|z\|^2)^{\eta_{n+1}}|p(z)|^2,$$

where $\eta_1\ge 0,\dots,\eta_{n+1}\ge 0$ *are integers and p is an analytic polynomial.*

As a matter of fact, we can restrict ourselves in (2.4′) to a smaller class of analytic polynomials. Let $\mathcal{T}_a$ *be as in the Introduction. The previous argument shows that we may assume that (2.4′) holds only for those* $p\in\mathcal{T}_a$.

(ii) If we seek a representing measure for a sequence γ *as above, whose support is in* $\partial K=\{z\in\mathbf{C}^n;|z_1|^2+\cdots+|z_n|^2=1\}$, *i.e., the unit sphere in* $\mathbf{C}^n$, *using the second part of Theorem 2.5 we find the conditions*

$$L_\gamma(p_{1,1}^{k_{1,1}}p_{1,2}^{k_{1,2}}\cdots p_{4,n}^{k_{4,n}}p_6^{k_6})\ge 0, \tag{2.5$'$}$$

for all integers $k_{1,1}\ge 0,\dots,k_{4,n}\ge 0,k_6\ge 0$, *and*

$$L_\gamma(p_{1,1}^{k_{1,1}}p_{1,2}^{k_{1,2}}\cdots p_{4,n}^{k_{4,n}}p_5^{k_5}p_6^{k_6})=0, \tag{2.5$''$}$$

for all integers $k_{1,1}\ge 0,\dots,k_{4,n}\ge 0,k_5>0,k_6\ge 0$. *According to a remark above, these two conditions can be replaced by*

$$L_\gamma((1-|z_1|^2)^{\eta_1}\cdots(1-|z_n|^2)^{\eta_n}|p(z)|^2)\ge 0 \tag{2.6$'$}$$

for all integers $\eta_1\ge 0,\dots,\eta_n\ge 0$, *and*

$$L_\gamma((1-|z_1|^2)^{\eta_1}\cdots(1-|z_n|^2)^{\eta_n})(1-\|z\|^2)^{\eta_{n+1}}|p(z)|^2)=0 \tag{2.6$''$}$$

for all integers $\eta_1\ge 0,\dots,\eta_n\ge 0,\eta_{n+1}>0$, *where* $p\in\mathcal{T}_a$ *is arbitrary.*

Example 2.7. *Let $K = \mathbf{D}^n = \{z \in \mathbf{C}^n; |z_j| < 1, j = 1, \dots, n\}$, i.e., the "unit" polydisc in $\mathbf{C}^n$. As in the proof of Theorem 3.7 from [22], we consider the following polynomials:*

$$p_{1,j}(z,\bar z) = \frac{1+\Re z_j}{2},\ p_{2,j}(z,\bar z) = \frac{1+\Im z_j}{2},$$
$$p_{3,j}(z,\bar z) = \frac{1-\Re z_j}{2},\ p_{4,j}(z,\bar z) = \frac{1-\Im z_j}{2},$$
$$p_{5,j}(z,\bar z) = 1-|z_j|^2,\ p_{6,j}(z,\bar z) = |z_j|^2,$$

for all $j = 1, \dots, n$.

Note that if $\mathcal{P} = \{1, p_{1,j}, p_{2,j}, p_{5,j}, j = 1, \dots, n\}$, then $K = K_{\mathcal{P}}$ and $\mathcal{P}$ generates the algebra $\Re P(\mathbf{C}^n)$. Moreover, $\hat{\mathcal{P}} = \{0, 1, p_{1,j}, \dots, p_{6,j}, j = 1, \dots, n\}$.

Let $\gamma = (\gamma_{\alpha,\beta})_{\alpha,\beta\in\mathbf{Z}_+^n}$ be a $2n$-sequence of complex numbers with $\bar\gamma_{\alpha,\beta} = \gamma_{\beta,\alpha}$ for all $\alpha, \beta \in \mathbf{Z}_+^n$ and $\gamma_{0,0} > 0$. By Theorem 2.5, γ is a K-moment sequence, with $K = \mathbf{D}^n$, if and only if

$$L_\gamma(p_{1,1}^{k_{1,1}} p_{1,2}^{k_{1,2}} \cdots p_{1,n}^{k_{1,n}} p_{2,1}^{k_{2,1}} \cdots p_{6,n}^{k_{6,n}}) \geq 0 \tag{2.7}$$

for all integers $k_{1,1} \geq 0, k_{1,2} \geq 0, \dots, k_{6,n} \geq 0$. As in Example 2.7, each expression $p_{1,1}^{k_{1,1}} p_{1,2}^{k_{1,2}} \cdots p_{6,n}^{k_{6,n}}$ can be written as a linear combination with positive coefficients of expressions of the form $(1-|z_1|^2)^{\eta_1} \cdots (1-|z_n|^2)^{\eta_n})|p(z)|^2$, where $\eta = (\eta_1, \dots, \eta_n) \in \mathbf{Z}_+^n$ and $p \in \mathcal{T}_a$. Consequently, if

$$L_\gamma((1-|z_1|^2)^{\eta_1} \cdots (1-|z_n|^2)^{\eta_n}|p(z)|^2) \geq 0 \tag{2.8}$$

for all $\eta = (\eta_1, \dots, \eta_n) \in \mathbf{Z}_+^n$ and all $p \in \mathcal{T}_a$, then γ is a K-moment sequence. Condition (2.8) is equivalent to

$$\sum_{\alpha,\beta\in\mathbf{Z}_+^n} \sum_{\xi\leq\eta} (-1)^{|\xi|} \binom{\eta}{\xi} c_\alpha \bar c_\beta \gamma_{\alpha+\xi,\beta+\xi} \geq 0 \tag{2.8$'$}$$

for all $\eta \in \mathbf{Z}_+^n$ and all finite sequences of complex numbers $(c_\alpha)_{\alpha\geq 0} \in \mathcal{C}_a$ (where $\mathcal{C}_a$ is defined as in the Introduction).

Condition (2.8) is obviously necessary.

For $n = 1$, a similar characterization is given in [3], where Hilbert space methods are used.

Example 2.8. *Let $\mathbf{T}^n = \{z \in \mathbf{C}^n; |z_j| = 1, j = 1, \dots, n\}$, i.e., the "unit" torus in $\mathbf{C}^n$. It follows from Theorem 2.5(ii), as well as from the previous example, that γ is a K-moment sequence, with $K = \mathbf{T}^n$, if and only if*

$$L_\gamma(|p(z)|^2) \geq 0 \tag{2.9}$$

and

$$L_\gamma((1-|z_1|^2)^{\eta_1} \cdots (1-|z_n|^2)^{\eta_n})|p(z)|^2) = 0, \tag{2.9$'$}$$

for all $\eta = (\eta_1, \dots, \eta_n) \in \mathbf{Z}_+^n$, $\eta \neq 0$, and all $p \in \mathcal{T}_a$. Condition (2.9$'$) can be considerably simplified. Namely, it can be replaced by the condition

$$L_\gamma((1-|z_j|^2)|p(z)|^2) = 0, \tag{2.9$''$}$$

for all $j = 1, \dots, n$ *and* $p \in \mathcal{T}_a$. *This follows from the fact that the set* $\mathcal{T}_a$ *is invariant under the multiplications by the coordinate functions.*

Remark 2.9. *Conditions* (2.9) *and* (2.9″) *can be used to recapture the classical solution to the trigonometric moment problem in one or several variables. Specifically, if* $\zeta = (\zeta_\alpha)_{\alpha \in \mathbf{Z}^n}$ *is an* n*-sequence of complex numbers with* $\zeta_0 > 0$ *and* $\zeta_{-\alpha} = \bar{\zeta}_\alpha$ *for all* $\alpha \in \mathbf{Z}^n$, *we define the* $2n$*-sequence* $\gamma = (\gamma_{\alpha,\beta})_{\alpha,\beta \in \mathbf{Z}^n_+}$ *by* $\gamma_{\alpha,\beta} = \zeta_{\alpha-\beta}$ *for all* $\alpha, \beta \in \mathbf{Z}^n_+$. *The* $2n$*-sequence* γ *has a representing measure concentrated in* $\mathbf{T}^n$ *(which clearly implies the existence of a representing measure for* ζ *on* $\mathbf{T}^n$*) if (i)* $L_\gamma(|p(z)|^2) \geq 0$ *and (ii)* $L_\gamma((1 - |z_j|^2)|p(z)|^2) = 0$ *for all* $j = 1, \dots, n$ *and* $p \in \mathcal{T}_a$. *Condition (i) is implied by the classical condition*

$$\sum_{\alpha,\beta} \zeta_{\alpha-\beta} \lambda_\alpha \bar{\lambda}_\beta \geq 0 \tag{2.10}$$

for each finite family $(\lambda_\alpha)_{\alpha \geq 0}$ *of complex numbers, while (ii) follows from the fact that* $\gamma_{\alpha+\theta,\beta+\theta} = \gamma_{\alpha,\beta}$ *for all* $\alpha, \beta, \theta \in \mathbf{Z}^n_+$.

Note also that in (2.10) *we may take only the finite families of complex numbers* $(\lambda_\alpha)_{\alpha \geq 0}$ *that are in the set* $\mathcal{C}_a$.

3. Moments on semi-algebraic compact sets with operator data

We can easily obtain some operator versions of the previous results. We start with the following general assertion, which is an operator version of Theorem 2.5 (and which is not explicitly stated in [22]).

Let $\Gamma = (\Gamma_{\alpha,\beta})_{\alpha,\beta \in \mathbf{Z}^n_+}$ be a $2n$-sequence of bounded operators acting in the Hilbert space $\mathcal{H}$. We set

$$L_\Gamma(z^\alpha \bar{z}^\beta) = \Gamma_{\alpha,\beta}, \ \alpha, \beta \in \mathbf{Z}^n_+,$$

and extend L_Γ to $P(\mathbf{C}^n)$ by linearity.

Theorem 3.1. *Let* $\mathcal{P} = \{p_0 = 1, p_1, \dots, p_m\}$ *be a finite family in* $\Re P(\mathbf{C}^n)$. *Suppose that* $K = K_{\mathcal{P}}$ *is compact and that the family* $\mathcal{P}$ *generates the algebra* $\Re P(\mathbf{C}^n)$.

Let also $\Gamma = (\Gamma_{\alpha,\beta})_{\alpha,\beta \in \mathbf{Z}^n_+}$ *be a sequence of bounded operators acting in the Hilbert space* $\mathcal{H}$, *such that* $\Gamma^*_{\alpha,\beta} = \Gamma_{\beta,\alpha}$ *for all* $\alpha, \beta \in \mathbf{Z}^n_+$ *and* $\Gamma_{0,0} = 1$.

(i) *The* $2n$*-sequence* $\Gamma = (\Gamma_{\alpha,\beta})_{\alpha,\beta \in \mathbf{Z}^n_+}$ *can be represented as*

$$\Gamma_{\alpha,\beta} = \int z^\alpha \bar{z}^\beta \, dF_\Gamma(z), \ \alpha, \beta \in \mathbf{Z}^n_+,$$

where F_Γ *is an operator-valued positive measure whose support lies in* K, *if and only if* L_Γ *is nonnegative on the set* $\Delta_{\mathcal{P}}$.

In addition, the operator-valued positive measure F_Γ *is uniquely determined.*

(ii) *Assume that* Γ *has a representing measure* F_Γ *whose support is in* K. *Also assume that there exists a polynomial* $r \in \Re P(\mathbf{C}^n)$ *such that* $L_\Gamma(rp) \geq 0$ *for all* $p \in \Delta_{\mathcal{P}}$. *Then*

$$\operatorname{supp}(F_\Gamma) \subset \{w \in K; r(w, \bar{w}) \geq 0\}.$$

If $L_\Gamma(rp) = 0$ *for some* $r \in P(\mathbf{C}^n)$ *and for all* $p \in \Delta_{\mathcal{P}}$, *then*

$$\mathrm{supp}(F_\Gamma) \subset \{w \in K; r(w, \bar{w}) = 0\}.$$

Proof. We fix an $x \in \mathcal{H}$, $x \neq 0$, and define $\gamma_{\alpha,\beta} = \langle \Gamma_{\alpha,\beta} x, x\rangle, \alpha, \beta \in \mathbf{Z}_+^n$. The sequence $\gamma = (\gamma_{\alpha,\beta})_{\alpha,\beta\in\mathbf{Z}_+^n}$ is a scalar K-moment sequence, via Theorem 2.2. Particularly, we can find a positive measure μ_x on K which is a representing measure for γ.

A standard polarization argument and the fact that μ_x is uniquely determined for each $x \in \mathcal{H}$ lead to the existence of a uniquely determined operator-valued positive measure F_Γ on K (see [5] for the general properties of such measures) satisfying $\Gamma_{\alpha,\beta} = \int_K z^\alpha \bar{z}^\beta \, dF_\Gamma(z)$, $\alpha, \beta \in \mathbf{Z}_+^n$, and so we have (i).

The assertion (ii) is obtained using the scalar measures $\mu_x = \langle F_\Gamma(*)x, x\rangle$, $x \in \mathcal{H}$, via the corresponding assertions from Theorem 2.5. □

The next result is an extension of Theorem 3.7 from [22] (see also [11], [13], [20] etc.).

Corollary 3.2. *Let* $\Gamma = (\Gamma_{\alpha,\beta})_{\alpha,\beta\in\mathbf{Z}_+^n}$ *be a sequence of bounded operators acting on* $\mathcal{H}$, *such that* $\Gamma^*_{\alpha,\beta} = \Gamma_{\beta,\alpha}$ *for all* $\alpha, \beta \in \mathbf{Z}_+^n$, *and* $\Gamma_{0,0} = 1$. *There exists an operator-valued positive measure* F_Γ *on* $\mathbf{D}^n = \{z \in \mathbf{C}^n; |z_j| \leq 1, j = 1, \ldots, n\}$ *such that*

$$\Gamma_{\alpha,\beta} = \int_{\mathbf{D}^n} z^\alpha \bar{z}^\beta \, dF_\Gamma(z), \alpha, \beta \in \mathbf{Z}_+^n$$

if and only if

$$\sum_{\alpha,\beta\in\mathbf{Z}_+^n} \sum_{\xi\leq\eta} (-1)^{|\xi|} \binom{\eta}{\xi} c_\alpha \bar{c}_\beta \Gamma_{\alpha+\xi,\beta+\xi} \geq 0 \tag{3.1}$$

for all $\eta \in \mathbf{Z}_+^n$ *and all collections of complex numbers* $(c_\alpha)_{\alpha\geq 0} \in \mathcal{C}_a$.

Proof. As in Example 2.7, condition (3.1) is equivalent to

$$L_\Gamma((1 - |z_1|^2)^{\eta_1} \cdots (1 - |z_n|^2)^{\eta_n})|p(z)|^2) \geq 0 \tag{3.2}$$

for all $\eta = (\eta_1, \ldots, \eta_n) \in \mathbf{Z}_+^n$ and all $p \in \mathcal{T}_a$. This implies the positivity of L_Γ on $\Delta_{\mathcal{P}}$. The existence of an operator-valued positive measure F_Γ on $\mathbf{D}^n$ with the desired properties is now a consequence of Theorem 3.1.

Condition (3.1) is clearly necessary. □

Proof of Theorem B. Let $\Gamma = (\Gamma_{\alpha,\beta})_{\alpha,\beta\in\mathbf{Z}_+^n}$ be given by $\Gamma_{\alpha,\beta} = \Theta_{\alpha-\beta}$. Note that $\Gamma^*_{\alpha,\beta} = \Gamma_{\beta,\alpha}$ for all $\alpha, \beta \in \mathbf{Z}_+^n$, and $\Gamma_{0,0} = 1$. By virtue of Theorem 3.1 and Example 2.8 (see also Remark 2.9), there exists an operator-valued positive measure F_Γ on $\mathbf{T}^n$ such that

$$\Gamma_{\alpha,\beta} = \int_{\mathbf{T}^n} z^\alpha \bar{z}^\beta \, dF_\Gamma(z), \ \alpha, \beta \in \mathbf{Z}_+^n,$$

if (and only if) we have (a) $L_\Gamma(|p(z)|^2) \geq 0$ and (b) $L_\Gamma((1-|z_j|^2)|p(z)|^2) = 0$ for all $j = 1, \ldots, n$ and $p \in \mathcal{T}_a$. Condition (a) is equivalent to

$$\sum_{\alpha,\beta \in \mathbf{Z}_+^n} c_\alpha \bar{c}_\beta \Theta_{\alpha-\beta} \geq 0, \ \{c_\alpha\}_{\alpha \geq 0} \in \mathcal{C}_a, \tag{3.3}$$

while condition (b) is automatically fulfilled, as a consequence of the equality $\Gamma_{\alpha+\theta,\beta+\theta} = \Gamma_{\alpha,\beta}$ for all $\alpha, \beta, \theta \in \mathbf{Z}_+^n$.

Proof of Theorem A. The idea of proof is well known. For the convenience of the reader, we give some details. With the notation $\Theta_\alpha = \Gamma(\alpha), \alpha \in \mathbf{Z}^n$, condition (3) from Theorem A becomes condition (2.3), showing that $\Gamma(\alpha) = \int_{\mathbf{T}^n} z^\alpha \, dF(z), \ \alpha \in \mathbf{Z}^n$, for some operator-valued positive measure F on $\mathbf{T}^n$. By the Naimark dilation theorem, there exist a Hilbert space $\mathcal{K} \supset H$ and a spectral measure E in $\mathcal{K}$ supported by $\mathbf{T}^n$, such that $F(*) = PE(*)|\mathcal{H}$, where P is the orthogonal projection of $\mathcal{K}$ onto $\mathcal{H}$. If $U_j = \int_{\mathbf{T}^n} z_j dE(z), \ j = 1, \ldots, n$, and $U = (U_1, \ldots, U_n)$, then U is a unitary dilation of T.

4. An application

Let $T = (T_1, \ldots, T_n)$ be a commuting multioperator from $\mathcal{L}(H)^n$ such that the joint spectrum of T is in $\mathbf{D}^n$, where $\mathbf{D} = \{z \in \mathbf{C}; |z| < 1\}$. We define the operators

$$C(T,w) = (1 - \bar{w}_1 T_1)^{-1} \cdots (1 - \bar{w}_n T_n)^{-1}, \ w \in \overline{\mathbf{D}}^n$$

and

$$P(T,w) = C(T,w)^* \Delta_T C(T,w), \ w \in \overline{\mathbf{D}}^n,$$

where $\Delta_T = \sum_{0 \leq \alpha \leq e} (-1)^{|\alpha|} T^{*\alpha} T^\alpha$, with $T^* = (T_1^*, \ldots, T_n^*)$. The operator functions $C(T,*)$ and $P(T,*)$ are called the Cauchy respectively the Poisson kernel associated to T. The map

$$P[f](T) = \int_{\mathbf{T}^n} f(w) P(T,w) d\sigma^n(w),$$

where σ^n is the normalized Haar measure on $\mathbf{T}^n$, is defined, in particular, for all continuous functions f on $\mathbf{T}^n$ and is called the Poisson transform of T at f (see [8] for details).

Theorem 4.1. *Let $T = (T_1, \ldots, T_n)$ be a commuting multioperator from $\mathcal{L}(H)^n$, whose joint spectrum lies in $\overline{\mathbf{D}}^n$. Suppose that the strong limit*

$$\Gamma(\alpha) = s - \lim_{r \to 1-} \int_{\mathbf{T}^n} w^\alpha P(rT, w) d\sigma^n(w)$$

exists for all $\alpha \in \mathbf{Z}^n$, and that $\sum_{\alpha,\beta \geq 0} c_\alpha \bar{c}_\beta \Gamma(\alpha - \beta) \geq 0$ for all collections of complex numbers $\{c_\alpha\}_{\alpha \geq 0} \in \mathcal{C}_a$. Then T has a unitary dilation.

Proof. Theorem 2.1 from [8] and the fact that the Poisson kernel is self-adjoint show that $\Gamma(*)$ is a $\mathbf{Z}^n$-extension of T. Thus the assertion follows from Theorem A.

Remark 4.2. *Theorem* 4.1 *applies to a class of commuting multioperators containing at least those satisfying Brehmer's condition* [19] *(see* [8] *for details).*

5. Final comments

Let $K \subset \mathbf{C}^n$ be a closed set and let μ be a positive measure on K such that $z^\alpha \in L^1(\mu)$, $\alpha \in \mathbf{Z}_+^n$. We may define the *analytic moments* $\zeta = (\zeta_\alpha)_{\alpha \in \mathbf{Z}_+^n}$ of μ, via the formula $\zeta_\alpha = \int_K z^\alpha d\mu(z)$. In addition, we may state the *analytic K-moment problem* (which is a truncated moment problem of a special type) by asking to characterize those n-sequences of complex numbers $\zeta = (\zeta_\alpha)_{\alpha \in \mathbf{Z}_+^n}$ which are the analytic moments of a positive measure on K.

When $K = K_{\mathcal{P}}$ is a compact semi-algebraic set with $\mathcal{P}$ as in Theorem 2.5, for proving that the given n-sequence $\zeta = (\zeta_\alpha)_{\alpha \in \mathbf{Z}_+^n}$ is an analytic K-moment sequence it suffices to find a $2n$-sequence of complex numbers $\gamma = (\gamma_{\alpha,\beta})_{\alpha,\beta \in \mathbf{Z}_+^n}$ with $\bar{\gamma}_{\alpha,\beta} = \gamma_{\beta,\alpha}$ for all $\alpha, \beta \in \mathbf{Z}_+^n$ and $\gamma_{0,0} > 0$, such that $\gamma_{\alpha,0} = \zeta_\alpha$, $\alpha \in \mathbf{Z}_+^n$. This operation is possible, at least in principle, because the inequalities $L_\gamma(p) \geq 0$, $p \in \Delta_{\mathcal{P}}$, may allow the determination of the unknown terms of γ, as the set $\Delta_{\mathcal{P}}$ is determined, provided the sequence ζ satisfies some compatibility conditions.

Particularly, if K is the unit disk in $\mathbf{C}$, given a sequence $\zeta = (\zeta_\alpha)_{\alpha \in \mathbf{Z}_+}$, to check that ζ is a an analytic moment sequence on K, we should find an appropriate sequence $\gamma = (\gamma_{\alpha,\beta})_{\alpha,\beta \in \mathbf{Z}_+}$ with $\bar{\gamma}_{\alpha,\beta} = \gamma_{\beta,\alpha}$ for all $\alpha, \beta \in \mathbf{Z}_+$ and $\gamma_{0,0} > 0$, such that $\gamma_{\alpha,0} = \zeta_\alpha$, $\alpha \in \mathbf{Z}_+$, satisfying the conditions

$$L_\gamma((1 - |z|^2)^\tau |z|^{2\theta} |z+1|^{2\alpha} |z-1|^{2\beta} |z+i|^{2\xi} |z-i|^{2\eta}) \geq 0,\ \tau, \theta, \alpha, \beta, \xi, \eta \in \mathbf{Z}_+,$$

which may allow the determination of the unknown terms of the sequence γ, also leading to compatibility conditions to be fulfilled by the sequence ζ.

More details concerning this type of problems, in one or several complex dimensions, will appear, we hope, in future work.

A similar discussion applies to analytic moment problems with operator data. According to this terminology, the problem of existence of a unitary dilation for a given commuting n-tuple of operators can be reduced to an operator-valued analytic $\mathbf{T}^n$-moment problem.

References

[1] T. Ando, *On a pair of commuting contractions,* Acta Sci. Math., **24** (1963), 88–90.

[2] A. Athavale and S. Pedersen, *Moment problems and subnormality*, J. Math. Anal. Appl. **146** (1990), 434–441.

[3] A. Atzmon, *A moment problem for positive measures on the unit disc*, Pacific J. Math. **59** (1975), 317–325.

[4] C. Berg, J.P.R. Christensen, P. Ressel, *Harmonic analysis on semigroups*, Springer-Verlag, New York/Berlin/Heidelberg/Tokyo, 1984.

[5] S.K. Berberian, *Notes on spectral theory*, Princeton, N.J., Van Nostrand, 1966.

[6] E. Becker and N. Schwartz, *Zum Darstellungssatz von Kadison-Dubois*, Arch. Math. **40** (1983), 421–428.

[7] G. Cassier,*Problème des moments sur un compact de* $\mathbf{R}^n$ *et décomposition de polynômes à plusieurs variables*, J. Functional Analysis, **58** (1984), 254–266.

[8] R. Curto and F.-H. Vasilescu, *Standard operator models in the polydisc*, Indiana Univ. Math. J. **42** (1993), 791–810.

[9] J.-L. Krivine, *Anneaux préordonnés*, J. Analyse Math., **12** (1964), 307–326.

[10] J.-L. Krivine, *Quelques propriétés des préordres dans les anneaux commutatifs unitaires*, C. R. Acad. Sci. Paris, **258** (1964), 3417–3418.

[11] J.S. Mac Nerney, *Hermitian Moment sequences*, Trans. Amer. Math. Soc.**103** (1962), 45–81.

[12] J.L. McGregor, *Solvability criteria for certain N-dimensional moment problems*, J. Approx. Theory, **30**, 315–333.

[13] F.J. Narcowich, *R-Operators II. On the Approximation of Certain Operator-Valued Analytic Functions and the Hermitian Moment Problem*, Indiana Univ. Math. J. **26** (1977), 483–513.

[14] S. Parrott, *Unitary dilations for commuting contractions*, Pacific J. Math., **34** (1970), 481–490.

[15] A. Prestel and Charles N. Delzell, *Positive Polynomials. From Hilbert's 17-th Problem to Real Algebra*, Springer Monographs in Mathematics, 2001.

[16] M. Putinar and F.-H. Vasilescu, *Problème des moments sur les compacts semi-algébriques* , C. R. Acad. Sci. Paris **323** (1996), Série I, 787–791.

[17] K. Schmüdgen, *The K-moment problem for compact semi-algebraic sets*, Math. Ann. **289** (1991), 203–205.

[18] J. Stochel and F.H. Szafraniec, *Unitary dilation of several contractions*, Operator Theory: Advances and Applications, Vol. **127** (2001), 585–598.

[19] B. Sz.-Nagy and C. Foias, *Analyse harmonique des opérateurs de l'espace de Hilbert*, Masson et Cie and Akadémiai Kiadó, 1967.

[20] B. Sz.-Nagy, *Sur les contractions de l'espace de Hilbert*, Acta Sci. Math., **15** (1953), 87–92.

[21] B. Sz.-Nagy, *Sur les contractions de l'espace de Hilbert. II, Acta Sci. Math.*, **18** (1957), 1–15.

[22] F.-H. Vasilescu, *Moment problems for multi-sequences of operators*, J. Math. Anal. Appl. **219** (1998), 246–259.

F.-H. Vasilescu
UFR de Mathématiques,
UMR du CNRS 8524,
Université de Lille I,
F-59550 Villeneuve d'Ascq Cedex, France
e-mail: `fhvasil@agat.univ-lille1.fr`

MIX
Papier aus verantwortungsvollen Quellen
Paper from responsible sources
FSC® C105338

If you have any concerns about our products, you can contact us on
ProductSafety@springernature.com

In case Publisher is established outside the EU, the EU authorized representative is:
Springer Nature Customer Service Center GmbH
Europaplatz 3, 69115 Heidelberg, Germany

Printed by Libri Plureos GmbH
in Hamburg, Germany